Phytopharmaceuticals and Biotechnology of Herbal Plants

Plant-based drugs play an important role in all cultures and have been crucial in maintaining health and fighting diseases, and by using contemporary methods of biotechnology, plants with explicit chemical compositions are propagated and genetically upgraded for the mining of pharmaceuticals. *Phytopharmaceuticals and Biotechnology of Herbal Plants* presents the latest advances in the development of transgenic plants, covering phototherapeutics, secondary metabolite production, metabolomics, metabolic engineering, bioinformatics, quality control of herbal plant products, crop improvement using various breeding techniques, tissue culture techniques, and the future of phytopharmaceuticals.

FEATURES

- Describes developments and applications of phytopharmaceuticals of herbal plants.
- Highlights the importance of phytopharmaceuticals of plants and potential applications in the food and pharma industries.
- Provides an overview of research on key anticancer drugs to elucidate the biotechnological approaches for their production in cell cultures.
- Places special emphasis on the biosynthetic pathway mapping and metabolic engineering.

Phytopharmaceuticals and Biotechnology of Herbal Plants will be an invaluable resource to academics and researchers associated with plant science, herbal drug manufacturers, and phytochemists.

Exploring Medicinal Plants

Series Editor
Azamal Husen
Wolaita Sodo University, Ethiopia

Medicinal plants render a rich source of bioactive compounds used in drug formulation and development; they play a key role in traditional or indigenous health systems. As the demand for herbal medicines increases worldwide, supply is declining as most of the harvest is derived from naturally growing vegetation. Considering global interests and covering several important aspects associated with medicinal plants, the Exploring Medicinal Plants series comprises volumes valuable to academia, practitioners, and researchers interested in medicinal plants. Topics provide information on a range of subjects including diversity, conservation, propagation, cultivation, physiology, molecular biology, growth response under extreme environment, handling, storage, bioactive compounds, secondary metabolites, extraction, therapeutics, mode of action, and healthcare practices.

Led by Azamal Husen, PhD, this series is directed to a broad range of researchers and professionals consisting of topical books exploring information related to medicinal plants. It includes edited volumes, references, and textbooks available for individual print and electronic purchases.

Omics Studies of Medicinal Plants
Ahmad Altaf

Traditional Herbal Therapy for the Human Immune System
Azamal Husen

Environmental Pollution and Medicinal Plants
Azamal Husen

Herbs, Shrubs and Trees of Potential Medicinal Benefits
Azamal Husen

Phytopharmaceuticals and Biotechnology of Herbal Plants
Sachidanand Singh, Rahul Datta, Parul Johri, and Mala Trivedi

Phytopharmaceuticals and Biotechnology of Herbal Plants

Edited by Sachidanand Singh, Rahul Datta,
Parul Johri, and Mala Trivedi

CRC Press
Taylor & Francis Group
Boca Raton London New York

CRC Press is an imprint of the
Taylor & Francis Group, an **informa** business

First edition published 2023
by CRC Press
6000 Broken Sound Parkway NW, Suite 300, Boca Raton, FL 33487–2742

and by CRC Press
4 Park Square, Milton Park, Abingdon, Oxon, OX14 4RN

CRC Press is an imprint of Taylor & Francis Group, LLC

Library of Congress Cataloging-in-Publication Data
Names: Singh, Sachidanand, editor.
Title: Phytopharmaceuticals and biotechnology of herbal plants / edited by Sachidanand Singh, Rahul Datta, Parul Johri, and Mala Trivedi.
Description: First edition. | Boca Raton, FL : CRC Press, 2023. | Series: Exploring medicinal plants ; book 4 | Includes bibliographical references and index. | Summary: "Plant-based drugs play an important role in all cultures and have been crucial in maintaining health and fighting diseases, and by using contemporary methods of biotechnology, plants with explicit chemical compositions are propagated and genetically upgraded for the mining of pharmaceuticals. Phytopharmaceuticals and Biotechnology of Herbal Plants presents the latest advances in the development of transgenic plants, covering phototherapeutics, secondary metabolite production, metabolomics, metabolic engineering, bioinformatics, quality control of herbal plant products, crop improvement using various breeding techniques, tissue culture techniques, and the future of phytopharmaceuticals"—Provided by publisher.
Identifiers: LCCN 2022030045 (print) | LCCN 2022030046 (ebook) | ISBN 9781032277752 (hbk) | ISBN 9781032277769 (pbk) | ISBN 9781003294023 (ebk)
Subjects: LCSH: Herbs—Therapeutic use. | Materia medica, Vegetable—Biotechnology. | Phytochemicals—Therapeutic use.
Classification: LCC RM666.H33 P495 2023 (print) | LCC RM666.H33 (ebook) | DDC 615.3/21—dc23/eng/20220802
LC record available at https://lccn.loc.gov/2022030045
LC ebook record available at https://lccn.loc.gov/2022030046

ISBN: 978-1-032-27775-2 (hbk)
ISBN: 978-1-032-27776-9 (pbk)
ISBN: 978-1-003-29402-3 (ebk)

DOI: 10.1201/b22917

Typeset in Times
by Apéx CoVantage, LLC

Contents

Preface ix
Editor Biographies xi
Contributors xiii

Chapter 1 Overview on Phytopharmaceuticals and Biotechnology of Herbal Plants 1

Priyanka Gangwar, Parul Johri, Rahul Datta, Sachidanad Singh, and Mala Trivedi

Chapter 2 Indian Spices: Treasure of Active Pharmaceutical Compounds 19

Priyanka Bhamu and Shaantanu Dabi

Chapter 3 Trends in Breeding Medicinal and Aromatic Plants: Challenges and Prospects 53

Deivanai Subramanian

Chapter 4 Controlled Environment Agriculture: An Opportunity for Customizing Secondary Metabolite Production 65

Ventris M. D'souza and Ritu Bhalla

Chapter 5 Advances in Micropropagation of Pharmaceutically Important Fruit Crops 81

Priti Sharma, Antara Das, and Maneesh Mishra

Chapter 6 Exploring the Diversity of Plant Alkaloids through a Multiomics Approach 109

Megha Rai and Mami Yamazaki

Chapter 7 Recent Investigations on Protective Mechanisms of Flavonoids against Sarcopenia 141

Ishita Pandey, Srishti Vashishtha, and Sonam Chawla

Chapter 8 Herbal Medicinal Compounds and Their Anti–Breast Cancer Actions: A Mechanistic Perspective 157

Daniel A. Gideon, Pushparaj Annadurai, Abhinav Parashar, Rachel Grace Murthy, Kandavel Dhandayuthapani, Sriram Srinivasan, Sivaramakrishnan R, and R. Jasmine

Chapter 9 Aromatic Plants as a Source of Natural Antioxidants and Their Potential to Reduce Disease Risk 185

Indira Mikkili, Krupanidhi Sreerama, Venkateswarulu TC, Abraham Peele Karlapudi, and Vidya Prabhakar Kodali

Chapter 10 Phytopharmaceuticals of Medicinal Plants for the Treatment of Common Illnesses 201

Venkata Kanaka Srivani Maddala, Asha S, and Sachidanand Singh

Chapter 11 A Disease-Driven Approach to Medicinal Natural Products........ 213

Jasmine R, Deena Priscilla H, and Sowmiya S

Chapter 12 Recent Insights into Cannabinoids: Functions, Mechanisms, Clinical Trials, and Technological Advancement in Cancer and Neurological Diseases 225

Tarun Pant, Matea Juric, and Zeljko J. Bosnjak

Chapter 13 Anti-Cancer Phytopharmaceuticals: A Green Approach Towards Treating Cancer 239

Priyanka Mudaliar, Apoorva Jagannath Nalawade, Priyamvada Arte, Akanksha Dhuliya Deepak Iyer, Kanchanlata Tungare, Mustansir Bhori, and Jyotirmoi Aich

Chapter 14 Synergism between Anticancer Drugs and Seaweed Metabolites for Cancer Treatment 255

Priscilla Arockiasamy, Sriram Srinivasan, Daniel A Gideon, Abhinav Parashar, Shivadharshini Ramachandran, and Janet Ann Netto

Chapter 15 Recent Perspectives on the Medicinal Potential of Plant-Based Adaptogens........ 267

Kajal Pandey, Manvi Arora, and Sonam Chawla

Chapter 16 Workflow and Methods of In Silico Analysis of Compounds Isolated from Medicinal Plants: An Overview 281

Rosita A. Sherlin, Pushparaj Annadurai, Thomas Jebastin, Sriram Srinivasan, Priscilla Arockiasamy, S. Danushri, Vijay Nirusimhan, Abhinav Parashar, and Daniel A. Gideon

Chapter 17 Basics of Photobiology and Clinical Indications for Phototherapy in Anti-Aging........ 307

Namrata Verma, Poornima Yadav, Pankhudi Gupta, Mahima Bhardwaj, Subhan Danish, Rahul Datta, and Sachidanand Singh

Chapter 18 Role of Nanotechnology in the Field of Phytopharmaceuticals for the Delivery of Herbal Drugs 317

Tanvi Jain, Kavita Srivastava, Rajnish Singh, M.J. Ansari, Sachidanand Singh, and P.K. Dutta

Chapter 19 Phytonanotechnology: Plant-Mediated Green Construction of MNPs and Their Biomedical Implementation against MDR Bacteria 331

Meenu Gupta, Neeraj Agarwal, and Abhidha Kohli

Chapter 20 Past, Present, and Future of Aromatic Plants........ 347

Vipin Kumar, Himashu Yadav, and Ramesh Kumar Srivastava

Index 361

Preface

The word "phytopharmaceutical" is a combination of two words: phyto, referring to plant, and pharmaceutical, referring to therapeutic drugs. Since prehistoric times medicinal plants have been used in traditional medicine practices. These plants contain an amalgamation of bioactive components—also referred as phytochemicals—residing in numerous plant parts ranging from stems to leaves, fruits, flowers, roots, etc. This book covers a range of topics for various biotechnological applications applied to the isolation of bioactive compounds of medicinal plants, their accessible sources, biochemistry, structural makeup and impending biological activities from a health perspective. The book also covers the photochemistry of different natural products; an innovative dietary approach for the management of a range of diseases and its therapeutic potential followed by applications in food and pharmacy industries. We foresee that this collection will be apposite for the scholars and researchers working in the academic field as well as those in the plant biotechnology, food, nutraceuticals and herbal trades. Incorporated in the twenty chapters of the book is the overall biotechnological application of various phytopharmaceuticals derived from herbal plants. Phytonutrients provide several health advantages and may be taken daily to lower risk factors for conditions such as high cholesterol, high blood pressure and diabetes. They are produced and assessed in the same manner that traditional therapeutic items are. The aromatic plants are of extraordinary value in terms of commercial crops. Hence, their compounds have the impending power to become new therapeutic drug molecules for human health. This book elaborates on the natural phytochemicals present in these plants, which can be further used as a potential lead molecule for the expansion of semisynthetic and synthetic derivatives and dicusses the aromatic plants, their classification, natural antioxidant bits and the role of antioxidants in preventing disease risk.

Seaweed or microalgae are rich in phytochemicals that exhibit various biological activities such as anticancer, antioxidant, anti-inflammatory and immunomodulatory functions. Purified bioactive compounds extracted from seaweeds have showed anticancer activity against drug-resistant cancer cell types, and the use of bioactive compounds along with anticancer drugs reduced the side effects caused by chemotherapy.

Plant-based adaptogens have been confirmed to show a large variation of medicinal properties like enhancing the immune system, stress regulation, treatment and inhibition of anxiety disorders, hypertension, memory impairment and in curing numerous neurodegenerative and cardiovascular ailments. Medicinal plants have abundant phytochemicals with diverse secondary metabolites like polyphenols (phenolic acids, anthocyanins, proanthocyanidins, flavonols, tannins), isoprenoids (sesqiterpenes, diterpenes, triterpenes, steroids, saponins), alkaloids (indole alkaloids, lysergic acid diethylamide, tropane alkaloids, ergot group) and fatty acids by which they impede or even scavenge the excess of free radicals produced in oxidative and neurotoxin-induced stresses in the nerve cells of the brain.

At present, essential oils are used by many people. Owing to its uses and benefits, the demand for essential oil in national and international markets has increased consistently. To fulfil the prerequisite and supply quality products, there is an increased scope of aromatic crop cultivators and their related trades. Finally, the book ends with an impression regarding the *in silico* attitudes used for analysis of compounds isolated from medicinal plants.

We are eternally grateful to Azamal Husen, the series editor, *Exploring Medicinal Plants*, for providing a constant support throughout the compilation of the book.

Sachidanad Singh
Rahul Datta
Parul Johri
Mala Trivedi

Preface

Editor Biographies

Dr. Sachidanand Singh, professor, Department of Biotechnology, Sankalchand Patel University, Visnagar, Gujrat. He completed his B.Tech in Biotechnology, M.Tech in Bioinformatics, and PhD in Bioinformatics. He has postdoctoral research experience from the College of Pharmacy, Ohio State University, USA. His specialization includes plant biotechnology, systems biology, and drug design. He has more than 13 years of research and academic experience and has served for a decade in the School of Biotechnology and Health Sciences, Karunya University, Coimbatore, Tamil Nadu, India. He also worked as a consultant for the College of Pharmacy, Ohio State University as a statistical geneticist in 2018. Later he joined Shri Ramswaroop Memorial University, Barabanki, Uttar Pradesh and was involved in different administrative and academic positions. He has published more than 45 research articles in peer-reviewed journals and completed one SERB, Government of India–funded project, and two university research grant projects. He has received several awards in his area of research and for his teaching strategies and has organized a number of conferences and workshops and received government funds for the same. He has shared his area of expertise with different colleges and universities as an invited speaker at different conferences and workshops. He has guided 36 B.Tech and 10 M.Tech projects. Presently four PhD students are enrolled under him, and two PhD students have completed research in the field of herbal products, drug design, and integration with network biology.

Dr. Rahul Datta is an assistant professor at Mendel University in Brno. He received his MSc and PhD from Mendel University, Czech Republic. He is a soil microbiologist and enzymologist, and his research is focused on the application of biochar, soil molecular, and microbiological techniques for the improvement of soil quality and crop productivity to attain sustainable agriculture and a solution to food security. He has had the opportunity as a visiting scientist to work with renowned scientists such as Prof. Dr. Richard Dick at Ohio State University, Ohio, USA, and Prof. Paolo Nannipieri at the University of Florence, Italy. To date, he has published about 110 research articles and review papers, two books with Springer, and one with Intech Open. He has reviewed more than 450 papers in peer-reviewed journals. He is academic editor of the journal *Open Agriculture, BMC Plant Biology, Soil Health* and is also hosting a special issue in the journal *Land, Agriculture and Sustainability.* At present, he is an honorary international advisor of the Asian PGPR Society

Dr. Parul Johri is an assistant professor, Department of Biotechnology, Dr. Ambedkar Institute of Technology for Handicapped, Kanpur, U.P. She completed her B.Tech in Biotechnology in 2006 from IET, Bundelkhand University, Jhansi, Uttar Pradesh, and her M.Tech in Bioinformatics from Padmashree Dr. D. Y. Patil University, Navi Mumbai, Maharashtra, India. She received her PhD in Biotechnology (specialization in Bioinformatics) from Amity Institute of Biotechnology, Amity University Uttar Pradesh Lucknow Campus. Dr. Johri has published about 18 research and review papers of national and international repute and more than 10 book chapters. She has been a recipient of the Vasant Rao Dada Saheb Gold Medal and is a merit holder in the top 0.1% of students across India in CBSE board exams. Dr. Johri received funds from DBT CTEP for conducting a popular lecture series in Biotechnology, Government of India, and also received funding from the National Commission of

Women (NCW) to conduct a webinar on women's health and rights. Dr. Johri has a wide teaching career of about 14 years and she has guided many postgraduate students for their final-year projects.

Mala Trivedi is a professor, Amity Institute of Biotechnology, Amity University Uttar Pradesh, Lucknow. She was a recipient of a double Gold Medal from the University of Allahabad during her M.Sc. (1993). She did her PhD in plant genetics and breeding at the Central Institute of Medicinal and Aromatic Plants (CIMAP) (CSIR), Lucknow and was a research associate at the Institute of Cell Biology, School of Life Sciences, Lanzhou University, Lanzhou, P.R. China. She has strong experience in plant transgenics. She has published more than 50 research papers in leading journals of national and international repute as well as a book and several book chapters with international publishers. She has five Indian and two U.S. patents to her credit. She is a member of many prestigious societies. She was also visiting faculty to the University of Lanzhou, Lanzhou PR China. Currently she is involved in mass multiplications of economically important plants (horticulture and medicinal plants), is the principal of a DBT-funded project, and has supervised 12 PhD scholars to date.

Contributors

Neeraj Agarwal
Department of Biotechnology
Noida Institute of Engineering and Technology
Uttar Pradesh, India

Jyotirmoi Aich
School of Biotechnology and Bioinformatics
DY Patil
Maharashtra, India

Pushparaj Annadurai
Department of Botany
Thanthai Periyar Government Arts and Science College (Autonomous), affiliated with Bharathidasan University
Tamil Nadu, India

M.J. Ansari
Department of Botany
Hindu College Moradabad (Mahatma Jyotiba Phule Rohilkhand University)
Bareilly, India

Priscilla Arockiasamy
Department of Biotechnology and Bioinformatics
Bishop Heber College (Autonomous), affiliated to Bharathidasan University
Tamil Nadu, India

Manvi Arora
Department of Biotechnology
Jaypee Institute of Information Technology
Uttar Pradesh, India

Priyamvada Arte
School of Biotechnology and Bioinformatics
DY Patil
Maharashtra, India

Syed Asha
Department of Biotechnology
Vignan's Foundation for Science Technology and Research
Andhra Pradesh, India

Ritu Bhalla
School of Applied Science
Republic Polytechnic
Singapore

Priyanka Bhamu
Department of Zoology
Govt R.R. Morarka College Nawalgarh
Jhunjhunu (Rajasthan), India

Mahima Bhardwaj
Department of Biotechnology
Seth Vishambhar Nath Institute of Engineering and Technology
Uttar Pradesh, India

Mustansir Bhori
School of Biotechnology and Bioinformatics
DY Patil
Maharashtra, India

Zeljko J. Bosnjak
Department of Medicine
and
Department of Physiology
Medical College of Wisconsin
Milwaukee, Wisconsin, USA

Sonam Chawla
Department of Biotechnology
Jaypee Institute of Information Technology
Uttar Pradesh, India

Ventris M. D'souza
School of Applied Science
Republic Polytechnic
Singapore

Shaantanu Dabi
Department of Zoology
Govt Lohia P.G. College
Churu (Rajasthan), India

Subhan Danish
Department of Soil Science
Faculty of Agricultural Sciences and Technology
Bahauddin Zakariya University
Multan, Pakistan

S. Danushri
Department of Biotechnology and Bioinformatics
Bishop Heber College (Autonomous), affiliated with Bharathidasan University
Tamil Nadu, India

Antara Das
ICAR-Central Institute for Subtropical Horticulture
Rehmankhera
Uttar Pradesh, India

Rahul Datta
Department of Geology and Pedology
Faculty of Forestry and Wood Technology
Mendel University in Brno
Brno, Czech Republic

Kandavel Dhandayuthapani
Department of Botany
Thanthai Periyar Government Arts and Science College (Autonomous), affiliated with Bharathidasan University
Tamil Nadu, India

Akanksha Dhuliya
School of Biotechnology and Bioinformatics
DY Patil
Maharashtra, India

P.K. Dutta
Department of Chemistry
Motilal Nehru National Institute of Technology
Prayagraj, India

Priyanka Gangwar
Amity Institute of Biotechnology
Amity University Uttar Pradesh, Lucknow Campus
Uttar Pradesh, India

Daniel A. Gideon
Department of Biotechnology and Bioinformatics
Bishop Heber College
Tamil Nadu, India

Daniel A. Gideon
Department of Biotechnology and Bioinformatics
Bishop Heber College (Autonomous) affiliated with Bharathidasan University
Tamil Nadu, India

Meenu Gupta
Department of Botany
J. D. Women's College Patna
Bihar, India

Pankhudi Gupta
Department of Biotechnology
Seth Vishambhar Nath Institute of Engineering and Technology
Uttar Pradesh, India

Deena Priscilla H
Department of Biotechnology & Bioinformatics
Bishop Heber College
Tiruchirappalli, India

Deepak Iyer
Laboratory Medicine Program
Toronto General Hospital
and
University Health Network
University of Toronto
Toronto, Canada

Tanvi Jain
Faculty of Biotechnology
Institute of Biosciences and Technology
Shri Ramswaroop Memorial University
Lucknow, India

R. Jasmine
Department of Biotechnology and Bioinformatics
Bishop Heber College (Autonomous), affiliated with Bharathidasan University
Tamil Nadu, India

Thomas Jebastin
Department of Biotechnology and Bioinformatics
Bishop Heber College (Autonomous), affiliated with Bharathidasan University
Tamil Nadu, India

Parul Johri
Department of Biotechnology
Dr. Ambedkar Institute of Technology for Handicapped
Uttar Pradesh, India

Matea Juric
Department of Biophysics
Medical College of Wisconsin
Milwaukee, Wisconsin, USA

Abraham Peele Karlapudi
Department of Biotechnology
Vignan's Foundation for Science, Technology and Research
Andhra Pradesh, India

Vidya Prabhakar Kodali
Department of Biotechnology
Vikrama Simhapuri University
Andhra Pradesh, India

Abhidha Kohli
Department of Biotechnology
IMS Engineering College
Uttar Pradesh, India

Vipin Kumar
CSIR-Central Institute of Medicinal and Aromatic Plants (CIMAP)
Lucknow, India

Venkata Kanaka Srivani Maddala
Department of Science and Humanities
Vignan's Foundation for Science Technology and Research
Andhra Pradesh, India

Indira Mikkili
Department of Biotechnology
Vignan's Foundation for Science, Technology and Research
Andhra Pradesh, India

Maneesh Mishra
ICAR-Central Institute for Subtropical Horticulture
Rehmankhera
Uttar Pradesh, India

Priyanka Mudaliar
School of Biotechnology and Bioinformatics
DY Patil
Maharashtra, India

Rachel Grace Murthy
Center for Nursing Education and Research
Baylor University Medical Center
Dallas, Texas, USA

Apoorva Jagannath Nalawade
School of Biotechnology and Bioinformatics
DY Patil
Maharashtra, India

Janet Ann Netto
Department of Biotechnology
Sastra University
Tamil Nadu, India

Vijay Nirusimhan
Department of Biotechnology and Bioinformatics
Bishop Heber College (Autonomous), affiliated with Bharathidasan University
Tamil Nadu, India

Ishita Pandey
Department of Biotechnology
Jaypee Institute of Information Technology
Uttar Pradesh, India

Kajal Pandey
Department of Biotechnology
Jaypee Institute of Information Technology
Uttar Pradesh, India

Tarun Pant
Department of Medicine
Medical College of Wisconsin
Milwaukee, Wisconsin, USA

Abhinav Parashar
Department of Biotechnology
Vignan's Foundation for Science, Technology and Research
Andhra Pradesh, India

Abhinav Parashar
Vignan's Institute for Science, Technology and Research, Guntur
Andhra Pradesh, India

Jasmine R
Department of Biotechnology & Bioinformatics
Bishop Heber College
Tiruchirappalli, India

Megha Rai
Graduate School of Pharmaceutical Sciences and
Plant Molecular Science Centre
Chiba University
Chiba, Japan

Shivadharshini Ramachandran
Department of Biotechnology
St. Joseph's College
Tamil Nadu, India

Priti Sharma
ICAR-Central Institute for Subtropical Horticulture
Rehmankhera
Uttar Pradesh, India

Rosita A. Sherlin
Department of Biotechnology and Bioinformatics
Bishop Heber College (Autonomous), affiliated with Bharathidasan University
Tamil Nadu, India

Rajnish Singh
Faculty of Biotechnology
Institute of Biosciences and Technology
Shri Ramswaroop Memorial University
Lucknow, India

Sachidanand Singh
Department of Biotechnology
Vignan's Foundation for Science, Technology and Research
Andhra Pradesh, India

Ramachandran Sivaramakrishnan
Department of Biotechnology
Faculty of Science
Chulalongkorn University
Bangkok, Thailand

S. Sowmiya
Department of Biotechnology & Bioinformatics
Bishop Heber College
Tiruchirappalli, India

Krupanidhi Sreerama
Department of Biotechnology
Vignan's Foundation for Science, Technology and Research
Andhra Pradesh, India

Sriram Srinivasan
Department of Biotechnology and Bioinformatics
Bishop Heber College
Tamil Nadu, India

Sriram Srinivasan
Department of Biotechnology and Bioinformatics
Bishop Heber College (Autonomous), affiliated with Bharathidasan University
Tamil Nadu, India

Kavita Srivastava
Faculty of Biotechnology
Institute of Biosciences and Technology
Shri Ramswaroop Memorial University
Lucknow, India

Ramesh Kumar Srivastava
CSIR-Central Institute of Medicinal and Aromatic Plants (CIMAP)
Lucknow, India

Deivanai Subramanian
School of Applied Science
Republic Polytechnic
Singapore

Mala Trivedi
Amity Institute of Biotechnology
Amity University Uttar Pradesh, Lucknow Campus
Uttar Pradesh, India

Kanchanlata Tungare
School of Biotechnology and Bioinformatics
DY Patil
Maharashtra, India

Srishti Vashishtha
Department of Biotechnology
Jaypee Institute of Information Technology
Uttar Pradesh, India

T.C. Venkateswarulu
Department of Biotechnology
Vignan's Foundation for Science, Technology and Research
Andhra Pradesh, India

Namrata Verma
Department of Biotechnology
Seth Vishambhar Nath Institute of Engineering and Technology
Uttar Pradesh, India

Himashu Yadav
CSIR-Central Institute of Medicinal and Aromatic Plants (CIMAP)
Lucknow, India

Poornima Yadav
Department of Biotechnology
Seth Vishambhar Nath Institute of Engineering and Technology
Uttar Pradesh, India

Mami Yamazaki
Graduate School of Pharmaceutical Sciences
and
Plant Molecular Science Centre
Chiba University
Chiba, Japan

1 Overview on Phytopharmaceuticals and Biotechnology of Herbal Plants

Priyanka Gangwar, Parul Johri, Rahul Datta, Sachidanad Singh, and Mala Trivedi

CONTENTS

1.1 Introduction ... 1
1.2 Complex Mixtures Made From Secondary Plant Metabolites ... 2
1.3 Phytotherapy and Homeopathy Are Not the Same Phenomena ... 3
1.4 Phytopharmaceuticals: Natural Chemicals Used to Treat Illness ... 3
1.5 High Tolerance and Strength ... 3
1.6 Phytotherapy With Clinical Evidence ... 4
1.7 Phytopharmaceuticals: Types and Medical Applications ... 4
1.7.1 Anthocyanins ... 4
1.7.2 Carotenoids ... 4
1.7.3 Lycopene ... 5
1.7.4 Flavonoids ... 5
1.7.5 Isoflavonoids ... 5
1.7.6 Polyphenols ... 10
1.7.7 Phytoestrogens ... 10
1.7.8 Resveratrol ... 10
1.7.9 Phytosterols ... 10
1.7.10 Terpenoids ... 11
1.8 Some Exotic Medicinal Plants ... 11
1.8.1 *Hypericum perforatum* ... 11
1.8.2 *Solanum lycopersicum* ... 11
1.8.3 *Digitalis purpurea* ... 12
1.8.4 *Filipendula ulmaria* ... 12
1.8.5 *Melissa officinalis* ... 12
1.9 Conclusion ... 12
Bibliography ... 13

1.1 INTRODUCTION

Herbal medical goods, often known as phytopharmaceuticals, are preparations derived from plants. As active components, they only comprise one or several herbal compounds or herbs, or a combination of them (EMA HMPC 2014). They are made up of a variety of natural chemicals

DOI: 10.1201/b22917-1

that stimulate or modify numerous specific processes in an organism, as opposed to synthetic medications that comprise chemically active molecules. The content of herbal medicinal preparations might vary greatly. As a result, standardized medicinal plant extracts are utilized to achieve consistent effectiveness and safety. Evidence-based phytotherapy is based on randomized controlled clinical trials that have scientifically shown effectiveness and safety. Herbal therapeutic solutions, on average, are better absorbed and have a higher benefit-to-risk ratio than synthetic medications (Karimi et al. 2015). As a result, evidence-based phytopharmaceuticals are becoming more commonly included in clinical practice recommendations as first-line therapy for a variety of disorders and indications.

The phrase "phytopharmaceutical" is a combination of phyto (meaning "plant") and pharmaceutical (meaning "medicinal drugs"). Thus, plant-derived substances with pharmacological properties are known as phytopharmaceuticals. They are defined as a pure and formalized portion, like an extract of either a medicinal herb or its part, that contains approximately four biologically active or phytochemical substances (qualitatively and quantitatively evaluated) for external or internal use by living organisms for disease detection, diagnosis, minimization, or containment (Bhatt 2016). In 2010, the Department of Ayush (formerly recognized as the Department of Indian Medication and Homeopathy) issued Rule 158 (B), which supported documents of effectiveness for a patent or proprietary Ayurveda Siddha and Unani (ASU) medication license (Singh 2018). Later, Good Clinical Practice (GCP) was created, which may be employed voluntarily for conducting clinical studies with copyrighted Ayurvedic, Siddha, and Unani medications.

The establishment of phytopharmaceutical medicines that could partly replace some traditional pharmaceuticals typically requires the import of raw ingredients and can be manufactured by drug manufacturers in underdeveloped countries under collaborative ventures. Clinically validated phytopharmaceuticals are held to the highest standards. Their effectiveness and safety must be demonstrated in randomized, double-blind, and controlled clinical studies. They are produced and assessed in the same manner that traditional therapeutic items are. Herbal treatment has long been seen as a viable alternative to contemporary allopathic medicine (Ahmad and Sharma 2020). But only a few herbs have been rigorously investigated for their efficiency in clinical therapy, despite their growing popularity.

1.2 COMPLEX MIXTURES MADE FROM SECONDARY PLANT METABOLITES

Phytopharmaceuticals are manufactured from phytometabolites and comprise complex mixes of several natural substances, unlike synthetic medications, which are single functional chemical molecules. Plants have a main metabolic process which creates and destroys amino acids, nucleotides, lipids, and carbohydrates. As a result, these synthesis routes are found in all plants and are fairly similar in various plant groups. The secondary metabolism is linked to the main metabolism since it employs primary metabolism building blocks to make a huge variety of specialized chemicals. Based on their synthetic backgrounds and chemical bonding, they are divided into a small number of categories (Bhusnure et al. 2019; Verma et al. 2021).

Amino acids are used to make alkaloids and amines. Polyketides, steroids, and phenylpropanoids are examples of secondary metabolites. Secondary higher plant metabolites have yielded over 80,000 distinct chemical structures (EMA HMPC 2014). Diverse plant families develop multiple secondary metabolites, which differ significantly in chemical structure, resulting in a large number of structurally similar compounds. Plants create a variety of secondary metabolites at various stages of growth. Secondary metabolites are essential for connecting with other species and dealing with the environment. To illustrate the influence of specific secondary metabolites as defensive barriers, tomatoes are a great example. Green tomatoes are protected from herbivores as well as fungus and lichen infestation by the glycoalkaloids tomatine and dehydrotomatine. As the tomatoes become older, they gradually deteriorate. Ripe tomatoes are safe to consume since they contain little or no tomatine and dehydrotomatine. When a plant is infected with microbial diseases, several secondary plant metabolites are generated, like phytoalexins, which are known to have antibacterial properties (Bhusnure et al. 2019; Obmann 2010).

1.3 PHYTOTHERAPY AND HOMEOPATHY ARE NOT THE SAME PHENOMENA

The German Medicines Act (AMG) establishes the definition of herbal medicines. Plant substances such as atropine and morphine, which can be extracted in pure form from plants, are classified as chemically defined compounds by the AMG and are hence classified as classical medicines. Phytopharmaceuticals, on the other hand, are always made up of whole plants, parts, or components. Based on scientific understanding, phytopharmaceuticals are utilized to cure ailments called phytotherapy. Phytopharmaceuticals, homeopathic, and anthroposophic medications are all classified as special therapy alternatives by the AMG; however, phytopharmaceuticals vary from homeopathic and anthroposophic drugs in terms of manufacture, approval, and therapeutic strategy (Büssing et al. 2011; Verma et al. 2021). Phytopharmaceuticals are made up of pharmacologically active chemicals which bind with protein elements in the human body, referred to as drug targets. As a result, they differ significantly from homeopathic treatments, which are extensively diluted to the point where little or no active component remains. Homeopathy, unlike phytotherapy, lacks substantial scientific proof and is thus considered an alternative or supplementary therapy option (Bhusnure et al. 2019). The premise of homeopathic products is radically different from that of contemporary pharmaceuticals.

1.4 PHYTOPHARMACEUTICALS: NATURAL CHEMICALS USED TO TREAT ILLNESS

Phytopharmaceuticals comprise natural medications whose effectiveness is attributed to one or more plant components or active elements. They've been utilized to treat illness from the beginning of time. Many medical treatments created from flora or parts of plants still rely on this ancient knowledge (Obmann 2010). For many years, herbal remedies have been made in Baden-Württemberg. Plants create a huge number of different natural chemicals. As a result, it's not strange that people make use of such a wide range of resources. The usage of medicinal herbs may be traced back to the Bronze Age, according to historical texts. Europe has a long history of employing medicinal plants, beginning with Hildegard von Bingen and continuing with Friedrich Sertürner, being the first to separate morphine in its purest form, and ending with present herbal medicine manufacture (Bhusnure et al. 2019).

1.5 HIGH TOLERANCE AND STRENGTH

In theory, phytopharmaceuticals have a high level of tolerability and safety, but they are subject to the same hazards as conventional treatments. They may have restrictions, like the danger of unwanted side effects and possible medication complications. But, on the other hand, they are generally well accepted and have a reduced risk of side effects than chemically synthesized treatments (Bhusnure et al. 2019). Phytopharmaceuticals have a wide therapeutic range and may be used to treat both simple and chronic diseases. They are highly suited for elderly individuals who take many drugs because of their good tolerance and limited interaction potential.

1.6 PHYTOTHERAPY WITH CLINICAL EVIDENCE

Clinically validated phytopharmaceuticals are held to the highest standards. Their effectiveness and safety must be established in randomized, double-blind, and controlled medical studies. They are produced and scientifically assessed in the same manner that traditional pharmaceuticals are. This is in contrast to conventional phytopharmaceuticals, which are generally used based on personal experiences, such as the prescription of tannin-containing black teas for diarrhea, St. John's wort as a therapy for depressive moods (Schrader 2000; Woelk 2000; EMA HMPC 2009; Linde et al. 2008), butterbur as a therapy for hay fever (Schapowal 2002), gingko as a therapy for declining mental performance (EMA HMPC 2015), black cohosh as a therapy for menopausal symptoms (Schellenberg et al. 2012), hawthorn as a therapy for cardiac symptoms (EMA HMPC 2016), and valerian and hops as a therapy for sleep disorders (Koetter et al. 2007)—these are just a few case studies of medically tested phytopharmaceuticals.

1.7 PHYTOPHARMACEUTICALS: TYPES AND MEDICAL APPLICATIONS

Polyphenols, terpenoids, alkaloids, phytosterols, and organosulfur compounds are all types of phytochemicals. Anthocyanidins, carotenoids, phytoestrogens, lycopene, flavonoids, terpenoids, limonoids, glucosinolates, polyphenols, isoflavonoids, phytosterols, omega-3 fatty acids, resveratrol, and probiotics are some of the most significant phytopharmaceuticals. In terms of human health, they have unique pharmacological effects (Singh et al. 2019). Several of these compounds have antibacterial, anti-inflammatory, anti-aging, anti-allergic, antioxidant, antispasmodic, antidiabetic, and a variety of other medicinal properties. Polyphenols are the most often encountered and researched phytochemical groups (Somani et al. 2015). Eight thousand polyphenolic substances have been found so far with antioxidant and prooxidant properties, depending on their concentrations (Ferguson 2001). Polyphenols have been demonstrated to have a role in inducing apoptosis, suppressing cell proliferation, migration, and the incursion of cancer cells at higher dosages, while their lower concentrations scavenge free radicals in cells. Polyphenols are chemical compounds that include one or more aromatic rings and one or more hydroxyl groups (Ferrazzano et al. 2011). They are divided into four groups: phenolic acids, isoflavones, flavonoids, and stilbenes based on their chemical structure.

1.7.1 Anthocyanins

Anthocyanins are the most abundant category of water-soluble pigments in plants. They are classified as possible antioxidants because they have been shown to offer medicinal value and disease control characteristics. Anthocyanin-rich diets have been linked to a lower risk of numerous illnesses, including atherosclerosis, dyslipidemia, and diabetes (Jaiswal et al. 2019). Anthocyanins can be found in the fruits, flowers, leaves, roots, and stems of higher plants. Vaccinium species, which include blueberry, red cabbage, blackberry, cranberry, and cherry, are high in anthocyanins. Pelargonidin, cyanidin, and delphinidin are the three primary anthocyanin molecules. Cyanidin and related glycosides are organic food colors that have been discovered to offer intriguing possible human health advantages, particularly in the diagnosis and intervention of diabetes mellitus (Khoo et al. 2017).

1.7.2 Carotenoids

Carotenoids are a category of over 700 phytochemicals identified in photosynthetic membranes that give plants and animals their colors. Only roughly 24 pigments out of 700 are found in human diets (Prakash and Gupta 2014). Zeaxanthin, lutein, β-carotene, α-carotene, and lycopene are the most studied carotenoids. Lycopene, β-carotene, lutein, and β-cryptoxanthin are the most common carotenoids found in foods. β-carotene and lycopene have been linked to a lower risk of cardiovascular disease and some malignancies, but lutein and zeaxanthin have been linked to eye issues (Tapiero 2004; Yasui et al. 2011). Lutein helps to prevent malignancies of the uterus, prostate, breast, colon, and lungs (Yang et al. 2013). They may also help to reduce the risk of stomach cancer. Carotenoids' antioxidant qualities have been proposed as the primary method by which they achieve their positive benefits.

1.7.3 Lycopene

Many fruits and vegetables contain lycopene, a carotenoid with no provitamin-A action. Consumption of lycopene-rich tomatoes and tomato products has been linked to a lower risk of chronic illnesses, including cancer and cardiovascular disease (Sesso et al. 2005). Lycopene levels in the blood and tissues have also been found to be inversely associated with the risk of chronic illness (Kristenson et al. 1997).

1.7.4 Flavonoids

Flavonoids are polyphenolic substances found in olive oil; legumes; chocolate; tea; peanuts; walnuts; spices; green vegetables; onion; and fruits such as grapes, apples, and berries (Panche et al. 2016). Antibacterial, antioxidant, antiviral, and analgesic effects are all found in flavonoids. Flavonoids, which comprise flavonols, minor flavonoids (flavanones and dihydroflavonols), and flavones, are a subclass of plant phenols. The most common flavonoids found in foods are flavonols. The primary members of this category include quercetin and kaempferol. They're usually found in low quantities of 15–30 mg/kg FW. Flavonols are abundant in onions, leeks, broccoli, and blueberries. Tomatoes and certain fragrant plants, such as mint (*Mentha piperita*), contain flavanones, but only citrus fruits have large quantities (Panche et al. 2016). The primary flavanones found in grapefruit, oranges, and lemons are naringenin, hesperetin, and eriodictyol. Bacteria, ulcers, tumors, viruses, free radical–mediated cellular signaling, allergies, inflammation, hepatotoxins, and platelet aggregation are all targets for flavonoids (Gupta and Prakash 2014).

1.7.5 Isoflavonoids

Another subtype of phenolic phytonutrients is isoflavonoids. Isoflavonoids are nearly solely generated by the Fabaceae (Leguminosae) family of plants. Tofu, soy cheese, soybean, and soy flour are the most common sources. Soybeans have a high concentration of isoflavones, such as genistein, and soy and daidzein are the major dietary suppliers of isoflavones. Because of their ability to adhere to the estrogen receptor family of substances, soy isoflavones have gotten a lot of attention (Liu et al. 2018). They constitute an activity of a group of phytochemicals known as phytoestrogens. Most of the cancer cells (hormone-dependent and -independent), particularly colon cancer cells, are inhibited by genistein in vitro. Isoflavones have gotten a lot of interest for their ability to prevent tumors and osteoporosis (Liu et al. 2018). Glucosinolates, found in cruciferous vegetables including broccoli, cauliflower, mustard, cabbage, and others, are liver detoxification enzyme activators. The pungent scent and bitter flavor of cruciferous vegetables are due to these compounds. The use of cruciferous vegetables protects against cancer and other kinds of reactive oxygen damage. Glucoraphanin is found in larger concentrations in the sprouts of various crucifers, such as broccoli and cauliflower, than in mature plants (Farnham et al. 2005; Sao et al. 2021). Table 1.1 includes a few phytopharmaceuticals along with their therapeutic advantages.

TABLE 1.1
A Few Phytopharmaceuticals and Their Therapeutic Benefits

S. No	List of Phytopharmaceuticals (Medicinal Plants/Aromatics Plants/Spices)	Plant Scientific Name/ Common Name	Economically Important Plant Part	Types of Metabolites (Tannin/Flavonoids, etc.)	Medicinal Importance	Biotechnological Interventions in These Plants
1.	Medicinal and aromatic	*Atropa belladonna/* Belladonna	Root	Atropine	Impaired memory retrieval	CRISPR/Cas9 system (Zeng et al. 2021)
2.	Medicinal and aromatic	*Coffea arabica, Camellia sinensis/*Coffee, tea	Seeds and leaves	Caffeine	CNS stimulant, decreased tonic immobility	Identify SNPs associated with caffeine content (Tran et al. 2018)
3.	Medicinal	*Erythroxylum coca/*Cocoa	Leaves	Cocaine	CNS stimulant	Genetic relationships among *Erythroxylum* species (Emche et al. 2011)
4.	Medicinal tree	*Ginkgo biloba/*Maidenhair tree	Leaves and seeds	Bilobalide and ginkgolides	Dementia, age-related cognitive impairment	LTR-RTs (Zou et al. 2021)
5.	Aromatic, culinary, and medical herb	*Melissa officinali/*Lemon balms	Leaves	Monoterpenoids and sesquiterpenes	Dementia, cognitive impairment	Transcriptome analysis (Mansouri and Mohammadi 2021)
6.	Medicinal herbs	*Panax ginseng/*Korean ginseng	Root	Ginsenosides	Cognitive performance	Genome and evolutionary study (Kim et al. 2018)
7.	Medicinal	*Podophyllum peltatu/*May apple	Root and underground stem (rhizome)	Podophyllotoxin	Anticancer	Germplasm conservation by synthetic seed technology (Lata et al. 2010)
8.	Aromatic annual herb	*Artemisia annu/*Sweet wormwood	Aerial parts and leaves	Artemisinin	Antimalarial	Chloroplast genome transformation (Kaushal et al. 2020)
9.	Medicinal	*Digitalis lanata/*Grecian foxglove	Leaves	Digoxin	Cardiotonic	Increasing productivity of digoxin (Bekhit 2009)
10.	Medicinal	*Mucuna prurien/*Cowitch	Seeds	(L)-dopa	Parkinsonism	De novo transcriptome assembly (Sathyanarayana et al. 2017)
11.	Medicinal	*Dioscorea deltoidea/*Nepal yam	Leaf, stem, tuber	Diosgenin	Antiproliferative	Genetic, and phytochemical fidelity assessment (Nazir et al. 2022)
12.	Medicinal	*Papaver somniferum/*Opium poppy	Seeds	Morphine	Analgesic	Genotyping-by-sequencing analysis (Hong et al. 2022)

13.	Medicinal	*Colchicum autumnale/* Autumn crocus	Bulb-like corms	Colchicine	Antitumor	Transcriptome analysis (Bass et al. 2019)
14.	Medicinal and aromatic	*Cinchona ledgeriana/*Peruvian bark	Bark	Quinine	Antimalarial	Transformed root cultures (Hamill et al. 1989)
15.	Medicinal	*Silybum marianum/*Milk thistle	Above-ground parts and seeds	Silymarin	Hepatic disorders	Gene expression analysis and metabolite profiling (Drouet et al. 2020)
16.	Medicinal shrub	*Rauwolfia serpentina/* Snakeroot	Root	Ajmaline	Antiarrhythmic	Hairy root cultures (Mehrotra et al. 2016)
17.	Medicinal fruit and food	*Berberis vulgaris/*Common barberry	Fruit, bark, root, and stem	Berberine	Antidiarrheal	Determining the efficacy and toxicity (Ghafourian et al. 2017)
18.	Medicinal	*Ephedra sinica/*Yellow horse, sea grape	Stem and branches	Ephedrine	Antiasthmatics	*N*-methyltransferase (Morris et al. 2018)
19.	Medicinal and aromatic	*Carapichea ipecacuanha/* Ipecac	Root	Ementine	Antiprotozoal	Clonal diversity and conservation genetics (Oliveira et al. 2010)
20.	Ornamental and medicinal plant	*Catharanthus roseus/* Bright eyes, Sada Bahar	Leaf, root, shoot, and stem	Vincristine, vinblastine	Antitumor	Cytogenetic characterization and genome size (Guimarães et al. 2012)
21.	Folk medicinal plant	*Scoparia dulcis/*Sweet broom	Herb	Terpenoids, flavonoids, tannins, alkaloids, hexacosonol, B-sitosterol, ketone, dulcitone, and amellin	Antitumoral, Antidiabetic activity, antimicrobial activity, antisickling activity, antihyperlipidemic activity	RNA-Seq Analysis to elucidate terpenoid metabolism (Yamamura et al. 2017)
22.	Medicinal	*Ajuga multiflora/* Bugleweed, ground pine	Aerial Part	Phytoecdysteroids, sesquiterpenoids, diterpenoids, triterpenoids, iridoids, withanolides	Rheumatic fevers, dysentery, malaria, hypertension, diabetes and gastrointestinal disorders	Genetic transformation for 20-hydroxyecdysone (Kim et al. 2005)
23.	Medicinal	*Drosera rotundifolia/* Round-leaved sundew or common sundew	Whole plant	Pharmacologically active compound 1,4-naphthoquinones	Treatment of asthma, coughs, and lung diseases	Genetic engineering of biochemical pathways (Hirsikorpi et al. 2002)

(Continued)

TABLE 1.1
(Continued)

S. No	List of Phytopharmaceuticals (Medicinal Plants/Aromatics Plants/Spices)	Plant Scientific Name/ Common Name	Economically Important Plant Part	Types of Metabolites (Tannin/Flavonoids, etc.)	Medicinal Importance	Biotechnological Interventions in These Plants
24.	Medicinal	*Eritrichium sericeum*	Callus and root culture	Rabdosiin and related caffeic acid metabolites	Anti-HIV and anti-allergic	Genetic transformation of *Rolc* gene (Bulgakov et al. 2005)
25.	Medicinal	*Aesculus glabra*/Ohio buckeye	Leaves	Flavonoids	Rheumatic pain and antimalarials	Somatic embryogenesis (Trick and Finer 1999)
26.	Medicinal	*Taxus bacata*/Common yew	Leaves and bark	Alkaloids, taxine A and B, ephedrine, taxiphyllin, A, cardiac glycoside	Anticancer drugs	Metabolic engineering for improving production of taxol (Kundu et al. 2017)
27.	Medicinal plant	*Tylophora indica*/ Antamool, damabuti	Aerial parts (mostly leaves), roots	Alkaloid, tylophorine	Immunosuppressive, anti-inflammatory, antitumor, anti-amoebic, and anticandidal properties	Genetic transformation (Chaudhuri et al. 2005)
28.	Woody medicinal shrub	*Rauvolfia micrantha*/ Small-flowered snakeroot	Roots	Alkaloids such as reserpine, reserpiline, serpentine, ajmaline, and ajmalicine	Antiarrhytmic	Hairy root cultures (Sudha et al. 2003)
29.	Aromatic and medicinal	*Lavandula intermedia*/ Hybrid lavender	Leaves and flower	Terpenoid substances such as linalyl acetate and linalool	Antiseptic, antibacterial, antifungal, anti-inflammatory, anticonvulsive, and antidepressant	RNAi study (Tsuro et al. 2019)
30.	Spices	*Trigonella foenum-graceum L.*/Fenugreek	Seeds	Saponins, flavonoids, alkaloids, and galactomannans	Antioxidant, anticancer, antidiabetic	Mutation breeding using sodium azide (Prabha et al. 2010)

31.	Spices	*Trachyspermum Ammi/* Ajwain	Flower, seeds, leaf	Thymol, carvacrol, alpha and beta-pinene, terpinene and para-cymene	Antifungal, antioxidant, antimicrobial, antinociceptive, cytotoxic activity	Transcriptome analysis to identify putative genes involved in thymol biosynthesis (Soltani Howyzeh et al. 2018)
32.	Spices	*Piper nigrum L./*Black pepper	Seeds	Alkaloid piperine	Reduction of insulin resistance, anti-inflammatory effects, and improvement of hepatic steatosis	Identification and characterization of piperine synthase (Schnabel et al. 2021)
33.	Spices	*Allium sativum/*Garlic	Root bulb	Sulfur compound allicin	Cardiovascular, antineoplastic, and antimicrobial properties	Antimicrobial gene identification1, role of diallyl sulfides and dipropyl sulfides (Casella et al. 2013)

1.7.6 Polyphenols

Polyphenols are natural substances that may be associated with a range of foods and drinks, including fruits, vegetables, cereals, and beverages. Polyphenols can also be found in legumes and chocolate. These compounds are secondary metabolites produced by plants and are used to protect against ultraviolet (UV) radiation and disease attack (Brglez Mojzer et al. 2016). Antioxidant-rich diets have been linked to a decreased risk of degenerative illnesses. Polyphenols may be found in apples, grapes, pears, cherries, and a variety of berries, with 200–300 mg per 100 g FW. However, there is little known regarding polyphenol absorption, both native and processed types, by the intestine's microorganisms (bacteria). Dietary phytochemicals (e.g., capsaicin, curcumin, β-carotene, catechins, resveratrol, and vitamins) are thought to reduce inflammation and to be preventive in cancer, neurological disorders, and cardiovascular illnesses, among other things (Zhang et al. 2015). A wide range of polyphenols has been found to offer significant UV-protective properties on the skin (Brglez Mojzer et al. 2016).

1.7.7 Phytoestrogens

Nonsteroidal phytoestrogens are phytochemicals that are structurally and functionally identical to the gonadal estrogen hormone. They provide an appealing alternative to hormone replacement treatment (HRT), with positive effects on the cardiovascular system and the potential to relieve menopausal symptoms (Estrada-Camarena et al. 2017). They might be a viable alternative to the synthetic selective estrogen receptor modulators used in HRT today. Phytoestrogens are categorized as flavonoids, lignans, isoflavonoids, coumestans, and stilbenes based on their chemical structure. They can be found in plants or their seeds. Soybeans are high in isoflavones, whereas soy sprouts are high in coumestrol, the main coumestan. Flavonoids have a structure that is similar to that of estrogens, and they can have both estrogenic and anti-estrogenic effects, as well as protect against bone loss and heart disease.

1.7.8 Resveratrol

Resveratrol is a phytoalexin produced naturally by plants in response to stress and pathogenic disease. Resveratrol is demonstrated to have neuroprotective properties along with positive impacts on the cardiovascular system at low doses. Its antioxidant capabilities are mostly responsible for these actions. Neuroprotective, antithrombogenic, cancer preventative, anti-inflammatory, therapeutic properties, cardioprotective, and anti-aging are the most obvious biological activities (Ranjan et al. 2021). Grapes and other foods have high levels of resveratrol. It's also found in eucalyptus, spruce, and lily trees, as well as blueberries, mulberries, strawberries, hops, peanuts, and their derivatives.

1.7.9 Phytosterols

Phytosterols are plant sterols and plant stanols. Phytosterols reduce total and low-densoity lipoprotein (LDL) blood cholesterol by inhibiting cholesterol intake from the gut, and they've been used to decrease blood cholesterol for almost a half-century (Cabral and Klein 2017). Plant sterols may be found in a variety of foods, including fruits, vegetables, nuts, and oils. Studies have shown that dosages of 2–3 g per day provide the best cholesterol-lowering results. As a result, it suggests the necessity for enhanced functional foods that offer sufficient phytosterol consumption, decreasing LDL cholesterol levels (Cabral and Klein 2017).

1.7.10 Terpenoids

Terpenes, also termed isoprenoids, are the most abundant phytonutrients found in plants and cereals. Higher plants, lichens, algae, mosses, and liverworts, along with insects, microorganisms, and marine creatures, contain these chemicals. Their relevance to plants stems from the need for them to fix carbon via photosynthetic processes involving photosensitizing pigments. Species have adapted to use these substances for hormonal and growth regulatory activities (vitamin A), and it is now known that the existence of such substances in organisms protects against several illnesses, particularly those involving growth dysregulation and chronic damage (Kim et al. 2020). Terpenes are also used to help penetrate the skin.

1.8 SOME EXOTIC MEDICINAL PLANTS

Because of the inability of contemporary therapies to treat chronic diseases and their very substantial side effects, the use of herbal medicines has grown in popularity in recent years. Various current medications and their synthetic counterparts have been developed from plant prototype molecules (Snijesh and Singh 2014). There is a wealth of research on the possible therapeutic use of plant isolates. Standardized extracts produced from *Commiphora guggul* and *Zingiber officinale* are two examples of phytopharmaceutical medications that are recommended by doctors and accessible on the Indian market (Savarikar et al. 2011; Levine et al. 1973).

1.8.1 *Hypericum perforatum*

St. John's wort is a great example of how secondary plant metabolites may have a wide variety of properties and effects. Hyperici herba (herbal medication) is made from dried *Hypericum perforatum* flowers and aerial parts. The formulation also contains additional flavonoids (hyperoside) and xanthones, as well as naphthodianthrone hypericin (antiviral action), the phloroglucinol derivative hyperforin (antibacterial effect), and other flavonoids (hyperoside). Thin-layer chromatography is used to identify the presence of all of these components in the completed pharmaceutical product, which is made from hydroalcoholic extracts (ethanol 50–60 percent, methanol 80 percent). Though hyperforin and hypericines are responsible for the herbal substance's antiviral and antibacterial properties, the exact mechanism of action of its antidepressant effect is still unknown. The whole extract is the active element in this situation since it has a medicinal impact (Obmann 2010).

1.8.2 *Solanum lycopersicum*

Tomatoes are renowned as a health-promoting fruit due to their unique combination of polyphenols. The principal bioactive chemicals found in ripened tomatoes are phenolics and carotenoids. A substantial quantity of lycopene is responsible for the red color of a ripe tomato (Marti et al. 2016; Perveen et al. 2015). The tomato fruit also has β-carotene, a provitamin A–like substance. Tomato carotenoids (lycopene) and vitamins (ascorbic acid and α-tocopherol) help to reduce oxidative stress and reduce the risk of cancer and cardiovascular disease. Lycopene levels in the blood are believed to be inversely related to the risk of cardiovascular disease (Sesso et al. 2004). Tomato consumption has been linked to a decreased risk of autoimmune disorders such as atherosclerosis (Hazewindus et al. 2014). LDL oxidative regulation is important for preventing atherosclerosis and cardiovascular disease (CVD). CVD is prevented by the influence of lipophilic tomato chemicals on LDLs on atherogenic mechanisms in endothelial cells (Viuda-Martos et al. 2014).

1.8.3 *Digitalis purpurea*

Digitalis purpurea L. (foxglove) is a member of the Scrophulariaceae family that is found across Europe. Cardiovascular glycosides, flavonoids, anthraquinones, and triterpenes were found in the phytochemical screening. The presence of several cardioactive glycosides such as digitoxin, which has shown to be one of the most essential medications in treating congestive heart failure, is the most important phytoconstituent supplied by *D. purpurea* (Haji and Movahed 2000). The glycosides of *D. purpurea* are thought to have wound healing, hepatoprotective, antioxidant, and cytotoxic properties, in addition to heart stimulant properties. In atrial fibrillation, atrial flutter, supraventricular tachycardia, and premature extrasystoles, many of them have a lowering effect on ventricular rate.

1.8.4 *Filipendula ulmaria*

Salicylic acid, derived from *Filipendula ulmaria* (often referred as meadowsweet), and acetylsalicylic acid, derived from willows, do have an ancient legacy of treating pain and fever but were made chemically for over a century (Kovaleva et al. 2018). The flowers are used in medications used to treat the common cold, infections, fever, peptic ulcer illness, gout, and rheumatism, among other conditions (Jarić et al. 2007; Šarić-Kundalić et al. 2011; Vogl et al. 2013). The traditional herbal medicinal products from meadowsweet flowers and *F. ulmariae* flowers have been authorized by the European Medicines Agency (Samardžić et al. 2018).

1.8.5 *Melissa officinalis*

The Lamiaceae family's *Melissa officinalis* (lemon balm) is a well-known herb that has historically been used in traditional medicine to cure a variety of ailments, and various research has been undertaken to identify the plant and its medicinal effects (Akhondzadeh et al. 2003). Because of its volatile natural substances and active compounds like terpenoids, quercetin, quercitrin, flavonoids, gallic acid, and rutin and strong antioxidant activity, the plant's extract can play a crucial role in protecting health and curing ailments (Zarei et al. 2015). Furthermore, a retrospective examination of the historical function of several European herbs in the enhancement of cognition, namely memory, suggests that *M. officinalis* and another labiatae family plant, *Salvia officinalis*, might be used as a natural therapy for Alzheimer's disease (Perry et al. 1998, 1999). *M. officinalis* possesses acetylcholine receptor activity in the central nervous system, including nicotinic and muscarinic binding capabilities, according to studies (Wake et al. 2000). Furthermore, recent research on healthy young volunteers found that this herb influences mood and cognitive function during acute administration (Kennedy et al. 2002).

1.9 CONCLUSION

Phytonutrients provide a number of health advantages and may be consumed daily to lower risk factors for conditions such as diabetes, high blood pressure, and high cholesterol. The food business has tremendous prospects to generate unique food items, including bioactive food components of both plant and animal origin, in the future of nutraceuticals. However, there is a need for investment and dedication in advancing phytopharmaceutical medication research and development. Furthermore, it is critical to promote items to pique consumers' interests and preferences. The current circumstances are favorable for the expansion of the phytopharmaceutical business, as long as the new set of rules is followed and all available chances are effectively utilized.

BIBLIOGRAPHY

Ahmad, A. S., & Sharma, R. (2020). Comparative analysis of herbal and allopathic treatment systems. *European Journal of Molecular & Clinical Medicine*, *7*(7), 2869–2876.

Akhondzadeh, S., Noroozian, M., Mohammadi, M., Ohadinia, S., Jamshidi, A. H., & Khani, M. (2003). Melissa officinalis extract in the treatment of patients with mild to moderate Alzheimer's disease: A double blind, randomised, placebo controlled trial. *Journal of Neurology, Neurosurgery & Psychiatry*, *74*(7), 863–866.

Bass, J. S., Gang, D. R., Kutchan, T. M., & Sivakumar, G. (2019). Gloriosa superba and Colchicum autumnale multi-tissue transcriptome analysis for colchicine pathway and rhizome development candidate gene identification. *Research Square*. https://doi.org/10.21203/rs.2.9594/v1

Bekhit, M. H. (2009). Increasing productivity of Digoxin content in *Digitalis lanata ehrh*. *Journal of Plant Production*, *34*(1), 333–345.

Bhatt, A. (2016). Phytopharmaceuticals: A new drug class regulated in India. *Perspectives in Clinical Research*, *7*(2), 59.

Bhusnure, O. G., Shinde, M. C., Vijayendra, S. S., Gholve, S. B., Giram, P. S., & Birajdar, M. J. (2019). Phytopharmaceuticals: An emerging platform for innovation and development of new drugs from botanicals. *Journal of Drug Delivery and Therapeutics*, *9*(3-s), 1046–1057.

Brglez Mojzer, E., Knez Hrnčič, M., Škerget, M., Knez, Ž., & Bren, U. (2016). Polyphenols: Extraction methods, antioxidative action, bioavailability and anticarcinogenic effects. *Molecules*, *21*(7), 901.

Bulgakov, V. P., Veselova, M. V., Tchernoded, G. K., Kiselev, K. V., Fedoreyev, S. A., & Zhuravlev, Y. N. (2005). Inhibitory effect of the Agrobacterium rhizogenes rolC gene on rabdosiin and rosmarinic acid production in Eritrichium sericeum and Lithospermum erythrorhizon transformed cell cultures. *Planta*, *221*(4), 471–478.

Büssing, A., Ostermann, T., Heusser, P., & Matthiessen, P. F. (2011). Usage of alternative medical systems, acupuncture, homeopathy and anthroposophic medicine, by older German adults. *Zhong xi yi jie he xue bao= Journal of Chinese Integrative Medicine*, *9*(8), 847–856.

Cabral, C. E., & Klein, M. R. S. T. (2017). Phytosterols in the treatment of hypercholesterolemia and prevention of cardiovascular diseases. *Arquivos brasileiros de cardiologia*, *109*, 475–482.

Casella, S., Leonardi, M., Melai, B., Fratini, F., & Pistelli, L. (2013). The role of diallyl sulfides and dipropyl sulfides in the in vitro antimicrobial activity of the essential oil of garlic, Allium sativum L., and leek, Allium porrum L. *Phytotherapy Research*, *27*(3), 380–383.

Chaudhuri, K. N., Ghosh, B., Tepfer, D., & Jha, S. (2005). Genetic transformation of Tylophora indica with Agrobacterium rhizogenes A4: Growth and tylophorine productivity in different transformed root clones. *Plant Cell Reports*, *24*(1), 25–35.

Drouet, S., Tungmunnithum, D., Lainé, É., & Hano, C. (2020). Gene expression analysis and metabolite profiling of silymarin biosynthesis during milk thistle (*Silybum marianum* (L.) Gaertn.) fruit ripening. *International Journal of Molecular Sciences*, *21*(13), 4730. https://doi.org/10.3390/ijms21134730

EMA HMPC (2009). Community herbal monograph on Hypericum perforatum L., herba (well-established medicinal use). EMA/HMPC/101304/2008.

EMA HMPC (2014). Questions & answers on the EU framework for (traditional) herbal medicinal products, including those from a 'non-European' tradition. EMA/HMPC/402684/2013.

EMA HMPC (2015). Community herbal monograph on *Ginkgo biloba* L., folium. EMA/HMPC/321097/2012.

EMA HMPC (2016). European Union herbal monograph on Crataegus spp., folium cum flore, EMA/HMPC/159075/2014.

Emche, S. D., Zhang, D., Islam, M. B., Bailey, B. A., & Meinhardt, L. W. (2011). AFLP phylogeny of 36 Erythroxylum species. *Tropical Plant Biology*, *4*(2), 126–133.

Estrada-Camarena, E., López-Rubalcava, C., Valdés-Sustaita, B., Azpilcueta-Morales, G. S., & González-Trujano, E. M. (2017). Use of phytoestrogens for the treatment of psychiatric symptoms associated with menopause transition. In Rodríguez-Landa, J. F. & Cueto-Escobedo, J. (Eds.), *A Multidisciplinary Look at Menopause* (pp. 81–109). Intech Open, London, UK.

Farnham, M. W., Stephenson, K. K., & Fahey, J. W. (2005). Glucoraphanin level in broccoli seed is largely determined by genotype. *HortScience*, *40*(1), 50–53.

Ferguson, L. R. (2001). Role of plant polyphenols in genomic stability. *Mutation Research/Fundamental and Molecular Mechanisms of Mutagenesis*, *475*(1–2), 89–111.

Ferrazzano, G. F., Amato, I., Ingenito, A., Zarrelli, A., Pinto, G., & Pollio, A. (2011). Plant polyphenols and their anti-cariogenic properties: A review. *Molecules*, *16*(2), 1486–1507.

Ghafourian, E., Sadeghifard, N., Pakzad, I., Valizadeh, N., Maleki, A., Jafari, F., . . . Ghafourian, S. (2017). Ethanolic extract of Berberis vulgaris fruits inhibits the proliferation of MCF-7 breast cancer cell line through induction of apoptosis. *Infectious Disorders-Drug Targets (Formerly Current Drug Targets-Infectious Disorders)*, *17*(3), 192–198.

Guimarães, G., Cardoso, L., Oliveira, H., Santos, C., Duarte, P., & Sottomayor, M. (2012). Cytogenetic characterization and genome size of the medicinal plant Catharanthus roseus (L.) G. Don. *AoB PLANTS*, *2012*, pls002. https://doi.org/10.1093/aobpla/pls002

Gupta, C., & Prakash, D. (2014). Phytonutrients as therapeutic agents. *Journal of Complementary and Integrative Medicine*, *11*(3), 151–169.

Haji, S. A., & Movahed, A. (2000). Update on digoxin therapy in congestive heart failure. *American Family Physician*, *62*(2), 409–416.

Hamill, J. D., Robins, R. J., & Rhodes, M. J. (1989). Alkaloid production by transformed root cultures of Cinchona ledgeriana. *Planta Medica*, *55*(4), 354–357.

Hazewindus, M., Haenen, G. R., Weseler, A. R., & Bast, A. (2014). Protection against chemotaxis in the anti-inflammatory effect of bioactives from tomato ketchup. *PLoS One*, *9*(12), e114387.

Hirsikorpi, M., Kämäräinen, T., Teeri, T., & Hohtola, A. (2002). Agrobacterium-mediated transformation of round leaved sundew (Drosera rotundifolia L.). *Plant science*, *162*(4), 537–542.

Hong, U. V. T., Tamiru-Oli, M., Hurgobin, B., Okey, C. R., Abreu, A. R., & Lewsey, M. G. (2022). Insights into opium poppy (Papaver spp.) genetic diversity from genotyping-by-sequencing analysis. *Scientific Reports*, *12*(1), 1–15.

Jaiswal, Y. S., Guan, Y., Moon, K. H., & Williams, L. L. (2019). Anthocyanins: Natural sources and traditional therapeutic uses. In *Flavonoids-A Coloring Model for Cheering Up Life*. IntechOpen, London, UK.

Jarić, S., Popović, Z., Mačukanović-Jocić, M., Djurdjević, L., Mijatović, M., Karadžić, B., . . . Pavlović, P. (2007). An ethnobotanical study on the usage of wild medicinal herbs from Kopaonik Mountain (Central Serbia). *Journal of Ethnopharmacology*, *111*(1), 160–175.

Karimi, A., Majlesi, M., & Rafieian-Kopaei, M. (2015). Herbal versus synthetic drugs; beliefs and facts. *Journal of Nephropharmacology*, *4*(1), 27.

Kaushal, C., Abdin, M. Z., & Kumar, S. (2020). Chloroplast genome transformation of medicinal plant Artemisia annua. *Plant Biotechnology Journal*, *18*(11), 2155.

Kennedy, D. O., Scholey, A. B., Tildesley, N. T., Perry, E. K., & Wesnes, K. A. (2002). Modulation of mood and cognitive performance following acute administration of Melissa officinalis (lemon balm). *Pharmacology Biochemistry and Behavior*, *72*(4), 953–964.

Khoo, H. E., Azlan, A., Tang, S. T., & Lim, S. M. (2017). Anthocyanidins and anthocyanins: Colored pigments as food, pharmaceutical ingredients, and the potential health benefits. *Food & Nutrition Research*, *61*(1), 1361779.

Kim, N. H., Jayakodi, M., Lee, S. C., Choi, B. S., Jang, W., Lee, J., . . . Yang, T. J. (2018). Genome and evolution of the shade-requiring medicinal herb Panax ginseng. *Plant Biotechnology Journal*, *16*(11), 1904–1917.

Kim, O. T., Manickavasagm, M., Kim, Y. J., Jin, M. R., Kim, K. S., Seong, N. S., & Hwang, B. (2005). Genetic transformation of Ajuga multiflora bunge with Agrobacterium rhizogenes and 20-hydroxyecdysone production in hairy roots. *Journal of Plant Biology*, *48*(2), 258–262

Kim, T., Song, B., Cho, K. S., & Lee, I. S. (2020). Therapeutic potential of volatile terpenes and terpenoids from forests for inflammatory diseases. *International Journal of Molecular Sciences*, *21*(6), 2187.

Koetter, U., Schrader, E., Käufeler, R., & Brattström, A. (2007). A randomized, double blind, placebo-controlled, prospective clinical study to demonstrate clinical efficacy of a fixed valerian hops extract combination (Ze 91019) in patients suffering from non-organic sleep disorder. *Phytotherapy Research: An International Journal Devoted to Pharmacological and Toxicological Evaluation of Natural Product Derivatives*, *21*(9), 847–851.

Kovaleva, T. Y., Ermakova, V. A., Aleksandrovna, D., Trashchenkova, E. A. D., Bokov, D. O., Shilova, I. V., & Samylina, I. A. (2018). Comparative study of the biologically active substances composition and content in meadowsweet (Filipendula ulmaria (l.) Maxim) crude herbal drugs (herb, leafs, flowers) of Russian origin. *International Journal of Pharmaceutical Quality Assurance*, *9*(3), 277–280.

Kristenson, M., Zieden, B., Kucinskiene, Z., Abaravicius, A., RazinkovienË, L., Elinder, L. S., . . . Olsson, A. G. (1997). Antioxidant state and mortality from coronary heart disease in Lithuanian and Swedish men: Concomitant cross sectional study of men aged 50. *Bmj*, *314*(7081), 629.

Kundu, S., Jha, S., & Ghosh, B. (2017). Metabolic engineering for improving production of taxol. *Transgenesis and Secondary Metabolism*, 463.

Lata, H., Moraes, R. M., Bertoni, B., & Pereira, A. (2010). In vitro germplasm conservation of Podophyllum peltatum L. under slow growth conditions. *In Vitro Cellular & Developmental Biology-Plant*, *46*(1), 22–27.

Levine, F. M., Spitalnik, R., & Dobos, C. (1973). Caudate nucleus effects on geriatric senility: effect of belladonna on learning and memory of geriatric patients. *Perceptual and Motor Skills*, *37*(3), 1003–1007.

Linde, K., Berner, M. M., & Kriston, L. (2008). St John's wort for major depression. *Cochrane Database of Systematic Reviews*, *4*.

Liu, E. Y., Xu, M. L., Jin, Y., Wu, Q., Dong, T. T., & Tsim, K. W. (2018). Genistein, a phytoestrogen in soybean, induces the expression of acetylcholinesterase via G protein-coupled receptor 30 in PC12 cells. *Frontiers in Molecular Neuroscience*, *11*, 59.

Mansouri, M., & Mohammadi, F. (2021). Transcriptome analysis to identify key genes involved in terpenoid and rosmarinic acid biosynthesis in lemon balm (Melissa officinalis). *Gene*, *773*, 145417.

Marti, R., Roselló, S., & Cebolla-cornejo, J. (2016). Tomato as a source of carotenoids and polyphenols targeted to cancer prevention. *Cancers*, *8*(6), 58.

Mehrotra, S., Srivastava, V., Goel, M. K., & Kukreja, A. K. (2016). Scale-up of agrobacterium rhizogenes-mediated hairy root cultures of *Rauwolfia serpentina*: A persuasive approach for stable reserpine production. In *Protocols for In Vitro Cultures and Secondary Metabolite Analysis of Aromatic and Medicinal Plants* (2nd ed., pp. 241–257). Humana Press, New York, NY.

Morris, J. S., Groves, R. A., Hagel, J. M., & Facchini, P. J. (2018). An *N*-methyltransferase from *Ephedra sinica* catalyzing the formation of ephedrine and pseudoephedrine enables microbial phenylalkylamine production. *The Journal of Biological Chemistry*, *293*(35), 13364–13376. https://doi.org/10.1074/jbc.RA118.004067

Nazir, R., Gupta, S., Dey, A., Kumar, V., Gupta, A. P., Shekhawat, M. S., . . . Pandey, D. K. (2022). In vitro tuberization, genetic, and phytochemical fidelity assessment of Dioscorea deltoidea. *Industrial Crops and Products*, *175*, 114174.

Obmann, A. (2010). Pharmakognosie—Phytopharmazie. *Scientia Pharmaceutica*, *78*(2), 381. https://doi.org/10.3797/scipharm.br-10-01

Oliveira, L. O. D., Venturini, B. A., Rossi, A. A. B., & Hastenreiter, S. S. (2010). Clonal diversity and conservation genetics of the medicinal plant Carapichea ipecacuanha (Rubiaceae). *Genetics and Molecular Biology*, *33*(1), 86–93.

Panche, A. N., Diwan, A. D., & Chandra, S. R. (2016). Flavonoids: An overview. *Journal of Nutritional Science*, *5*.

Perry, E. K., Pickering, A. T., Wang, W. W., Houghton, P., & Perry, N. S. (1998). Medicinal plants and Alzheimer's disease: Integrating ethnobotanical and contemporary scientific evidence. *The Journal of Alternative and Complementary Medicine*, *4*(4), 419–428.

Perry, E. K., Pickering, A. T., Wang, W. W., Houghton, P. J., & Perry, N. S. (1999). Medicinal plants and Alzheimer's disease: From ethnobotany to phytotherapy. *Journal of Pharmacy and Pharmacology*, *51*(5), 527–534.

Perveen, R., Suleria, H. A. R., Anjum, F. M., Butt, M. S., Pasha, I., & Ahmad, S. (2015) Tomato (Solanum lycopersicum) carotenoids and lycopenes chemistry; metabolism, absorption, nutrition, and allied health claims—a comprehensive review. *Crit Rev Food Sci Nutr*, *55*(7), 919–929.

Prabha, R., Dixit, V., & Chaudhary, B. R. (2010). Sodium azide-induced mutagenesis in Fenugreek (Trigonella foenum graecum Linn). *Legume Research*, *33*, 235–241.

Prakash, D., & Gupta, C. (2014). 12 carotenoids: Chemistry and health benefits. In *Phytochemicals of Nutraceutical Importance* (p. 181). CAB International, Oxfordshire, UK.

Ranjan, S., Dinda, S. C., & Verma, A. (2021). Resveratrol as anti-cancer and cardio protective agent. *Oncology and Radiotherapy*, *15*(12), 1–12.

Samardžić, S., Arsenijević, J., Božić, D., Milenković, M., Tešević, V., & Maksimović, Z. (2018). Antioxidant, anti-inflammatory and gastroprotective activity of Filipendula ulmaria (L.) Maxim. and Filipendula vulgaris Moench. *Journal of Ethnopharmacology*, *213*, 132–137.

Sao, P., Chand, Y., Kumar, A., & Singh, S. (2021). Potential drug target identification in porphyromonas gingivalis using in-silico subtractive metabolic pathway analysis. *Bangladesh Journal of Medical Science*, *20*(4), 887–896.

Šarić-Kundalić, B., Dobeš, C., Klatte-Asselmeyer, V., & Saukel, J. (2011). Ethnobotanical survey of traditionally used plants in human therapy of east, north and north-east Bosnia and Herzegovina. *Journal of Ethnopharmacology*, *133*(3), 1051–1076.

Sathyanarayana, N., Pittala, R. K., Tripathi, P. K., Chopra, R., Singh, H. R., Belamkar, V., Bhardwaj, P. K., Doyle, J. J., & Egan, A. N. (2017). Transcriptomic resources for the medicinal legume Mucuna pruriens: De novo transcriptome assembly, annotation, identification and validation of EST-SSR markers. *BMC Genomics*, *18*(1), 409. https://doi.org/10.1186/s12864-017-3780-9

Savarikar, S. S., Barbhind, M. M., Halde, U. K., & Kulkarni, A. P. (2011). Pharmaceutical and analytical evaluation of triphalaguggulkalpa tablets. *Journal of Ayurveda and Integrative Medicine*, *2*(1), 21.

Schapowal, A. (2002). Randomised controlled trial of butterbur and cetirizine for treating seasonal allergic rhinitis. *BMJ*, *324*(7330), 144.

Schellenberg, R., Saller, R., Hess, L., Melzer, J., Zimmermann, C., Drewe, J., & Zahner, C. (2012). Dose-dependent effects of the Cimicifuga racemosa extract Ze 450 in the treatment of climacteric complaints: A randomized, placebo-controlled study. *Evidence-Based Complementary and Alternative Medicine*, *2012*.

Schnabel, A., Athmer, B., Manke, K., Schumacher, F., Cotinguiba, F., & Vogt, T. (2021). Identification and characterization of piperine synthase from black pepper, Piper nigrum L. *Communications Biology*, *4*(1), 1–10.

Schrader, E. (2000). Equivalence of St John's wort extract (Ze 117) and fluoxetine: A randomized, controlled study in mild-moderate depression. *International Clinical Psychopharmacology*, *15*(2), 61–68.

Sesso, H. D., Buring, J. E., Norkus, E. P., & Gaziano, J. M. (2004). Plasma lycopene, other carotenoids, and retinol and the risk of cardiovascular disease in women. *Am J Clin Nutr*, *79*(1), 47–53.

Sesso, H. D., Buring, J. E., Norkus, E. P., & Gaziano, J. M. (2005). Plasma lycopene, other carotenoids, and retinol and the risk of cardiovascular disease in men. *The American Journal of Clinical Nutrition*, *81*(5), 990–997.

Singh, A. (2018). *The future of phytopharmaceuticals*. www.pharmabiz.com/NewsDetails.aspx?aid=110880 &sid=9

Singh, A., Sengar, R. S., Singh, A. K., Singh, S., Kumar, A., & Ahmad, K. (2019). Impact of genetic engineering on agriculture food crops. *Biotech Today: An International Journal of Biological Sciences*, *9*(1), 17–24.

Snijesh, V. P., & Singh, S. (2014). Molecular modeling and network based approach in explaining the medicinal properties of nyctanthes arbortristis, lippia nodiflora for rheumatoid arthritis. *Journal of Bioinformatics and Intelligent Control*, *3*(1), 31–38.

Soltani Howyzeh, M., Sadat Noori, S. A., Shariati, J. V., & Amiripour, M. (2018). Comparative transcriptome analysis to identify putative genes involved in thymol biosynthesis pathway in medicinal plant Trachyspermum ammi L. *Scientific Reports*, *8*(1), 1–19.

Somani, S. J., Modi, K. P., Majumdar, A. S., & Sadarani, B. N. (2015). Phytochemicals and their potential usefulness in inflammatory bowel disease. *Phytotherapy Research*, *29*(3), 339–350.

Sudha, C. G., Reddy, B. O., Ravishankar, G. A., & Seeni, S. (2003). Production of ajmalicine and ajmaline in hairy root cultures of Rauvolfia micrantha Hook f., a rare and endemic medicinal plant. *Biotechnology Letters*, *25*(8), 631–636.

Tapiero, H. (2004). Influence of alcohol consumption and smoking habits on human health. *Biomedicine & Pharmacotherapy*, *58*(2), 75–76.

Tran, H. T., Ramaraj, T., Furtado, A., Lee, L. S., & Henry, R. J. (2018). Use of a draft genome of coffee (Coffea arabica) to identify SNP s associated with caffeine content. *Plant Biotechnology Journal*, *16*(10), 1756–1766.

Trick, H. N., & Finer, J. J. (1999). Induction of somatic embryogenesis and genetic transformation of Ohio buckeye (Aesculus glabra Willd.). *In Vitro Cellular & Developmental Biology-Plant*, *35*(1), 57–60.

Tsuro, M., Tomomatsu, K., Inukai, C., Tujii, S., and Asada, S. (2019). RNAi targeting the gene for 1,8-cineole synthase induces recomposition of leaf essential oil in lavandin (*Lavandula* × intermedia Emeric). *Vitr. Cell. Dev. Biol. Plant*, *55*, 165–171. doi: 10.1007/s11627-018-09949-z

Verma, N., Sao, P., Srivastava, A., & Singh, S. (2021). Physiological and molecular responses to drought, submergence and excessive watering in plants. In *Harsh Environment and Plant Resilience* (pp. 305–321). Springer, Cham.

Viuda-Martos, M., Sanchez-Zapata, E., Sayas-Barberá, E., Sendra, E., Pérez-Álvarez, J. A., & Fernández-López, J. (2014) Tomato and tomato byproducts. Human health benefits of lycopene and its application to meat products: A review. *Crit Rev Food Sci Nutr, 54*(8), 1032–1049.

Vogl, S., Picker, P., Mihaly-Bison, J., Fakhrudin, N., Atanasov, A. G., Heiss, E. H., . . . Kopp, B. (2013). Ethnopharmacological in vitro studies on Austria's folk medicine—An unexplored lore in vitro anti-inflammatory activities of 71 Austrian traditional herbal drugs. *Journal of Ethnopharmacology, 149*(3), 750–771.

Wake, G., Court, J., Pickering, A., Lewis, R., Wilkins, R., & Perry, E. (2000). CNS acetylcholine receptor activity in European medicinal plants traditionally used to improve failing memory. *Journal of Ethnopharmacology, 69*(2), 105–114.

Woelk, H. (2000). Comparison of St John's wort and imipramine for treating depression: Randomised controlled trial. *BMJ, 321*(7260), 536–539.

Yamamura, Y., Kurosaki, F., & Lee, J. B. (2017). Elucidation of terpenoid metabolism in Scoparia dulcis by RNA-seq analysis. *Scientific Reports, 7*(1), 1–14.

Yang, Y., Kim, B., & Lee, J. Y. (2013). Astaxanthin structure, metabolism, and health benefits. *J. Hum. Nutr. Food Sci, 1*(1003), 1–1003.

Yasui, Y., Hosokawa, M., Mikami, N., Miyashita, K., & Tanaka, T. (2011). Dietary astaxanthin inhibits colitis and colitis-associated colon carcinogenesis in mice via modulation of the inflammatory cytokines. *Chemico-Biological Interactions, 193*(1), 79–87.

Zarei, A., Changizi-Ashtiyani, S., Taheri, S., & Hosseini, N. (2015). A brief overview of the effects of Melissa officinalis L. extract on the function of various body organs. *Zahedan Journal of Research in Medical Sciences, 15*, 29–34.

Zeng, L., Zhang, Q., Jiang, C., Zheng, Y., Zuo, Y., Qin, J., . . . Deng, H. (2021). Development of Atropa belladonna L. Plants with high-yield hyoscyamine and without its derivatives using the CRISPR/Cas9 system. *International Journal of Molecular Sciences, 22*(4), 1731.

Zhang, Y. J., Gan, R. Y., Li, S., Zhou, Y., Li, A. N., Xu, D. P., & Li, H. B. (2015). Antioxidant phytochemicals for the prevention and treatment of chronic diseases. *Molecules, 20*(12), 21138–21156.

Zou, K., Liu, X., Hu, Q., Zhang, D., Fu, S., Zhang, S., Huang, H., Lei, F., Zhang, G., Miao, B., Meng, D., Jiang, L., Liu, H., Yin, H., & Liang, Y. (2021). Root endophytes and *Ginkgo biloba* are likely to share and compensate secondary metabolic processes, and potentially exchange genetic information by LTR-RTs. *Front. Plant Sci.*, 12, 704985. doi: 10.3389/fpls.2021.704985

2 Indian Spices
Treasure of Active Pharmaceutical Compounds

Priyanka Bhamu and Shaantanu Dabi

CONTENTS

2.1 Introduction 19
2.2 Spices: As Per Ayurveda 20
2.3 Plant Secondary Metabolites as Active Pharmaceutical Compounds 21
2.3.1 Cinnamon (*Cinnamomum zeylanicum*) 23
2.3.2 Black Pepper (*Piper nigrum*) 23
2.3.3 Clove (*Syzygium aromaticum*) 24
2.3.4 Cumin (*Cuminum cyminum*) 25
2.3.5 Garlic (*Allium sativum*) 25
2.3.6 Ginger (*Zingiber officinale*) 26
2.3.7 Fennel (*Foeniculum vulgare*) 26
2.3.8 Fenugreek (*Trigonella foenum-graecum*) 27
2.3.9 Saffron (*Crocus sativus*) 28
2.3.10 Turmeric (*Curcuma longa*) 28
2.3.11 Nutmeg and Mace (*Myristica fragrans*) 29
2.3.12 Carom Seeds (*Trachyspermum ammi*) 30
2.3.13 Asafoetida 30
2.3.14 Cardamom 31
2.3.15 Star Anise (*Illicium verum*) 32
2.4 Spices and COVID-19 33
2.4 Bibliography 43

2.1 INTRODUCTION

Spices and condiments have been used since ancient times to enhance the taste and flavor of cooked food. Spices are materials of herbal origin that possess peculiar, often tangy, fragrances used in trace amounts in food. Spices have significantly impacted world history. The aromas and flavors of spices are an important source of information and play a significant role in the development of human civilization. In addition to enhancing culinary experiences, spices are used to treat a variety of ailments. Their applications in many parts of the world have been mentioned as trade commodities, religious items, preservatives, narcotics and stimulants (Nascimento et al., 2020).

Spices are plant parts other than leaves (rhizome, buds, fruits, seeds, bark and onions) used for aromatic and medicinal purposes. The word "spice" comes from the Latin word "species", which means "a certain type". The term "herb" is used to describe parts of a plant that have similar benefits but are derived from leaf or soft flowering areas. The two words refer to the same plant, where fresh leaves are used as herbs, while dried parts are used as spices, such as cilantro, mint and bay leaf. Spices can be used more than just for food; they can also be used to preserve food and provide nutritional and health benefits (Nazeem, 1995).

DOI: 10.1201/b22917-2

As discussed earlier, various parts or organs of the plant, except leaves, can be an aromatic spice that is full of many pharmaceutically active compounds *viz*, ginger and turmeric (rhizome or roots), peppers, carom seeds (fruits), saffron and clove (flower part), garlic and onion (bulbs), cinnamon (bark), nutmeg, coriander, mace, fennel, fenugreek, cumin and cardamom (seeds), asafoetida (latex) and most importantly oregano, basil, mint, bay leaf and fenugreek (dried leaves) (Nascimento et al., 2020).

In traditional medicines, these spices are used for their antioxidant, antimicrobial, anti-inflammatory, antiseptic, antihelminthic, analgesic, diuretic, stimulant, carminative and many other properties. In the current scenario, antibiotics and many other drugs available in the market show unwanted toxic symptoms and side effects and the emergence of resistant pathogenic microorganisms. As a result, herbal extracts and pharmacologically active substances isolated from various plant species used in traditional remedies have attracted great attention. Traditional medicine is becoming increasingly important as a major source of health care for the world population.

The pharmacological components of spices are affected by their ripening time, drying, processing and storage. For each step, specific criteria are met to achieve a specific composition, quantity and quality of active chemicals. These active chemicals are alkaloids, glycosides, flavonoids, stimulants, saponins, tannins and terpenoids, which provide spices with distinctive medicinal effects. Spices are embedded in the matrix of carbohydrates, protein and fiber. When spices are processed, the cell matrix shrinks and volatile components are released (Raghavan, 2007).

2.2 SPICES: AS PER AYURVEDA

Ayurveda, the world's oldest complete natural medicine system, originated in India over 3,000 years ago. The Sanskrit words *Ayur* (life) and *Veda* (knowledge) combine to form the term *Ayurveda* (knowledge or wisdom). It considers that the five fundamental elements *viz.* water, space, earth, fire and air are the basic components of all living beings on this planet. When these elements come together in pairs, three body humors are formed: air or *vata* (space and air), bile or *pitta* (fire and water) and phlegm or *kapha* (water and earth). Any imbalance in these humors will lead to three different types of *doshas* that mean imbalance. These *doshas* are responsible for the different energies found in every cell in the body and the activities that take place therein. Ayurveda divides all diseases into three categories: *vata*, *pitta* and *kapha*. Spices can balance a specific defect, or *dosha*, with the help of their phytochemical components.

Individuals with *vata dosha* should eat smaller portions and avoid mixing different recipes. Allowing enough time to eat and chew thoroughly facilitates the release of pre-existing digestive enzymes in the saliva, making food easier to digest. Warm spices such as pepper, ginger and cardamom are especially good for this type of body humor.

Individuals with *pitta dosha* have a strong digestive system and can eat almost anything. However, the meal should not be too spicy, salty or sour, as this will intensify the beta fire. Bitter, astringent and sweet flavors reduce this imbalance, as it has a narrowing, reducing and cooling effect on the digestive system. Sweet and light spices such as mint and cinnamon are especially beneficial in *pitta dosha*.

Kapha dosha–affected individuals should increase their metabolism and therefore eat more warm food for digestion. Pungent and bitter tastes stimulate digestion and reduce *kapha dosha*. Spices that have a lot of heat, such as pepper, turmeric and ginger, are especially beneficial in *kapha dosha*.

Since the oxidative processes in the body are associated with a variety of metabolic diseases and age-related degenerative disorders, the use of herbs and spices as sources of antioxidants to prevent oxidation deserves further research.

Each spice is associated with a range of pharmacological activities which is due to its active pharmaceutical molecules. These bioactive molecules are secondary metabolites that can be further classified as per their chemical structure.

2.3 PLANT SECONDARY METABOLITES AS ACTIVE PHARMACEUTICAL COMPOUNDS

Secondary metabolites are organic substances released by bacteria, fungi and plants that play no direct function in their regular growth, development or reproduction.

These by-products of primary metabolites play a vital role in ecological processes such as defense and pigment production. The control of secondary metabolite synthesis is more complicated than that of primary metabolites. Apart from genetic considerations, different environmental conditions such as salinity, drought and temperature have been discovered to have a major impact on secondary metabolite synthesis.

Only the plant kingdom has approximately 50,000 different kinds of secondary metabolites. The activities of herbal medicine, spices and many contemporary medications are mostly dependent on secondary plant metabolites. Almost all active pharmacological chemicals and bioactive molecules are secondary metabolites, which are present in trace amounts in a variety of plants. The majority of these chemicals are hydrophobic and insoluble.

We'll talk about pharmacologically active chemicals in this chapter and how everyday culinary spices may be a treasure trove of these bioactive magic molecules.

Although secondary metabolites come in a variety of forms, the following bioactive molecules are the most common:

1. Alkaloids
2. Terpenes and terpenoids
3. Phenolic compounds

1. Terpenes are polymers of five-carbon isoprene units C_5H_8 (2-methyl-1,3-butadiene), and terpenoids are derivatives of terpenes, containing other functional groups such as oxygen. e.g. abscisic acid, curcumin, menthol, squalene, humulene, lycopene, (α-, β-, γ-) carotene, xanthophyll and vitamin A (Lalonde, 2005).
2. Alkaloids contain a nitrogen atom in a heterocyclic ring and may have carbon, hydrogen oxygen, sulfur and other elements. Examples include quinine, caffeine, piperine, nicotine and theobromine (Ziegler and Facchini, 2008). Alkaloids in almost all plants are toxic or poisonous and are synthesized as a defense mechanism against other species. Alkaloids are natural compounds that are usually produced from plant sources.
3. The phenolic group, which is one of the aromatic ring and hydroxyl groups, describes the third category of the most abundant secondary metabolites, phenolic substances. Phenolic acids, flavonoids, and tannins are examples of phenolic compounds. These substances are characterized by their strong antioxidant activity. (Giada, 2013). Polyphenols are derivatives of phenylalanine or shikimic acid, which are common precursors and classified under the following four categories *viz.* phenolic acids, stillbean, flavonoids and lignans. Among these groups, flavonoids have been the most researched and are mainly responsible for the attractive colors of fruits, flowers and leaves (Pandey and Rizvi, 2009).

Although all types of secondary metabolites have been attributed with a very diverse range of pharmacological effects, their general effects can be summarized as per Table 2.1 and various pharmacological properties of various spices are depicted in Figure 2.1.

TABLE 2.1
Types of Active Phyto-Constituents Present in Spices with Their General Effects

No.	Phytoconstituents	General Effects
1.	Alkaloids	Affect central nervous system, addictive
2.	Glycosides	Affect heart contractions, antispasmodic, sedative, affect respiration, acrid, stimulant, antibiotic
3.	Flavonoids	Antiseptic, anti-inflammatory, antispasmodic
4.	Saponins	Anti-inflammatory, diuretic, stimulant
5.	Tannins	Astringent, antiseptic
6.	Terpenes and terpenoids	Antioxidant, aids in digestion
7.	Resins	Acrid, astringent, antiseptic, healing properties
8.	Volatile oils	Aromatic, antiseptic, fungicidal, irritant and stimulant

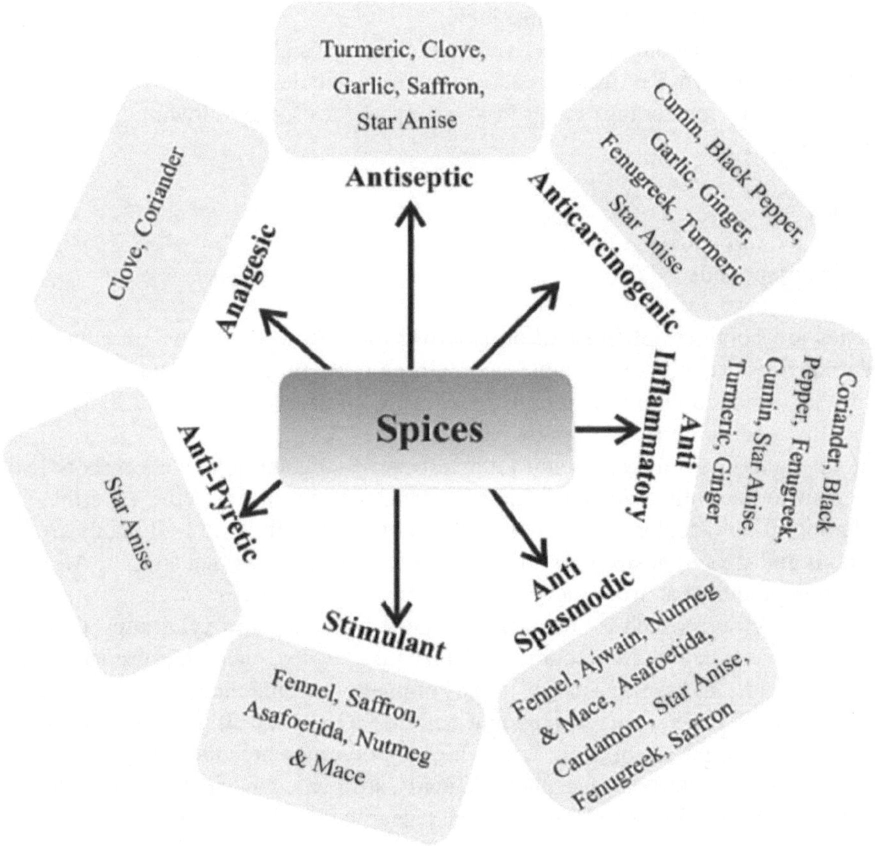

FIGURE 2.1 Various pharmacological properties of spices.

In this chapter, we are presenting the pharmaceutical applications of bioactive compounds of commonly used Indian spices that have been applied in traditional medicine and food.

2.3.1 Cinnamon (*Cinnamomum zeylanicum*)

Cinnamon, often called "Dalchini" in India, is the dried bark of the cinnamon tree. It belongs to the Lauraceae family and is known as the perennial tree of tropical medicine. It is one of the oldest and most important spices and has a unique taste and aroma that are widely used in Indian and Oriental cuisine. In total, there are more than 250 species of cinnamon known in this genus. Cinnamon is a native plant of India, Sri Lanka, Ceylon, Indonesia and Myanmar. It is made by drying the middle part of the bark and sold as quills or powder.

Cinnamon has been a part of the Ayurvedic system for thousands of years. In Ayurveda it is called "tvak", which means skin or bark, and is used to treat various ailments like cough, flu and indigestion. According to Ayurveda, it has a pungent and sweet taste. It is warm and is often prescribed to those with a *vata*, or air imbalance. It acts as a pain reliever and detoxifier (Thomas and Duethi, 2001).

Due to its aroma, cinnamon is widely used in aromatherapy along with the preparation of diverse meals, perfumes and pharmaceuticals. It is an antioxidant, anti-inflammatory, antidiabetic, antimicrobial, anticancer, lipid-lowering chemical that reduces cardiovascular disease by modulating different signaling pathways (Ranasinghe et al., 2013). Cinnamon has also been linked to the treatment of neurological diseases such as Parkinson's disease and Alzheimer's disease (Angelopoulou et al., 2021; Anderson et al., 2013). Cinnamon has been reported to protect against the toxicity of chemotoxins in the main vital organs of the human body because of its antioxidant and free radical scavenging effect.

The main bioactive compounds found in real cinnamon are cinnamaldehyde, cinnamic acetate, cinnamic acid, camphor and eugenol (Nascimento et al., 2020). Cinnamon is also found to have a polymer called methyl hydroxychalcone (MHCP), which lowers blood sugar levels by mimicking insulin and helps in the treatment of diabetes (Thomas and Duethi, 2001).

Essential oils include cinnamaldehyde and *trans*-cinnamaldehyde (t-CIN), which are responsible for the aroma and various biological functions. E-cinnamaldehyde, one of the major constituents of essential oil produced from *C. zeylanicum*, exhibits antityrosinase properties, making it a useful skin whitening agent (Chou et al., 2013). Cinnamon bark also contains flavonoid compounds, catechins and their polymer procyanidins which are involved in zinc homeostasis.

2.3.2 Black Pepper (*Piper nigrum*)

The most common and most widely used spice in the world is black pepper, commonly known as peppercorns. It is the dried fruit of *Piper nigrum*. The historical homeland of this precious spice is southwestern India, especially the western coasts of Southern Peninsular India and the Malabar coast. Due to its strong, bitter taste, it is widely used in culinary applications, including as a flavoring agent and preservative for ready-made food products.

According to the *Charak Samhita*, an ancient Indian treatise on Ayurveda, black pepper is said to increase internal heat. It is used in traditional medicine to treat respiratory and digestive problems, as well as to increase appetite. It has antioxidant and astringent effects. Pepper promotes the free delivery of oxygen to the brain while increasing the bioavailability of nutrients in the body by increasing absorption and circulation (Kesarwani et al., 2013).

It also has vitamins A, C, E and K; choline; folic acid; pyridoxine; riboflavin; thiamine; and niacin, as well as minerals including copper, calcium, magnesium, zinc and phosphorus. Piperine is the main alkaloid in pepper, which gives it a strong pungent flavor, and it has antiproliferative, antineoplastic, antioxidant, antidiabetic, anti-obesity, cardioprotective, antimicrobial, anti-aging and immunomodulatory properties. Hepatoprotective, anti-allergic, anti-inflammatory and neuroprotective activities of piperine have also been studied (Haq et al., 2021).

Piperine, α-pinene and β-pinene, as well as myrcene, α-phellandrene, limonene, linalool, methyl propanol, 2- and 3-methyl butanol, butyric acid and 3-methyl butyric acid are compounds behind pepper's distinctive odor (Lee et al., 2020). Terpenes are the main volatile flavor compounds in black pepper (Clery et al., 2014). Terpenes, 2,3-diethyl-5-methyl pyrazine and 2-isopropyl-3-methoxy pyrazine, give pepper its strong flavor (Jagella and Grosch, 1999). E-caryophyllene and β-thujene of black pepper are the other major chemical constituents of the essential oil (Lee et al., 2020). E-caryophyllene aids in the treatment of age-related bone loss by enhancing the creation of a mineralized extracellular matrix during the osteogenic differentiation process (Shan et al., 2017). Yu et al. 2018 found that D-limonene has anticancer properties. α-terpinene is associated with trypanocidal activities (Baldissera et al., 2016). β-caryophyllene, δ-3-carene and sabinene are also present in different percentages in *P. nigrum* oil (Dosoky et al., 2019).

2.3.3 Clove (*Syzygium aromaticum*)

Cloves are stunted flower buds of a tree belonging to the Myrtaceae family, known as Mother Nature's antiseptic. It is called "Lavang" in Ayurveda. Clove, according to Ayurveda, increases circulation and digestion and helps with stomach problems like gas, bloating and nausea. It is an excellent pain reliever. It is also used to treat coughs and sore throats due to its antibacterial, antioxidant and anti-inflammatory effects and also helps in lowering blood sugar levels. Clove essential oil is used in oral hygiene products to promote dental health and keep the mouth fresh. It also has mosquito repellent properties.

Cloves originally belong to the Maluku Islands of Indonesia, and in India, they thrive in the humid tropics of Kerala, Tamil Nadu and Karnataka. In food, cloves are used whole or ground to intensify the taste.

Eugenol is the main ingredient (72–90%) that gives clove its distinctive aroma. It is the most important phytochemical, having antibacterial and analgesic properties (Chaieb et al., 2007). It is commonly used as a bactericide in mouthwashes, as well as in dental plasters, fillers and cement. Clove gel is also used for local anesthesia due to its analgesic properties. Eugenol has been shown to have anticancer properties in the treatment of melanoma and leukemia (Khalil et al., 2017; Haro-González et al., 2021) along with colon, stomach, breast, prostate and skin cancers. Apart from eugenol, acetyl eugenol, β-caryophyllene, vanillin, crategolic acid, tannins, gallotannic acid, methyl salicylate, eugenin, kaempferol, rhamnetin, eugenitin and triterpenoids like oleanolic acid, stigmasterol and campesterol are important constituents present in the essential oil of clove, which makes it an important therapeutic agent. (D'Souza et al., 2017).

Flavonoids (quercetin), hydroxybenzoic acids, hydroxycinnamic acids, hydroxyphenyl propane, gallic acid, caffeic acid, ferulic acid, ellagic acid and salicylic acids are major phenolic compounds present in clove (Cortés-Rojas et al., 2014).

Clove essential oil has been shown to have antiviral, antibacterial, antifungal, anticancer, antioxidant, anti-inflammatory effects (Han and Parker, 2017).

The essential oil has shown aphrodisiac, antipyretic, appetite, hypnotic, anxiolytic, antiemetic, analgesic, decongestant, antimicrobial, anti-epileptic, myorelaxant, anti-inflammatory and expectorant properties (Elwakeel et al., 2007). Tannins, ellagic acid, gallic acid, flavonoids and their glycosides isolated from alcoholic and aqueous clove bud extracts show antithrombotic, antiprotozoal, hypoglycemic, anti-inflammatory, gastroprotective and aphrodisiac properties (Tajuddin et al., 2004; Issac et al., 2015; Johannah et al., 2015).

2.3.4 Cumin (*Cuminum cyminum*)

Cumin is an ancient spice with a strong aroma. Ripe cumin is the fruit (seed) belonging to the Apiaceae family. It is a spice widely used in whole and ground form in many dishes of different civilizations. Cumin is a spice that originates in the Eastern Mediterranean and extends to South Asia. India now produces 70% of the world's cumin.

Cumin is called "jeeraka" in Sanskrit, which is derived from the word "jeerna", which means digestion, so cumin has been a famous remedy for gastrointestinal problems like flatulence (Reza et al., 2015). Cumin has a bitter taste according to Ayurveda. It has the power to cure *vata* (air) and *kapha* (phlegm) *dosha*s. Cumin seeds and distilled extract of cumin work as a stimulant, antispasmodic, carminative and antimicrobial. It is often used as a flatus-relieving remedy in veterinary medicine (Amin, 2001).

Cumin is rich in amino acids, essential fatty acids, vitamins and minerals. It is also a great source of iron. The major bioactive components in cumin essential oil are cuminaldehyde, cymene and terpenoids (Allahghadri et al., 2010). The distinctive aroma and taste of cumin are attributed to the aldehydes present in the essential oil of cumin, such as cuminaldehyde or cumenol.

Cumin oil contains less than 1% fraction each of α-pinene, myrcene, limonene, 1–8-cineole, p-menth-3-en-7-ol, caryophyllene and β-phellandrene. Additionally, β-bisabolene, p-mentha-1,3-dien-7-ol, β-pinene, β-farnesene, P-cymene and cuminyl alcohol are present in comparatively high proportions in cumin oil. D-terpinene, the cuminic aldehyde, is the most abundant phytochemical present in its essential oil (Nadeem and Riaz, 2012).

Cumin aldehyde, the principal component of cumin, has been reported for its antimicrobial and antibiofilm properties and as a potent lipoxygenase (LOX) inhibitor. Recently cumin extract has been linked with skin whitening (Wang et al., 2021) and antitumor activity (Aqil et al., 2021).

2.3.5 Garlic (*Allium sativum*)

Garlic is the oldest authenticated aromatic spice related to the member of the Amaryllidaceae family. It has been used for flavoring foods during the cooking process and has been used for medicinal and other purposes for thousands of years. Presently it is the most widely studied spice used to treat numerous ailments like cold, influenza, diabetes and hypertension as well as a remedy for snake bites.

In Charak Samhita and Sushrut Samhita (both are one of the foundation texts of Ayurveda), garlic is known as "lasuna" or "rasona". They described garlic as *snigdha* (unctuous), *ushna* (hot), *katu* (pungent), *guru* (heavy), *tikshna* (sharp), *picchila* (slimy), *svadurasa* (sweet) and *sara* (laxative). Because of these properties in traditional medicines, garlic is used as an immunomodulator or immune booster, to treat cardiovascular disease and for the quality formation of sperm cells. As well, it is used for therapeutic purposes including the treatment of lung disorders, whooping cough, stomach disorders, cold and earache (Badal et al., 2019). From ancient times it has been used to treat digestive disorders, respiratory and urinary tract infections and cardiac disorders, as it shows carminative, antipyretic, sedative, antibiotic, aphrodisiac and diuretic effects (Souza et al., 2011).

The active compounds present in *Allium* species have antimicrobial, antifungal, antiprotozoal, antioxidant, anti-inflammatory, anti-aging and anticancer properties (Batiha et al., 2020). Allicin (s-2-propenyl-2-propene-1-sulfinothionate) is the major lipid-soluble sulfur-containing bioactive molecule present in garlic, which is responsible for its smell and taste (Rahman, 2007; Slusarenko et al., 2008). Alliin (S-allyl-L-cysteine sulfoxide) is the precursor of allicin, which is transformed into allicin by alliinase enzymes (Kaye et al., 2000).

Along with these, the bulb of *Allium sativum* is reported to have numerous other phytochemicals, including the sulfur-containing compounds like ajoenes, vinyl dithiins (2-vinyl-4H-1,3-dithiin,

3-vinyl-4H-1, 2-dithiin) and sulfides (diallyl sulfides, diallyl disulfide and diallyl trisulfide) (Al-Snafi et al., 2013). S-propyl-cystein sulfoxide and S-methyl-cysteine sulfoxide are the key odorous compounds of freshly crushed garlic homogenates (Zeng et al., 2017).

Consumption of a few garlic cloves early in the morning has been associated with good cardiovascular health for ages in India. Garlic plays a vital role in modulating systolic as well as diastolic blood pressure (Varshney and Budoff, 2016). Garlic is popularly known to prevent and mitigate cardiovascular disorders such as arrhythmia, thrombosis, hyperlipidemia, atherosclerosis and hypertension (Rahman, 2001).

2.3.6 Ginger (*Zingiber officinale*)

Ginger is the tropical part of the Zingiberacceae family. The consumable part of the ginger is a *rhizome*. It is the horizontal stem from which the root is grown. For centuries, it has been used to treat various diseases like cough, rhinitis, arthritis and hypertension. In Sanskrit, it was called *Srngaveram*, meaning "horn root" because of its appearance. In Ayurveda, dried ginger is known as "Sunthi". Ginger assumes a significant role in the antiquated customs of Ayurveda. In Ayurvedic medication, the decoction of the ginger rhizome is generally utilized (Agrahari et al., 2015). It has an excellent safety profile. It is a *kapha* and *vatta* sedative and a good stamina booster (PV, 1981; KRS, 2005).

It is utilized internally as well as topically. It also recommended for patients experiencing nausea from pregnancy, motion sickness or chemotherapy. Ginger has a peppery and somewhat sweet flavor, as well as a powerful and pungent scent. Therefore, it is preferably used in teas, soups, decoctions and for enhancing the flavor of dishes (Isha Kumari et al., 2015). Fresh ginger juice is used by pregnant women at the time of childbirth, as it helps in easy childbirth (Moghaddasi and Kashani, 2012).

The rich phytochemistry of ginger makes this spice a significant health booster. Both fresh and dry forms of ginger rhizome have diverse phytochemical constituents (Langner, 1998). *Gingerols* are the main bioactive compound present in fresh ginger (Wohlmuth et al., 2005). It possesses antiviral and antifungal properties that promote saliva formation and reduce oral candidiasis, whereas *shogaols*, the dehydrated form of gingerols, are the major compounds in dry ginger (Bhattarai and Duke, 2001).

The phytochemical components of ginger are divided into two types: volatile and nonvolatile (Zhan et al., 2008). Ginger has a unique aroma due to its volatile oil content. The monoterpenoid hydrocarbons and sesquiterpene are the most volatile oils of ginger. Geraniol, curcumin β-phellandrene, (+)-camphene, 1,8-cineole, citral, terpineol, borneol, linalool and neral are examples of monoterpenoid hydrocarbons present in ginger. Sesquiterpenoids (terpenoid made up of three isoprene units) are represented by zerumbone, α-zingiberene, β-sesquiphellandrene, β-bisabolene, (E-E)-α-farnesene, arcurcumene and zingiberol. Among nonvolatile pungent constituents of ginger, gingerols, shogaols, paradols and gingeron contribute to its intense flavor (Jiang et al., 2006; Langner et al., 1998).

2.3.7 Fennel (*Foeniculum vulgare*)

Fennel is an aromatic dried ripe fruit or seed of a perennial plant classified under the Apiaceae (Umbellifereae) family. It has high fiber content, which improves digestion. In India, people commonly chew fennel seeds post-meal because its volatile oil reduces inflammation and stimulates bile production to allow food to move faster through the body.

It is described as a stimulant, carminative, emmenagogue (medicine that promotes menstrual discharge) and galactagogue (induces milk secretion) (Badgujar et al., 2014; Akbar, 2018; Dymock et al., 1890). In Indian traditional medicine, fennel water is recommended for improving eyesight, to open liver and spleen obstructions (Kabeeruddin, 1937), to relieve flatulent colic and as a diuretic and diaphoretic in children (Khory and Katrak, 1984; Nadkarni, 1954; Saha and Kasinathan, 1961).

Major phytocomponents in fennel's essential oil are anethole, α-thuzone, limonene, phenytoin, estrogol (methyl chavicol), camphon, α-penene, fenchyl alcohol and anisaldehyde (Hausner et al., 2008). Trace amounts of δ-α-phellandrene, dipentene and ρ-hydroxy phenyl acetone have also been reported in fennel. Fennel is also rich in minerals like calcium, phosphorus, salt and potassium.

Petrocylic, oleic, linoleic and palmitic acids are the basic components in stationary oil (Farrell, 1999). Among the active chemicals in fragrance are anethole, estragole, (+)-fenchone and P-anisaldehyde, which have been shown to have estrogenic, acaricide and anticoagulant properties (Rather et al., 2016).

The chewing of fennel seeds has been reported to increase salivary pH, therefore making it a suitable anticarcinogenic agent (Manohar et al., 2020). Studies indicate that fennel's major bioactive molecule trans-anethole can play a hepatoprotective role, especially against diabetes-induced liver damage, probably through its hypoglycemic effect. (Samadi-Noshahr et al., 2020).

2.3.8 Fenugreek (*Trigonella foenum-graecum*)

Fenugreek is an ancient legume-based medicinal spice. Fenugreek seeds, whole or powdered, are edible, contain a strong aroma and have a pleasantly bittersweet taste. They are commonly used for flavoring foods in different cuisines. The leaves and seeds of the plant are of high medicinal and nutritional value.

India is said to be one of the natives of fenugreek; presently it occupies a prime position among fenugreek-growing countries in the world. According to historical facts and *Charak Samhita* and *Sushrut Samhita*, the classical texts of Ayurveda, fenugreek was used as a blood cleanser, as a detoxifier and as a diaphoretic because it can bring on a sweat.

Fenugreek is well known as a principal source of soluble fiber. It has been shown that fenugreek has antidiabetic, anticancer, hypercholesterolemic, anti-inflammatory, antioxidant and chemopreventive activity due to its useful phytoconstituents (Khorshidian et al., 2016).

The chemical components of fenugreek are said to provide excellent functional and therapeutic properties. Fenugreek seeds contain slimy fiber, which binds to bile acids and reduces cholesterol and fat absorption (Al-Asadi, 2014). Fenugreek is known for its hypoglycemic and hypocholesterolemic properties since its seeds contain galactomannan, which inhibits the absorption of bile salts in the colon and slows the digestion and absorption of carbohydrates (Madar and Shomer, 1990; Mathern et al., 2009; Sharma et al., 1990; Ajabnoor and Tilmisany, 1998). Apart from that, it has been reported to contain 4-hydroxy isoleucine, a free inhibitory amino acid that enhances insulin sensitivity and glucose uptake in diabetic individuals (Sauvaire et al., 1998; Baquer et al., 2011).

Fenugreek has choline, gentianine, carpain and trigonelline as alkaloids; apigenin, luteolin, orientin, quercetin, vitexin and isovitexin as flavonoids; and diosgenin, gitogenin and yamogenin as steroidal sapogenin as the main bioactive ingredients responsible for its various biological effects (Trivedi et al., 2007; Snehlata and Payal, 2012; Yoshikawa et al., 1997). Phytoestrogens and diosgenin of fenugreek act as galactagogues and are most frequently used to stimulate breast milk secretion (Turkyılmaz et al., 2011).

Trigonelline, a key alkaloid constituent of fenugreek, has hypoglycemic, hypolipidemic, neuroprotective, antimigraine, sedative, memory-enhancing, antibacterial, antiviral and antitumor activities. Evidence also suggests that it can alleviate diabetic auditory neuropathy and platelet aggregation through β-cell regeneration, promoting insulin secretion, glucose metabolism modulation, neutralizing reactive oxygen species and affecting axonal extension and neuron excitability (Zhou et al., 2012).

Diosgenin, a sterol saponin of fenugreek, has been reported to cure and prevent lung, liver, colon, breast and prostate cancers by altering cellular pathways (Arya and Kumar, 2021).

Apigenin is a flavonoid polyphenol that has been reported to have strong antioxidative effects by scavenging free radicals, especially in the peripheral nervous system, so it can play a neuroprotective role to postpone peripheral neurodegenerative disease, such as cerebral ischemia (Kim et al., 2019). Yamogenin, which is a diastereomer of diosgenin, is a powerful antihyperlipidemic compound and tends to inhibit lipid accumulation through gene suppression in lipogenesis in hepatocytes (Moriwaki et al., 2014).

2.3.9 Saffron (*Crocus sativus*)

Saffron is a member of the *Crocus sativus* L. family of Iridaceae. Derived from the dried stigma of its flower, it is one of the most expensive spices in the world. It is commonly known as "*kesar*" and hails from India's Himalayan region of Kashmir. For generations, saffron has been used for flavoring and fragrance in cuisines and skincare for its glow-enhancing properties for centuries. According to Ayurveda, saffron is "tridoshic", which means it balances all three types of humors. It is a powerful antioxidant with calming anti-inflammatory capabilities.

Saffron can treat asthma and cough, strengthen the heart, relieve an enlarged liver and help with menstrual problems. It is an analgesic, antihistamine, antiseptic, antispasmodic, aphrodisiac, balsamic, cardiotonic, diaphoretic, flatus reliever, expectorant, soporific, stimulant and emmenagogue. It has antibacterial, antiseptic and antifungal properties and is used to treat cataracts, night blindness and vision impairment (Hosseinzadeh and Nassiri-Asl, 2013).

According to phytochemical research, saffron contains at least four active ingredients: crocin, crocetin, picrocrocin and safranal (Alavizadeh and Hosseinzadeh, 2014).

Cis and trans crocins and crocetins are the main components that give saffron its color. Crocins are extraordinarily water-soluble carotenoids and hold different healing properties, especially on the nervous system. Crocins can relieve dementia and depression and can act as an anxiolytic and aphrodisiac. Furthermore, it has antioxidant, cardioprotective and antitumor properties (Alavizadeh and Hosseinzadeh, 2014; Ben Salem et al., 2016). Saffron is formed during the hydrolysis and dehydration of picrocrocin, which causes the saffron odor (Hossaini et al., 2018).

Picrocrocins are produced by enzymatic degradation of protocrocin and are responsible for its bitter taste. Upon hydrolysis and dehydration, picrocrocins are converted in safranal, which is the main chemical behind the aroma of saffron (Hosseini et al., 2018).

Like crocins, safranal has been found to have effective antidepressant and neuroprotective effects (Khazdair et al., 2015). It has a beneficial effect on serotonin, a neurotransmitter in the brain (Hausenblas et al., 2013). It acts as a radical scavenger, protects the brain from oxidative damage, increases the diffusion of oxygen in tissues and enhances memory (Assimopoulou et al., 2005; Farahmand et al., 2013; Mashmoul et al., 2013).

Saffron also contains xanthophylls, carotenes, lycopene and zeaxanthin, all of which are important phytocomponents (Siracusa et al., 2011). It contains more vitamin B_2 than known sources. Since picrocrocin and crocin are easily oxidized by direct contact with oxygen and moisture, saffron should be stored at dry, low humidity and low temperature.

2.3.10 Turmeric (*Curcuma longa*)

The powerful golden-yellow root has been used for thousands of years as a medicine, spice and culinary flavoring agent. It belongs to the ginger family. Apart from being a fragrance, it has also been used in the past as a natural coloring agent (food, cosmetics and textiles), as well as an antibacterial and insect repellent (Moghadamtousi et al., 2014). Turmeric is used by Indians for Ayurvedic uses as well as for cleansing the blood and treating skin problems. In many parts of India, pre-wedding bridal skin is painted with yellow turmeric paste, which is believed to brighten the skin and keep harmful germs away from the body. This yellow spice is now used in the composition of many sunscreens. Turmeric is also very popular due to its medicinal properties such as thermogenic, softening, pain-killing, carminative, stomachic and antiseptic properties.

Curcumin, demethoxycurcumin (DMC) and bis-demethoxycurcumin (BDMC), collectively known as curcuminoids, are the major bioactive polyphenol components present in turmeric (Majeed et al., 1999). Curcumin (1,7-bis-4-hydroxy-3-methoxyphenyl-1,6 heptadiene-3,5-dione), or diferuloylmethane, has anti-inflammatory, antiviral, antioxidant, antivenom, antibacterial, anti-asthmatic, anticancer, anti-arthritis, antidiabetic, wound healing and anti-obesity properties (Rathore et al., 2020; Goel et al., 2008; Araújo and Leon, 2001). Treatment of curcumin in diabetic rats has been shown to lower blood glucose and hemoglobin levels (Arun and Nalini, 2002). Recently, curcumin has been established as an antiaging compound since its treatment enhanced superoxide dismutase (SOD) activity and also decreased malondialdehyde (MDA) and lipofuscin levels (Zia et al., 2021). Due to inadequate absorption, rapid metabolism and rapid elimination, curcumin has poor bioavailability (Hewlings et al., 2017), although when consumed with black pepper, its piperine has been shown to significantly increase the bioavailability of curcumin by 2000%.

The essential oil of turmeric has been a storehouse of numerous compounds, although *ar*-turmerone, α-turmerone and β-turmerone are its main constituents, then followed by significant quantities of α-zingiberene, curlone, *ar*-curcumene, α-santalene, santalenone, β-sesquiphellandrene, (*Z*)-β-ocimene, β-bisabolene, β-caryophyllene, α-phellandrene, (*Z*)-β-farnesene, humulene oxide, β-selinene, caryophyllene oxide, (*E*)-γ-atlantone, 1,8-cineole and terpinolene (Braga et al., 2003; Dosoky and Setzer, 2018).

Turmeric essential oil has the potential to protect against cardiovascular diseases. The essential oil was reported to have antihyperlipidemic effects on high-fat diet (HFD)–induced hyperlipidemia in rats (Ling et al., 2012). *ar*-turmerone exhibited a strong inhibitory effect on platelets. It has also shown antimutagenic, powerful hypoglycemic, neuroprotective, cytotoxic, anti-inflammatory, antivenom and insect-repellent activities (Dosoky and Setzer, 2018).

2.3.11 Nutmeg and Mace (*Myristica fragrans*)

Nutmeg and mace are different parts of a common fruit of the nutmeg tree, which belongs to the Myristicaceae family. The kernel of the dried seed is nutmeg, while the dried aril around the seed is mace. The nutmeg tree originated in the Banda Islands in the Moluccas. Nutmeg and mace are mostly produced in Indonesia, second only to Grenada. Kerala and Tamil Nadu are the largest producers of this fragrant spice in India. The two spices have the same taste, but nutmeg is often used in cooking due to its somewhat sweeter taste. Nutmeg and mace are commonly used in savory foods, food processing, cosmetics and the pharmaceutical industry, either whole or in the powdered form. This is called "Jatiphala" in Ayurveda. It is used as a heart tonic and blood purifier and in traditional medicine for the treatment of gastrointestinal problems, impotence, insomnia and skin diseases (Garg et al., 2007). Melasma is also treated with it (Debbarma et al., 2020). Nutmeg oil is used as a tonic, while nutmeg ointment is used as an anti-irritant. (Sharma et al., 2002). Ayurveda recommends nutmeg and mace to balance *vata* (air) and *kapha* (phlegm) humors of the body (Somani et al., 2008).

A wide variety of phytochemicals are present in *Myristica*, namely, lignans along with neolignans, diphenyl alkanes, phenylpropanoids, terpenoids, alkanes, fatty acids and its esters. Steroids, saponins, triterpenoids and flavonoids are also present in trace amounts.

The phytochemicals in nutmeg and mace essential oils contribute to their medicinal properties. Monoterpene hydrocarbons, aromatic ethers, oxygenated monoterpenes, alkanes, aromatic monoterpenes and organic acids are the major components in spice oil (Abourashed and El-Alfy, 2016). Pinene, camphene, p-cymene and sabinene monoterpene hydrocarbons are recognizable, while myristicin (1-allyl-5-methoxy-3,4-methylenedioxybenzene), elemicin, isoelimicin and safrol are aromatic ethers that contribute significantly to the aroma and taste of the spice (Phulsagar et al.,

2014). Myristicin, which is a major phytochemical present in the spice, is found to be involved in its antioxidant, anti-inflammatory, antiproliferative, antimicrobial, insecticidal and larvicidal activities (Seneme et al., 2021).

Myristicin and elemicin both act as stress relievers and are responsible for the hallucinogenic effects (Kumari et al., 2021a). Nutmeg is traditionally used to treat diarrhea, flatulence, abdominal pain, nausea, vomiting, rheumatism, sciatica pain, malaria and leprosy.

Nutmeg extract has been found to prevent hepatotoxicity through its anti-apoptotic, antioxidant and anti-inflammatory properties (Dkhil et al., 2019).

2.3.12 Carom Seeds (*Trachyspermum ammi*)

Carom is also known as bishop weed or thymol seed, and in India, it is popular as "ajwain". The seeds, fruits and leaves of the plant are used. The seed-like fruit is actually a schizocarp, which is bitter and pungent-tasting. Although it is now widely used in cuisine around the world, it is native to Egypt. Carom is widely grown in Gujarat and Rajasthan and is widely used to aid digestion in fried meals such as fritters.

It is also known as "Amanashak" in Ayurveda, as it helps in reducing the development of toxins in the digestive system and relieving constipation. Many other studies have proved ajwain as an antifungal, antioxidant, antimicrobial, antinociceptive, cytotoxic, hypolipidemic, antihypertensive, antispasmodic, bronchodilator, urinative, abortion inducer, cough suppressor, nematicidal, anthelminthic and antifilarial (Bairwa et al., 2012). A teaspoon of seeds mixed with a pinch of salt is a common home remedy for indigestion caused by a change in diet (Malhotra and Vijay, 2004).

Carrom seed's essential oil is composed of thymol, carvacrol, limonene, dillapiole, *p*-cymene, γ-terpinene, β-pinene, terpinen-4-ol, oleic acid, linoleic acid, palmitic acid and xylene (Malhotra and Vijay, 2004). Aromatic monoterpenes (thymol, p-cymene and **γ**-terpinene), the major (90%) components of the oil, are small-molecular-weight compounds and are highly polarizing. These phytochemicals are among the most potent components of essential oils and have antioxidant, antibacterial and antidiabetic properties. Thymol, a powerful oil-derived antiseptic, is used in a variety of skin ointments, powders, deodorants, mouthwashes, toothpaste and gargles (Malhotra and Vijay, 2004).

Essential oil of carom seeds has been successfully tested for its antimicrobial properties against *Lactobacillus acidophilus, Bacillus cereus, Saccharomyces cerevisiae, Mycoderma* sp. and *Aspergillus niger* (Meena and Sethi, 1994). Carom seeds have been reported to provide relief from bloating, abdominal pain, indigestion, diarrhea, cholera, hysteria and bowel diseases (Pruthi, 2001; Latif and Rahman, 1999).

2.3.13 Asafoetida

Asafoetida is a dry latex derived from the rhizome of the ferula plant, belonging to the Apiaceae family; it is often referred to as "food of the gods" or "devil's excrement" due to its foul-smelling and repellent taste. Ferula is indigenous to Iran and Afghanistan and the Kashmir region of India. The whole plant radiates a strong sulfurous odor. The pungent, resinous gum is widely used in Indian cuisines and is considered one of the important remedial plants in Ayurveda.

It is used to treat diseases of the gastrointestinal tract and is a good remedy for flatulence and hysteria in traditional medicine. Ferula is used in pickles to aid digestion and as a spice in meals. It is also used to treat nerve, gastrointestinal, reproductive and skin disorders and is usually recommended in whooping cough, asthma, bronchitis, intestinal parasites and influenza (Takeoka et al., 2001; Lee et al., 2009; Al-Ja'fari et al., 2013; Amalraj and Gopi, 2017).

Additionally, it has stimulant, carminative, expectorant and antispasmodic properties. Ferula is also used to treat infertility, unwanted miscarriages, prematurity, pain, difficult and abnormal menstruation and leukorrhea, among other gynecological disorders, and also stimulates progesterone production (Choudhary et al., 2021). These include antioxidants (Milani et al., 2007), antimicrobials (Dehpour et al., 2009), antifungals (Rani et al., 2009), antitumor (Saleem et al., 2001) and antidiabetic (Abu-Zaiton, 2010) and neuroprotective activities due to their phytochemical contents (Tayeboon et al., 2013).

Ferula has a resin content of 64%, gum content of 25% and oil content of 11%. Vanillin, 3,4-dimethoxycinnamyl-3-(3,4-diacetoxyphenyl) acrylate, picealactone C, and 7-oxocallitristic acid are phenolic compounds and diterpenes found in the plant. Ferulic acid and its esters, coumarin, sesquiterpene coumarins and other terpenoids, are found in the resin.

Glucose, galactose, L-arabinose, rhamnose, glucuronic acid, polysaccharides and glycoproteins are found in the gums, and the volatile oil contains 2-butyl 1-propenyl disulfide, 1- (methyl thio)-2-propenyl disulfide, 2-methyl-2-propanethiol, 2-butyl 3-(methylthio)-2-propenyl disulfide, 2,3-dimethylthiirane, S-methylpropanethioate, dimethyl trisulfide, dipropyl disulfiefoetisulfide A and C, monoterpenes and other volatile terpenoids, which are responsible for its various biological activities (Mahran et al., 1975).

The active ingredients like α-pinene, α-terpineol, diallyl disulfide, ferulic acid, isopimpinellin, luteolin, umbelliferone and vanillin have shown anticancer effects in many in vitro and in vivo studies. Anti-inflammatory effects of α-pinene, azulene, β-pinene, ferulic acid, isopimpinellin, luteolin and umbelliferone have been documented (Mahendra and Bisht, 2012).

Ferula, called "hing" in Indian cuisine, is known for its antispasmodic properties that are attributed to the presence of azulene, ferulic acid, luteolin, umbelliferone and valeric acid. In addition, α-pinene, α-terpineol, azulene, diallyl-disulfide, diallyl-sulfide, ferulic-acid and luteolin are antibacterial compounds found in hing. α-pinene, diallyl-disulfide, ferulic acid, luteolin and vanillin have been shown to have antiviral effects. Ferula contains antioxidants such as ferulic acid, luteolin and vanillin (Mahendra and Bisht, 2012).

In addition to the pharmacological activities already noted, asafoetida has been shown to have lipoxygenase inhibitory, hepatoprotective, tranquilizing and cytotoxic effects (Mahendra and Bisht, 2012). It contains large amounts of calcium, as well as phosphorus, iron, carotene, riboflavin and niacin. A disulfide $C_{11}H_{20}S_2$ is said to be the main cause of the unpleasant odor of ferula oil (Shankaranarayana et al., 1982; Rajanikanth et al., 1984).

2.3.14 Cardamom

Cardamom is the dried seeds of perennial herbaceous plants of the ginger root family, which are valued for their pleasant perfume and flavor. Green cardamom belongs to the Western Ghats of the evergreen rainforests of Southern India, from which it spreads to other tropical countries. Black cardamom, on the other hand, is native to the Eastern Himalayas.

Cardamoms are broadly divided into two categories:

1. **Small cardamom:** Green cardamom, or "*chotti elaichi*" (*Elettaria cardamomum*). Popularly called the queen of spices.
2. **Large cardamom:** Black cardamom, or "*badi elaichi*" (*Amomum subulatum Roxb*).

Cardamom is one of the spices found in almost every kitchen on the planet. It is found in tea, coffee, snacks, desserts and practically every food in the world. Their seeds are also used in traditional medicine to treat digestive problems, heartburn, intestinal pains, irritable bowel syndrome, colds, coughs, bronchitis and urinary problems. The antidote to snake and scorpion venom is also made using cardamom (Rajathi et al., 2017). Cardamom oil is used in perfumes, alcoholic beverages, as a flavoring agent and as an additive carminative. Some cardamom seeds are recommended for chain smokers who want to quit their addiction. Black cardamom pods are used as a spice similar to green cardamom pods, but the taste is very different.

The major constituents of green cardamom are α-pinene, β-pinene, sabinene, myrcene, α-phellandrene, limonene, 1,8-cineole, γ-terpinene, p-cymene, terpinolene, linalool, linalyl acetate, α-terpineol, α-terpinyl acetate, citronellol, geraniol, methyl eugenol and trans-nerolidol (Lawrence, 1979; Korikanthimath, 2001).

When cardamom is chewed, it has an astringent and spicy flavor. Cardamom volatile oil contains small amounts of mono and sesquiterpenes, which give it its distinctive aroma. Ingredients in volatile oils such as 1,8-cineole, terpinene, terpineol, sabinine, α-pinene and limonene act as heart and liver tonics, appetite enhancers, increase bile excretion and help reduce liver congestion (Rajathi et al., 2017). Cardamom essential oil is distinguished by the presence of ether, 1,8-cineole and esters, terpinyl and linalyl acetates. Different types of cardamom contain different ratios of 1,8-cineole, camphor or borneol, which are responsible for its peculiar aroma (Sengupta and Bhattacharjee, 2009). Esters and 1,8-cineole from the diverse types of cardamoms are responsible for the total flavor variations (Lewis et al., 1966; Salzer, 1975; Wijesekera and Jayawardena, 1973; Korikanthimath et al., 1997).

The main ingredient in cardamom essential oil is 1,8-cineole (65–80%); a small fraction of terpenyl acetate is also present (traces up to 5%). Limonene, sabeinene terpenes and pinene monoterpenes are important components in hydrocarbon formation (Madhusoodanan and Saideshwara Rao, 2004). Unani physicians use large cardamoms to treat digestive problems (Jafri et al., 2001).

2.3.15 Star Anise (*Illicium verum*)

Star anise is a star-shaped fruit that grows on the fragrant evergreen tree of the Illiciaceae family. Illicium is a native Chinese plant whose name comes from the Latin word "alluring", which means "fragrant". Since star anise is a cold-sensitive plant, it thrives only in warm subtropical climates.

This star-shaped fruit is considered one of the most important spices in Oriental cuisine. In countries like China and Vietnam, it is considered to be the most valuable medicinal plant with a wide variety of healing properties, as well as a widely used spice.

Ayurveda, ancient Indian herbal medicine, recommends these starry fruits for indigestion, flatulence, spastic colon pain, diarrhea, cough, asthma, rheumatoid arthritis, facial paralysis and many other conditions. Star anise is the main ingredient in the famous Indian garam masala (a powdered mixture of many spices) and is often used as a seasoning. It also has a wide range of pharmaceutical applications, as it is a storehouse of various types of bioactive compounds.

This botanical fruit is a major component of Chinese medicinal herbs and is famously known for its antiviral properties. In addition to antiviral properties, star anise has been shown to have antioxidant, antimicrobial, antifungal, anthelmintic, antiseptic, antisecretory, antipain, anti-inflammatory, expectorant, spasmolytic and estrogenic capabilities (Patra et al., 2020).

In most cases, Chinese star anise (*Illicium verum*) is found to be contaminated with Japanese star anise (*Illicium anisatum*). The unusual neurotoxin sesquiterpene, dilactone anisatin, found in Japanese star anise can cause seizures, hallucinations and nausea.

The essential chemicals in the essential oil of this spice are α-pinene, β-pinene, myrcene, ρ-cymene, α-phellandrene, limonene, γ-terpineol, linalool, α-terpineol, estragole, cis and trans-anethole, cubebene, caryophyllene oxide, α- humulene and foeniculin (Aly et al., 2016; Luís et al., 2019; Wang et al., 2011).

Due to the high concentration of cis anethole (Chouksey et al., 2010; Marinov and Valcheva-Kuzmanova, 2015), the spice has antiviral, antimicrobial, antifungal, antihelminthic, antioxidant, anti-inflammatory, gastroprotective and antiseptic properties. Shikimic acid has anticoagulant, antibacterial, anti-inflammatory and analgesic properties (Boota et al., 2018; Chouksey et al., 2010; Estevez and Estevez, 2012). Bisabolene is used as a fragrance and has antiepileptic properties. Foeniculin has antiseptic properties, and limonene is an important antitumor, anticancer and anti-inflammatory compound found in star anise (Chouksey et al., 2010; de Souza et al., 2015). α-pinene is an important biochemical molecule that has gastroprotective, selective anti-inflammatory, anticatabolic, antinociceptive, antimicrobial, neuroprotective, and cytoprotective properties (Chouksey et al., 2010; de Almeida et al., 2015; Him et al., 2008; Rufino et al., 2014; Da Silva et al., 2012).

Star anise has gained scientific interest due to shikimic acid, which has antiviral properties and is currently used in the production of the avian influenza drug oseltamivir (Tamiflu) (Patra et al., 2020).

2.4 SPICES AND COVID-19

COVID-19 is now the most serious health problem in the world. The virus causing severe acute respiratory syndrome is also known as a coronavirus (SARS-COV-2) (Zhu et al., 2020). COVID-19 was declared a global pandemic by the World Health Organization in March 2020, and no effective treatment or cure was available at an early stage due to the novelty of the virus. In the face of this uncertainty, boosting the natural immunity of the human body was the best option available to tackle this global problem. Traditional immune-boosting herbs and spices were suddenly in the highest demand globally. In this unusual condition global interest was renewed in Ayurveda.

The majority of allopathic medicines used in the treatment of coronaviruses have formed resistant strains of viruses, as well as a wide variety of adverse effects. As a result, there is an urgent need for a safe and effective new antiviral drug. It is crucial to combat the COVID-19 epidemic by improving our diet with natural antibiotics produced from a variety of plant-based ingredients, spices and herbs.

Spices, herbal remedies and their essential oils are rich in active phytochemicals that can be used to develop new antiviral drugs. These active plant components stimulate the immune system and can interfere with the viral life cycle, including viral entry, replication, assembly, exit and virus-specific host targets (Boukhatem et al., 2020). It can also stimulate cytokine inhibitors, natural killer cells, macrophages and lymphocytes to protect before and after exposure. As a result, botanicals reduce the adverse effects of antiviral drugs by reducing the amount needed, while improving treatment and outcomes by reducing inflammation and respiratory symptoms (Khubber et al., 2020).

SARS-COV-2 contains a barbed protein, which facilitates virus entry into host cells and releases virus cells by binding to the human cell surface angiotensin-converting enzyme II (ACE II) that accelerates viral transcription and replication (Schoeman and Fielding, 2019). A new molecule paromomycin emerged as a potential dual target drug since it has shown its inhibitory role for spike and main protease of COVID-19 virus (Tariq et al., 2020).

In addition, the main protease (M^{pro}) is responsible for viral polyprotein degradation, viral genome replication and transcription (Huynh et al., 2020; Meyer-Almes, 2020; Sankar et al., 2021). Consequently, it is a good therapeutic target against COVID-19 (Gao et al., 2020; Cavasotto et al., 2021).

The active phytochemicals found in ginger, cumin, black pepper, turmeric and garlic, the plant sterols β-sitosterol, β-elemene, sesquiterpene and β-chlorogenin saponins, have been shown to have a significant affinity with the active site of S.-ACE II and Mpro proteins (Sankar et al., 2021).

Organosulfur compounds such as allicin and alliin, as well as flavonoids such as quercetin present in garlic, have been shown to inhibit viral replication (Khubber et al., 2020). Similarly, the active compounds in cinnamon, tenuifolin and pavetannin C1, have a strong affinity for the coronavirus spike protein (Prasanth et al., 2021).

Spices such as ginger, cumin, cloves, black cumin, bay leaf, fenugreek, asafoetida, coriander, nutmeg, mace, turmeric, garlic, pepper, carom seeds and star anise, which are discussed in this chapter, are rich sources of immunity-boosting phytochemicals like polyphenols, flavonoids, saponins and alkaloids (Palai et al., 2020). The Ministry of AYUSH, Government of India, supports these immune-boosting spices in various decoctions (Palai et al., 2020).

As a result, researchers in a variety of medical fields are increasingly interested in researching active plant extracts derived from spices to develop safer, more effective and less expensive antiviral drugs.

In short, the ingredients and essential oils found in spices and condiments are powerful antioxidants that also act as natural anti-inflammatory agents. Table 2.2 provides a summary of various spices discussed in this chapter. Chemical structures of all bioactive compounds mentioned in this chapter are given in Table 2.3.

Although they are natural compounds, they can cause allergic reactions when taken in large quantities (D'Souza et al., 2017). As a result, according to the U.S. Food and Drug Administration (FDA) and the World Health Organization, the use of these active plant substances should be limited.

TABLE 2.2
Summary of Spices

S. No.	Spice	Botanical Name	Family	Part Used	Origin
1.	True cinnamon	*Cinnamomum zeylanicum Blume*	Lauraceae	Dried bark	Sri Lanka, the Malabar coast of India and Myanmar (Burma)
2.	Black pepper	*Piper nigrum L.*	Piperaceae	Fruit	Malabar coast of India
3.	Clove	*Syzygium aromaticum*	Myrtaceae	Flower buds	The Maluku Islands in Indonesia
4.	Cumin	*Cuminum cyminum*	Umbelliferae/ Apiaceae	Dried seed	Iran and the Mediterranean region
5.	Garlic	*Allium sativum*	Amaryllidaceae	Bulb of the plant	Central Asia
6.	Ginger	*Zingiber officinale*	Zingiberaceae	Rhizome	Southeast Asia
7.	Fennel	*Foeniculum vulgare*	Umbelliferae/ Apiaceae	Seeds	The Mediterranean region
8.	Fenugreek	*Trigonella foenum-graecum*	Fabaceae	Seeds and leaves	Southeastern Europe and West Asia
9.	Saffron	*Crocus sativus*	Iridaceae	Stigma of flower	Iran
10.	Turmeric	*Curcuma longa*	Zingiberaceae	The underground stem, or rhizome	Southern India and Indonesia
11.	Nutmeg and mace	*Myristica fragrans*	Myristicaceae	Shelled, dried seed, the dried net-like covering of the shell of the nutmeg seed	Banda Islands of Indonesia
12.	Carom seeds	*Trachyspermum ammi*	Umbelliferae/ Apiaceae	Fruit	Egypt
13.	Asafoetida	*Ferula assa-foetida*	Umbelliferae/ Apiaceae	Dried latex (gum oleoresin)	Eastern Iran to Afghanistan
14.	Cardamom	*Elettaria cardamom*um and *Amomum subulatum*	*Zingiberaceae*	Seed pods	The Eastern Himalayas
15.	Star aniseed	*Illicium verum*	Illiciaceae	Fruit	China

TABLE 2.3
Structure of Major Bioactive Compounds

S. No.	Spice	Active Pharmaceutical Compounds	Structures of Major Bioactive Compounds Present
1.	True cinnamon	Cinnamaldehyde, (E)-cinnamyl acetate, camphor, eugenol, cinnamic acid, ρ-coumaric acid	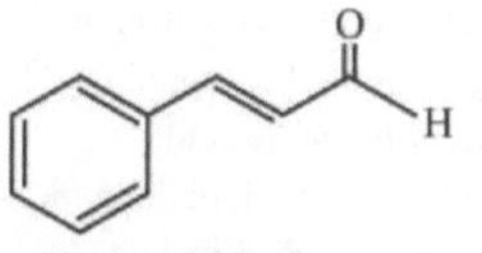 Cinnamaldehyde 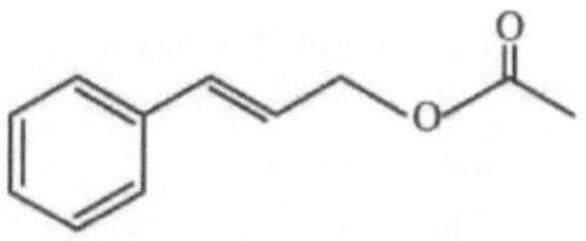 Cinamyl -Acetate 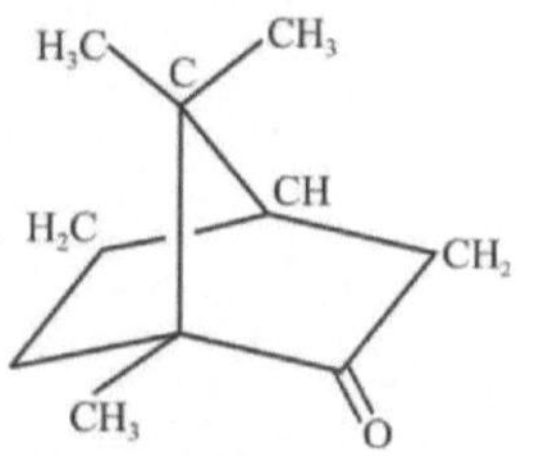 Camphore
2.	Black pepper	Piperine, α-and β-pinene, myrcene, α-phellandrene, limonene, linalool, E-caryophyllene, β-thujene, terpenes, 2,3 diethyl 5 methyl pyrazine and 2 isopropyl 3 methoxy pyrazine	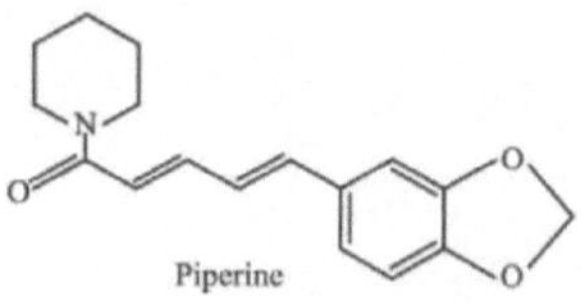 Piperine 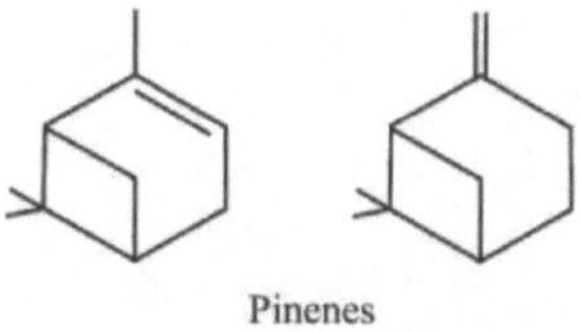 Pinenes 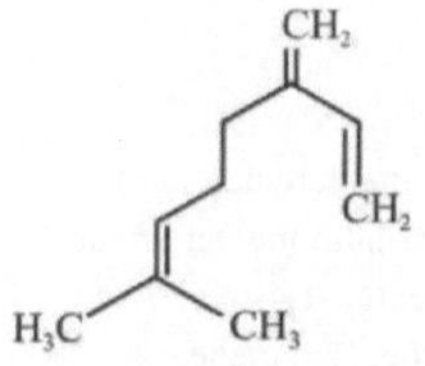 Myrcene 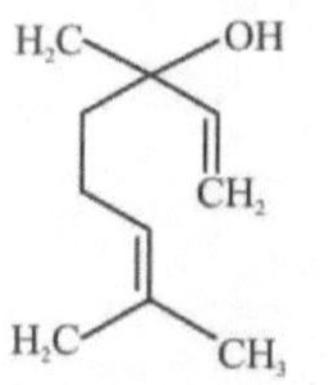 Linalool

(Continued)

TABLE 2.3
(Continued)

S. No.	Spice	Active Pharmaceutical Compounds	Structures of Major Bioactive Compounds Present
3.	Clove	Eugenol, eugenol acetate, β-caryophyllen, tannins, vanillin, gallic acid, crategolic acid, gallotannic acid, methyl salicylate, sesquiterpenes, eugenin, kaempferol, rhemnetin, eugenitin, oleanolic acid, stigmasterol and campesterol	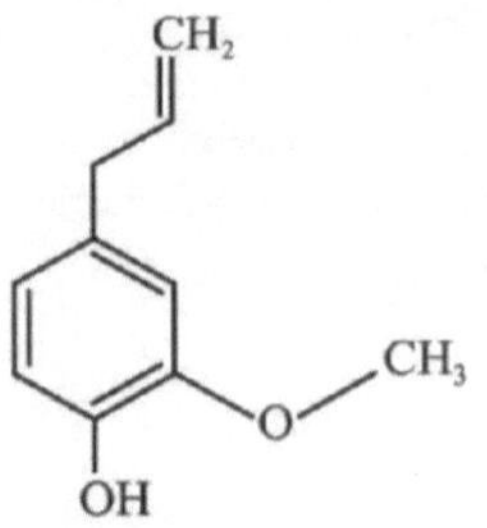 Eugenol 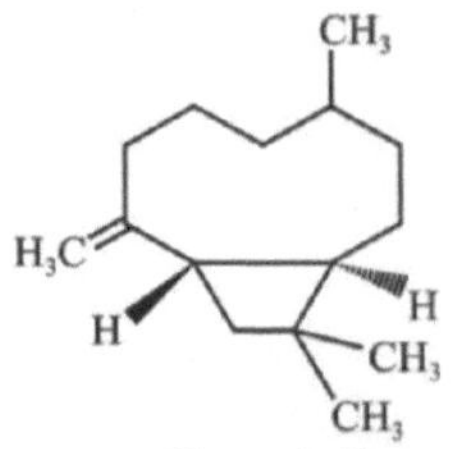 Caryophyllene 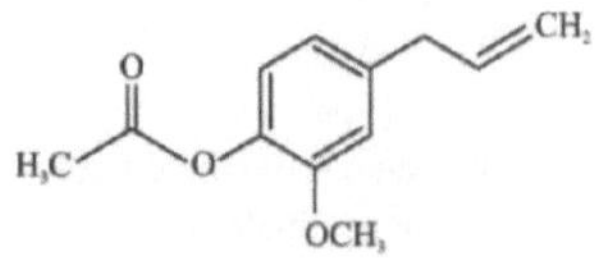 Eugenyl acetate 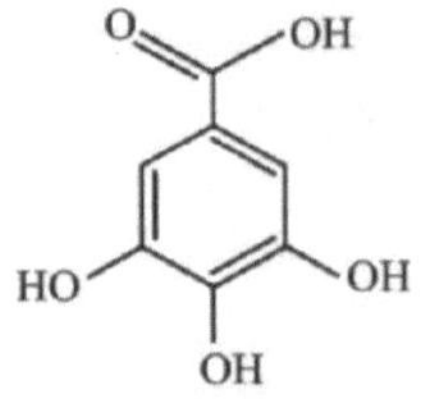 Gallic acid
4.	Cumin	Cuminaldehyde, thymoquinone, terpinene, ρ-mentha-1,4-dien-7-ol, pinene, P-cymene, myrcene, limonene, caryophyllene	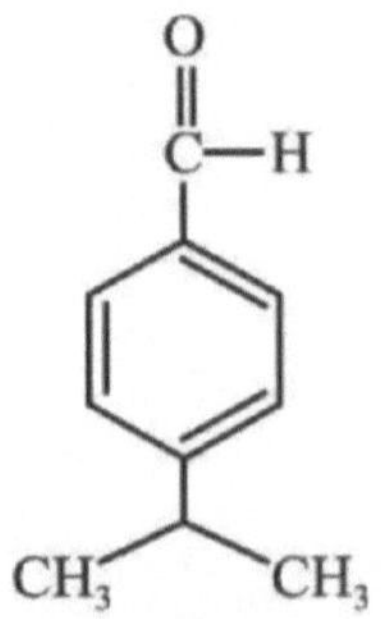 Cuminaldehyde

S. No.	Spice	Active Pharmaceutical Compounds	Structures of Major Bioactive Compounds Present
5.	Garlic	Allicin, alliin, ajoenes, vinyldithiins, sulfides (diallyl sulfide, diallyl disulfide and diallyl trisulfide), S-propyl-cystein sulfoxide and S-methyl-cysteine sulfoxide	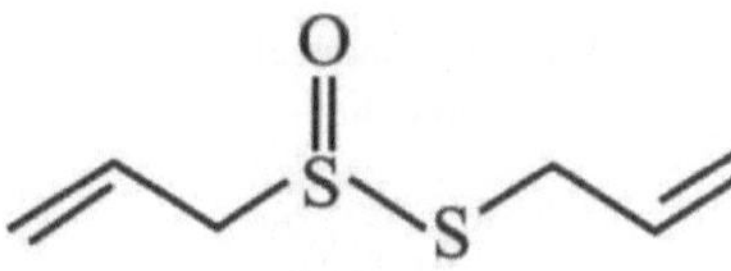 Allicin 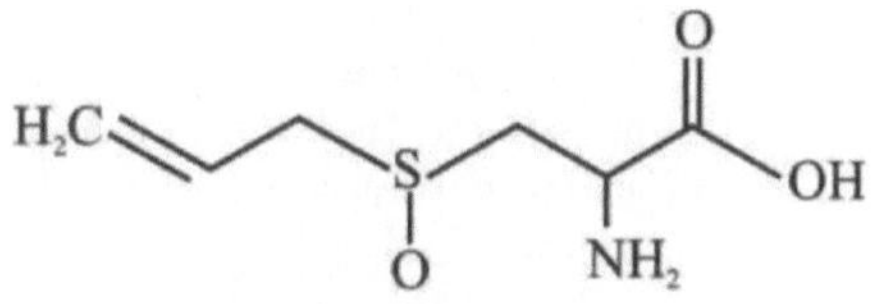 Alliin 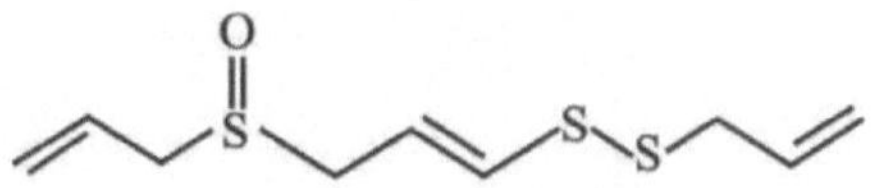 Ajoene
6.	Ginger	Gingerols, shogaols, paradols, zingerone, geraniol, curcumene, β-Phellandrene, (+)-camphene, 1,8-Cineole, citral, terpineol, borneol, linalool, zerumbone, α-zingiberene, β-sesquiphellandrene, β-bisabolene, (E-E)-afarnesene, arcurcumene, zingiberol	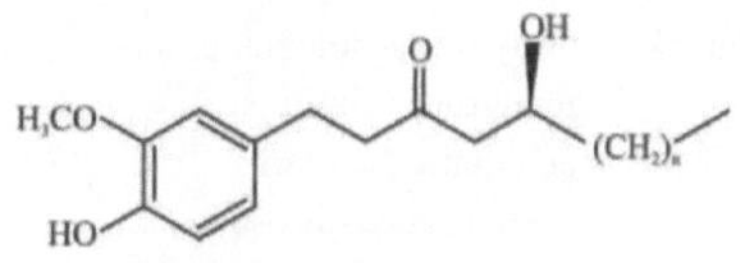 Gingerols 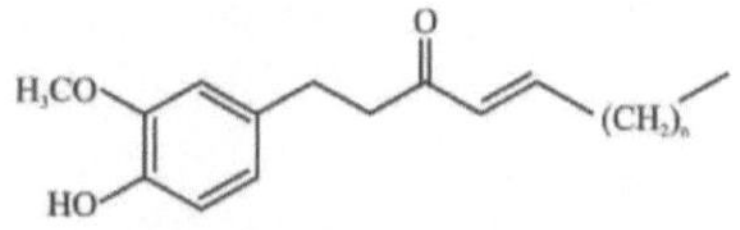 Shogaols 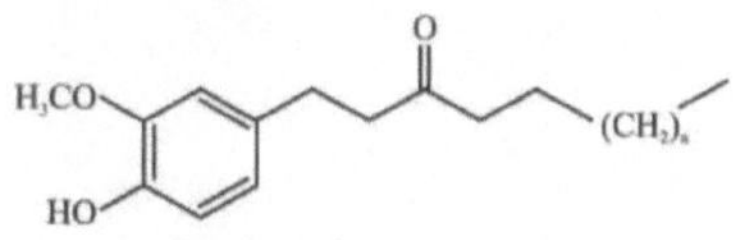 Paradols 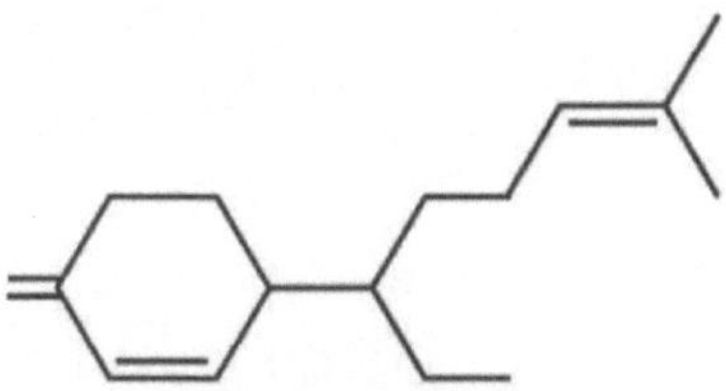 β-Sesquiphelandrene

(Continued)

TABLE 2.3
(Continued)

S. No.	Spice	Active Pharmaceutical Compounds	Structures of Major Bioactive Compounds Present
7.	Fennel	Anethole, α-thujone, limonene, fenchone, estragole (methyl chavicol), camphene, α-pinene, fenchyl alcohol, anisaldehyde. δ-α-phellandrene, dipentene, and ρ-hydroxy phenyl acetone	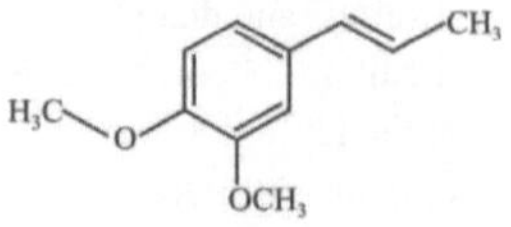 trans-Anethole 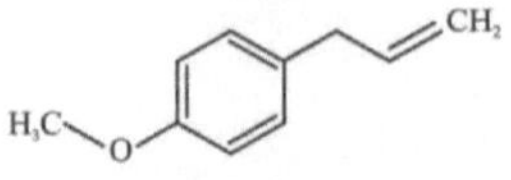 Estragol 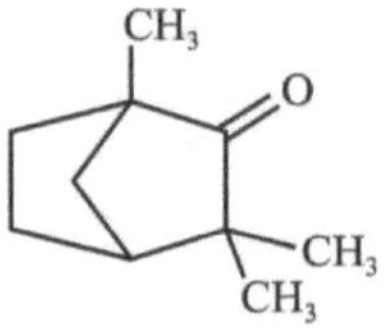 Fenchone
8.	Fenugreek	4-hydroxy isoleucine, fenugrin-B choline, gentianine, carpain, trigonolline, spirostanols, furastanols like diosgenin, gitogenin and yamogenin, rhaponticin, isovitexin	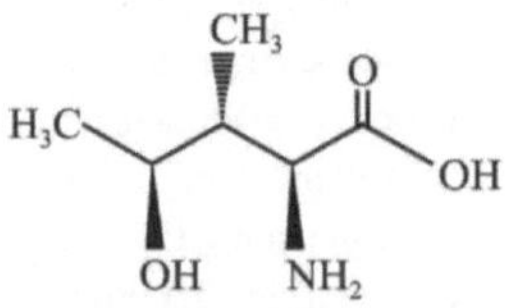 4-Hydroxy Isoleucine 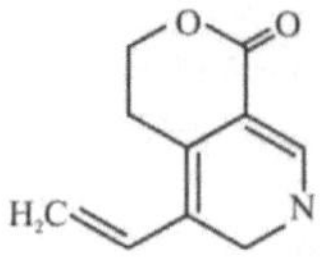 Gentianine 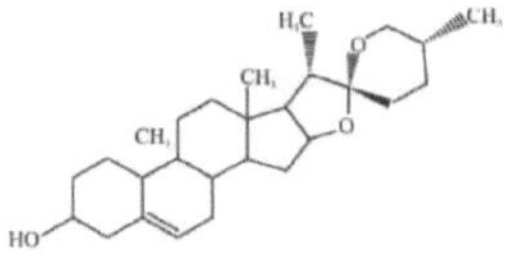 Diosgenin 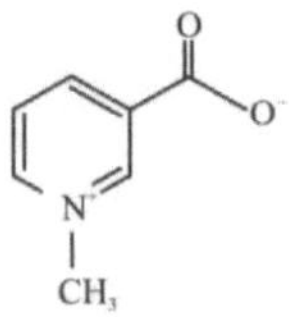 Trigonelline 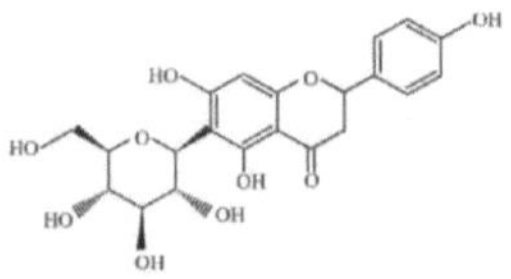 Isovitexin

S. No.	Spice	Active Pharmaceutical Compounds	Structures of Major Bioactive Compounds Present
9.	Saffron	Crocins, crocetins, picrocrocins, safranal, xanthophylls, carotenes, lycopene and zeaxanthin	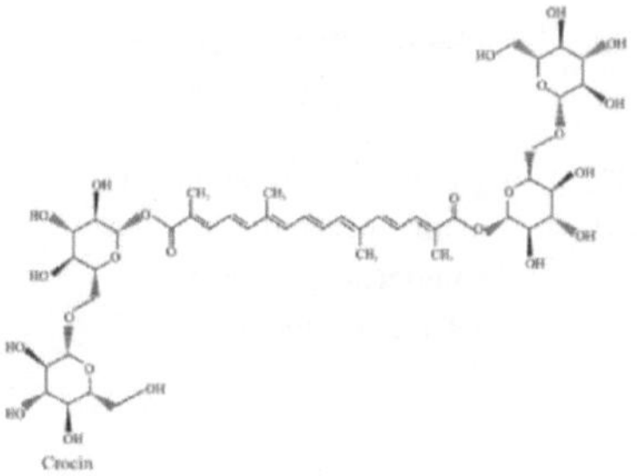 Crocin 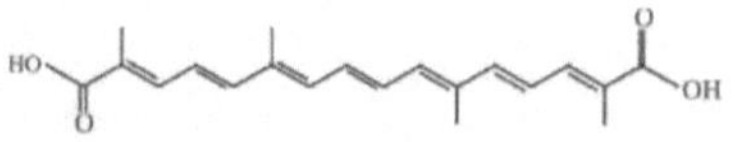 Crocetin 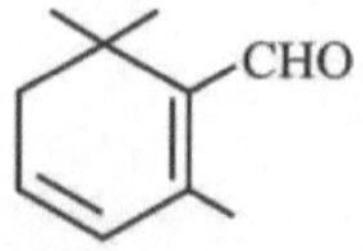 Safranal 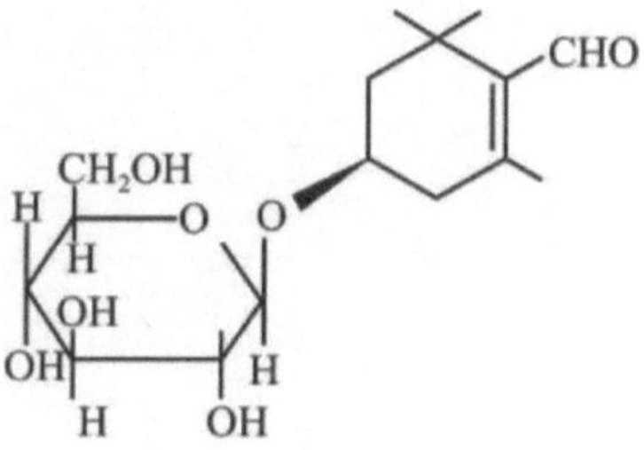 Picrocrocins
10.	Turmeric	Curcumin, demethoxycurcumin (DMC), bisdemethoxycurcumin (BDMC), turmerones and sesquiterpenoids	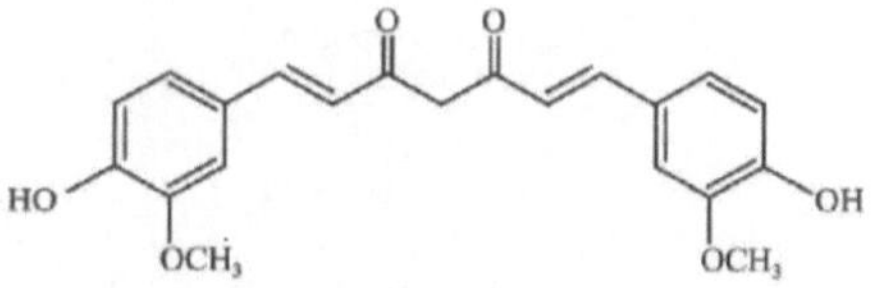 Curcumin 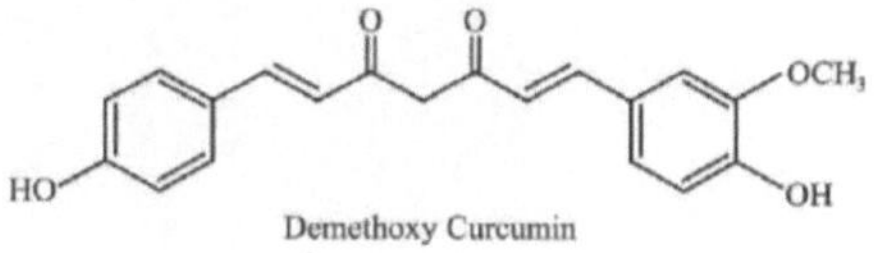 Demethoxy Curcumin 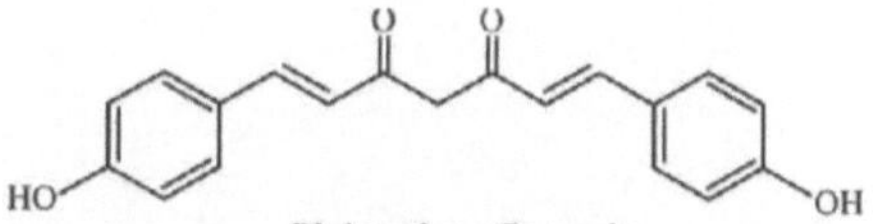 Bisdemethoxy Curcumin

(Continued)

TABLE 2.3
(Continued)

S. No.	Spice	Active Pharmaceutical Compounds	Structures of Major Bioactive Compounds Present
11.	Nutmeg and mace	Myristicin, elemicin, iso elemicin, sabinene, pinene, camphene, p-cymene, safrole, phellandrene	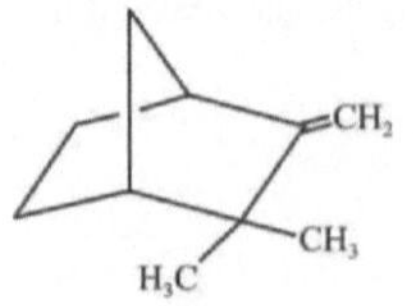 Camphene 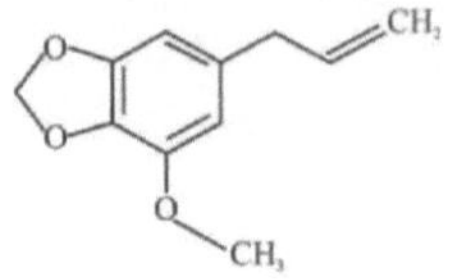 Mytristicin 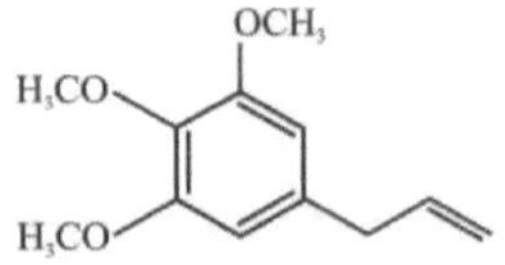 Elimicin 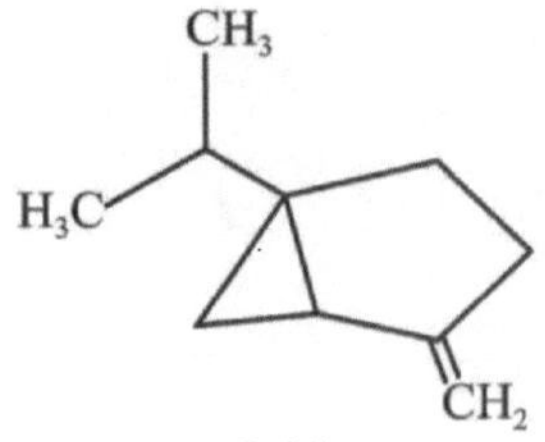 Sabinene
12.	Carom seeds	Thymol, p-cymene γ-terpinene, pinenes, carvacrol, α-terpinene	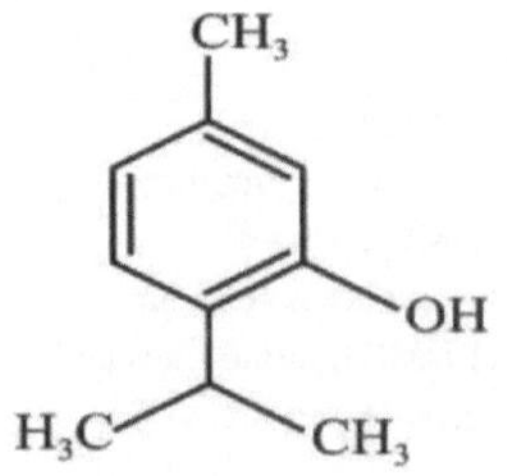 Thymol 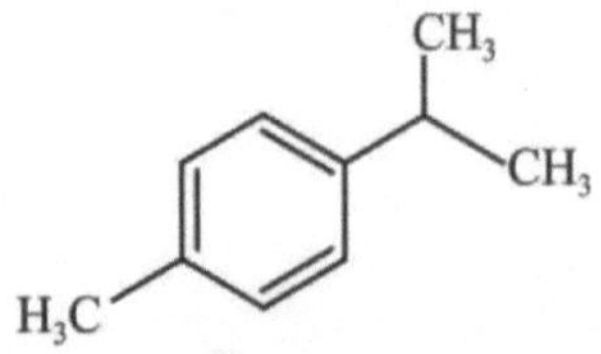 p-Cymene

S. No.	Spice	Active Pharmaceutical Compounds	Structures of Major Bioactive Compounds Present
13.	Asafoetida	Ferulic acid, azulene, umbelliferone	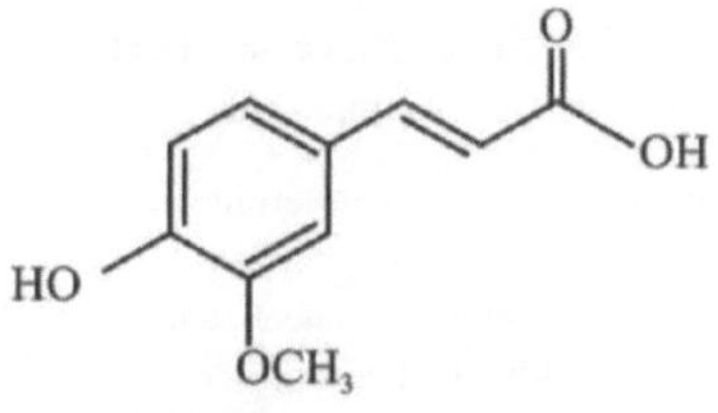 Ferulic acid 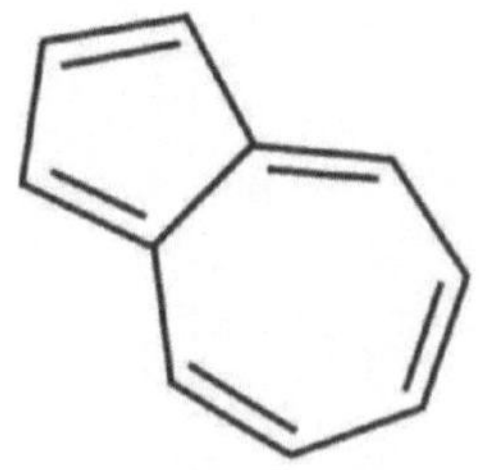 Azulene 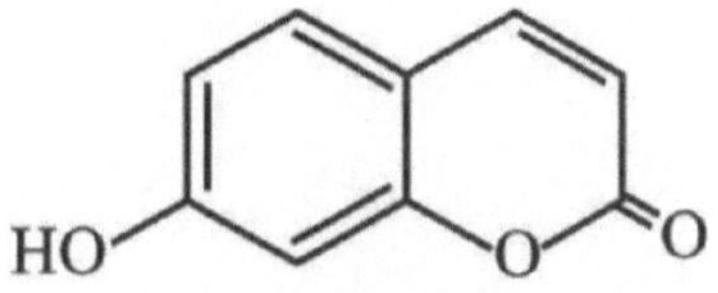 Umbelliferone

(Continued)

TABLE 2.3
(Continued)

S. No.	Spice	Active Pharmaceutical Compounds	Structures of Major Bioactive Compounds Present
14.	Cardamom	1,8-cineole, α-terpineol, terpinyl and linalyl -acetates, sabinene, and β-linalool, camphor or borneol	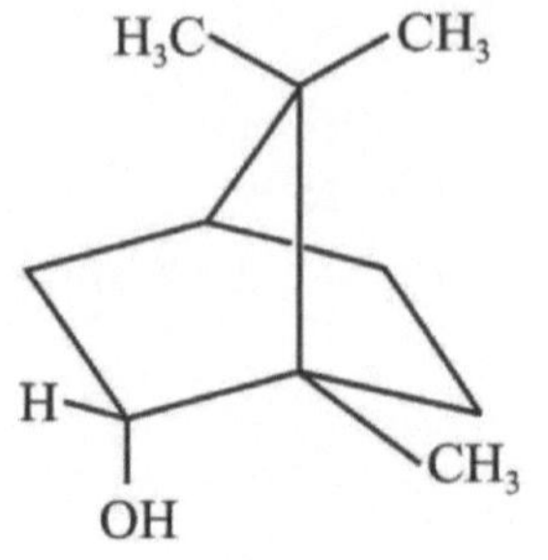 Borneol 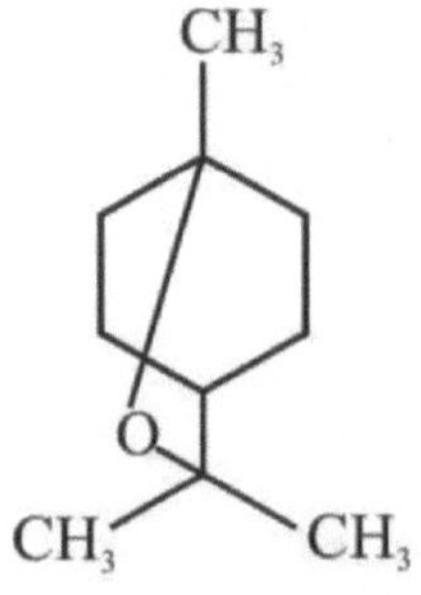 1,8 -Cineole 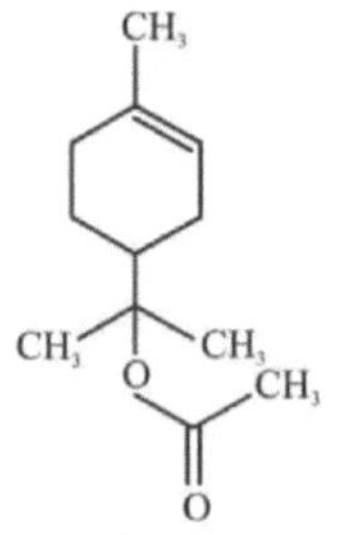 α-Terpinyl Acetate 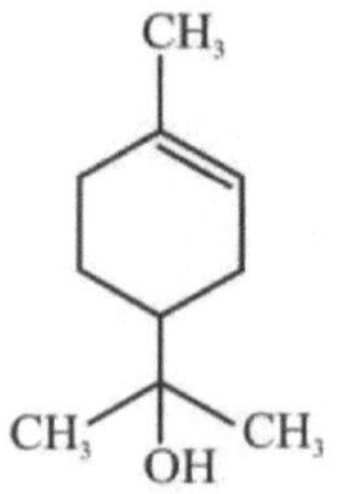 α-Terpineol

S. No.	Spice	Active Pharmaceutical Compounds	Structures of Major Bioactive Compounds Present
15.	Star anise	α-pinene, β-pinene, myrcene, P-cymene, α-phellandrene, limonene, gama terpineol, linalool, α-terpineol, estragole, cis and trans anethole, α-cubebene, caryophyllene oxide, α-humulene, shikimic acid	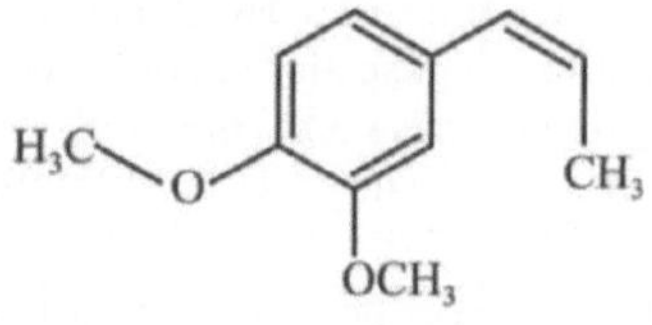 cis-Anethole 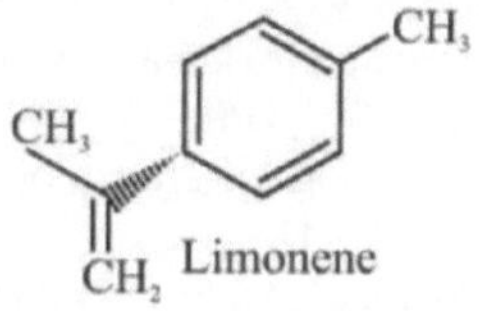 Limonene 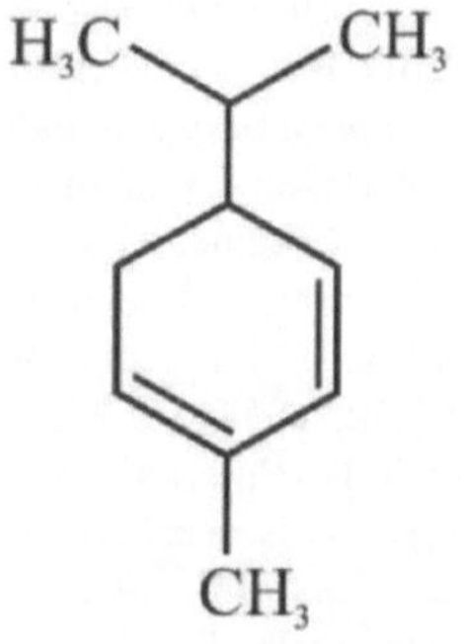 α-Phellandrene 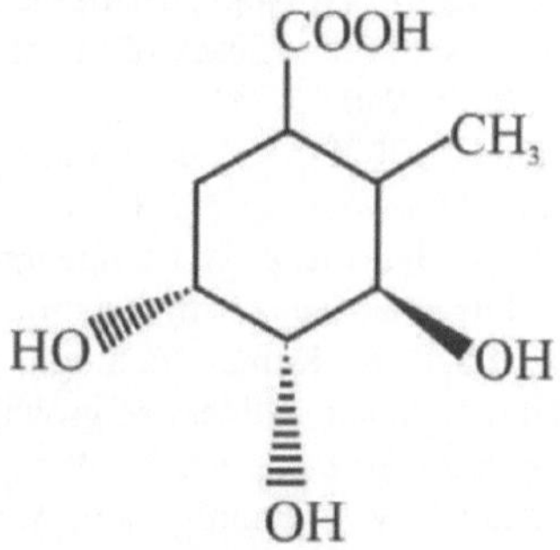 Shikimic Acid

BIBLIOGRAPHY

Abourashed, E. A., & El-Alfy, A. T. 2016. Chemical diversity and pharmacological significance of the secondary metabolites of nutmeg (*Myristica fragrans* Houtt.). *Phytochem Rev*, 15, 1035–1056.

Abu-Zaiton, A. S. 2010. Anti-diabetic activity of Ferula asafoetida extract in normal and alloxan-induced diabetic rats. *Pakistan Journal of Biological Sciences: PJBS*, 1;13(2), 97–100.

Agrahari, P., Panda, P., Verma, N., Khan, W., & Darbari, S. 2015. A brief study on zingiber officinale-a review. *J. Drug Discov. Ther.*, 3, 20–27.

Ajabnoor, M. A., & Tilmisany, A. K. 1998. Effect of Trigonella foenum-graecum on blood glucose levels in normal and alloxan-diabetic mice. *J Ethnopharmacol*, 22, 45–49.

Akbar, S. 2018. Fennel (*Foeniculum vulgare Mill.*): A common spice with unique medicinal properties. *Annals of Complementary and Alternative Medicine*, 1(1) 1001:1–9.

Akram, M., Shahab-uddin, Ahmed A., Khan, U., Hannan, A., Mohiuddin, E., & Asif, M. 2010. Curcuma Longa and Curcumin: A review article, *Rom. J. Biol—Plant Biol.*, 55(2), 65–70.

Al-Asadi, J. N. 2014. Therapeutic uses of Fenugreek (*Trigonella foenum-graecum L.*). *American Open Access Journal Of Social Issues and Humanities*, 21–36.

Alavizadeh, S. H., & Hosseinzadeh, H. 2014. Bioactivity assessment and toxicity of crocin: A comprehensive review. *Food Chem Toxicol.*, 64, 65–80.

Al-Ja'fari, A. H, Vila, R., Freixa, B., Costa, J., & Cañigueral, S. 2013. Antifungal compounds from the rhizome and roots of Ferula hermonis. *Phytotherapy Research*, 27(6):911–915.

Allahghadri, T., Rasooli, I., Owlia, P., Nadooshan, M. J., Ghazanfari, T., Taghizadeh, M., & Astaneh, S. 2010. Antimicrobial property, antioxidant capacity, and cytotoxicity of essential oil from cumin produced in Iran. *J. Food Sci.*, 75, H54—H61.

Al-Snafi, A. 2013. Pharmacological effects of Allium species grown in Iraq: An overview. *Int. J. Pharm. Health Care Res.*, 1, 132–147.

Aly, S. E., Sabry, B. A., Shaheen, M. S., & Hathout, A. S. 2016. Assessment of anti mycotoxigenic and antioxidant activity of star anise (*Illicium verum*) in vitro. *J Saudi Soc Agri Sci*, 15(1), 20–27.

Amalraj, A., & Gopi, S. 2017. Biological activities and medicinal properties of Asafoetida: A review. *Journal of Traditional and Complementary Medicine*, 7, 347–359.

Amin, G. H. 2001. Tehran University of Medical Sciences. In Peter, K. V. (Ed.), *Handbook of Herbs and Spices*. Cambridge, UK: Woodhead Publishing Limited.

Anderson, R. A., Qin, B., Canini, F., Poulet, L., & Roussel, A. M. 2013. Cinnamon counteracts the negative effects of a high fat/high fructose diet on behavior, brain insulin signaling and Alzheimer-associated changes. *PLoS One*, 8(12), e83243.

Angelopoulou, E., Paudel, Y. N., Piperi, C., & Mishra, A. 2021. Neuroprotective potential of cinnamon and its metabolites in Parkinson's disease: Mechanistic insights, limitations, and novel therapeutic opportunities. *Journal of Biochemical and Molecular Toxicology*, 35, 4.

Aqil, F., Jeyabalan, J., Munagala, R., Ahmad, I., Schultz, D. J., & Gupta, R. C. 2021. Cumin prevents 17β-estradiol-associated breast cancer in ACI rats. *Int. J. Mol. Sci.*, 22, 6194.

Araújo, C., & Leon, L. 2001. Biological activities of *Curcuma longa L. Mem. Inst. Oswaldo Cruz*, 96, 723–728.

Arun, N., & Nalini, N. 2002. Efficacy of turmeric on blood sugar and polyol pathway in diabetic albino rats. *Plant Foods Hum. Nutr.*, 57, 41–52.

Arya, P., & Kumar, P. 2021. Diosgenin a steroidal compound: An emerging way to cancer management. *Journal of Food Biochemistry*, 45(12), e14005.

Assimopoulou, A. N., Sinakos, Z., & Papageorgiou, V. P. 2005. Radical scavenging activity of *Crocus sativus L.* extract and its bioactive constituents. *Phytother Res* 19:997–1000.

Badal, D. S., Dwivedi, A. K., Kumar, V., Singh, S., Prakash, A., Verma, S., & Kumar, J. 2019. Effect of organic manures and inorganic fertilizers on growth, yield and its attributing traits in garlic (*Allium sativum L.*). *J. Pharmacogn. Phytochem.*, 8, 587–590.

Badgujar, S. B., Patel, V. V., & Bandivdekar, A. H. 2014. *Foeniculum vulgare mill*: A review of its botany, phytochemistry, pharmacology, contemporary application, and toxicology. *BioMed Research International Volume*, 32.

Bairwa, R., Sodha, R. S., & Rajawat, B. S. 2012. Trachyspermum ammi. *Pharmacognosy Reviews*, 6(11), 56–60.

Baldissera, M. D., Grando, T. H., Souza, C. F., Gressler, L. T., Stefani, L. M., Da Silva, A. S., & Monteiro, S. G. 2016. In vitro and in vivo action of terpinen-4-ol, γ-terpinene, and α-terpinene against *Trypanosoma evansi. Exp. Parasitol.*, 162, 43–48.

Baquer, N. Z., Kumar, P., Taha, A., Kale, R. K., Cowsik, S. M., & Mclean, P. 2011. Metabolic and molecular action of *Trigonella foenum-graecum* (fenugreek) and trace metals in experimental diabetes tissues. *J Biosci*, 36: 383–396.

Batiha, G. E. S., Alkazmi, L. M., Wasef, L. G., Beshbishy, A. M., Nadwa, E. H., & Rashwan, E. K. 2020. *Syzygium aromaticum L.* (Myrtaceae): Traditional uses, bioactive chemical constituents, pharmacological and toxicological activities. *Biomolecules*, 10: 202.

Ben Salem, I., Boussabbeh, M., Kantaoui, H., Bacha, H., & Salwa, A. 2016. Crocin, the main active saffron constituent, mitigates dichlorvos-induced oxidative stress and apoptosis in HCT-116 cells. *Biomedicine & Pharmacotherapy*, 82, 65–71.

Bhattarai, S., & Duke, C. C. 2001. The stability of gingerol and shogaol in aqueous solutions. *Journal of Pharmaceutical Sciences*, 9010, 1658–1664.

Boota, T., Rehman, R., Mushtaq, A., & Kazerooni, E. G. 2018. *Star Anise:* A review on benefits, biological activities and potential uses. *International Journal of Chemical and Biochemical Sciences*, 14, 110–114.

Boukhatem, M. N., & Setzer, W. N. 2020. Aromatic herbs, medicinal plant-derived essential oils, and phytochemical extracts as potential yherapies for Coronaviruses: Future perspectives. *Plants*, 9, 800.

Braga, M. E., Leal, P. F., Carvalho, J. E., & Meireles, M. A. 2003. Comparison of yield, composition, and antioxidant activity of turmeric (Curcuma longa L.) extracts obtained using various techniques. *Journal of Agricultural and Food Chemistry*, 51(22), 6604–6611.

Cavasotto, C. N., Lamas, S. M., & Maggini, J. 2021. Functional and druggability analysis of the SARS-CoV-2 proteome. *Eur. J. Pharmacol.* 890, 173705.

Chaieb, K., Hajlaoui, H., Zmantar, T., Kahla-Nakbi, A. B., Rouabhia, M., Mahdouani, K., & Bakhrouf, A. 2007. The chemical composition and biological activity of clove essential oil, Eugenia caryophyllata (*Syzygium aromaticum L. Myrtaceae*): A short review. *Phytother. Res.*, 21, 501–506.

Chou, S. T., Chang, W. L., Chang, C. T., Hsu, S. L., Lin, Y. C., & Shih, Y. 2013. *Cinnamomum cassia* essential oil inhibits α-MSH-induced melanin production and oxidative stress in murine B16 melanoma cells. *Int J Mol Sci.*, 14, 19186–19201.

Choudhary, S., Walia, B., & Chaudhary, G. 2021. *Ferula asafetida* (Hing): A review based upon its ayurvedic and pharmacological properties. *Int. J. Pharm. Sci. Rev. Res.*, 68(2); 06, 31–39.

Chouksey, D., Sharma, P., & Pawar, R. 2010. Biological activities and chemical constituents of *Illicium verum* hook fruits (Chinese star anise). *Der Pharmacia Sinica*, 1(3), 1–10.

Clery, R. A., Hammond, C. J., & Wright, A. C. 2014. Nitrogen-containing compounds in black pepper oil (*Piper nigrum L.*). *J Essent Oil Res.*, 18: 1–3.

Cortés-Rojas, D. F., Regina, C., De-Souza, F., & Oliveira, W. P. 2014. Clove (*Syzygium aromaticum*): A precious spice. *Asian Pac J Trop Biomed*, 4(2): 90–96.

Da Silva, A. C. R., Lopes, P. M., de Azevedo, M. M. B., Costa, D. C. M., Alviano, C. S., & Alviano, D. S. 2012. Biological activities of a-pinene and β-pinene enantiomers. *Molecules*, 17(6), 6305–6316.

De Almeida Pinheiro, M., Magalhaes, R. M., Torres, D. M., Cavalcante, R. C., Mota, F. S. X., Coelho, E. M. A. O., & Cardoso, J. H. L. 2015. Gastroprotective effect of alpha-pinene and its correlation with antiulcerogenic activity of essential oils obtained from Hyptis species. *Pharmacognosy Magazine*, 11(41), 123.

Debbarma, R., GV, R., & MB, B. 2020. A clinical evaluation of Jatiphala Churna Lepa in Vyanga wsr to Melasma. *International Journal of Ayurveda and Pharma Research*, 28, 17–26.

Dehpour, A. A., Ebrahimzadeh, M. A., Fazel, N. S., & Mohammad, N. S. 2009. Antioxidant activity of the methanol extract of *Ferula asafoetida* and its essential oil composition. *Grasas y aceites*, 30;60(4), 405–412.

De Sousa, D. P., Mesquita, R. F., de Araújo Ribeiro, L. A., & de Lima, J. T. 2015. Spasmolytic activity of carvone and limonene enantiomers. *Natural Product Communications*, 10(11), 1893–1896.

Dkhil, M. A., Abdel Moneim, A. E., Hafez, T. A., Mubaraki, M. A., Mohamed, W. F., Thagfan, F. A., & Al-Quraishy, S. 2019. Myristica fragrans kernels prevent paracetamol-induced hepatotoxicity by inducing anti-apoptotic genes and Nrf2/HO-1 pathway. *International Journal of Molecular Sciences*, 20(4), 993.

Dosoky, N. S., Satyal, P., Barata, L. M., da Silva, J. K. R., & Setzer, W. N. 2019. Volatiles of Black Pepper Fruits (*Piper nigrum* L.). *Molecules*, 24(23): 4244.

Dosoky, N. S., & Setzer, W. N. 2018. Chemical composition and biological activities of essential oils of curcuma species. *Nutrients*, 10(9), 1196.

D'Souza, S. P., Chavannavar, S. V., Kanchanashri, B., & Niveditha, S. B. 2017. Pharmaceutical perspectives of spices and condiments as alternative antimicrobial remedy. *Journal of Evidence-Based Complementary & Alternative Medicine*, 22(4): 1002–1010.

Dymock, W., Warden, C. J. H., & Hooper, D. 1890. Pharmacographia Indica. Karachi, Pakistan: Reprinted by The Institute of Health and Tibbi Research. *Hamdard National Foundation*, 1972;2, 124–126.

El-Saber Batiha, G., Magdy Beshbishy, A. G., Wasef, L., Elewa, Y. H. A., A Al-Sagan, A., Abd El-Hack, M. E., Taha, A. E. M., Abd-Elhakim, Y., & Prasad Devkota, H. 2020. Chemical constituents and pharmacological activities of garlic (*Allium sativum* L.): A review. *Nutrients*, 24;12(3), 872.

Elwakeel, H., Moneim, H., Farid, M., & Gohar, A. A. 2007. Clove oil cream: A new effective treatment for chronic anal fissure. *Colorectal Dis.*, 9, 549–552.

Estevez, A. M., & Estevez, R. J. 2012. A short overview on the medicinal chemistry of (-)-shikimic acid. *Mini Reviews in Medicinal Chemistry*, 12(14), 1443–1454.

Farahmand, S. K., Samini, F., Samini, M., & Samarghandian, S. 2013. Safranal ameliorates antioxidant enzymes and suppresses lipid peroxidation and nitric oxide formation in aged male rat liver. *Biogerontology*, 14, 63–71.

Farrell, K. T. 1999. *Spices, Condiments and Seasonings.* Westport, CT: AVI Publishing; New York: Van Nostrand.

Gao, J., Zhang, L., Liu, X., Li, F., Ma, R., Zhu, Z., et al. 2020. Repurposing low—molecular-weight drugs against the main protease of severe acute respiratory syndrome coronavirus 2. *J. Phys. Chem. Lett*, 11, 7267–7272.

Garg, A., Kumar, A., Nair, A., & Reddy, A. 2007. Analysis of some Indian medicinal herbs by INAA. *Journal of Radioanalytical and Nuclear Chemistry*, 271(3), 611–619.

Giada, M. 2013. Food Phenolic Compounds: Main Classes, Sources and Their Antioxidant Power. In (Ed.), *Oxidative Stress and Chronic Degenerative Diseases – A Role for Antioxidants.* London, UK: IntechOpen.

Goel, A., Kunnumakkara, A. B., & Aggarwal, B. B. 2008. Curcumin as "Curcumin": From kitchen to clinic. *Biochem. Pharmacol.*, 75, 787–809.

Han, X., & Parker, T. L. 2017. Anti-inflammatory activity of clove (*Eugenia caryophyllata*) essential oil in human dermal fibroblasts. *Pharm. Biol.*, 55, 1619–1622.

Haq, I. U., Imran, M., Nadeem, M., Tufail, T., Gondal, T. A., & Mubarak, M. S. (2021). Piperine: A review of its biological effects. *Phytotherapy Research: PTR*, 35(2), 680–700.

Haro-González, J. N., Castillo-Herrera, G. A., Martínez-Velázquez, M., & Espinosa-Andrews, H. 2021. Clove essential oil (*Syzygium aromaticum L. Myrtaceae*): Extraction, chemical composition, food applications, and essential bioactivity for human health. *Molecules*, 26, 6387.

Hausenblas, H. A., Saha, D., Dubyak, P. J., & Anton, S. D. 2013. Saffron (*Crocus sativus L.*) and major depressive disorder: A meta-analysis of randomized clinical trials. *J Integr Med.*, 11(6), 377–383.

Hausner, H., Bredie, W. L., & Molgaard, C. 2008. Differential transfer of dietary flavour compounds into human breast milk. *Physiol Behav.*, 95, 118–124.

Hewlings, S. J., & Kalman, D. S. 2017. Curcumin: A review of its effects on human health. *Foods*, 6, 92.

Him, A., Ozbek, H., Turel, I., & Oner, A. C. 2008. Antinociceptive activity of alpha-pinene and fenchone. *Pharmacology*, 3, 363–369.

Hosseini, A., Razavi, B. M., & Hosseinzadeh, H. 2018. Pharmacokinetic properties of saffron and its active components. *European Journal of Drug Metabolism and Pharmacokinetics*, 43(4), 383–390.

Hosseinzadeh, H., & Nassiri-Asl, M. 2013. Avicenna's (Ibn Sina) the canon of medicine and saffron (*Crocus sativus*): A review. *Phytother Res.*, 27(4), 475–483.

Huynh, T., Wang, H., & Luan, B. 2020. In silico exploration of the molecular mechanism of clinically oriented drugs for possibly inhibiting SARS-CoV-20 s main protease. *J. Phys. Chem. Lett.*, 11, 4413–4420.

Issac, A., Gopakumar, G., Kuttan, R., Maliakel, B., & Krishnakumar, I. M. 2015. Safety and anti-ulcerogenic activity of a novel polyphenol-rich extract of clove buds (*Syzygium aromaticum L). Food Funct.*, 6, 842–852.

Jafri, M. A., Farah, J. K., & Singh, S. 2001. Evaluation of the gastric antiulcerogenic effect of large cardamom fruits of (*Amomum subulatum Roxb.*). *Journal of Ethnopharmacology*, 75(2–3), 89–94.

Jagella, T., & Grosch, W. 1999. Flavour and off-favour compounds of black and white pepper (*Piper nigrum L.*) II. Odour activity values of desirable and undesirable odorants of black pepper. *Eur Food Res Technol.*, 209(1), 22–26.

Jiang, H., Xie, Z., Koo, H. J., McLaughlin, S. P., Timmermann, B. N, & Gang, D. R. 2006. Metabolic profiling and phylogenetic analysis of medicinal Zingiber species: Tools for authentication of ginger (Zingiber officinale Rosc.). *Phytochemistry*, 19(20), 2957–2964.

Johannah, N., Renny, R., Gopakumar, G., Balu, M., Sureshkumar, D., Krishnakumar, I. M. 2015. Beyond the flavour: A de-flavored polyphenol-rich extract of clove buds (Syzygium aromaticum L) as a novel dietary antioxidant ingredient. *Food Funct.*, 6, 3373–3382.

Kabeeruddin, M. 1937. Kitabul-Advia. In *Vol. II Makhzan-al-Mufradat*. New Delhi: Aligarh Barqi Press, pp. 57–58.

Kaye, A. D., De Witt, B. J., Anwar, M., Smith, D. E., Feng, C. J., Kadowitz, P. J., & Nossaman, B. D. 2000. Analysis of responses of garlic derivatives in the pulmonary vascular bed of the rat. *J. Appl. Physiol.*, 89, 353–358.

Kesarwani, K., Gupta, R., & Mukerjee, A. 2013. Bioavailability enhancers of herbal origin: An overview. *Asian Pacific Journal of Tropical Biomedicine*, 3(4), 253–266.

Khalil, A. A., Rahman, U., Khan, M. R., Sahar, A., Mehmood, T., & Khan, M. 2017. Essential oil eugenol: Sources, extraction techniques and nutraceutical perspectives. *RSC Adv.*, 7, 32669–32681.

Khazdair, M. R., Boskabady, M. H., Hosseini, M., Rezaee, R. M., & Tsatsakis, A. 2015. The effects of *Crocus sativus* (saffron) and its constituents on nervous system: A review. *Avicenna J Phytomed.*, 5(5), 376–391.

Khorshidian, N., Yousefi Asli, M., Arab, M., Mirzaie, A. A., & Mortazavian, A. M. 2016. Fenugreek: Potential applications as a functional food and nutraceutical. *Nutrition and Food Sciences Research*, 3(1), 5–16.

Khory, R. N., & Katrak, N. N. 1984. *Materia Medica of India and Their Therapeutics*. Delhi-52: Neeraj Publishing House, p. 292.

Khubber, S., Hashemi Fesharaki, R., Mohammadi, M., & Gharibzahedi, S. M. T. 2020. Garlic (*Allium sativum L.*): A potential unique therapeutic food rich in organosulfur and flavonoid compounds to fight with COVID19. *Nutrition Journal*, 19, 124.

Kim, M., Jung, J., Jeong, N. Y., & Chung, H. J. 2019. The natural plant flavonoid apigenin is a strong antioxidant that effectively delays peripheral neurodegenerative processes. *Anatomical Science International*, 94(4), 285–294.

Korikanthimath, V. S., Mulge, R., & Zachariah, J. T. 1997. Variation in yield and quality characters of cardamom clones. *J. Medicinal and Aromatic Plant Sciences*, 19(4), 1024–7,

Korikanthimath, V. S. 2001. Indian Institute of Spices Research, Karnataka. In Peter, K. V. (Ed.), *Handbook of Herbs and Spices* (Vol. 1). Woodhead Publishing Ltd.

KRS, M. 2005. *Sushruta samhita (700 BC)*. Varanasi: Choukhamba Orientalia.

Kumari, I., Kaurav, H., & Chaudhary, G. 2021a. *Myristica fragrans* (Jaiphal): A significant medicinal herbal plant. *International Journal for Research in Applied Sciences and Biotechnology*, 8(2).

Kumari, I., Madhusudan, S., Walia, B., & Chaudhary, G. 2021b. *Zingiber Officinale* (GINGER): A review based upon its ayurvedic and modern therapeutic properties. *International Journal of Current Research Vol.*, 13(3), 16583–16587.

LaLonde, R. T. 2005. *Terpenes and Terpenoids, Van Nostrand's Encyclopedia of Chemistry*. New York: John Wiley & Sons, Ltd.

Langner, E., Greifenberg, S., & Gruenwald, J. 1998. Ginger: History and Use. *Advances in Therapy*, 15(1), 25–44.

Latif, A., & Rahman, S. Z. 1999. Medicinal use of spices for skin care in Unani medicine. In. Proc. Golden Jubilee National Symposium on Spices, Medicinal and Aromatic Plants—biochemistry conservation and utilization. *IISR Calicut*, 10–12, 274–281.

Lawrence, B. M. 1979. Major tropical spices—cardamom (*Elettaria cardamomum*). In *Essential Oils*. Allured Publ., Wheaton, III., 104.

Lee, C. L., Chiang, L. C., Cheng, L. H., Liaw, C. C., Abd El-Razek, M. H., Chang, F. R., & Wu, Y. C. 2009. Influenza A (H1N1) antiviral and cytotoxic agents from *Ferula assafoetida*. *Journal of Natural Products*, 25;72(9), 1568–1572.

Lee, J. G., Chae, Y., Shin, Y., & Kim, Y. J., 2020. Chemical composition and antioxidant capacity of black pepper pericarp. *The Korean Society for Applied Biological Chemistry*, 63, 35.

Lewis, Y. S., Nambudiri, E. S., & Philip, T. 1966. Composition of cardamom oils, perfume. *Essent. Oil Rec.*, 57, 623.

Ling, J., Wei, B., Lv, G., Ji, H., & Li, S. 2012. Anti-hyperlipidaemic and antioxidant effects of turmeric oil in hyperlipidaemic rats. *Food Chem.*, 130, 229–235.

Luís, Â., Sousa, S., Wackerlig, J., Dobusch, D., Duarte, A. P., Pereira, L., & Domingues, F. 2019. Star anise (*Illicium verum* Hook. F.) essential oil: Antioxidant properties and antibacterial activity against *Acinetobacter baumannii. Flavor Frag J*, 34(4), 260–270.

Madar, Z., & Shomer, I. 1990. Polysaccharide composition of a gel fraction derived from fenugreek and its effect on starch digestion and bile acid absorption in rats. *J Agr Food Chem.*, 38(7), 1535–1539.

Madhusoodanan, K. J., & Saideswara Rao, Y. 2004. Indian Cardamom Research Institute, Kerala. In Peter, K. V. (Ed.), *Handbook of herbs and spices* (Vol. 2). Cambridge, UK: Woodhead Publishing Ltd.

Mahendra, P., & Bisht, S. 2012. *Ferula asafoetida*: Traditional uses and pharmacological activity. *Pharmacogn Rev.*, 6(12), 141–146.

Mahran, G. H., El-Alfy, T. S., & Ansary, H. A. 1975. A phytochemical study of the gum and resin of Afghanian asafoetida. *Bull Fac Pharm.*, 12, 119–32.

Mahulette, A. S., Hariyadi, Yahya, S., & Wachjar, A. 2020. Physico-chemical properties of clove oil from three forest clove accession groups in Maluku. *IOP Conf. Series: Earth and Environmental Science*, 418, 012028.

Majeed, M., Murray, F., & Badmaev, V. 1999. Turmeric and the healing curcuminoids. *McGraw-Hill Education*, 122–127.

Malhotra, S. K., & Vijay, O. P. 2004. National Research Centre on Seed Spices, India. In Peter, K. V. *Handbook of spices*. Cambridge, UK: Woodhead Publishing Limited.

Manohar, R., Ganesh, A., Abbyramy, N., Abinaya, R., Balaji, S. K., & Priya, S. B. 2020. The effect of fennel seeds on pH of saliva: A clinical study. *Indian Journal of Dental Research: Official Publication of Indian Society for Dental Research*, 31(6), 921–923.

Marinov, V., & Valcheva-Kuzmanova, S. 2015. Review on the pharmacological activities of anethole. *Scripta Sci Pharma*, 2(2), 14–19.

Mashmoul, M., Azlan, A., Khaza'Ai, H., Yusof, B. N. M., & Noor, S. M. 2013. Saffron: A natural potent anti-oxidant as a promising anti-obesity drug. *Antioxidants*, 2, 293–308.

Mathern, J. R., Raatz, S. K., Thomas, W., & Slavin, J. L. 2009. Effect of fenugreek fiber on satiety, blood glucose and insulin response and energy intake in obese subjects. *Phytother Res.*, 23(11), 1543–1548.

Meena, M. R., & Sethi, V. 1994. Anti-microbial activity of essential oils from spices. *J. Food. Sci. Tech.*, 31(1), 68–78.

Meyer-Almes, F. J. 2020. Repurposing approved drugs as potential inhibitors of 3CL-protease of SARS-CoV-2: Virtual screening and structure based drug design. *Comput. Biol. Chem.*, 88, 107351.

Milani, J. M., Rezaii, K., Safari, M., Ghanbarzadeh, B., & Phillips, G. O. 2007. Extraction and physicochemical properties of Barijeh (*Ferula galbaniflua*) gum. *Journal of Food and Bioprocess Engineering (JFBE)*, 9, 20.

Moghadamtousi, S. Z., Kadir, H. A., Hassandarvish, P., Tajik, H., Abubakar, S., & Zandi, K. 2014. A review on antibacterial, antiviral, and antifungal activity of curcumin. *BioMed Res. Int.*, 186864.

Moghaddasi, M. S., & Kashani, H. H. 2012. Ginger (*Zingiber officinale*): A review. *Journal of Medicinal Plants Research*, 6, 4255–4258.

Moriwaki, S., Murakami, H., Takahashi, N., Uemura, T., Taketani, K., Hoshino, S., Tsuge, N., Narukami, T., Goto, T., & Kawada, T. 2014. Yamogenin in fenugreek inhibits lipid accumulation through the suppression of gene expression in fatty acid synthesis in hepatocytes. *Bioscience, Biotechnology, and Biochemistry*, 78(7), 1231–1236.

Nadeem, M., & Riaz, A. 2012. Cumin (*Cuminum cyminum*) as a potential source of antioxidants. *Pakistan Journal of Food Science*, 22, 101–107.

Nadkarni, A. K. 1954. *Indian Materia Medica*. 3rd edn. Bombay: Popular Book Depot and Panval Dhootapapeshwar Parakashan Ltd.

Nascimento, L. D. D., Moraes, A. A. B. D., Costa, K. S. D., Galúcio, J. M. P., Taube, P. S., Costa, C. M. L., Cruz, J. N., Andrade, E. H. A., & Faria, L. J. G. 2020. Bioactive natural compounds and antioxidant activity of essential oils from spice plants: New findings and potential applications. *Biomolecules*, 10, 988.

Palai, S., Dehuri, M., & Patra, R. 2020. Spices boosting immunity in COVID-19. *Annals of Phytomedicine*, 9(2): 80–96.

Pandey, K. B., & Rizvi, S. I. 2009. Plant polyphenols as dietary antioxidants in human health and disease. *Oxidative Medicine and Cellular Longevity*, 2(5), 270–8.

Patra, J. K., Das, G., Bose, S., Banerjee, S., Vishnuprasad, C. N., Rodriguez-Torres, M. D. P., & Shin, H. S. 2020. Star anise (Illicium verum): Chemical compounds, antiviral properties, and clinical relevance. *Phytotherapy Research*, 1–20.

Phulsagar, S., Dundi, M., Bhagwat, S., & Girigaon, Y. 2014. An inside review of *Myristica Fragrans Houtt.*—a potential medicinal plant of India. *International Journal of Medical Science and Clinical Inventions*, 1(9), 500–513.

Prasanth, D. S. N. B. K., Murahari, M., Chandramohan, V., Panda, S. P., Atmakuri, L. R., & Guntupalli, C. 2021. *In silico* identification of potential inhibitors from Cinnamon against main protease and spike glycoprotein of SARS CoV-2. *Journal of Biomolecular Structure and Dynamics*, 39(13), 4618–4632.

Pruthi, J. S. 2001. Minor spices and condiments. *ICAR, New Delhi*, 124–133, 659–660.

PV, S. 1981. *Charaka samhita*. Varanasi: Choukhamba Orientalia.

Raghavan, S. 2007. *Handbook of Spices, Seasonings, and Flavorings*. 2nd edn. Boca Raton: CRC Press, Taylor & Francis Group.

Rahman, K. 2001. Historical perspective on garlic and cardiovascular disease. *J. Nutr.*, 131, 977S–979S.

Rahman, M. S. 2007. Allicin and other functional active components in garlic: Health benefits and bioavailability. *Int. J. Food Prop.*, 10, 245–268.

Rajanikanth, B., Ravindranath, B., & Shankaranarayana, M. L. 1984. Volatile polysulphides of asafoetida. *Phytochemistry.*, 23, 899–900.

Rajathi, A. A., Sundarraj, A. A., Leslie, S., & PragalyaaShree, M. M. 2017. Processing and medicinal uses of cardamom and ginger—a review. *J. Pharm. Sci. & Res.*, 9(10), 2117–2122.

Ranasinghe, P., Pigera, S., Premakumara, G. S., Galappaththy, P., Constantine, G. R., & Katulanda, P. 2013. Medicinal properties of 'true' cinnamon (*Cinnamomum zeylanicum*): A systematic review. *BMC Complementary and Alternative Medicine*, 13(1), 275.

Rani, A., Jain, S., & Dureja, P. 2009. Synergistic fungicidal efficacy of formulations of neem oil, nicotinic acid and Ferula asafoetida with α, β-unsaturated carbonyl compounds against ITCC 5226 Sclerotium rolfsii & ITCC 0482 Macrophomina phaseolina. *Journal of Pesticide Science*, 0908030091.

Rather, M. A., Dar, B. A., Sofi, S. N., Bhat, B. A., & Qurishi, M. A. 2016. *Foeniculum vulgare:* A comprehensive review of its traditional use, phytochemistry, pharmacology, and safety. *Arabian Journal of Chemistry*, 9(2), S1574–S1583.

Rathore, S., Mukim, M., Sharma, P., Devi, S., Nagar, J. C., & Khalid, M. 2020. Curcumin: A review for health benefits. *International Journal of Research and Review*, 7, 1.

Rawat, I., Verma, N., & Joshi, K. 2020. Cinnamon (*Cinnamomum zeylanicum*). *Medicinal Plants in India: Importance and Cultivation*, 164–177.

Reza, Z. M., Atefeh, J. Y., & Faezeh, F. 2015. Effect of γ-irradiation on the antibacterial activities of *Cuminum cyminum* L. essential oils in vitro and in vivo systems. *J. Essent. Oil Bear. Plants*, 18, 582–591.

Rufino, A. T., Ribeiro, M., Judas, F., Salgueiro, L. G., Lopes, M. C., Cavaleiro, C., & Mendes, A. F. 2014. Anti-inflammatory and chondroprotective activity of (+)-α-pinene: Structural and enantiomeric selectivity. *Journal of Natural Products*, 77(2), 264–269.

Saha, J. C., & Kasinathan, S. 1961. Ecbolic properties of Indian medicinal plants II. *Indian J Med Res.*, 49, 1094–1098.

Saleem, M., Alam, A., & Sultana, S. 2001. Asafoetida inhibits early events of carcinogenesis: A chemopreventive study. *Life Sciences*, 9, 68(16), 1913–1921.

Salzer, U. J. 1975. Analytical evaluation of seasoning extracts (oleoresins) and essential oils from seasonings. *I. International Flavours and Food Additives*, 6(3), 151.

Samadi-Noshahr, Z., Hadjzadeh, M. A., Moradi-Marjaneh, R., & Khajavi-Rad, A. 2020. The hepatoprotective effects of fennel seeds extract and trans-Anethole in streptozotocin-induced liver injury in rats. *Food Science & Nutrition*, 9(2), 1121–1131.

Sankar, M., Ramachandran, B., Pandi, B., Mutharasappan, N., Ramasamy, V., Prabu, P. G., Shanmugaraj, G., Wang, Y., Muniyandai, B., Rathinasamy, S., Chandrasekaran, B., Bayan, M. F., Jeyaraman, J., Halliah, G. P., & Ebenezer, S. K. 2021. *In silico* screening of natural phytocompounds towards identification of potential lead compounds to treat COVID-19. *Frontiers in Molecular Biosciences*, 8, 637122.

Sauvaire, Y., Petit, P., Broca, C., Mantehgetti, A. K., Baissac, Y., Fernandez-Alvarez, J., Gross, R., Roye, M., Leconte, A., Gomis, R., & Ribes, G. 1998. 4-hydroxyleucine: A novel amino acid potentiator of insulin secretion. *Diabetes*, 47, 206–210.

Schoeman, D., & Fielding, B. C. 2019. Coronavirus envelope protein: Current knowledge. *Virol. J.*, 16, 69.

Seneme, E. F., Dos Santos, D. C., Silva, E., Franco, Y., & Longato, G. B. 2021. Pharmacological and therapeutic potential of myristicin: A literature review. *Molecules*, 26(19), 5914.

Sengupta, A., & Bhattacharjee, S. 2009. Cardamom (*Elettaria cardamomum*) and its active constituent, I,8-cineole. 10.1142/9789812837912_0003.

Shan, J., Chen, L., & Lu, K. 2017. Protective effects of trans-caryophyllene on maintaining osteoblast function. *IUBMB Life*, 69, 22–29.

Shankaranarayana, M. L., Raghavan, B., & Natarajan, C. P. 1982. Odorous compounds of asafetida. VII. Isolation and identifi cation. *Indian Food Pack*, 36, 65–76.

Sharma, P. C., Yeine, M. B., & Dennis, T. J. 2002. Database on medicinal plants used in ayurveda, Vol-IV, New Delhi; Central Council for Research in Ayurveda and Siddha. Dept. of ISM and H Ministry of Health and Welfare, Govt. of India: 217.

Sharma, R. D., Raghuram, T. C., & Rao, N. S. 1990. Effects of fenugreek seeds on blood glucose and serum lipid in type I diabetes. *Eur J clin nutr*, 44, 301–306.

Singh, N., Rao, A. S., Nandal, A., Kumar, S., Yadav, S. S., Ganaie, S. A., & Narasimhan, B. 2021. Phytochemical and pharmacological review of *Cinnamomum verum*. J. Presl-a versatile spice used in food and nutrition. *Food Chemistry*, 338, 127773.

Siracusa, L., Gresta, F., & Ruberto, G. 2011. Saffron *(Crocus Sativus L.)* apocarotenoids: A review of their biomolecular features and biological activity perspectives. *Carotenoids: Properties, Effects and Diseases*, 145–178.

Slusarenko, A. J., Patel, A., & Portz, D. 2008. Control of plant diseases by natural products: Allicin from garlic as a case study. *Eur. J. Plant Pathol.*, 121, 313.

Snehlata, H. S., & Payal, D. R. 2012. Fenugreek (*Trigonella foenumgraecum L.*): An overview. *IJCPR*, 2(4), 169–187.

Somani, R., Karve, S., Jain, D., Jain, K., & Singhai, A. 2008. Phytochemical and pharmacological potential of *Myristica fragrans Houtt:* A comprehensive review. *Pharmacognosy Reviews*, 2(4), 68.

Souza, G. A., Ebaid, G. X., Seiva, F. R., Rocha, K. H., Galhardi, C. M., Mani, F., & Novelli, E. L. 2011. N-acetylcysteine, an Allium plant compound improves high-sucrose diet-induced obesity and related effects. *Evid. Based Complement. Altern. Med.*, 2011, 643269.

Tajuddin, Ahmad S., Latif, A., & Qasmi, I. A. 2004. Effect of 50% ethanolic extract of Syzygium aromaticum (L.) Merr. & Perry. (clove) on sexual behaviour of normal male rats. *BMC Complementary and Alternative Medicine*, 4, 17.

Takeoka, G. 2001. Instrumental analysis of food flavors-4 volatile constituents of Asafoetida. In *ACS symposium series* (794: 33–44). Washington, DC: American Chemical Society, 1974.

Tariq, A., Mateen, R. M., Afzal, M. S., & Saleem, M. 2020. Paromomycin: A potential dual targeted drug effectively inhibits both spike (S1) and main protease of COVID-19. *Int. J. Infect. Dis.*, 98, 166–175.

Tayeboon, G. S., Tavakoli, F., Hassani, S., Khanavi, M., Sabzevari, O., & Ostad, S. N. 2013. Effects of cymbopogon citratus and Ferula assafoetida extracts on glutamate-induced neurotoxicity. *In Vitro Cellular & Developmental Biology-Animal*, 1;49(9):706–715.

Thomas, J., & Duethi, P. P. 2001. *Handbook of Herbs and Spices* (Vol. 1), Peter, K. V. (Ed.). Sawston, Cambridge: Woodhead Publishing Ltd., pp. 143–152.

Trivedi, P. D., Pundarikakshudu, K., Rathnam, S., & Shah, K. S. 2007. A validated quantitative thin-layer chromatographic method for estimation of diosgenin in various plant samples, extract, and market formulation. *J Aoac Int.*, 90(2), 358–363.

Turkyılmaz, C., Onal, E., Hirfanoglu, I. M., Turan, O., Koç, E., Ergenekon, E., et al. 2011. The effect of galactagogue herbal tea on breast milk production and short-term catch-up of birth weight in the first week of life. *J Altern Complem Med.*, 17(2), 139–142.

Varshney, R., & Budoff, M. J. 2016. Garlic and Heart Disease. *J. Nutr.*, 146, 416S–421S.

Wang, G. W., Hu, W. T., Huang, B. K., & Qin, L. P. 2011. *Illicium verum*: A review on its botany, traditional use, chemistry and pharmacology. *Journal of Ethnopharmacology*, 136(1), 10–20.

Wang, Y., Du, G. Y., Guo, T., Zou, H. M., & Jia, D. (2021). Skin-whitening mechanism of cumin (*Cuminum cyminum* L.) extract. *Pakistan Journal of Pharmaceutical Sciences*, 34(1), 77–84.

Wijesekera, R. O. B., & Jayawardena, A. L. 1973. Recent developments in the production of spices and their essential oils in Ceylon, *Proc. Conf. Spices, Tropical Products Institute, London, 159.*

Wohlmuth, H., Leach, D. N., Smith, M. K., & Myers, S. P. 2005. Gingerol content of diploid and tetraploid clones of ginger (*Zingiber officinale Roscoe*). *Journal of Agricultural and Food Chemistry*, 53(14), 5772–5778.

Yoshikawa, M., Murakami, T., Komatsu, H., Murakami, N., Yamahara, J., & Matsuda H. 1997. Medicinal foodstuffs: IV. Fenugreek seeds (1): Structures of trigoneosides Ia, Ib, IIb, IIIa and IIIb new furostanol saponins from the seeds of Indian. *Trigonella foenumgraecum L. Chem Pharmacol Bull*, 45, 81–87.

Yu, X., Lin, H., Wang, Y., Lv, W., Zhang, S., Qian, Y., Deng, X., Feng, N., Yu, H., & Qian, B. 2018. D-limonene exhibits antitumor activity by inducing autophagy and apoptosis in lung cancer. *Onco. Targets. Ther.*, 11, 1833–1847.

Zeng, Y., Li, Y., Yang, J., Pu, X., Du, J., Yang, X., Yang, T., & Yang, S. 2017. Therapeutic role of functional components in Alliums for preventive chronic disease in human being. Evid. Based Complement. *Altern. Med.*, 9402–849.

Zhan, K., Wang, C., Xu, K., & Yin, H. 2008. Analysis of volatile and non-volatile compositions in ginger oleoresin by gas chromatography-mass spectrometry. Se. pu. *Chinese Journal of Chromatography*, 26(6), 692–696.

Zhou, J., Chan, L., & Zhou, S. 2012. Trigonelline: A plant alkaloid with therapeutic potential for diabetes and central nervous system disease. *Current Medicinal Chemistry*, 19(21), 3523–3531.

Zhu, N., Zhang, D., Wang, W., Li, X., Yang, B., Song, J., Zhao, X., Huang, B., Shi, W., Lu, R., Niu, P., Zhan, F., Ma, X., Wang, D., Xu, W., Wu, G., Gao, G. F., Tan, W., & China Novel Coronavirus Investigating and Research Team. 2020. A novel coronavirus from patients with Pneumonia in China, 2019. *The New England Journal of Medicine*, 382(8), 727–733.

Zia, A., Farkhondeh, T., Pourbagher-Shahri, A. M., & Samarghandian, S. 2021. The role of curcumin in aging and senescence: Molecular mechanisms. *Biomedicine & Pharmacotherapy*, 134, 111119.

Ziegler, J., & Facchini, P. J. 2008. Alkaloid biosynthesis: metabolism and trafficking. *Annual Review of Plant Biology*, 59, 735–769.

3 Trends in Breeding Medicinal and Aromatic Plants
Challenges and Prospects

Deivanai Subramanian

CONTENTS

3.1 Introduction ... 53
3.2 Traditional Breeding Approach ... 54
3.2.1 Artificial Selection ... 54
3.2.2 Hybridization Technique ... 55
3.2.3 Mutation Breeding ... 56
3.2.4 Polyploidy Breeding ... 56
3.3 Biotechnology Approach ... 58
3.3.1 Plant Tissue Culture ... 58
3.3.2 Genetic Engineering ... 58
3.3.3 Genome Editing ... 59
3.3.4 Marker-Assisted Selection ... 59
3.4 Conclusions and Perspectives ... 60
Bibliography ... 60

3.1 INTRODUCTION

Plants with therapeutic and aromatic properties have been used to relieve pain and heal various ailments since time immemorial and are often referred to as medicinal and aromatic plants (MAPs). They are an essential part of conventional medical systems followed by oral traditions in various cultures in managing health and well-being. These plants consist of an array of active compounds which form a major source of numerous valuable drugs (Farnsworth 1990; Bodeker et al. 2005). Many affluent countries have accepted the use of MAPs as a means of primary health care, and the vast majority of them are collected from wild resources (Hamilton 2004; Leamann 2011). It is estimated that nearly 50,000–80,000 plant species are used as medicaments worldwide (Chen et al. 2016a, 2016b). In 2014, the global trade value of medicinal plants was accounted at 702,813 tons worth USD 3.60 billion. "Countries like China and India from Asia; Egypt and Morocco from Africa; Poland, Bulgaria and Albania from Europe; Chile and Peru from South America are the lead suppliers while USA, Japan and Europe are the main consumers of the world" (Lange 2004; Vasisht et al. 2016). In Asia, Europe, and North America, the demand for MAPs is expected to increase at a pace of 8–15% per year (Chen et al. 2016a, 2016b). Consequently, medicinal plants are under great pressure due to the substantial demand and consumption that cause overexploitation of resources (Nalawade et al. 2003). Furthermore, unsustainable harvesting practice has reduced the regeneration and survival of medicinal plants, causing severe irreplaceable habitat loss (Nishteswar 2014; Ouarghidi 2017) and eventually leading to extinction. It is evident that some 25,000 plant species are at risk of extinction in Asia and the Pacific (UNEP 2011). For example, medicinal plants like American ginseng (*Panax quinquefolius*), black cohosh (*Cimicifuga racemosa*), goldenseal (*Hydrastis canadensis*), slippery elm (*Ulmus rubra*), and yew (*Taxus* species) have been

DOI: 10.1201/b22917-3

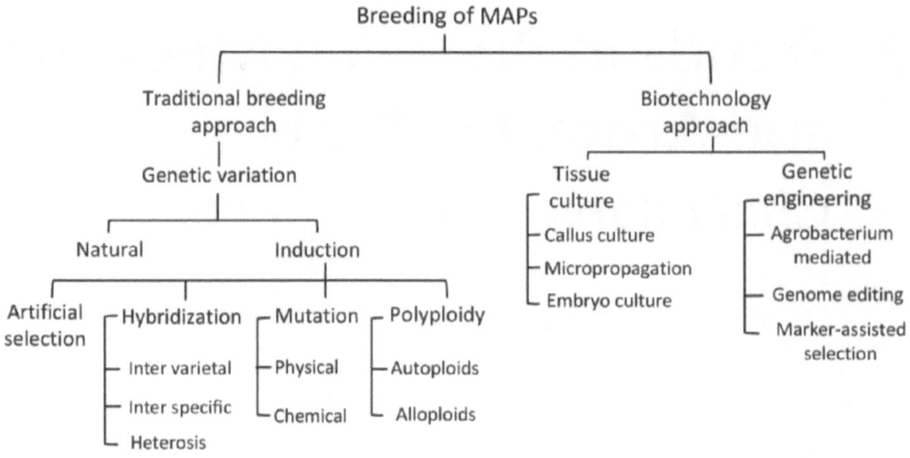

FIGURE 3.1 Schematic representation of breeding methods in MAPs.

seriously depleted (Emily 2008). Several initiatives have been taken for sustainable harvesting and protection against existing threats—cultivation of MAPs is one such promising alternative which enables preservation of endangered species. It is expected that cultivation of MAPs under a formal agricultural system may offer optimum yield with a high-quality product. Some MAP species are grown for commercial production worldwide in this regard, for instance, 20 species in India, 250 species in China, 40 species in Hungary, 100 in Germany, and 140 in Europe (Schippmann et al. 2002; Chen et al. 2014). However, factors such as low germination rate, slow growth, and habitat specificity limit the cultivation of some species. Further, a lack of suitable planting materials also causes an issue in cultivation because most ex situ conservation works were focused on interspecific diversity rather than intraspecific, and few works were undertaken in selection and improvement of seed stock for propagation. Though cultivation of MAPs occupies a small area when compared to agriculture crops, breeding techniques open an opportunity to resolve the inherent issues and expand its distribution. A bibliometric survey on MAPs has revealed that most of the research in last two decades was focused on the search for new medicines or active compounds rather than on breeding. This chapter reviews breeding methods (Figure 3.1), challenges, recent advances, and development prospects in terms of MAP cultivation.

3.2 TRADITIONAL BREEDING APPROACH

The first attempt at cultivation consists of collecting plant materials or seeds from wild populations and growing them in a permanent location. Artificial selection with desirable characteristics then follows to domesticate the wild progenitors.

3.2.1 Artificial Selection

Availability of a diverse germplasm is the basis for selection and can be exploited by collecting specimens from the wild; further, it offers an opportunity for conservation and breeding of MAPs with wider adaptation to the natural environment (Kumar et al. 2016; Ndjiondjop et al. 2017). However, administration of genetic resources in field collections is tedious and expensive due to the requirement of large areas and adverse environmental effects. Hence, a core collection of germplasm is established which uses both phenotypic traits and molecular data as representative of gene pools

(Yuan et al. 2010; Guzmán et al. 2017; Liu et al. 2019). When devising a core collection for MAPs, it's important to evaluate the presence of secondary metabolites because synthesis and accumulation are influenced by various genetic and environmental factors. Further, the quality of medicinal plants is defined by the level of active ingredient, which is distinct from crop plants and often difficult to breed using traditional methods (Wang et al. 2020). Although breeding of MAPs is still in its initial stage, success has been achieved through a few cycles of selection based on visible phenotypes. For instance, MAP such as chamomile (*Matricaria recutita* L.), peperina (*Minthostachys verticillate* Griseb.), rosemary (*Rosmarinus officinalis* L.), hemp (*Cannabis sativa* L.), Greek sage (*Salvia fruticosa* Mill.), oregano (*Origanum vulgare* L.), horehound (*Marrubium vulgare* L.), sweet flag (*Acorus calamus* L.), wilga (*Geijera parviflora* Lindl.,), gidgee fuchsia (*Eremophila dalyana* F. Muell.), thymus species, tansy (*Tanacetum vulgare* L.), pennywort (*Centella asiatica* L.), etc., were selected from the collection for chemotype diversity. Likewise, desert ginseng (*Cistanche herba*) was chosen for better adaptability and rosemary (*Rosmarinus officinalis* L.) for biomass yield. Thus, several medicinal plants are selected based on their wider adaptability to varying environments and high yield of valuable metabolites. Despite constant efforts to improve MAPs, only a few of them are cultivated on a large scale due to degradation in genetic diversity relative to their wild counterparts. The presence of low genetic diversity in a species increases the risk of adaptability and long-term survival (Linde and Selmes 2012).

Breeding techniques such as hybridization, mutation, and polyploidy increase the intraspecific genetic diversity of the participating populations, and it is accomplished between carefully chosen parents. However, the success of breeding relies on clear objectives of the traits of interest to be improved, and these should be specific, measurable, and achievable. Some typical breeding objectives of MAPs as stated by Pank (2006) and Faehnrich et al. (2021) are "i) high and stable yield, ii) homogeneous raw material, iii) high content of desired compounds and absence of harmful substances, iv) resistance to biotic stress (pests and diseases) and abiotic stress (climatic adaptation to the cultivation region), and v) high functional value and safety".

3.2.2 Hybridization Technique

This technique uses deliberate interbreeding of closely or distantly related parental lines to produce a new cultivar or lines with desirable traits in a span of five or more years. The new cultivars should fit into the technological processes in agriculture and industry with reduced cost for sustainable production. To date, hybridization technique in MAPs has yielded a number of promising cultivars, for example, 1) a strong lemon fragrant hybrid of Lao basil (*Ocimum citriodorum*) derived from basil (*Ocimum basilicum*) and African basil (*Ocimum americanum*) (Darrah 1980); 2) a high-morphine-content opium poppy hybrid obtained from an intervarietal cross between Dhola Chhotta Gotia and Kantia Pink (Kaicker and Choudhury 1983); 3) a high-citral-oil lemon grass hybrid (*Cymbopogan flexuosus*) CKP 25 developed between *C. khasianum* and *C. pendulus* (Farooqi and Sreeramu 2001); 4) a promising medicinal and ornamental valued foxglove hybrid achieved from an interspecific hybridization between *Digitalis purpurea* and *D. lanata* (Ardelean et al. 2006); 5) a chemotype thyme hybrid (*Thymus vulgaris* L.), Varico 3, rich in thymol was recommended for cultivation in Switzerland and middle Europe (Carlen et al. 2010); 6) a high-yielding German chamomile variety, P15 (*Matricaria recutita* L.), with high α-bisabolol content showed favorable agronomic traits which facilitated mechanical harvesting (Albrecht et al. 2016; Fejér and Salamon 2016; Chauhan et al. 2022); 7) high-yield red sage cultivars ZD1and ZD2 obtained from a cross between a male sterile line and inbred lines within the same species of *Salvia miltiorrhiza* (Chen et al. 2016a, 2016b); and 8) a hybrid Amerindian yam (*Dioscorea trifida*) developed between local landraces and cultivated yam showed high phenotypic variability for tuber quality parameters (Beyerlein et al. 2019). Interspecific and intraspecific crossing between nine species of *Salix* such as *alba* (white willow), *daphnoides*, *humboldtiana* (Chile willow), *lasiandra*, *nigra* (black willow), *pentandra* (bay willow), *purpurea* (purple willow), *S.* x *rubens*, and *viminalis* (common osier) produced chemotypes with higher salicylates (Förster et al. 2021). In addition, hybridization also acts as a potential tool to exploit the phenomenon of heterosis or hybrid vigor or outbreeding enhancement, whereby a progeny outperforms its parents in terms of growth rate, reproductive success, and yield (Lippman and Zamir 2007). Though the heterosis phenomenon is widely explored in agricultural production to achieve increased yield in crop plants, it is still at an early stage in MAP breeding. Some promising hybrids are developed through heterosis breeding, for example, *Artemisia annua* F_1 hybrid for the antimalarial compound artemisinin (Townsend et al. 2013), an opium poppy hybrid −1 × 4 was released as a concentrated poppy straw variety "CIMAP AJAY" (Lal et al. 2014), ginseng F_1 progeny obtained from an interspecific reciprocal cross between *Panax ginseng* and *Panax quinquefolius* for bigger roots and greater ginsenosides contents (Kim et al. 2016), and two hybrids "TMOT × Ofis-8" and "TMOT × 2010" of poppy (*Papaver somniferum* L.) were selected for capsule, total alkaloid, and morphine content as well as seed yield (Valizadeh et al. 2017). Two hybrids of *Withania somnifera* (JA-20 and JA-134) were selected for root yield and quality traits (Dhuri et al. 2017). Despite the progress, cross breeding often is associated with problems of cross-incompatibility, hybrid sterility and hybrid inviability that restricts interspecific or intergeneric gene transfer (Allendorf et al. 2001). Asexual propagation can be used to overcome issues in a hybridization program, hence combining vegetative mass propagation with the breeding program helps to maintain genetic integrity and ensure uniformity of clones. However, studies have shown that in some vegetative propagation, development of adventitious root was poor. For example, *Aconitum heterophyllum* and *Saraca asoca* exhibited high variability in the rooting ability (Waman et al. 2019).

3.2.3 Mutation Breeding

Genetic variation is also induced through the process of mutagenesis using physical (gamma radiations, ultraviolet [UV], and x-ray) and chemical (methyl methane-sulfonate [MMS], or ethyl methane-sulfonate [EMS], sodium azide, N-methyl-N-nitrosourea [MNU] and hydrogen fluoride [HF]) mutagens to develop new improved varieties. The choice of suitable plant material, optimal dose rate to cause 50% lethality, and properties of the mutagen play a key role in mutagenesis (Roychowdhury and Tah 2013). Mutation breeding involves two major steps *viz*., 1) screening and selection of a putative mutant from a large mutated population and 2) confirmation of putative mutants in a replicated trial under a controlled environment (Mba 2013; Oladosu et al. 2016) for cultivar development. Generally, screening for quantitatively inherited characters starts in the M_3 generation onwards, and a stable homozygous mutant will be selected and multiplied directly. Some studies have shown that MAPs like *Isatis indigotica*, *Belamcanda chinensis*, *Gentiana manshurica*, *Melilotus officinalis*, *Melilotus dentatus*, and *Melilotus albus* mutation breeding have been attempted. Furthermore, in a review on the role of mutation in MAPs by Shashikala et al. (2018), Agrawal and Kumar (2021) listed commercial mutant varieties as the achievements of mutagenesis in medicinal plants (Table 3.1). Like other breeding techniques, mutation breeding is also limited by poor permeability in plant tissues, low reproducibility, and less desired dominant alleles, which necessitate screening a large population for the selection of desired mutants, which in turn is ineffective in inducing large chromosomal variations (Shu and Lagoda 2007).

3.2.4 Polyploidy Breeding

Polyploidy is important in the evolution of speciation, biodiversity, and adaptation to environmental change (Wood et al. 2009). Polyploidy's occurrence is common in nature, and polyploidy has occurred in 50–70% of angiosperms during their evolutionary process (Chen et al. 2007). Antimitotic agents such as colchicine, oryzalin, and trifluralin are used to induce polyploids in

TABLE 3.1
List of Commercial MAP Varieties Developed from Mutation Breeding

Plant Species	Family	Type of Mutagen	Traits Improved
Medicinal Plants			
Catharanthus roseus	Apocynaceae	Chemical mutagen (EMS)	High alkaloid content, resistant to stem gall and salt tolerant
Coleus forskohlii	Lamiaceae	Physical mutagen (gamma irradiation)	High forskolin content and increased tuber yield
Curcuma longa	Zingiberaceae	Chemical mutagens (EMS)	High curcumin content, short duration, resistance to bacterial wilt and rhizome rot, drought tolerant
Cymbopogon citratus	Poaceae	Physical mutagen (gamma irradiation)	High essential oil
Datura innoxia	Solanaceae	Chemical mutagens (EMS)	Increased alkaloid yield (scopdamine)
Hyoscyamus niger	Solanaceae	Physical mutagen (gamma irradiation)	High yield and alkaloid content
Matricaria recutita	Asteraceae	Physical mutagen (gamma irradiation)	More flowers and high oil yield
Papaver somniferum	Papaveraceae	Chemical mutagens (EMS)	High seed yield and morphine content
Phyllanthus niruri	Phyllanthaceae	Physical mutagen (gamma irradiation)	High phyllanthin content
Plantago ovata	Plantaginaceae	Physical mutagen (gamma irradiation)	Increased seed yield
Rauvolfia serpentine	Apocynaceae	Physical mutagen (gamma irradiation)	High alkaloid (reserpine) content
Solanum viarum	Solanaceae	Physical mutagen (gamma irradiation)	High solasodine content
Aromatic Plants			
Coriandrum sativum	Apiaceae	Chemical mutagens (EMS)	Resistant to stem gall
Cuminum cyaminum	Apiaceae	Physical mutagen (gamma irradiation)	Increased seed yield and resistant to wilt disease
Cymbopogon martinii	Poaceae	Physical mutagen (gamma irradiation)	High essential oil
Cymbopogon winterianus	Poaceae	Physical mutagen (x-ray irradiation)	High essential oil
Foenicum vulgare	Apiaceae	Physical mutagen (gamma irradiation)	Resistant to root-knot nematodes and powdery mildew
Mentha citrata	Lamiaceae	Physical mutagen (gamma irradiation)	High essential oil with linalool, linalyl acetate content.
Mentha piperita	Lamiaceae	Physical mutagen (gamma irradiation)	High oil yield with menthol and menthone content
Mentha spicata	Lamiaceae	Physical mutagen (gamma irradiation)	High essential oil
Zingiber officinale	Zingiberaceae	Physical mutagen (gamma irradiation)	Increased rhizome size with yellow flesh

plants. These agents disrupt mitosis resulting in either duplication of a whole set of chromosomes or alterations in the genome (Chen et al. 2020), commonly referred as an autopolyploid and allopolyploid, respectively. According to Urwin (2014), induced autopolyploids are characterized by "larger leaves, flowers and seeds, better photosynthetic properties, and increased biomass production than diploids". An investigation on artificial hybridization between the diploid *Salix purpurea* L. and tetraploids of *S. miyabeana* Seemen has improved biomass yield in triploid genotypes (Serapiglia et al. 2014). Similarly, the aneuploid plant *Echinacea purpurea* L., when crossed with a triploid female parent, resulted in increased production of aneuploid individuals, while a diploid female parent produced ha igher percentage of diploid progenies. The study on *Echinacea purpurea* L. revealed that in artificial hybridization, the efficacy of polyploid induction is dependent on the parent combination used (Li et al. 2017a, 2017b). The number of tetraploid types in *Eucommia ulmoides* Oliver, *Agastache foeniculum* Gronov, *Cannabis sativa* L. var. *sativa*, *Jatropha curcas* L, *Paulownia tomentosa* Thunb., *Salix viminalis* L, *Echinacea purpurea* L., etc., were generated by the application of colchicine at an optimal dose to plant propagules *viz*., seeds, shoot tips, meristematic tissue, axillary buds, explants of leaf, petiole, and nodal segments (Trojak-Goluch et al. 2021). Polyploidy sometimes causes degeneration of endosperm and results in embryo death (Guo et al. 2011), which can be overcome by embryo culture.

3.3 BIOTECHNOLOGY APPROACH

The traditional breeding approach for MAPs is long-term, which takes considerable time and resources to achieve the desired results due to a high level of heterozygosity, extended juvenile periods, and self-incompatibility. Biotechnology approaches such as tissue culture, micropropagation, marker-based approaches, etc., offer unique opportunities to improve the genetic makeup and efficient manipulation of the gene of interest with precision. The two basic techniques in plant biotechnology are 1) plant tissue culture and 2) genetic engineering.

3.3.1 Plant Tissue Culture

Plant tissue culture techniques are the most frequently used tool in biotechnology for basic and applied investigation in the process of plant development. It offers an integrated approach to produce disease-free plants, homogenous plantlets, breaking dormancy, generating transgenic lines, and synthesizing secondary metabolites of interest (Moraes et al. 2021). It consists of in vitro aseptic culturing of cells, tissues, organs, or the whole plant under controlled environmental and nutritional conditions (Brown and Thorpe 1995). This technique is also successfully employed to conserve endangered, threatened, and rare species which are difficult to regenerate by conventional methods. Using callus culture, *Celosia argentea* (Var.) cristata was successfully regenerated using Murashige and Skoog (MS) medium for conservation (Bakar et al. 2014). There have been reports on callus culture of *Scrophularia striata* Boiss (Khanpour-Ardestani et al. 2015), *Plumbago zeylanica* (Roy and Bharadvaja 2017), *Styrax benzoin* Dryander (Nurwahyuni et al. 2020), and *Bletilla striata* (Pan et al. 2020), which form the foundation for micropropagation studies. Recently, micropropagation has turned out to be a promising tissue culture technique to produce rapid multiplication of high-quality plants from the excised tissue. It is also used to restore and expand wild habitats in *Aegle marmelos, Acorus calamus, Celastrus paniculatus, Commiphora mukul, Peganum harmala, Prosopis cineraria, Simmondsia chinensis, Spilanthes acmella, Stevia rebaudiana, Sapindus mukorossi. Artemisia annua, Catharanthus roseus, Campotheca acuminata, Leucojum aestivum, Narcissus* sp., *Hyoscyamus niger, Pilocarpus* sp., *Podophyllum emodi, Podophyllum peltatum, Rauwolfia serpentina,* and *Taxus* sp. (Yadav et al. 2012; Moraes et al. 2021). Furthermore, *Cephaelis ipecacuanha* was efficiently propagated using micropropagation methods—it grew uniformly in the field and was reported to have a high alkaloid content (Yoshimatsu 2008). Though in vitro technology provided a good number of identical plants, it is extremely limited by acclimatization, which necessitates substantial hardening of the in vitro plantlets to prevent high mortality after transfer to in situ conditions (Ziv and Chen 2008).

3.3.2 Genetic Engineering

Genetic engineering technology alters the plant genotypes and brings an intended targeted change to improve resistance against pest, diseases, and various abiotic stresses. It is also used to increase biomass yield and bioactive compounds. Agrobacterium-mediated transformation is the most common way to introduce heterologous DNA into a host plant to achieve stable and high productivity secondary metabolites of the MAP. Though the random integration of single or multiple copies of the DNA causes undesirable effects, there have been some remarkable studies on the genetic transformation of medicinal plant families of "Apocynaceae, Araceae, Araliaceae, Asphodelaceae, Asteraceae, Begoniaceae, Crassulaceae, Fabaceae, Lamiaceae, Linaceae, Papaveraceae, Plantaginaceae, Scrophulariaceae and Solanaceae" (Bandurska et al. 2016). In some species such as *Artemisia* (mugwort), *Echinacea* (coneflower), *Digitalis* (foxglove), *Scrophularia* (figwort), *Taxus* (yew), and *Thalictrum* (meadow rues), the *Agrobacterium tumefaciens*–based genetic transformation system is well established (Canter et al. 2005). Recently, a novel and efficient in planta protocol was developed using ovary injection and shoot apical meristem to overcome the obstacles in *Catharanthus roseus* transformation program. (Bahari et al. 2020). In another study, embryogenic callus of *Veratrum dahuricum* L. produced the pharmaceutically important steroid alkaloids cyclopamine, jervine, and veratramine (Ma et al. 2020). Similarly, zygotic embryos explants of *Trachyspermum ammi* increased thymol synthesis in the aerial part (Nomani and Tohidfar 2021). Although in vitro plant regeneration has been established for many MAPs, some specie remains intractable; hence a careful planning and optimization of the transformation method mediated by *A. tumefaciens* for the species decides the efficiency of transgenic events.

3.3.3 Genome Editing

"Genome editing (GE) is a technique consists of transcription activator-like effector nucleases (TALENs), zinc-finger nucleases, (ZFN) and the most recently developed clustered regularly interspaced short palindromic repeats /CRISPR-associated protein 9 (CRISPR/Cas9 system) methods" (Gaj et al. 2016) that can produce user-designed medicinal plants. The technique combines a synthetic guide RNA molecule with CRISPR-associated (Cas-9) proteins to edit DNA with unprecedented ease and precision (Liu et al. 2017a, 2017b). It is a flexible tool developed to alter metabolic pathways; improve crop quality; and develop drugs through gene mutation, gene silencing, and transcriptional regulation. It has a variety of applications *viz*., in agricultural crops CRISPR/Cas-editing was successful in improving crop yield and stress tolerance, while in MAP assembly of genomes, it is challenging due to high repetitiveness, extreme genome sizes, and the polyploid nature (Cheng et al. 2021). However, very few medicinal plants have been edited using CRISPR/Cas for secondary metabolites, for example, benzyl isoquinoline alkaloid (BIA) biosynthesis in *Papaver somniferum* L (Alagoz et al. 2016), the flavone pathway of *Salvia miltiorrhiza* (Li et al. 2017a, 2017b), the lignocellulose biosynthesis pathway in *Dendrobium officinale* (Yan et al. 2015; Kui et al. 2017), improved fatty acid composition in *Camelina sativa* seeds (Jiang et al. 2017), and editing of farnesyl pyrophosphate synthase gene coding for squalene in *Dioscorea zingiberensis* (Feng et al. 2018). These studies provided a deep understanding of the biosynthetic pathways of MAPs; however, insufficiency of sequence information limits its application.

3.3.4 Marker-Assisted Selection

Marker-assisted selection (MAS) is deployed to accelerate phenotypic selection and improves selection accuracy by reducing the population size in breeding for crop improvement. The MAS breeding of MAPs is relatively new, and the efficacy of MAS depends on the reliability of DNA markers used in constructing a genetic linkage map. To identify markers that are closely associated with target traits with monogenic or oligogenic inheritance, molecular markers ranging from amplified fragment length polymorphism (AFLP), random amplified polymorphic DNA (RAPD), sequence-related amplified polymorphism (SRAP), and simple sequence repeat (SSR) to single-nucleotide polymorphism (SNP) have been used. There have been very few reports on MAS in MAPs, for example, "Miaoxiang Kangqi 1" is the first root rot and rust rot disease-resistant cultivar of *Panax notoginseng* (Dong et al. 2017), a new *Perilla frutescens* variety "Zhongyan Feishu No. 1" was developed for high yield, and resistance (Shen et al. 2017), and new chemical compounds of 5-hydroxymethylfurfural (5-HMF) and 2,3-dihydro-3,5-dihydroxy-6-methyl-4H-pyran-4-one (DDMP) have been developed from *Moringa oleifera* gamma irradiation (Mansour et al. 2018).

3.4 CONCLUSIONS AND PERSPECTIVES

MAPs play a significant role in the history of human well-being, and the majority of world's population still relies on it for their primary health care. Growing demand for MAPs has led to overexploitation and resource depletion; hence several ex situ and in situ conservation strategies are recommended for sustainable use of MAPs. Systematic cultivation as a conservation strategy aims to alleviate pressure on the collection of wild MAP populations and retain natural genetic variability. However, most MAPS are considered a minor crop, and successful large-scale cultivation depends on the availability of superior genotypes. Despite the effort, only a small portion has been cultivated because factors such as low germinability, slow growth rate, soil types, susceptibility to pests, etc., limit the progress. Therefore, to improve the cultivation, agronomic variables associated with uniformity, stability, growth and development, and resistance to biotic and abiotic stresses should be taken into consideration. Several breeding methods have been attempted to improve the previously mentioned attributes ranging from traditional techniques such as crossbreeding, mutation, and polyploid breeding, to recent biotechnology approaches including transgenic technology, genome editing, and MAS. These techniques can create custom-designed medicinal plants with different secondary metabolite profiles. Although advances have been made in the breeding of some MAPs, application of these technologies is still in its initial stage because of inadequate planting materials, limited knowledge on agronomic practices, self-incompatibility, environmental fluctuation, etc. Furthermore, the bioactive profile of most MAPs is yet to be identified and is often under the influence of the environment. Change in the climate frequently generates unpredictable variations in the quality and quantity of secondary metabolites, and the cues are mainly species specific. Therefore, a multidisciplinary approach integrating the current state of the art with ecology may promote development and utilization of medicinal plant resources.

BIBLIOGRAHPY

Agrawal, L., and Kumar, M. 2021. Improvement in ornamental, medicinal, and aromatic plants through induced mutation. *Journal of Applied Biology and Biotechnology* 9(4): 162–169.

Alagoz, Y., Gurkok, T., Zhang, B., and Unver, T. 2016. Manipulating the biosynthesis of bioactive compound alkaloids for next-generation metabolic engineering in opium poppy using CRISPR-Cas 9 genome editing technology. *Scientific Reports* 6: 30910.

Albrecht, S., Sonnenschein, M., and Plescher, A. 2016. P 15: Breeding of a high yielding chamomile variety (*Matricaria recutita* L.) with improved traits for machine harvesting. 6th International Symposium Breeding Research on Medicinal and Aromatic Plants, BREEDMAP 6, Quedlinburg, Germany, June 19–23.

Allendorf, F. W., Leary, R. F., Spruell, P., and Wenburg, J. K. 2001. The problems with hybrids: Setting conservation guidelines. *Trends in Ecology & Evolution* 16(11): 613–622.

Ardelean, M., Costea, A. M., and Cordea, M. 2006. Breeding foxglove (Digitalis sp.) for ornamental and/or medical purposes. Symposium on prospects for the 3rd millennium agriculture. *Bulletin of the University of Agricultural Sciences and Veterinary Medicine* 63: 22–31.

Bahari, Z., Sazegari, S., Niazi, A., and Afsharifar, A. 2020. The application of an Agrobacterium-mediated in planta transformation system in a *Catharanthus roseus* medicinal plant. *Czech Journal of Genetics and Plant Breeding* 56: 34–41.

Bakar, D. A., Ahmed, B. A., and Taha, R. M. 2014. In vitro callus induction and plant regeneration of Celosia argentea—an important medicinal plant. *Brazilian Archives of Biology and Technology* 57: 860–866.

Bandurska, K., Berdowska, A., and Król, M. 2016. Transformation of medicinal plants using *Agrobacterium tumefaciens. Postepy Hig Med Dosw* (Online) 70: 1220–1228.

Beyerlein, P., Mendes, A. M. S., and Pereira, H. S. 2019. Floral phenology, germination and hybrid plants of the Amerindian yam (*Dioscorea trifida*). *Acta Amazonica* 49(3): 167–172.

Bodeker, C., Bodeker, G., Ong, C. K., Grundy, C. K., Burford, G., and Shein, K. 2005. *WHO global atlas of traditional, complementary and alternative medicine*. Geneva, Switzerland: World Health Organization.

Brown, D. C. W., and Thorpe, T. A. 1995. Plant tissue culture: Current status and opportunities. *World Journal of Microbiology &Biotechnology* 11: 409–415.

Canter, P. H., Thomas, H., and Ernst, E. 2005. Bringing medicinal plants into cultivation: Opportunities and challenges for biotechnology. *Trends in Biotechnology* 23(4): 180–185.

Carlen, C., Schaller, M., Carron, C. A., Vouillamoz, J. F., and Baroffio, C. A. 2010. The new *Thymus vulgaris* L. hybrid cultivar 'Varico 3' compared to five established cultivars from Germany, France and Switzerland. *Acta Horticulturae* 860: 161–166.

Chauhan, R., Singh, S., Kumar, V., Kumar, A., Kumari, A., Rathore, S., Kumar, R., and Singh, S. (2022). A comprehensive review on biology, genetic improvement, agro and process technology of German Chamomile (*Matricaria chamomilla* L.). *Plants* 11: 29.

Chen, C., Li, P., Wang, R. H, Schaal, B. A., and Fu, C. X. 2014. The population genetics of cultivation: Domestication of a traditional Chinese medicine, *Scrophularia ningpoensis* Hemsl. (Scrophulariaceae). *PLoS One* 9(8): e105064.

Chen, J. T., Coate, J. E., and Meru, G. 2020. Editorial: Artificial polyploidy in plants. Front. *Plant Science* 11: 621849.

Chen, L., Lou, Q., Zhuang, Y., Chen, J., Zhang, X., and Wolukau, J. N. 2007. Cytological diploidization and rapid genome changes of the newly synthesized allotetraploids Cucumis× hytivus. *Planta* 225: 603–614.

Chen, M., Yang, C., Sui, C., Jin, Y., and Wei, J. 2016a. Zhongdanyaozhi no. 1 and Zhongdanyaozhi no. 2 are hybrid cultivars of *Salvia miltiorrhiza* with high yield and active compounds content. *PLoS One* 11(9): e0162691.

Chen, S. L., Yu, H., Luo, H. M., Wu, Q., Li, C. F., and Steinmetz, A. 2016b. Conservation and sustainable use of medicinal plants: problems, progress, and prospects. *Chinese Medicine* 11, 37.

Cheng, Q. Q., Ouyang, Y., Tang, Z. Y., Lao, C. C., Zhang, Y. Y., Cheng, C. S., and Zhou, H. 2021. Review on the development and applications of medicinal plant genomes. *Frontiers in Plant Science* 12.

Darrah, H. 1980. The cultivated basils. Thomas Buckeye, Independence, Missouri, 40pp.

Dhuri, H., Joshi, P. K., Tirkey, A., Devi, H., and Porte, S. S. 2017. Heterosis for root yield and quality characters in Ashwagandha. *International Journal of Current Microbiology and Applied Sciences* 6(11), 2268–2278.

Dong, L. L., Chen, Z. J., Wang, Y., Wei, F. G., Zhang, L. J., Xu, J., Wei, G. F., Wang, R., Yang, J., Liu, W. L., Li, X. W., Yu, Y. Q., Chen, S. L. 2017. [DNA marker-assisted selection of medicinal plants (I). Breeding research of disease-resistant cultivars of *Panax notoginseng*]. *Zhongguo Zhong Yao Za Zhi* 42(1), 56–62.

Emily, R. 2008. Medicinal planta at risk. In Nature's Pharmacy, Our Treasure Chest: Why We Must Conserve Our Natural Heritage, A Native Plant Conservation Campaign Report. Center for Biological Diversity P.O. Box 710 Tucson, AZ.

Faehnrich, B., Franz, C., Nemaz, P., and Kaul, H. P. 2021. Medicinal plants and their secondary metabolites–State of the art and trends in breeding, analytics and use in feed supplementation—with special focus on German chamomile. *Journal of Applied Botany and Food Quality* 94: 61–74.

Farnsworth, N. R. 1990. *The role of ethno pharmacology in drug development.* Ciba Foundation Symposium 154. Bioactive Compounds from Plants. Chichester, England: John Wiley & Sons, Baffins Lane, 2–21.

Farooqi, A., and Sreeramu, B. S. 2001. *Cultivation of medicinal and aromatic crops.* Hyderabad: Universities Press (India) Ltd., 506.

Fejér, J., and Salamon, I. 2016. Breeding of German chamomile, Matricaria recutita L., with a high content of α-bisabolol. *Acta Horticulturae*: 287–292.

Feng, S., Song, W., Fu, R., Zhang, H., Xu, A., Li, J. 2018. Application of the CRISPR/Cas9 system in *Dioscorea zingiberensis. Plant Cell, Tissue and Organ Culture (PCTOC)* 135: 133–141.

Förster, N., Antoniadou, K., Zander, M., Baur, S., Mittermeier-Kleßinger, V. K., Dawid, C., Ulrichs, C., and Mewis, I. 2021. Chemoprofiling as breeding tool for pharmaceutical use of Salix. *Frontiers in Plant Science*: 12.

Gaj, T., Sirk, S. J., Shui, S. L., & Liu, J. 2016. Genome-editing technologies: Principles and applications. *Cold Spring Harbor Perspectives in Biology* 8(12): a023754.

Guo, Y., Zhao, Y., Li, K., Liu, Z., Lin, H., Guo, X., and Li, C. 2011. Embryo rescue of crosses between diploid and tetraploid grape cultivars and production of triploid plants. *African Journal of Biotechnology* 10: 19005–19010.

Guzmán, L. F., Machida-Hirano, R., Borrayo, E., Cortés-Cruz, M., Espíndola-Barquera, M. D., and Heredia García, E. 2017. Genetic structure and selection of a core collection for long term conservation of Avocado in Mexico. *Frontiers in Plant Science* 8: 243.

Hamilton, A. C. 2004. Medicinal plants, conservation and livelihoods. *Biodiversity Conservation* 13: 1477–1517.

Jiang, W. Z., Henry, I. M., Lynagh, P. G, Comai, L., Cahoon, E. B., and Weeks, D. P. 2017. Significant enhancement of fatty acid composition in seeds of the allohexaploid, Camelina sativa, using CRISPR/Cas9 gene editing. *Plant Biotechnology Journal* 15(5): 648–657.

Kaicker, U. S., and Choudhury, B. 1983. Hybrids of opium poppy (*Papaver somniferum* L.) with changed morphine alkaloid content. *Acta Horticulturae* 132: 197–202.

Khanpour-Ardestani, N., Sharifi, M., and Behmanesh, M. 2015. Establishment of callus and cell suspension culture of *Scrophularia striata* Boiss.: An in vitro approach for acteoside production. *Cytotechnology* 67(3): 475–485.

Kim, Y. J, Silva, J., Zhang, D., Shi, J., Joo, S. C., Jang, M. G., Kwon, W. S., and Yang, D. C. 2016. Development of interspecies hybrids to increase ginseng biomass and ginsenoside yield. *Plant Cell Report* 35(4): 779–790.

Kui, L., Chen, H., Zhang, W., He, S., Xiong, Z., Zhang, Y., et al. 2017. Building a genetic manipulation tool box for orchid biology: Identification of constitutive promoters and application of CRISPR/Cas9 in the orchid. *Dendrobium officinale. Frontiers in Plant Science* 7: 2036.

Kumar, S., Ambreen, H., Variath, M. T., Rao, A. R., Agarwal, M., Kumar, A., et al. 2016. Utilization of molecular, phenotypic, and geographical diversity to develop compact composite core collection in the oilseed crop, safflower (*Carthamus tinctorius* L.) through maximization strategy. *Frontiers in Plant Science* 7: 1554.

Lal, R. K., Gupta, V., Gupta, P., Sarkar, S., and Singh, S. 2014. Genetics and inheritance pattern of heterosis in impact of genetic divergence in opium poppy (*Papaver somniferum* L.). *Journal of Herbs, Spices & Medicinal Plants* 20(1): 70–82.

Lange, D. 2004. Medicinal and aromatic plants: Trade, production, and management of botanical resources. *Acta Horticulturae* 629: 177–197.

Leamann, D. J. 2011. Newsletter of the medicinal plant specialist group of the IUCN species survival commission. Volume 14.

Li, B., Cui, G., Shen, G., Zhan, Z., Huang, L., Chen, J., and Qi, X. 2017a. Targeted mutagenesis in the medicinal plant *Salvia miltiorrhiza*. *Scientific Reports* 7: 43320.

Li, Q., Jiang, W., Ren, Y., Chen, R., Li, X., Yang, Y., and Wu, H. 2017b. In vitro cloning potential and phytochemical evaluations of aneuploid individuals produced from reciprocal crosses between diploid and triploid in Echinacea purpurea L. *Acta Societatis Botanicorum Poloniae* 86: 3556.

Linde, C. C., and Selmes, H. 2012. Genetic diversity and mating type distribution of Tuber melanosporum and their significance to truffle cultivation in artificially planted truffieres in Australia. *Applied and Environmental Microbiology* 78: 6534–6539.

Lippman, Z. B., and Zamir, D. 2007. Heterosis: Revisiting the magic. *Trends in Genetics* 2: 60–66.

Liu, F. M., Zhang, N. N., Liu, X. J., Yang, Z. J., Jia, H. Y., and Xu, D. P. 2019. Genetic diversity and population structure analysis of *Dalbergia odorifera* germplasm and development of a core collection using microsatellite markers. *Genes* 10: 281.

Liu, X., Wu, S., Xu, J., Sui, C., and Wei, J. 2017. Application of CRISPR/Cas9 in plant biology. *Acta Pharmaceutica Sinica B* 7(3): 292–302.

Ma, R., Yu, Z., Cai, Q., Li, H., Dong, Y., Oksman-Caldentey, K. M., and Rischer, H. 2020. Agrobacterium-mediated genetic transformation of the medicinal plant veratrum dahuricum. *Plants* 9: 191.

Mansour, H. M., Hamideldin, N., Abdel-Tawab, F. M., and Fahmy, E. M. 2018. Marker-assisted selection of new compounds in *Moringa oleifera* Lam induced by gamma irradiation. *Egyptian Journal of Radiation Sciences and Applications* (Online) 31(1): 49–62.

Mba, C. 2013. Induced mutations unleash the potentials of plant genetic resources for food and agriculture. *Agronomy* 3(1): 200–231.

Moraes, R. M., Cerdeira, A. L., and Lourenço, M. V. 2021. Using micropropagation to develop medicinal plants into crops. *Molecules* (Basel, Switzerland) 26(6): 1752.

Nalawade, S. M., Sagare, A. P., Lee, C. Y., Kao, C. L., and Tsay, H. S. 2003. Studies on tissue culture of Chinese medicinal plant resources in Taiwan and their sustainable utilization. *Botanical Bulletin of Academia Sinica* 44: 79–98.

Ndjiondjop, M. N., Semagn, K., Gouda, A. C., Kpeki, S. B., Dro Tia, D., Sow, M., et al. 2017. Genetic variation and population structure of Oryza glaberrima and development of a mini-core collection using DArTseq. *Frontiers in Plant Science* 8: 1748.

Nishteswar, K. 2014. Depleting medicinal plant resources: A threat for survival of Ayurveda. *Ayu* 35(4): 349–350.

Nomani, M., and Tohidfar, M. 2021. Plant regeneration and transformation of *Trachyspermum ammi* using *Agrobacterium tumefaciens* and zygotic embryos. *Journal of Genetic Engineering and Biotechnology* 19: 68.

Nurwahyuni, I., Situmorang, M., and Sinaga, R. 2020. Plant regeneration through Callus cultures from leaf explant of Sumatra Benzoin (Styrax benzoin). *International Journal of Forestry Research* 7.

Oladosu, Y., Rafii, M. Y., Abdullah, N., Hussin, G., Ramli, A., Rahim, H. A., Miah, G., Usman, M. 2016. Principle and application of plant mutagenesis in crop improvement: A review. *Biotechnology & Biotechnological Equipment* 30(1): 1–16.

Ouarghidi, A., Powell, B., Martin, G. J., and Abbad, A. 2017. Traditional sustainable harvesting knowledge and distribution of a vulnerable wild medicinal root (a. pyrethrum var. pyrethrum) in Ait M'hamed valley. *Morocco Economic Botany* 71: 83–95.

Pan, Y., Li, L., Xiao, S., Chen, Z., Sarsaiya, S., Zhang, S., et al. 2020. Callus growth kinetics and accumulation of secondary metabolites of Bletilla striata Rchb.f. using a callus suspension culture. *PLoS One* 15(2): e0220084.

Pank, F. 2006. Adaptation of medicinal and aromatic plants to contemporary quality and technological demands by breeding: Aims, methods and trends. *Revista Brasileira de Plantas Medicinais* 8: 39–42.

Roy, A., and Bharadvaja, N. 2017. Establishment of the shoot and callus culture of an important medicinal plant Plumbago zeylanica. *Advances in Plants & Agriculture Research* 7(5): 407–409.

Roychowdhury, R., and Tah, J. 2013. Mutagenesis—a potential approach for crop improvement. In: Hakeem, K. R., Ahmad, P., Ozturk, M., editors. *Crop improvement: New approaches and modern techniques.* New York, NY: Springer, 149–187.

Schippmann, U., Leaman, D. J., and Cunningham, A. B. 2002. Impact of cultivation and gathering of medicinal plants on biodiversity: Global trends and issues. In: *Biodiversity and the ecosystem approach in agriculture, forestry and fisheries.* Rome: FAO Document Repository of United Nations, 1–21.

Serapiglia, M. J., Gouker, F. E., and Smart, L. B. 2014. Early selection of novel triploid hybrids of shrub willow with improved biomass yield relative to diploids. *BMC Plant Biology* 14: 74.

Shashikala, S. K., Sadashiv, N., Shridevi, A. J., Lakshmana, D., Hanumanthappa, M., and Gangaprasad, S. 2018. Role of mutation breeding in improvement of medicinal and aromatic crops: Review. In: National conference on Conservation, Cultivation and Utilization of medicinal and Aromatic plants, College of Horticulture, Mudigere Karnataka, 2018. *Journal of Pharmacognosy and Phytochemistry* SP3: 425–429.

Shen, Q., Zhang, D., Sun, W., Zhang, Y. J., Shang, Z. W., and Chen, S. L. 2017. [Medicinal plant DNA marker assisted breeding (Ⅱ) the assistant identification of SNPs assisted identification and breeding research of high yield *Perilla frutescens* new variety]. *Zhongguo Zhong Yao Za Zhi* 42(9): 1668–1672.

Shu, Q. Y., and Lagoda, P. J. L. 2007. Mutation techniques for gene discovery and crop improvement. *Molecular Plant Breeding* 5: 193–195.

Townsend, T., Segura, V., Chigeza, G., Penfield, T., Rae, A., Harvey, D., et al. 2013. The use of combining ability analysis to identify elite parents for Artemisia annua F1 hybrid production. *PLoS One* 8(4): e61989.

Trojak-Goluch, A., Kawka-Lipińska, M., Wielgusz, K., and Praczyk, M. 2021. Polyploidy in industrial crops: Applications and perspectives in plant breeding. *Agronomy* 11: 2574.

United Nations Environment Programme. (2011). *UNEP 2010 annual report.* https://wedocs.unep.org/20.500.11822/7915

Urwin, N. A. 2014. Generation and characterization of colchicine-induced polyploid Lavandula × intermedia. *Euphytica* 197: 331–339.

Valizadeh, N., Arslan, N., and Khawar, K. M. 2017. Heterosis and heterobeltiosis studies on yield and yield components of some Turkish poppy hybrids (*Papaver somniferum* L.), *Journal of Applied Research on Medicinal and Aromatic Plants* 6: 41–51.

Vasisht, K., Sharma, N., and Karan, M. 2016. Current perspective in the international trade of medicinal plants material: An update. *Current Pharmaceutical Design* 22(27):4288–4336.

Waman, A. A., Smitha, G. R., and Bohra, P. 2019. A review on clonal propagation of medicinal and aromatic plants through stem cuttings for promoting their cultivation and conservation. *Current Agriculture Research Journal* 7(2): 122–138.

Wang, W., Xu, J., Fang, H., Li, Z., and Li, M. 2020. Advances and challenges in medicinal plant breeding. *Plant Science* 298: 110573.

Wood, T. E., Takebayashi, N., Barker, M. S., Mayrose, I., Greenspoon, P. B., and Rieseberg, L. H. 2009. The frequency of polyploid speciation in vascular plants. *Proceedings of the National Academy of Sciences. USA* 106: 13875–13879.

Yadav, K., Singh, N., and Verma, S. 2012. Plant tissue culture: A biotechnological tool for solving the problem of propagation of multipurpose endangered medicinal plants in India. *Journal of Agricultural Technology* 8(1): 305–318.

Yan, L., Wang, X., Liu, H., Tian, Y., Lian, J., Yang, R., et al. 2015. The genome of dendrobium officinale illuminates the biology of the important traditional chinese orchid herb. *Molecular Plant* 8(6): 922–934.

Yoshimatsu, K. 2008. Tissue culture of medicinal plants: Micropropagation, transformation and production of useful secondary metabolites. *Studies in Natural Products Chemistry* 34: 647–752.

Yuan, Q. J., Zhang, Z. Y., Hu, J., Guo, L. P., Shao, A. J., and Huang, L. Q. 2010. Impacts of recent cultivation on genetic diversity pattern of a medicinal plant, *Scutellaria baicalensis* (Lamiaceae). *BMC Genetics* 11: 29.

Ziv, M., and Chen, J. 2008. The anatomy and morphology of tissue cultured plants. In: George, E. F., Hall, M. A., De Klerk, G. J., editors. *Plant propagation by tissue culture* (3rd ed.), The Background, vol. 1. Dordrecht: Springer, 465–477.

4 Controlled Environment Agriculture

An Opportunity for Customizing Secondary Metabolite Production

Ventris M. D'souza and Ritu Bhalla

CONTENTS

4.1 Introduction 65
4.2 Controlled Environment Agriculture 66
4.3 Environmental Factors Influencing Secondary Metabolite Production 69
4.3.1 Temperature 69
4.3.2 Carbon Dioxide (CO_2) Concentration 71
4.3.3 Light 72
4.3.4 Other Factors (Hydration, Humidity, Nutrient Concentration) 74
4.4 Synergism Among Different Parameters 75
4.5 Limitations 76
4.6 Conclusion and Future Perspectives 76
Acknowledgments 77
Bibiliography 77

4.1 INTRODUCTION

Plants produce an enormous variety of phytochemical compounds that are classified into functionally overlapping categories, namely primary metabolites, secondary metabolites, and hormones. Primary metabolites (e.g., carbohydrates, lipids, proteins, nucleic acids, and pigments) are highly conserved, as they are necessary for growth and development processes across all plants. Secondary metabolites (e.g., phenolics, terpenoids, and nitrogen-containing compounds) originate from either primary metabolites or from intermediates in the biosynthesis of primary metabolites, and their distribution in the plant kingdom can vary, based on the taxonomic group. They play multifunctional roles in the regulation of plant growth; defense against pests, predators, and pathogens; and plant adaptation to environmental stresses (Erb & Kliebenstein, 2020). Plant secondary metabolites are responsible for the distinctive odors, tastes, and colors in plants and therefore represent unique resources for dietary supplements, flavors, and fragrances. They are also of medicinal importance, as they have potential health-promoting properties such as antioxidant, anti-inflammatory, anticancer, antihypertensive, and antimicrobial activities (Hussein & El-Anssary, 2018). Artemisinin from *Artemisia annua* used for the treatment of malaria, silymarin from the seeds of *Silybum marianum* used for treatment of digestive problems and liver diseases, digoxin from *Digitalis lanata* used for the treatment of various heart ailments, and alkaloids from *Catharanthus roseus* used as anticancer agents are some of the numerous examples of plant-derived drugs.

DOI: 10.1201/b22917-4

Traditional methods of cultivating plants to produce secondary metabolites of human interest are time-consuming, susceptible to biotic and abiotic contamination, may yield variable/low amounts of product, are subject to environmental and seasonal constraints, and face a risk of extinction of some native plant populations due to overharvesting (Chandran, Mukesh, Barupal, & Sharma, 2020). Large-scale plant cell culture technologies provide an opportunity for the reliable and continuous year-round production of bioactive secondary metabolites, unaffected by climate and soil conditions that affect soil-grown plants. However, plant cell culture is costly because of the slow growth rate and low productivity. In addition, this method would not be applicable to the production of secondary metabolites that are synthesized or sequestered only in specialized, differentiated plant tissues (Hussain et al., 2012). Understanding the pathways involved in the biosynthesis of the desired metabolite in plants would also be necessary for increasing the productivity of cultured plant cells, for example, by the addition of appropriate precursors.

Growing plants under controlled environment agriculture (CEA) represents an attractive proposition for optimizing their production of secondary metabolites (Fonseca, Rushing, Rajapakse, Thomas, & Riley, 2006). The term CEA refers to the growth of plants within closed systems using a combination of plant science, engineering technologies, and data analysis to monitor and regulate plant growing conditions. Kozai has pointed out several advantages of closed systems such as fast and effective growth due to optimized growth conditions, improved yield and quality of desired plant components, increased productivity arising from greater planting density and reduction in plant pests or disease, shorter production cycle, and better control of plant growth and development (Kozai, Niu, & Takagaki, 2019). Secondary metabolites affect leaf and flower color (e.g., anthocyanins, betalains, and carotenoids), nutrition (e.g., phytosterols, precursors of vitamins), flavor (e.g., esters, sulfur compounds, and essential oils), and aromas (volatile organic compounds) in crops. Since flux through metabolic pathways varies based on the growing environment, precise control of the environment can help in the growing of crops with consistent flavors and nutrition.

This chapter aims to highlight the enhanced production of secondary metabolite production in plants by modulation of environmental parameters essential for plant growth.

4.2 CONTROLLED ENVIRONMENT AGRICULTURE

The market share for CEA is estimated to reach USD 172.16 billion in 2025 (Insights). The recent surge of interest in CEA has been in response to continuing growth of the world's population; impact of weather and climate change on conventional agriculture; consumer demand for produce that is sustainably and locally grown, free of disease, pesticide-free, and available all year-around; and the need for stable food supplies. Crops commonly grown using CEA are those that are profitable, high in demand, quick-growing, and possess compact growth morphologies. These include leafy greens; herbs; microgreens; medicinal crops; vegetables such as tomatoes, peppers, and cucumbers; strawberries; and mushrooms.

CEA systems can range from simple shade structures to greenhouses to fully indoor or vertical farms. Figure 4.1 shows snapshot views of a range of CEA facilities.

The degree to which the environment is controlled increases in complexity from the use of simple sensors to the most advanced system that is fully insulated from its surrounding environment with automated monitoring and regulation of all variables such as lighting, water, and humidity. The advantages and shortcomings of some of these systems are listed in Table 4.1.

Table 4.2 provides an overview of some of the methods adopted in CEA. The most widely adopted method is soilless farming in which an inert substrate (of either organic or inorganic origin) holds plants in place while water, oxygen, and nutrition are provided to the plants in a controlled manner. Besides growing crops for nutrition, CEA has been successful in improving crop yield; biomass production; and the quality, purity, and uniformity of bioactive compounds obtained for several medicinal herb and food crops (Hayden, 2006; Maggini, Kiferle, Guidi, Pardossi, & Raffaelli, 2012).

a. Large scale commercial agriculture under netting cover

b. Greenhouse with controlled environment and Internet of Things (IoT) for monitoring growth conditions

c. Modern glasshouse using building integrated photovoltaic (BIPV) system. Insert: Closeup of solar panels and supplemental grow lights

d. Modern greenhouse with automated vents and fans for temperature and humidity control

e. Plant growth chamber for experiments on controlled and consistent growth conditions

f. Container farming with precise growing conditions for temperature and humidity, producing consistent harvests all year round

g. Indoor vertical farm using artificial lights for controlled agriculture

h. Plant factory for large-scale commercial production under controlled conditions

FIGURE 4.1 Overview of controlled environment agriculture facilities for research and commercial production.

TABLE 4.1
Types of Growing Environments and Methods Practiced in Controlled Environment Agriculture

Type of Growing Environment for CEA	Characteristics	Advantages	Disadvantages
Outdoor cultivation, e.g., tunnel farming	Small metal or PVC hoops pushed into the soil over a bed of crops and covered with plastic sheeting that can be raised	Easy to construct; warms the soil, protects plants from frost and wind	Susceptibility to pests, least control over environmental conditions compared to other systems
Greenhouse	An outdoor structure consisting of a simple frame covered by materials such as acrylic, PVC, double-layered polyethylene, polycarbonate, plastic, fiberglass, or glass	Uses sunlight, which can be supplemented with artificial lighting; plants are protected from extreme weather, rodents, and insects	Less controlled than closed systems, as exchange of heat energy, CO_2, and water may occur between its interior and the external environment Accordingly, may need ventilation through exhaust fans on hot summer days or heating in winter; harvest season is controlled by the climate; the amount of light diffusion depends on the material of the greenhouse
Growth chambers/ rooms	Closed cabinet or dedicated room with multiple, adjustable tiers, insulated from the exterior environment, constructed to create ideal circumstances for plant growth	Programmable control of temperature, illumination, humidity, CO_2; good reliability of experiments; good for simulation experiments (e.g., greenhouse models for space stations, polar regions, plant growth under prehistoric climatic conditions)	Too high cost per individual plant, cannot accommodate meaningful plant population sizes
Indoor farms (containers, plant factories/vertical farms, warehouses)	Commercial facility for growing plants in indoor spaces by harnessing technology to grow plants in stacks or multilevels Consists of a cultivation system, nutrient solution system, temperature and humidity control system, CO_2 gas supply system	Automated control of temperature, light, humidity, irrigation, and nutrients; free of soil-borne disease; year-round, high-quantity and -highquality commercial production possible; high productivity per unit area; can grow closer to the point of consumption	Plant height limitation to less than 30 cm; high set-up and running costs

Production of secondary metabolites is influenced by many factors such as nutrients, temperature, light irradiation (duration, intensity, quality), humidity, etc. (Yang, Wen, Ruan, Zhao, & Wang, 2018). Carefully controlled environments enable the optimization of cultivation conditions to grow high-quality medicinal plants in an efficient and sustainable manner to maximize the yield of beneficial plant secondary metabolites of commercial interest.

TABLE 4.2
Types of Growing Methods Practiced in CEA

System	Characteristics	Advantages	Disadvantages
Soil-based	Limited to outdoor farming and greenhouse growing	Natural environment, lower costs	Slower growing cycle, pests, difficult to optimize nutrients
Aeroponics	Soilless growth of plants with periodic spraying of roots with a water and nutrient solutions	Uses less water compared to all other growing approaches	Highly susceptible to power outages, expensive to set up automated nutrient monitoring and back-up power supply
Hydroponic deepwater culture (DWC)	Plants are constrained to a net pot suspended from the lid with roots emerging into a reservoir containing nutrient solution	Easy and simple to set up and scale up Low maintenance costs and inexpensive to maintain	Requires monitoring of concentration of nutrient solution and pH and aeration to oxygenate the nutrient
Hydroponic drip system	Plants in a soilless medium obtain nutrient solution as it drips through nozzles at a slow rate, from a reservoir Extra solution may be recirculated or drained out	Promotes more control over water and nutrient supply	Dependence on electricity, drip emitters become clogged easily due to algae growth or nutrient mineral build-up, can fluctuate in pH and nutrients, especially in recovered solution
Ebb and flow hydroponic system	Water and nutrients flow over plant roots in a grow tray during the flow phase During the drain phase, the nutrient solution drains back into the reservoir A pump based on a timer controls the inflow	Relatively easy to use and manage	Need to optimize timings for nutrients and water to drain in and out, sensitive to power failure, challenging to keep nutrient and pH levels constant
Hydroponic nutrient film technique (NFT)	The container with plants is at an angle, and nutrient flow solution flows over the plant roots and drains at the opposite end to return to the reservoir	Saves water, easy, cheap, and efficient	Sensitive to pump failure, channels can be clogged with plant roots
Hydroponic wick system	Nutrients from a reservoir move into the root system by capillary action	Simplest, easiest, and least expensive; does not require electricity	Poor oxygenation of plant roots, can have toxic build-up of nutrients in growing media

4.3 ENVIRONMENTAL FACTORS INFLUENCING SECONDARY METABOLITE PRODUCTION

4.3.1 Temperature

Temperature represents an abiotic stress that triggers alterations in plant metabolism, since biochemical reactions central to plant physiological processes are sensitive to temperature. Plants use various defense mechanisms for coping with elevated temperatures (Hasanuzzaman, Nahar, Alam, Roychowdhury, & Fujita, 2013). For example, when elevated temperatures lead to the production of free radicals like superoxide ($O_2^{\cdot-}$) and hydroxyl radicals ($HO^{\cdot}$) and nonradical molecules such

TABLE 4.3
Effect of Temperature on Secondary Metabolites in Plants Grown under Controlled Environment Conditions

Optimum Temperature	Plant Species	Secondary Metabolites Affected	Alterations in Levels	Reference
35°C for 15 days	*Hypericum perforatum*	Hypericin, pseudohypericin, pseudohypericin	Increased	(Zobayed, Afreen, & Kozai, 2005a)
30°C over the growth period	*Triticum* spp.	Total phenolics, total flavonoids	Increased	(Shamloo et al., 2017)
40°C for 2 hours	*Camptotheca acuminata*	Camptothecin, 10-hydroxycamptothecin	Increased	(Zu et al., 2003)
32°C day/23°C night for 7 days	*Cordeauxia edulis*	Monoterpenes, sesquiterpenes, and total isoprenoids	Increased	(Egigu et al., 2014)
35°C day/25°C night for 7 days	*Stevia rebaudiana*	Ketones, aldehydes, and alcohols	Increased	(Tursun, Jabran, Gurkan, & Telci, 2021)
4°C for 7 days	*Withania somnifera*	Withanolides	Increased	(Mir et al., 2015)

as hydrogen peroxide (H_2O_2) and singlet oxygen ($^1O_2$) (all of which can cause oxidative damage to biomolecules such as proteins, DNA, and lipids), plants produce antioxidants such as polyphenols which act as electron and hydrogen donors to neutralize these reactive oxygen species (ROS). Under controlled environment conditions, temperature can be modulated during the cultivation of plants to enhance the generation of secondary metabolites. However, the influence of temperature on plant growth and accumulation of secondary metabolites has been found to be species-specific. Table 4.3 provides examples of some effects of temperature on secondary metabolite production in plants grown under controlled environment conditions.

Jochum et al. (Jochum, Mudge, & Thomas, 2007) found that *Panax quinquefolius*, an understory herb, grown at 5°C higher day and night temperatures for 3 months, displayed decreased photosynthesis (52%), stomatal conductance (60%), total biomass (28%), and root biomass (33%) as compared to plants grown at lower temperatures. However, storage-root ginsenosides were found to be 49% higher in the plants grown at high temperatures compared to plants grown at low temperatures, suggesting that reduction of growth due to temperature stress led to partitioning of carbon from growth to secondary metabolite production. In contrast, studies done using *V. vinifera* L. cv. Cabernet Sauvignon grapevines grown under controlled environment conditions (Mori, Goto-Yamamoto, Kitayama, & Hashizume, 2007) revealed that high temperatures (around 35°C) resulted in grape berries with less than half of the total anthocyanin content compared to the control berries (grown at around 25°C). This was attributed to a decrease in transcription of the genes for anthocyanin biosynthesis as well as increased degradation of anthocyanin at high temperatures. Anthocyanins are the plant secondary metabolites that give red, purple, and blue plant tissues their rich coloring. In addition to their usage as natural dyes, anthocyanins are potential pharmaceutical ingredients with numerous beneficial health effects. Studies using cell lines, animal models, and human clinical studies have indicated that anthocyanins attenuate oxidative stress, have antimicrobial properties, enhance visual and neurological health, and protect against various inflammatory or age-related diseases (Khoo, Azlan, Tang, & Lim, 2017; Lila, 2004).

In addition to elevated temperatures, lowering the temperature can be a cause of abiotic stress. Shriya et al. (Shriya, Sharma, Kumari, & Choudhury, 2019) exposed seedlings of *Datura stramonium* to five temperature conditions (50°C, 40°C, 30°C, 20°C, and 10°C), with seedlings grown at 25°treated as the control. Alkaloid and flavonoid contents were upregulated in plants treated at high temperatures, as well as lower temperatures compared to the control. Leaves of wheat (*Triticum aestivum*) and alfalfa (*Medicago sativa*), when exposed to cold temperatures, have been reported to have altered levels of polyamines, mainly putrescine, spermidine, and spermine (Nadeau, Delaney, & Chouinard, 1987). These variations were found to be reversible upon restoration of initial growing conditions, suggesting that these secondary metabolites were involved in the biochemical process of cold hardening.

Molecular, physiological, and biochemical changes that enable plants to survive and resist injury at high or low temperatures (acclimation) can have significant impact on plant secondary metabolite content (Verma & Shukla, 2015). However, altering temperatures steeply can also affect plant growth and development. This calls for a careful fine-tuning of the controlled environment conditions for enhanced production of secondary metabolites of human interest.

4.3.2 Carbon Dioxide (CO_2) Concentration

Plants as photosynthetic autotrophs make their own food by photosynthesis, wherein they use light energy from the sun to combine inorganic carbon dioxide and water from their environment to synthesize carbohydrates and oxygen, which is released to the atmosphere. The energy trapped in organic compounds like carbohydrates is released during processes such as aerobic respiration in which oxygen is consumed and carbon dioxide and water are produced as by-products. Photosynthesis and cellular respiration are thus the reverse of one another. Plants carry out cellular respiration (to obtain energy for growth) and light-dependent reactions of photosynthesis during the day and cellular respiration and light-independent reactions of photosynthesis during the night. Heterotrophs cannot make their own food so subsist directly (herbivores) or indirectly (carnivores) on autotrophs.

Assuming all other resources (nutrients, water, sunlight) are not limiting, raising levels of CO_2 in the field has been shown to result in greater uptake of CO_2 by plants, leading to increased rates of photosynthesis, enhanced growth, and yield; decreased uptake of water; and lowered concentrations of tissue nitrogen and protein (Taub, 2010). The levels of CO_2 in controlled environments can vary based on the degree of ventilation, the presence of human workers, plant species, variable uptake by plants depending on the plant development stage, etc. Decrease in CO_2 levels below the optimal can limit photosynthesis and growth of plants. Growers therefore monitor CO_2 levels using infrared gas analyzers and regulate CO_2 levels via methods such as ventilation, CO_2 generation, and liquid CO_2 supplementation.

The effect of elevated CO_2 on diverse secondary metabolites in plants growing in CAEs has been studied by many researchers. The reported outcomes have been varied, with secondary metabolites increasing in concentrations in some plant species and decreasing in others. Of the two theories to explain increased production of secondary metabolites, one hypothesizes that the decrease of the nitrogen levels in vegetative plant tissues under elevated CO_2 concentrations could result in the reallocation of carbon to the synthesis of non-nitrogenous secondary metabolite compounds (Ibrahim, Jaafar, Karimi, & Ghasemzadeh, 2014). The second theory proposes that inhibition of growth due to limiting resources (such as water, nutrients, and light) results in secondary metabolism being promoted (Herms & Mattson, 1992). Examples of enhanced production of secondary metabolites in plants grown under elevated CO_2 conditions are presented in Table 4.4.

Consistent production of plant secondary metabolites possessing pharmacological activities is of commercial importance. Controlled environment facilities have the potential for reproducibly growing medicinal plants that are high-yielding and that produce high-value metabolites with minimized microbial and insect contamination. Plants belonging to the *Scutellaria* genus, for example, are harvested from the wild for use as common herbal remedies in systems of alternative medicine ranging from traditional Chinese medicine (TCM) to Native American herbalism, as they contain many flavonoids that exhibit a range of pharmacological activities. Some species of *Scutellaria*

TABLE 4.4
Effect of CO_2 on Secondary Metabolites in Plants Grown under Controlled Environment Conditions

Maximum CO_2 (μmol mol^{-1}) Tested	Plant Species	Secondary Metabolites Affected	Alterations in Levels	Reference
800	*Camellia sinensis*	Polyphenols, catechins	Increased	(Li et al., 2017)
1200	*Labisia pumila*	Phenolics, flavanoids	Increased	(Ibrahim, Jaafar, Karimi, & Ghasemzadeh, 2014)
820	*Brassica oleracea*	Glucosinolates	Increased	(Schonhof, Klaring, Krumbein, & Schreiner, 2007)
1000	*Lactuca sativa* L.	Flavonol glycosides, phenolic compounds	Increased	(Becker & Klaring, 2016)
800	*Zingiber officinale*	Flavonoids (quercetin, rutin, catechin, epicatechin, kaempferol, naringenin, fisetin, and morin) and phenolic acid (gallic acid, vanillic acid, ferulic acid, tannic acid, cinnamic acid, and salicylic acid)	Increased	(Ghasemzadeh, Jaafar, & Rahmat, 2010)
1200	*Elaeis guineensis*	Total flavonoids, total phenolics	Increased	(Ibrahim & Jaafar, 2012)
1050	*Hypericum perforatum*	Hypericin, pseudohypericin, pseudohypericin	Increased	(Zobayed & Saxena, 2004)
750	*Hypericum perforatum*	Flavonoids	Increased	(Save, de Herralde, Codina, Sanchez, & Biel, 2007)

are becoming rare because of increasing human population, environmental pollution, and habitat destruction. Stutte and Eraso (Stutte, Eraso, & Rimando, 2008) reported enhanced growth rate and flavonoid content in two *Scutellaria* species grown under controlled environment conditions with CO_2 enrichment. Bioactive flavonoids in species of *Scutellaria lateriflora*, for example, were found to increase 2.4 times when grown in control environments enriched to 1200 μmol mol^{-1} CO_2 and 5.9 times when grown in control environments enriched to 3000 μmol mol^{-1} CO_2.

4.3.3 Light

Energy is transported through the air through vibrations of electric and magnetic waves having different frequencies and wavelengths. Human eyes capture visible light between 400 and 700 nm wavelength, with the shortest wavelengths at around 400 nm being perceived as blue color and the longest wavelengths in the 600 nm range seen as red light. The areas of the spectrum that are most effective in driving photosynthesis are the red region (600–700 nm) followed by the blue region (400–500 nm) and lastly the green region (500–600 nm). Hence the range between 400 and 700 nm is referred to as photosynthetically active radiation (PAR). Other relevant regions of the solar spectrum are ultraviolet light (UV-A, 315–400 nm), UV-B (280–320 nm), UV-C (200–280 nm), and

TABLE 4.5
Effect of Lighting on Secondary Metabolites in Plants Grown under Controlled Environment Conditions

Light Conditions	Plant Species	Secondary Metabolites Affected	Alterations in Levels	Reference
UV-B (0.063 and 0.359 W m^{-2}) for 14 days	*Rosmarinus officinalis*	Rosmarinic acid, carnosic acid, total phenolics	Increased	(Luis, Perez, & Gonzalez, 2007)
UV-B (0.127 W m^{-2}) 1 h, d^{-1} for 5 days	*Salvia verticillata*	Chlorogenic acid, caffeic acid, rosmarinic acid	Increased	(Rizi, Azizi, Sayyari, Mirzaie-Asl, & Conti, 2021)
UV-B (0.35 W m^{-2}) 30 mins d^{-1} for 15 days	*Prunella vulgaris*	Phenolic compounds, flavonoids, rosmarinic acid, caffeic acid, hyperoside	Increased	(Zhang et al., 2017)
Combined red and blue light 1R (40 μmol m^{-2} s^{-1}):2B (80 μ mol m^{-2} s^{-1}) for 17 days	*Ocimum basilicum*	Rosmarinic acid, caffeic acid	Increased	(Lobiuc et al., 2017)
Combined red and blue light (7R:3B), 300 μmol m^{-2} s^{-1} for 60 days	*Salvia miltiorrhiza* Bunge	Rosmarinic acid, salvianolic acid B	Increased	(Zhang et al., 2020)
Growth under 100% blue light (150 μmol m^{-2} s^{-1}); 20–28 days	*Crocus sativus*	Carotenoids	Increased	(Moradi et al., 2021)
16–33% blue light (total PPFD 150 μmol m-2 s-1) 10–14 days	*Brassica juncea, Beta vulgaris, Petroselinum crispum* microgreens	Tocopherols, carotenoids	Increased	(Samuoliene et al., 2017)

far-red light (700–800 nm). The energy of light is captured by different molecules such as chlorophylls (primarily red and blue light), cryptochromes and phototropins (blue light), phytochromes (red, far-red light), and carotenoids (violet and blue-green light). Light thus absorbed is used for photosynthesis, which provides energy for biosynthetic pathways, biomass production, and reproductive processes. Activities such as stomatal regulation, phototropism, germination, flowering, and stem elongation also rely on light signals perceived by the plant. Examples of enhanced production of secondary metabolites in plants grown under varied light conditions are presented in Table 4.5.

The biosynthesis of secondary metabolites such as terpenoids, polyphenols, and alkaloids is significantly affected by light quality.

For example, while high intensity of UV radiation may interfere with plant growth and productivity, low levels of UV-B and UV-A radiation improve the stress tolerance in plants by eliciting the biosynthesis of several kinds of metabolites. As summarized by Zhang et al. (Zhang et al., 2021), light-activated photoreceptor proteins undergo structural changes at specific wavelengths and interact with downstream signaling proteins such as transcription factors that regulate expression of genes encoding transcription factors or key enzymes involved in secondary metabolite biosynthesis pathways. Some of the metabolites, such as polyphenols, play important roles in plant acclimation to UV-B radiation, likely due their high antioxidant capacity. Yoshida et al. (Yoshida, Nishikawa, Hikosaka, & Goto, 2021) have reported that continuous UV-B irradiation of the medicinal plant *Perilla frutescens* at 0.05 W m^{-2} for 3 hours resulted in a two-fold increase of the polyphenol rosmarinic acid compared to its levels in control plants. Blue- and red-light wavelengths have also been shown to have a stimulatory effect on accumulation of secondary metabolites in plants. Hornyak et al. (Hornyak et al., 2022) found that buckwheat grown in controlled environments under light-emitting diode (LED) lighting, emitting mainly blue and red waves, had significantly higher phenolic and total flavonoid content than control plants.

Lighting is of particular importance in vertical farms where plants are grown in stacked layers to increase the output within small areas. Artificial lights must be provided at each level for all plants to be exposed to adequate light. LED lights are currently the most favored for indoor farming since they make it possible to modulate light quality and quantity. In addition, they possess characteristics of durability, long lifespan, energy efficiency, and low heat emission. While red and blue LED lights (matching the absorption curve of chlorophyll) were initially used, there is now increasing awareness that productivity and concentrations of metabolites of interest can be increased by customizing lighting (in terms of both spectra and light intensity) to illuminate crops (Urrestarazu, Najera, & del Mar Gea, 2016). In a study by Amaki et al., sweet basil seedlings were cultivated under one of four LED lights (blue, blue-green, green, and red) with growth under white light used as control. After treatment over a 16-hr photoperiod at 50 µmol m^{-2} s^{-1} PPFD for 70 days, the essential oil content (sum total of seven essential oil components) was found to be 1.2–4.4 times higher when plants were grown under blue light compared to those grown under white light, and lowest contents of essential oil were obtained under red light (Amaki, Yamazaki, Ichimura, & Watanabe, 2011).

CEA using customized lighting has been proposed to supplement field production to meet the increasing market demand for high-quality herbs. Research into the secondary metabolite compositions of various herb species and their response to varying light treatments involving monochromatic LED lights (used singly or in combination) may be advantageous not only to obtain greater harvests but also to produce molecules that contribute to a desirable flavor profile.

4.3.4 Other Factors (Hydration, Humidity, Nutrient Concentration)

Drought stress impairs normal plant functionality as the reduction in leaf water potential and stomatal and mesophyll conductance results in lower CO_2 availability, decrease in photosynthesis, growth inhibition, and increased ROS production, among other morphological and biochemical changes. Plant responses to water stress can affect the biosynthesis, concentration, distribution, and storage of both primary and secondary metabolites. The medicinal plant St. John's wort (*Hypericum perforatum*) was grown in soil under controlled environment conditions and drought stress imposed on 50-day-old plants for 12 days before harvesting (Zobayed, Afreen, & Kozai, 2007). The concentrations of some secondary metabolites (hypericin, pseudohypericin) were found to be reduced significantly under water stress conditions and remained low even after recovery. However, concentration of the major secondary metabolite, hyperforin, in the leaf tissue was found to be double the amount found in the control, suggesting an association with the high antioxidant potential detected in water-stressed plants.

Vapor pressure deficit (VPD), an accurate measure of humidity in controlled environments, is the difference, in terms of pressure, between the water vapor in saturated air and the water vapor in the air being measured, at the same temperature. A high VPD (relatively dry air) will increase water loss through the leaves (transpiration), increasing the water and nutrient uptake through the roots. If too high, the evaporation rate from the leaf's stomata might increase, causing the plant to close its stomata to prevent evaporation, in turn stopping the intake of CO_2 and reducing growth. Low VPD would reduce uptake of water and fertilizer from the growing medium, slowing plant metabolism and development and increasing susceptibility to rot and diseases. Removal of excess water vapor in controlled environments (dehumidification) is achieved by methods such as air conditioning or using dehumidification methods involving air circulation, heat pumps, and desiccants. Air temperature and humidity are thus important stress factors in controlled environments, with the potential for influencing the levels of bioactive components in plants. For example, research on the medicinal herb *Thymus transcaucasicus*, grown under controlled environment conditions, has revealed that a 20°C air temperature and low air humidity of 50% were optimal for the recovery of carotenoids and polyphenols in this thyme species, while a second subset of other polyphenols were at a maximum level at a high air humidity of 90% (Manukyan, 2019). The hypothesis is that synthesis of bioactive compounds in *T. transcaucasicus* are an essential biological adaptation in the plant to deal with stresses such as environmental temperature and humidity.

Cultivation of plants in a controlled environment such as a plant factory requires optimization of the nutrient solution formulation as a first step, as it affects the efficiency of water and nutrient acquisition by the roots. Determining the optimal nutrient solution concentration, for accumulation of the secondary metabolite of interest, is the next step. The electrical conductivity (EC) of the nutrient solution serves as an indicator of the amount of macro- and micro-element ions available in the root zone of plants. Excessively low or high EC values result in the plants experiencing nutrient deficiency or salinity stress, respectively. The challenge lies in selecting a nutrient solution having an EC value that maximizes the production of bioactive compounds without decreasing plant growth. In the case of the aromatic herb *Agastache rugosa*, for example, the optimal EC for both growth and major bioactive compound accumulation (tilianin, rosmarinic acid) was found to be 4.0 dS m^{-1} (Lam, Kim, & Park, 2020), hence obviating the necessity of making a choice between yields of biomass versus metabolites of interest. In contrast, lycopene concentration of tomato fruit (*Lycopersicon esculentum*) grown under controlled environment conditions increased significantly in four cultivars under high EC conditions (4.5 dS m^{-1}) when compared to fruit from plants grown under lower EC (2.4 dS m^{-1}) conditions (Kubota, Thomson, Wu, & Javanmardi, 2006). It was suggested that ethylene production induced by saline stress triggered the earlier fruit maturation and accumulation of lycopene. However, as ECs climb higher, fruit yield is reduced, first due to a reduction in fruit size followed by decreases in both number and size of fruit (Dorais, Papadopoulos, & Gosselin, 2001), necessitating a trade-off between quality and productivity.

4.4 SYNERGISM AMONG DIFFERENT PARAMETERS

While the majority of studies on medicinal plants cultivated in controlled environments tend to focus on a single environmental factor, multiple stress-producing conditions may have synergistic or antagonistic effects.

Both temperature and photoperiod influenced sensory and phytochemical profiles of broccoli florets, as the highest aliphatic glucosinolate content was obtained under low temperature and long photoperiod conditions, while the indolic glucosinolate content was favored by a high temperature and short photoperiod (Molmann et al., 2015). This illustrates how tweaking the environment to promote increased accumulation of one class of secondary metabolites could adversely affect the production of other key metabolites. This was also the conclusion by Fonseca et al. (Fonseca, Rushing, Rajapakse, Thomas, & Riley, 2006), who found that the feverfew plant (*Tanacetum parthenium*) subjected to water stress under controlled environment conditions accumulated higher levels of the key metabolite, parthenolide (PRT), compared to control plants receiving water daily. In addition, plants under water stress produced increased amounts of PRT when exposed to UV A+B irradiation before harvest. However, plants that received continuous irrigation had higher phenolic contents and lower PRT amounts when subjected to UV irradiation.

A synergistic impact of elevated night-time temperatures and cultivation period was found to produce lettuce rich in polyphenols under controlled greenhouse conditions (Jeong et al., 2015). Lettuce leaves were cultivated under different night-time temperatures (4°, 12°, and 20°C) and cultivation durations (5, 15, and 20 days). Lettuce grown under high night-time temperatures (12° and 20°C) and durations (20 days) were found to be richer in polyphenols compared to lettuce grown under low temperatures (4°C).

The highly nutritious sprouts of *Vigna radiata* (mung bean), extensively consumed as a food-upplement, are commonly grown in the dark and under high relative humidity to provide moisture during germination. In a study by Amitrano et al. (Amitrano, Arena, De Pascale, & De Micco, 2020), it was found that germinating mung bean using a combination of light and low relative humidity resulted in increased polyphenol and anthocyanin content, as well as higher antioxidant capacity, compared to sprouts obtained under dark conditions.

Since achievement of high yield of secondary metabolites at enriched concentrations in plants of interest is desirable, it has been recommended that plants be provided optimal physicochemical conditions initially to obtain enhanced growth rate and biomass production followed by subjecting plants to environmental stressors several days prior to harvest to positively influence the production of phytonutrient components (Dou, Niu, Gu, & Masabni, 2017). This requires research into the adjustment of various growth parameters (temperature, humidity, carbon dioxide, fertilization, light) to elicit the desired responses. Such production systems are more easily developed under closed environment conditions than in open fields.

4.5 LIMITATIONS

Plant secondary metabolites are economically important sources of drugs, food additives, flavors, and other industrial components. However, sourcing these secondary metabolites by conventional means involves the risk of biotic (e.g., bacterial and fungal) and abiotic (heavy metal) contamination, adulteration with weeds or other closely related plant species, variation in active constituents due to field-growing conditions/genetic variability of wild plant species/climatic fluctuations, and risk of extinction of wild populations of plants of interest due to overharvesting. Producing secondary metabolites under sanitary and controlled conditions has less environmental interference, is season-independent, has very few risks of contamination by undesired toxic compounds, and is therefore highly advantageous. Nevertheless, CEA has its own challenges, beginning with the necessity of selection of a superior germplasm that will generate a high biomass and increased levels of secondary metabolite production. This stage is followed by the optimization of cultivation protocols which may require extensive research into the appropriate nutritional and environmental parameters such as lighting, temperature, relative humidity, CO_2, irrigation, etc. (Zobayed, Afreen, & Kozai, 2005b). The levels of secondary metabolite produced may need to be enhanced by subjecting the plants to different stresses (irradiation, temperature, water stress, etc.) to elicit the production of the secondary metabolite to a desirable level. This is in addition to the high set-up and operation costs involved in CEA and need for specialized technologies and labor, compared to the relatively cheaper alternative of growing medicinal plants conventionally.

4.6 CONCLUSION AND FUTURE PERSPECTIVES

Secondary metabolites have complex chemical compositions and perform different physiological roles as plants adapt to a variety of environmental stresses. Environmental variables that affect plant growth (light, temperature, humidity, water, minerals, and CO_2) influence their ability to synthesize secondary metabolites. Extended use of these metabolites in pharmaceutical industries, cosmetics, fragrances, flavors, dyes, and dietary supplements has accelerated a need to focus on increasing their production via CEA. Controlled-environment plant production systems enable stable temperature, hydration, and light conditions that make it possible to obtain high-quality plants with desirable, health-promoting secondary metabolites that can be harvested more frequently. However reliable production and maximizing yields are important factors for the commercial-level production of secondary metabolites. The challenge for the future lies in constant innovation in technologies and adoption of pioneering mindsets to create the best possible environment for plant development that results in improvement in the yield and quality of phytochemicals.

ACKNOWLEDGMENTS

The authors would like to thank the following individuals for providing the photographs in Figure 4.1: Mr. Yezdi Darukhanawalla for taking the photographs (a-g); Genesis One Tech farm for providing the photograph shown in (h). We would also like to thank LivFresh Pte Ltd for access to the facility shown in image d and Republic Polytechnic, Singapore for access to facilities shown in images b-c and e-g.

BIBILIOGRAPHY

Amaki, W., Yamazaki, N., Ichimura, M., & Watanabe, H. (2011). Effects of light quality on the growth and essential oil content in sweet basil. *Acta Horticulturae*, *907*, 91–94. https://doi.org/10.17660/ActaHortic.2011.907.9

Amitrano, C., Arena, C., De Pascale, S., & De Micco, V. (2020). Light and low relative humidity increase antioxidants content in mung bean (*Vigna radiata* L.) sprouts. *Plants*, *9*(9), 1093. https://doi.org/10.3390/plants9091093

Becker, C., & Klaring, H.-P. (2016). CO_2 enrichment can produce high red leaf lettuce yield while increasing most flavonoid glycoside and some caffeic acid derivative concentrations. *Food Chemistry*, *199*, 736–745. https://doi.org/10.1016/j.foodchem.2015.12.059

Chandran, H., Mukesh, M., Barupal, T., & Sharma, K. (2020). Plant tissue culture as a perpetual source for production of industrially important bioactive compounds. *Biotechnology Reports (Amsterdam, Netherlands)*, *26*(e00450). https://doi.org/10.1016/j.btre.2020.e00450

Dorais, M., Papadopoulos, A. P., & Gosselin, A. (2001). Influence of electric conductivity management on greenhouse tomato yield and fruit quality. *Agronomie*, *21*(4), 367–383. https://doi.org/10.1051/agro:2001130

Dou, H., Niu, G., Gu, M., & Masabni, J. G. (2017). Effects of light quality on growth and phytonutrient accumulation of herbs under controlled environments. *Horticulturae*, *3*(2), 36. https://doi.org/10.3390/horticulturae3020036

Egigu, M. C., Ibrahim, M. A., Riikonen, J., Yahya, A., Holopainen, T., Julkunen-Tiito, R., & Holopainen, J. K. (2014). Effects of rising temperature on secondary compounds of Yeheb (*Cordeauxia edulis* Hemsley). *American Journal of Plant Sciences*, *5*(5), 517–527. https://doi.org/10.4236/ajps.2014.55066

Erb, M., & Kliebenstein, D. J. (2020). Plant secondary metabolites as defenses, regulators, and primary metabolites: The blurred functional trichotomy. *Plant Physiology*, *184*(1), 39–52. https://doi.org/10.1104/pp.20.00433

Fonseca, J. M., Rushing, J. W., Rajapakse, N. C., Thomas, R. L., & Riley, M. B. (2006). Potential implications of medicinal plant production in controlled environments: The case of feverfew (*Tanacetum parthenium*). *HortScience*, *41*(3), 531–535. https://doi.org/10.21273/HORTSCI.41.3.531

Ghasemzadeh, A., Jaafar, H. Z., & Rahmat, A. (2010). Elevated carbon dioxide increases contents of flavonoids and phenolic compounds, and antioxidant activities in malaysian young ginger (*Zingiber officinale* Roscoe.) varieties. *Molecules*, *15*(11), 7907–7922. https://doi.org/10.3390/molecules15117907

Hasanuzzaman, M., Nahar, K., Alam, M. M., Roychowdhury, R., & Fujita, M. (2013). Physiological, biochemical, and molecular mechanisms of heat stress tolerance in plants. *International Journal of Molecular Sciences*, *14*(5), 9643–9684. https://doi.org/10.3390/ijms14059643

Hayden, A. L. (2006). Aeroponic and hydroponic systems for medicinal herb, rhizome, and root crops. *Hortscience*, *41*(3), 536–538. https://doi.org/10.21273/HORTSCI.41.3.536

Herms, D. A., & Mattson, W. J. (1992). The dilemma of plants: To grow or defend. *The Quarterly Review of Biology*, *67*(3), 283–335. https://doi.org/10.1086/417659

Hornyak, M., Dziurka, M., Kula-Maximenko, M., Pastuszak, J., Szczerba, A., Szklarczyk, M., & Plazek, A. (2022). Photosynthetic efficiency, growth and secondary metabolism of common buckwheat (*Fagopyrum esculentum* Moench) in different controlled-environment production systems. *Scientific Reports, 12*(257). https://doi.org/10.1038/s41598-021-04134-6

Hussain, M. S., Fareed, S., Ansari, S., Rahman, M. A., Ahmad, I. Z., & Saeed, M. (2012). Current approaches toward production of secondary plant metabolites. *Journal of Pharmacy & Bioallied Sciences, 4*(1), 10–20. https://doi.org/10.4103/0975-7406.92725

Hussein, R. A., & El-Anssary, A. A. (2018). Plants secondary metabolites: The key drivers of the pharmacological actions of medicinal plants. In P. F. Builders (Ed.), *Herbal Medicine* (pp. 11–30). London, UK: IntechOpen. http://dx.doi.org/10.5772/intechopen.76139

Ibrahim, M. H., & Jaafar, H. Z. (2012). Impact of elevated carbon dioxide on primary, secondary metabolites and antioxidant responses of *Eleais guineensis* Jacq. (Oil Palm) seedlings. *Molecules, 17*, 5195–5211. https://doi.org/10.3390/molecules17055195

Ibrahim, M. H., Jaafar, H. Z., Karimi, E., & Ghasemzadeh, A. (2014). Allocation of secondary metabolites, photosynthetic capacity, and antioxidant activity of Kacip Fatimah (*Labisia pumila* Benth) in response to CO_2 and light intensity. *The Scientific World Journal, 2014*. https://doi.org/10.1155/2014/360290

Jeong, S. W., Kim, G.-S., Lee, W. S., Kim, Y.-H., Kang, N. J., Jin, J. S., Lee, G. M., Kim, S. T., Kim, S. T., Abd El-Aty, A. M., Shim, J.-H., & Shin, S. C. (2015). The effects of different night-time temperatures and cultivation durations on the polyphenolic contents of lettuce: Application of principal component analysis. *Journal of Advanced Research, 6*(3), 493–499. https://doi.org/10.1016/j.jare.2015.01.004

Jochum, G. M., Mudge, K. W., & Thomas, R. B. (2007). Elevated temperatures increase leaf senescence and root secondary metabolite concentrations in the understory herb *Panax quinquefolius* (Araliaceae). *American Journal of Botany, 94*(5), 819–826. https://doi.org/10.3732/ajb.94.5.819

Khoo, H. E., Azlan, A., Tang, S. T., & Lim, S. M. (2017). Anthocyanidins and anthocyanins: Colored pigments as food, pharmaceutical ingredients, and the potential health benefits. *Food & Nutrition Research, 61*(1). https://doi.org/10.1080/16546628.2017.1361779

Kozai, T., Niu, G., & Takagaki, M. (Eds.). (2019). *Plant Factory: An Indoor Vertical Farming System for Efficient Quality Food Production*. (2nd ed.). Academic Press.

Kubota, C., Thomson, C. A., Wu, M., & Javanmardi, J. (2006). Controlled environments for production of value-added food crops with high phytochemical concentrations: Lycopene in tomato as an example. *HortScience, 41*(3), 522–525. doi:10.21273/hortsci.41.3.522

Lam, V. P., Kim, S. J., & Park, J. S. (2020). Optimizing the electrical conductivity of a nutrient solution for plant growth and bioactive compounds of *Agastache rugosa* in a plant factory. *Agronomy, 10*(1), 76. https://doi.org/10.3390/agronomy10010076

Li, X., Zhang, L., Ahammed, G. J., Li, Z.-X., Wei, J.-P., Shen, C., Yan, P., Zhang, L.-P., & Han, W.-Y. (2017). Stimulation in primary and secondary metabolism by elevated carbon dioxide alters green tea quality in *Camellia sinensis* L. *Scientific Reports*, 7 Article Number: 7937. https://doi.org/10.1038/s41598-017-08465-1

Lila, M. A. (2004). Anthocyanins and human health: An in vitro investigative approach. *Biomed Research International, 2004*(article ID 673916), 306–313. https://doi.org/10.1155/S111072430440401X

Lobiuc, A., Vasilache, V., Oroian, M., Stoleru, T., Burducea, M., Pintilie, O., & Zamfirache, M.-M. (2017). Blue and red LED illumination improves growth and bioactive compounds contents in acyanic and cyanic *Ocimum basilicum* L. microgreens. *Molecules, 22*(12), 2111. https://doi.org/10.3390/molecules22122111

Luis, J. C., Perez, R. M., & Gonzalez, F. V. (2007). UV-B radiation effects on foliar concentrations of rosmarinic and carnosic acids in rosemary plants. *Food Chemistry, 101*(3), 1211–1215. https://doi.org/10.1016/j.foodchem.2006.03.023

Maggini, R., Kiferle, C., Guidi, L., Pardossi, A., & Raffaelli, A. (2012). Growing medicinal plants in hydroponic culture. *Acta Horticulturae, 952*, 697–704. doi: 10.17660/ActaHortic.2012.952.88

Manukyan, A. (2019). Secondary metabolites and their antioxidant capacity of Caucasian endemic thyme (*Thymus transcaucasicus* Ronn.) as affected by environmental stress. *Journal of Applied Research on Medicinal and Aromatic Plants, 13*, 100209. https://doi.org/10.1016/j.jarmap.2019.100209

Mir, B. A., Mir, S. A., Khazir, J., Tonfack, L. B., Cowan, D. A., Vyas, D., & Koul, S. (2015). Cold stress affects antioxidative response and accumulation of medicinally important withanolides in *Withania somnifera* (L.) Dunal. *Industrial Crops and Products, 74*, 1008–1016. https://doi.org/10.1016/j.indcrop.2015.06.012

Molmann, J. A., Steindal, A. L., Bengtsson, G. B., Seljasen, R., Lea, P., Skaret, J., & Johansen, T. J. (2015). Effects of temperature and photoperiod on sensory quality and contents of glucosinolates, flavonols and vitamin C in broccoli florets. *Food Chemistry, 172*, 47–55. https://doi.org/10.1016/j.foodchem.2014.09.015

Moradi, S., Kafi, M., Aliniaeifard, S., Salami, S. A., Shokrpour, M., Pedersen, C., Moosavi-Nezhad, M., Wrobel, J., & Kalaji, H. M. (2021). Blue light improves photosynthetic performance and biomass partitioning toward harvestable organs in saffron (*Crocus sativus* L.). *Cells*, *10*(8), 1994. https://doi.org/10.3390/cells10081994

Mori, K., Goto-Yamamoto, N., Kitayama, M., & Hashizume, K. (2007). Loss of anthocyanins in red-wine grape under high temperature. *Journal of Experimental Botany*, *58*(8), 1935–1945. https://doi.org/10.1093/jxb/erm055

Nadeau, P., Delaney, S., & Chouinard, L. (1987). Effects of cold hardening on the regulation of polyamine levels in wheat (*Triticum aestivum* L.) and alfalfa (*Medicago sativa* L.). *Plant Physiology*, *84*(1), 73–77. https://doi.org/10.1104/pp.84.1.73

Rizi, M. R., Azizi, A., Sayyari, M., Mirzaie-Asl, A., & Conti, L. (October, 2021). Increased phenylpropanoids production in UV-B irradiated *Salvia verticillata* as a consequence of altered genes expression in young leaves. *Plant Physiology and Biochemistry*, *167*, 174–184. https://doi.org/10.1016/j.plaphy.2021.07.037

Samuoliene, G., Virsile, A., Brazaityte, A., Jankauskiene, J., Sakalauskiene, S., Vastakaite, V., Novickovas, A., Viskeliene, A., Sasnauskas, A., & Duchovskis, P. (2017). Blue light dosage affects carotenoids and tocopherols in microgreens. *Food Chemistry*, *228*, 50–56. https://doi.org/10.1016/j.foodchem.2017.01.144

Save, R., de Herralde, F., Codina, C., Sanchez, X., & Biel, C. (2007). Effect of atmospheric carbon dioxide fertilization on biomass and secondary metabolites of some plant species with pharmacological interest under greenhouse conditions. *Affinidad*, *64*(528), 237–241.

Schonhof, I., Klaring, H.-P., Krumbein, A., & Schreiner, M. (2007). Interaction between atmospheric CO_2 and glucosinolates in broccoli. *Journal of Chemical Ecology*, *33*, 105–114. https://doi.org/10.1007/s10886-006-9202-0

Shamloo, M., Babawale, E. A., Furtado, A., Henry, R. J., Eck, P. K., & Jones, P. J. (2017). Effects of genotype and temperature on accumulation of plant secondary metabolites in Canadian and Australian wheat grown under controlled environments. *Scientific Reports*, *7*, 9133. https://doi.org/10.1038/s41598-017-09681-5

Shriya, Sharma, M., Kumari, A., & Choudhury, S. (2019). Temperature stress mediated consequences on physiology and secondary metabolites of *Datura stramonium* (L.). *International Journal of Pharmaceutical Sciences and Research*, *10*(6), 3085–3091. http://dx.doi.org/10.13040/IJPSR.0975-8232.10(6).3085-91

Stutte, G. W., Eraso, I., & Rimando, A. M. (2008). Carbon dioxide enrichment enhances growth and flavonoid content of two *Scutellaria* species. *Journal of the American Society for Horticultural Science*, *133*(5), 631–638. https://doi.org/10.21273/JASHS.133.5.631

Taub, D. R. (2010). Effects of rising atmospheric concentrations of carbon dioxide on plants. *Nature Education Knowledge*, *3*(10), 21.

Tursun, A. O., Jabran, K., Gurkan, H., & Telci, I. (2021). Effects of elevated temperature and carbon dioxide concentrations on aromatic compounds of Stevia rebaudiana. *Sugar Tech Volume*, *23*(4), 941–948. doi: 10.1007/s12355-020-00932-w

Urrestarazu, M., Najera, C., & del Mar Gea, M. (March, 2016). Effect of the spectral quality and and intensity of light-emitting diodes on several horticultural crops. *HortScience*, *51*(3). https://doi.org/10.21273/HORTSCI.51.3.268

Verma, N., & Shukla, S. (December, 2015). Impact of various factors responsible for fluctuation in plant secondary metabolites. *Journal of Applied Research on Medicinal and Aromatic Plants*, *2*(4), 105–113. https://doi.org/10.1016/j.jarmap.2015.09.002

Yang, L., Wen, K.-S., Ruan, X., Zhao, Y.-X., & Wang, Q. (2018). Response of plant secondary metabolites to environmental factors. *Molecules*, *23*(4), 762. https://doi.org/10.3390/molecules23040762

Yoshida, H., Nishikawa, T., Hikosaka, S., & Goto, E. (2021). Effects of nocturnal UV-B irradiation on growth, flowering, and phytochemical concentration in leaves of greenhouse-grown red perilla. *Plants*, *10*(6), 1252. https://doi.org/10.3390/plants10061252

Zhang, S., Ma, J., Zou, H., Zhang, L., Li, S., & Wang, Y. (2020). The combination of blue and red LED light improves growth and phenolic acid contents in *Salvia miltiorrhiza* Bunge. *Industrial Crops and Products*, *158*, 112959. https://doi.org/10.1016/j.indcrop.2020.112959.

Zhang, S., Zhang, L., Zou, H., Qiu, L., Zheng, Y., Yang, D., & Wang, Y. (10 Dec, 2021). Effects of light on secondary metabolite biosynthesis in medicinal plants. *Frontiers in Plant Science*, *12*. https://doi.org/10.3389/fpls.2021.781236

Zhang, X.-R., Chen, Y.-H., Guo, Q.-S., Wang, W.-M., Liu, L., Fan, J., Cao, L.-P., & Li, C. (2017). Short-term UV-B radiation effects on morphology, physiological traits and accumulation of bioactive compounds in

Prunella vulgaris L. *Journal of Plant Interactions*, *12*(1), 348–354. https://doi.org/10.1080/17429145.2017.1365179

Zobayed, S. M., Afreen, F., & Kozai, T. (2005a). Temperature stress can alter the photosynthetic efficiency and secondary metabolite concentrations in St. John's wort. *Plant Physiology and Biochemistry*, *43*(10–11), 977–984. https://doi.org/10.1016/j.plaphy.2005.07.013

Zobayed, S. M., Afreen, F., & Kozai, T. (2005b). Necessity and production of medicinal plants under controlled environments. *Environmental Control in Biology*, *43*(4), 243–252. doi: 10.2525/ecb.43.243

Zobayed, S. M., Afreen, F., & Kozai, T. (2007). Phytochemical and physiological changes in the leaves of St. John's wort plants under a water stress condition. *Environmental and Experimental Botany*, *59*(2), 109–116. https://doi.org/10.1016/j.envexpbot.2005.10.002

Zobayed, S. M., & Saxena, P. K. (2004). Production of St. John's wort plants under controlled environment for maximizing biomass and secondary metabolites. *In Vitro Cellular & Developmental Biology. Plant*, *40*(1), 108–114. https://doi.org/10.1079/IVP2003498

Zu, Y.-G., Tang, Z.-H., Yu, J.-H., Liu, S.-G., Wang, W., & Guo, X.-R. (2003). Different responses of camptothecin and 10-hydroxycamptothecin to heat shock in *Camptotheca acuminata* seedlings. *Acta Botanica Sinica*, *45*(7), 809–814.

5 Advances in Micropropagation of Pharmaceutically Important Fruit Crops

Priti Sharma, Antara Das, and Maneesh Mishra

CONTENT

5.1 Introduction ... 81
5.2 Pharmaceutical Importance of Fruit Crops ... 82
5.3 Advances in Micropropagation of Fruit Crops ... 85
5.3.1 Mango (*Mangifera indica* L.) ... 85
5.3.2 Guava (*Psidium guajava* L.) ... 90
5.3.3 Anola (*Emblica officinalis* Gaertn) ... 91
5.3.4 Bael (*Aegle marmelos* Correa) ... 95
5.3.5 Jamun (*Syzizium cumini* Skeel) ... 97
5.3.6 Banana (*Musa* spp.) ... 97
5.3.7 Papaya (*Carica papaya* L.) ... 98
5.5 Conclusion and Future Perspective ... 100
Bibliography ... 101

5.1 INTRODUCTION

India ranks second in terms of fruit production in the world after Brazil. Horticulture has become a well-organized sector that is attracting enormous private investments. Horticulture products have become more economically significant due to increasing domestic and international demand. Most fruit crops are rich in bioactive compounds and used in the pharma industries. The role of aonla, bael, jamun, and mango is well documented in terms of nutraceutical properties. However, the productivity of fruits is still abysmally low (11.7 tons) (ICAR-CISH-Vision, 2050). The commercial orcharding of many important fruit crops is still restricted due to a lack of clonal planting material. Many superior cultivars have been released in mango, guava, aonla, bael, papaya, banana, grape, litchi, and pomegranate. Still, a severe shortage of planting material has hindered their expansion, and in some crops like citrus, papaya, and banana, viral and wilt diseases severely reduced production. It is, therefore, essential to produce the disease-free planting material for such crops. Micropropagation is a necessary and efficient means for commercial production of high-productivity and disease-free quality planting material of fruit crops.

Propagation is mainly done through seed in fruit plants to increase the rootstocks required for grafted plants. Papaya fruit plants are propagated through seeds, but dormancy, poor germination, low seed viability, and unfavorable environmental conditions are the major restrictive factors. Technologies such as plant tissue culture have provided new methods for the rapid production of high-quality and disease-free planting material. This technique is beneficial for proper and valuable planting material production, virus control, and management of genetic resources. Nowadays, the range of the usual plant tissue culture system has opened new vistas through somatic embryogenesis, somatic hybridization, in vitro mutagenesis, virus elimination, embryo rescue techniques, anther

DOI: 10.1201/b22917-5

TABLE 5.1
List of Horticultural Plants Where Micropropagation Has Been Commercialized

Aechmea	Dieffenbackia	Narcissus	Strawberry
Actinida	*Ficus benjamia*	Orchids	Streptocarpus
Alstromeria	*F. decora*	Oilpalm	Syngonium
Anthurium	*F. lyrata*	Papaya	Taragon
Apple	Fuchsia	Poppy	Tomato
Arum	Gerbera	Periwinkle	Tulips
Asparagus	Gladiolus	Peach	Vanilla
Banana	Grapes	Pelargonium	*Viburnum*
Basil	Gypsophila	Pear	Black pepper
Begonia	Hemerocallis	Potato	Large cardamom
Camellia	Hydrangea	Primrose	*Costus pictus*
Cardamom	Iris	Prunus	*Aegle marmelos*
Camomille	Kalanchoe	Raspberry	Safed muesli
Chrysanthemum	Lettuce	Rhododendrons	Ginger
Cordyline	Lily	Ribes	Turmeric
Dahlia	Melons	Saintpaulia	Pineapple
Datepalm	Mint	Spathyphyllum	Guava
Fig (*Ficus carica*)	Ber	Bamboo	Populus

Source: Updated from Chadha et al. (2000)

culture, protoplast culture, somatic fusion, etc. In the methods mentioned earlier, micropropagation is the most common technique for the large-scale production of fruit plants. Meristem culture and micrografting techniques were standardized to produce virus-free forms in various fruit plants (e.g., banana, citrus). In the past few years, attention has been paid to the potential beneficial effects of in vitro plant culture of microorganisms. Recently, bioimmunized tissue culture technology in bananas has been developed at the ICAR-Central Institute of Subtropical Horticulture, Lucknow, India, to produce wilt disease (*Fusarium oxysporum* Sp. *Cubense* Tropical Race-4)–tolerant plants (ICAR-CISH Annual Report, 2021).

Among the tropical and subtropical fruit crops, the banana is being exploited commercially on a large scale. Papaya and apple rootstocks are also being micropropagated commercially in many countries. Grapes and citrus have also been reported to be regenerated from axillary shoots in large quantities. However, commercial exploitation of these techniques of fruit crops is still upmarket. Micropropagation techniques have also been successfully developed for pineapple, guava, pomegranate, oil palm, date palm, coconut, and some nut crops. Clonal micropropagation of papaya, guava, aonla, and bael has recently been reported (Mishra et al., 2006, 2010; Pati et al., 2008). However, except pomegranate, date palm, papaya, and pine apple, none of them have reached a commercial level. Despite much early work, relatively reliable techniques and plenty of research and development (R&D) initiatives, commercial micropropagation of fruit crops is still at an infancy stage. Chadha et al. (2000) have thoroughly reviewed the commercialization of several horticultural crops (Table 5.1).

5.2 PHARMACEUTICAL IMPORTANCE OF FRUIT CROPS

As an excellent source of essential minerals and fiber, the fruits contain antioxidants, including flavonoids and polyphenols. Secondary metabolites are the significant factors rendering fruits for pharmaceutical uses, which provide antibacterial, antifungal, antiprotozoal, antiviral, antitumor, antidiabetic, anti-inflammatory, and other biological properties. A fruit-rich diet lowers blood pressure; checks the risk of heart disease, eye problems, digestion problems, and certain types of cancer; and also causes a positive effect on blood sugar, which manages appetite.

Mango (*Mangifera indica*), belonging to the Anacardiaceae family, is native to southern Asia, especially eastern India, Burma, and the Andaman Islands. Known as the king of fruits, mango is rich in vitamins and minerals apart from its unique and delicious taste. It is incredibly healthy due to its macronutrient and micronutrient composition and the presence of bioactive compounds. Mango is loaded with vitamins A, B, and C and soluble/insoluble fiber. The soluble fiber may help prevent heart disease and improve gastrointestinal health. Mango fruit contains polyphenols (gallic acid, gallatin, quercetin, isoquercetin, mangiferin, ellagic acid, and beta-glucosamine) that have potent antioxidant activity and other potential therapeutic effects. Mango contains the biologically active compound mangiferin, which the plant synthesizes as a chemical defense compound. Mango fruit pulp is beneficial due to its high concentration of antioxidant nutrients and carotenoids (phytochemicals). These phytochemicals are antioxidants that slow down the oxidative process, thereby preventing damage to cells in the body. Mangoes also contain beta-carotene, lutein, and zeaxanthin, which are essential for maintaining eye health and preventing macular degeneration. Mango bark extract is reported to be antibacterial, antiviral, anti-inflammatory, and antifungal due to catechins, amino acids, and phenolic and triterpenoid compounds. Various plants of the Anacardiaceae family contain a compound called xanthonmangiferin, which is beneficial in terms of antitumor, antidiabetic, and antimicrobial actions.

Guava (*Psidium guajava*), often called a 'super-fruit,' is a tropical fruit that belongs to the Myrtaceae family. It is native to tropical America extending from Mexico to Peru. Guava contains essential vitamins, minerals, and polyphenolic and flavonoid compounds that slow down the process of cancer and aging. Besides being low in calories and fat, guava has anti-inflammatory, antiviral, antiplaque, antinociceptive, and antimutagenic activities. Guava readily fits into the new functional food group due to its distinctive taste, flavor, and health-promoting properties and is rich in high-profile nutrients. This fruit is rich in vitamin C, which helps protect against oxidative stress and can boost immune system function. The lycopene present in the pink pulp of guava protects the skin from damage caused by ultraviolet (UV) rays and protects against prostate cancer. Guavas are also a good source of soluble dietary fiber, which binds to cancer-causing chemicals in the colon to help protect the colon's mucous membrane and reduce the time it takes for exposure to toxins. The potassium in guava helps control heart rate and blood pressure, and copper is essential for producing red blood cells. Guava leaf tea may help prevent spikes in blood sugar and improve symptoms of type 2 diabetes. The ethanolic extract of guava can be used to treat infertile men, as it increases the quality and quantity of sperm.

Aonla (*Emblica officinalis*), or Indian gooseberry, native to tropical Southeast Asia, is an underutilized fruit. It is also called 'amritphal' due to its medicinal and nutritional properties and culinary uses. Ayurvedic and Unani medicines contain Indian gooseberry due to its excellent dietary (minerals, vitamins) and therapeutic (phenolics and tannins) values. Apart from having high amounts of vitamin C, it has astringent, diuretic, cooling and laxative properties. It helps cure bleeding, diarrhea, dysentery, anemia, jaundice, dyspepsia, and cough. Aonla fruit is widely used in medicinal and cosmetic products. It contains alkaloids, tannins, polyphenols, multivitamins, inorganic compounds, and organic chemical constituents such as gallic acid, ellagic acid, ascorbic acid, quercetin, etc.

Jamun (*Syzygium cumini*), or black plum, of the Myrtaceae family is an underutilized tropical fruit of commercial value. It is native to India and Burma but has naturalized throughout Southeast Asia and the Pacific Islands. In Unani medicine, the plant is used as a liver tonic to enrich the blood, strengthen teeth and gums, and make lotions to cure ringworm infection of the scalp (Ayyanar and Babu, 2012). Four compounds, *viz*., lupeol, 12-oleanen-3-of-3í-acetate, stigmasterol, and ísitosterol from the n-hexane fraction of jamun leaf extract were isolated by Alam et al. (2012), which showed potential antidiabetic activities. Jamun is a wonderful fruit due to numerous curative benefits. A potent phytonutrient, anthocyanin, present in the dark skin color of the fruit, provides the body with sufficient antioxidants, which defend the cells from harm caused by oxidative stress and free radicals. It is suitable for hot summer, as glucose and water content are present abundantly. It converts starch in the body into energy that keeps the blood sugar levels low. It has a low glycemic index (GI) and can be safely consumed by diabetic people to control frequent thirst and urination symptoms. Jamun is packed with vitamin C, iron, and fiber, apart from minerals, sugars, and proteins. The iron acts as a blood purifier and helps invigorate the red blood cells. It is also loaded with compounds containing glucoside, anthocyanins, isoquercetin, ellagic acid, myricetin, and kaempferol. Jamun also has a beneficial plant compound called polyphenol, which guards the body against potential cancer-causing substances. It has diuretic properties, astringent properties, antibacterial properties, and antidiabetic effects. Jamun seeds have antioxidant competence that works to combat free radicals such as hydroxyl, lipid peroxide, and superoxide. Jamun contains beta-carotene, an antioxidant that inhibits auto-oxidation and protects the human body from many harmful diseases.

Bael (*Aegle marmelos*) is native to India and is a significant member of the Rutaceae family. This fruit contains the highest riboflavin content among fruits, apart from having chemicals like tannins, flavonoids, and coumarins, which help reduce inflammation. Also, some of these chemicals help to lower blood sugar. Marmelosin is the most preferred therapeutically active compound of bael fruit. The astringent taste of bael fruit is due to phenolic compounds. The medicinal properties of bael have been dealt with in *Charaka Samhita* (Aiyer, 1956). Bael fruit has curative, worm controlling, and nutritional properties. Every part of the bael plant contains many coumarins, alkaloids, sterols, and essential oils. Therefore it controls inflammation, parasites, viral diseases, pyria, etc. It is laxative and claimed to be aphrodisiac. Fresh leaves and fruits are beneficial in curing beriberi disease.

Banana (*Musa paradisiaca* L.) belongs to family Musaceae and was first domesticated in tropical Indo-Malaya and Australia. Rich in nutrients, bananas are a natural antioxidant source and a functional food. Bananas have many biological properties due to secondary compounds like gallocatechin and dopamine. Bananas are high in potassium and pectin. It contains a serotonin protein, which helps cure depression and is called the 'happy hormone.' Banana possesses antimicrobial, antioxidant, and antiulcer activities, as it has many nutritional health-promoting phytochemicals like potassium, carbohydrates, fiber, vitamin C, flavonoids, carotenoids, phenolics, biogenic amines, and sterols. Every part of the banana has medical importance. The banana flowers contain tannins, saponins, sterols, triterpenes, and reducing and nonreducing sugars. At the same time, the pulp is rich in antioxidants, vitamins, carotenoids, and phenolic compounds; the peel contains cellulose, hemicelluloses, and amino acids; the leaves contain polyphenols like epigallocatechin gallate (EGCG); and the fruit contains essential vitamins (vitamin C, riboflavin, folate), carbohydrate, and phytonutrients, including phenolic compounds, iron, phosphorus, potassium, calcium, carotenoid, and oligosaccharides.

Papaya (*Carica papaya*) belongs to the family Caricaceae and is originally from Central America. Papaya plants are medicinally useful due to natural metabolites present in the bark, leaf, and stem with antitumor and pesticide properties (Priyadarshi and Ram, 2018). Papayas contain vitamins A, C, and E apart from high antioxidants. Diets rich in antioxidants reduce the risk of heart disease by preventing the oxidation of cholesterol. In addition, papaya's high fiber content lowers cholesterol levels, thus reducing heart disease risk. The folic acid present in papaya is essential for converting the amino acid homocysteine to a less harmful amino acid. Papaya peel, pulp, and seeds contain a variety of phytochemicals such as lycopene, chlorogenic acid, benzyl isothiocyanates, and benzyl glucocyanates, beta-carotenoid, caffeic acid, quercetin, protocatechuic acid, etc. Papaya seeds also have prunasin, which is a cyanogenic substance. Papaya leaf contains papain, a compound that can break down large proteins into smaller, easier-to-digest proteins and amino acids, and karpain, which kills bacteria causing gastritis. Recently, Soni et al. (2022) have highlighted the role of papaya leaf in the management of dengue. A large number of papaya leaf products are already on the market for increasing the platelet count in the blood (Soni et al., 2022)

5.3 ADVANCES IN MICROPROPAGATION OF FRUIT CROPS

Micropropagation remains a popular choice for bulking up planting material. The technology of micropropagation was started in 1952 in a research laboratory via shoot meristem culture, which resulted in virus-free plants. Murashige and Skoog (1962) developed a tissue culture medium which changed the whole perspective of tissue culture and led to the spread of micropropagation for a large number of plant species which ranges from annuals to perennials, orchid to fruit, flower to plantation crops. Micropropagation was further fine-tuned by incorporating different tools and techniques in several horticultural crops. The development of transgenic crops necessitates a robust tissue culture protocol. Several transgenic horticultural crops were developed, and some were commercialized. The first transgenic papaya variety, Rainbow, was developed for resistance against papaya ring spot virus disease owing to a highly repetitive somatic embryogenesis protocol. The limitation in the development of a transgenic protocol for a large number of hardy woody perennial horticultural crops is lack of a robust tissue culture protocol in place.

The Netherlands in the early 1980s took the plunge in the field of commercial tissue culture and produced 20 million plantlets in a year. By end of the 1990s, it produced 50 million plants through tissue culture. A big boom of tissue culture industries was seen in the United States, Europe, and Israel. Commercial production started booming in leaps and bounds. South America, Southeast Asia, and Eastern Europe entered the clonal propagation market with a low-cost workforce and personnel by the end of the 1980s. The major tissue culture products were houseplants, cut flowers, and orchids in the market. Europe had around 500 tissue culture industries, producing 80–100 million plants by 1990s, whereas the United States produced around 120 million tissue culture plants during same time. Israel also focused on commercial tissue culture and started producing 20–25 million tissue culture plants despite being a small country. India entered in commercial tissue culture a little late; however, owing to availability of cheap and skilled manpower, the industry is doing fairly well, and a range of products from banana, papaya, and pomegranate to foliage and ornamental plants are being micropropagated.

5.3.1 MANGO (*MANGIFERA INDICA* L.)

Mango is also known as the king of fruits in India and has a significant consumer preference due to its diverse shape and size, attractive color, and aroma. It is one of the cultivated members of the Anacardiaceae family. Southeast Asia is the center of origin for the mango and the center of diversity for its genus, *Mangifera*. It is the most cultivated species of this genus. Mango occupies nearly 42% of the fruit-cultivated area in India. Mango is highly heterozygous due to its cross-pollination. Vegetative propagation is the only way to get the true-to-type planting materials. It is being propagated through veneer grafting, wedge grafting, epicotyle grafting, or inarching. These processes are season dependent and time- and labor-consuming. Multiplication under protected conditions has not only extended the time of propagation but it has also improved the success of grafting.

The usefulness of the micropropagation technique in mango can improve and increase the planting material production, which fulfills the orchardists' demand. Mango appears to be a challenging plant material for tissue culture due to its poor response to it; various attempts at micropropagation mango production have been made. Still, the success rate is relatively low and cultivar specific. Several difficulties like phenol leaching, alteration to axenic cultures, and browning of the explant are the most common during the mango tissue culture; all these factors subsequently inhibit the conversion rate, resulting in poor somatic embryoids and pitiful root and shoot development. Nevertheless, the endeavor is continuing to defeat these troubles.

The success of a prompt gene delivery/transformation system depends on an efficient, standardized in vitro regeneration protocol. Moreover, the in vitro propagation method likewise eases the fast procreation of top-notch clones within a shortened duration. Polyembryonic mango cultivars are used as rootstocks for their preferable traits, like being more tolerant of abiotic and biotic stresses; due to their polyembryonic nature, the nucellar embryos are the clones of the mother plant, and they always carry the same genome as the mother genotype. Still, the numbers of clones are limited and season-dependent and durable. So, the previous discussion clearly shows that micropropagation is the best tool to produce many identical true-to-type plants. The standard micropropagation methods can quickly make such a large number of clonal rootstocks.

In Vitro Propagation

One of the most streamlined means for clonal micropropagation is performed via in vitro nucellar embryogenesis in mango. In this in vitro process, nucellar tissue is used as an explant, and somatic embryos are developed by culturing them on a nutrient medium supplemented with different growth hormones. Clonal propagation of horticultural plants via induced nucellar embryogenesis in in vitro conditions has mainly been exploited in various fruit trees, which includes *Musa* spp., *Ribes rubrum* (Zatyko et al., 1975), *Mangifera indica* (Mishra et al., 2010), *Malus domestic* (Eichholtz et al., 1979), *Vitis vinifera* (Mullins and Srinivasan, 1976), *Syzygium cumini* (Yadav et al., 1990), *Eriobotrya japonica* (Litz, 1985), and *Theobroma cacao.*

Mango is one of the recalcitrant plant species in terms of tissue culture. For the past two decades, several groups in India, Australia, and the United States kept their focus on regenerating mango through a somatic embryogenesis pathway. Some authors (Rao et al., 1981; Raghuvanshi and Srivastava, 1995) reported initial success in culturing cotyledon or leaf explants for callus culture. However, no such efforts converted into reality. Excessive necrosis and in vitro oxidative browning are major bottlenecks in the embryogenesis pathways. It was determined that the browning activity during in vitro culture was increased by the enzyme activity of polyphenol oxidase (Litz and Vijayakumar, 1988). In order to reduce oxidative browning of tissue, different changes in media composition were tried like altering the concentration of the primary salt (lowering the level gives better results) and vitamins (increased level of vitamins) with additional components from organic sources like coconut water and yeast extracts. Use of polyvinyl pyrrilidon (PVP), ascorbic acid, and citric acid in the medium also reduces browning to a great extent. These changes, with an altered incubation environment, helped improve the nucellar embryogenesis of this fruit crop (Litz et al., 1986; Patena et al., 2002). Another problem with in vitro nucellar embryogenesis is the fasciation of embryos in specific polyembryonic mango cultivars. Nonseparable pluricotyly aggregated seeds are found. Many times this causes difficulties for in vitro culture. Ethylene acts as an inducer for flaccidity, vitrification, embryonal fasciation, and other abnormalities with in vitro embryogenesis. Ethylene inhibitors like ancymidol, salicylic acid, AVG, and $AgNO_3$ were used to inhibit embryos' fasciation problems and boost individual embryo formation and growth (Litz and Yurgalevitch, 1997; Chi et al., 1991).

Litz et al. (1982) used 40- to 60-day-old fruits after anthesis from some polyembryonic cultivars to extract nucellar tissues from the endocarp and used these as explants for embryogenesis. (Litz, 1984). A higher frequency of somatic embryogenesis was recorded in 2- to 3-cm-long fruits. Somatic embryos developed through the embryogenic pathway have high proliferability. Globular to heart to cotyledonary and torpedo stages of embryos have been recorded in mangos. The morphogenetic potentiality of the nucellus is the primary factor that enhances somatic embryo induction. Adventitious embryos are mainly developed from competent cells of the nucellus in polyembryonic cultivars (Litz, 2003). Though various developmental retardation is observed like fasciation, polycotyledony, lack of bipolarity, and several secondary embryos formation from hypocotyls, all these features are inhibited by the average embryo's maturity (Dewald et al., 1989b). These retarded germination problems and forward-moving necrosis also prevent the prompt retrieval of mango plantlets (Dewald et al., 1989a). To resolve these troubles, Dewald et al. (1989a) looked into the appropriate conditions for induction of embryogenic cultures and maintained the optimum condition throughout the maturation phase of somatic embryos. Several researchers from across the globe perfected somatic embryogenesis in polyembryonic (Litz et al., 1982, 1998; Dewald et al., 1989a; Patena et al., 2002) as well as monoembryonic cultivars of mango (Litz, 1984; Jana et al., 1994; Pliego-Alfaro et al., 1996; Laxmi et al., 1999; Thomas, 1999; Ara et al., 2000b; Deore et al., 2000; Chaturvedi et al., 2004a, 2004b; Rivera-Dominguez et al., 2004). The somatic embryogenesis pathway could be gainfully utilized for genetic transformation of mango (Mathews et al., 1993).

Somatic embryogenesis includes induction of pro-embryonic calli from nuvellus tissues, development of globular embryos, proliferation of embryoids and conversion of torpedo embryos in to shootlets, and finally rooting of in vitro shoots. A solid medium created problems in successful embryogenesis. Dewald et al. (1989b) conceptualized the idea of liquid media with a shaking option for mango embryogenic culture, which causes washing out of the browning agent from the media of the explants, and every time changing the media will reduce the problem of browning. Afterward, media optimization was performed by adding 20% coconut water and 6% sucrose to the modified B5 media. This media was not equally successful for other cultivars like Anwar Rataul and Chausa. The intensity of browning depends upon the phenolic exudation, and the phenol level varies from cultivar to cultivar. It was caused by overweening phenolic emission from nucellus tissues to the medium (Usman et al., 2005). A media was standardized for cultivars like Kurakkan with the composition of B5 media (B5 major salts, little salts with vitamin and iron sources) along with 1.0 mg l^{-1} 2, 4-D, 400 mg l^{-1} glutamine, 100 mg l^{-1} ascorbic acid, 1.0 mg l^{-1} spermidine, and 500 mg/l malt extract, which was the best media for somatic embryogenesis via induction and proliferation for the Kurakkan mango cultivar (Mishra et al., 2010). Other researchers used a comparable medium without ascorbic acid. The addition of Fe-EDTA was used by Ara et al. (2000b) for nucellar induction to promote somatic embryogenesis. The response of nucellar tissue from various developmental stages under the induction media differed. Chaturvedi et al. (2004a) recorded the results and standardized the same nutrient medium with an alteration of the hormone level for proper embryogenesis. The same media composition with 0.25 mg/l BAP and 1.0 mg l^{-1} NAA was required for the youngest fruits (2.5 cm). Only 0.5 mg l^{-1} 2iP was sufficient to fulfill the hormonal demand for the oldest fruits (5.0 cm), and both the nucellar explants were converted to somatic embryos. Different in-plant synthesized hormonal levels in different developmental stages might be due to different in-plant synthesized hormonal levels. Temperature has a direct relationship with oxidative browning. The nucellar tissues were incubated in a new media at a temperature of 25°C in a dark chamber. The tissues were intermittently transferred to fresh media until browning caused by phenol exudation of explant stopped entirely (Litz, 2003). In 2002, Patena et al. altered the basal media to Barba and Patena's formulation, also known as BP medium, from B5 media. Browning was controlled for eight germplasm of Carabao and two anonymous varieties of mango in the altered media. Lad et al. (1997) performed standardization of embryogenic competency for Carabao via exposing the explants at least for 28 days to 2,4-D. The explants were transferred to the induction media mentioned earlier for 30 days. From various experiments, it was concluded that embryogenic response is cultivar dependent. Litz et al. (1998) classified the cultivars depending on their embryogenesis success rate; some cultivars with a higher embryogenic response, polyembryonic cultivars like Hindi and Parris, were identified as highly embryogenic. Monoembryonic cultivars Lippens and Tommy Atkins are moderately embryogenic, and cultivars had the least or null embryogenic response known as difficult-to-regenerate like the polyembryonic cultivar Nam Doc Mai. Manzanilla Ramiriez et al. (2000) also performed a similar experiment to study the embryogenic response of three cultivars. The polyembryonic cultivar 'Ataulfo' was much more embryogenic than the monoembryonic cultivar 'Tommy Atkins' and monoembryonic cultivar 'Haden,' respectively. Litz and Yurgalevitch (1997) recommended that differential expression of enzymes for spermidine and ethylene biosynthesis pathways also regulate mango's embryogenic response factors. These might be the significant contributive components in controlling induction of somatic embryogenesis in in vitro conditions and confirmed the positive influence of spermidine level for embryogenesis in mangos (Table 5.2).

An importantly higher proliferation rate of somatic embryos was found in conditions like liquid media suspension in uninterrupted cultures. Browning and darkening were low in liquid media, and nutrient availability was also increased in this suspension condition, increasing the proliferation rate (Litz, 2003; Dewald et al., 1989a). Production of spherical embryogenic masses is one of the essential steps during the embryogenesis process; both cytokinin and auxin levels are important factors for their growth and development in mango (Litz, 2003). The different role of cytokinin was studied in depth by Deore et al. (2000), and it was found to be not responsible for induction. Still, it played an essential role in stimulating the organizational pattern for the maturation of the apical meristem. Better callus induction was found in media incremented with kinetin (0.1 mg l^{-1}) + 2, 4-D (2.0 mg l^{-1}). Dewald et al. (1989a) studied liquid suspension media supplemented with 2, 4-D (4.0–8.0 μm), and kinetin (4.6 μM) induced an advanced and higher number of somatic embryos (heart-shaped) for the cultivar "Parris." Ara et al. (2004) recommended that the development and maturity of the pro-embryogenic cultures (PECs) be increased by adding 2, 4-D (1.0 mg l^{-1}) and NAA (1.0 mg l^{-1}) with or without kinetin (1.0 mg l^{-1}) in any form of media (liquid or solid). A suspension culture with cultivar-specific media composition enhances the embryo proliferation of secondary embryos from pro-embryonic masses (PEMs), though gradually, the morphogenesis potentiality was released by the embryogenic culture (Dewald et al., 1989a; Litz et al., 1993, 2000).

TABLE 5.2
Somatic Embryogenesis in Mango

Explant	Cultivars	Reference
		Litz et al. (1982)
Nucellus	Golden Brooks, Brodes, Keitt, Irwin, Kent, Van Dyk, Ruby, Tommy Atkins, and Sensation,	Litz (1984)
Ovulas halves	Saigon and Parris	Dewald et al. (1989a)
Ovulas halves	Hindi	
Leaves	Amrapali	Raghuvanshi and Srivastava (1995)
Nucellus	Hindi, Tommy Atkins, Lippens, and Nam Doc Mai.	Litz et al. (1998)
Microshoots of nuceller somatic embryos	Amrapali	Ara et al. (1998)
Nucellus	Amrapali	Ara et al. (1999)
Nucellus	Amrapali	Laxmi et al. (1999)
Nucellus	Amrapali and Chausa	Ara et al. (2000a, 2000b)
Nucellus	Carabao and Manila super	Patena et al. (2002), Xiao et al.
Immature zygotic embryos		(2004)
Nucellus	Ambalavi	Chaturvedi et al. (2004a)

Zimmerman (1993) recommended that the advancement from the spherical to the heart shape of the somatic embryos is induced solely by auxin deficiency. Induction of maturity of the somatic embryos is required to pass two critical morphogenetic changes like the development of globular embryos bipolarity followed by cotyledons which need stimulation by transferring the heart-shaped embryos to maturation media. Mishra et al. (2010) recommended brief exposure to 15°C temperatures to embryoids helps them mature and convert to plantlets. Laxmi et al. (1999) determined that maximal embryo production was seen in half-MS add-on with BAP (20.0 µM) and absent 2,4-D. Slow necrosis is a significant problem during somatic embryo maturation; it slowly pushes the cotyledon and hypocotyl toward mortification. ABA stirred up an increased somatic embryo amount (Dewald et al., 1989b). Singh et al. (2001) recommended that the add-on of 100 µM ABA to regeneration media (hormone-free) enhances the embryogenesis frequency and induces a higher number of normal cotyledonary embryos. Cell reprogramming and rearrangement of apical meristem in the globular embryo were quicker in maturation media add-on with kinetin or BAP (1.0 to 3.0 mg l^{-1}) (Litz et al., 1993; Lad et al., 1997). Maturation media supplemented with kinetin (4.65 µM) or benzyl adenine (4.44 µM) enhanced the cotyledon's development (Litz et al., 2000). The effect of necrosis on somatic embryos in various developmental stages would retard the growth of globular heart-shaped embryos, developed cotyledon, or hypocotyl structure, so due to this disorder, the embryos never reach their maturity (Mathews et al., 1992; Monsalud et al., 1995). Another restricting cause for necrosis was the absence of highly embryogenic media in suspension culture mode. Pliego-Alfaro et al. (1996) resolved these issues by adding ABA (750–1750 µM), mannitol (7.5–12.5%), and their mixture to the existing standard media. Recurrence of hyperhydricity in somatic embryos was recorded by Monsalud et al. (1995); two ways were standardized partial desiccation of heart-shaped seeds under harnessed conditions with higher relative humidity for 24–48 hours, and secondly by exploring the gelling gum concentration to 2–6%. Another major problem, precocious germination in embryogenesis of somatic cells of mango, was repressed by supplementation of ABA (100 µM) by Pliego-Alfaro et al. (1996). It has been discovered that the length of cotyledonary embryos also influences germination efficiency; 1.0- to 1.5-cm-long embryos displayed a higher germination rate than the smaller ones. It was recommended that generate and phytagel are the most suitable gelling agents for somatic embryo germination over other gelling sources (Dewald et al., 1989a, 1989b; Laxmi et al., 1999). Chaturvedi et al. (2004b) restated that liquid suspension media with all nutrients and hormones is the best for induction, germination, maturation, and plantlet formation of cotyledonary embryos. Xiao et al. (2004) studied the role of kinetin (23 µM) with the growth media for enhancing somatic embryogenesis and regeneration of plantlets from immature zygotic embryos. Ara et al. (2000a) standardized and performed a well-developed protocol for somatic embryogenesis and regeneration of plantlets for the cultivar Amrapali. The first nucellar region was cultured in a liquid suspension of media, and PEMs were developed from the nucellar tissues. It is tough to induce rooting in in vitro conditions for natural mango shoots as explants due to higher phenolic emission (Krishna et al., 2008) and more systemic contaminants caused by pests and diseases (De Andrade et al., 2009), which are problems for culture formation. First-time rooting was elicited from callus from mango cotyledons in MS media enriched with NAA and kinetin (Rao et al., 1981). A two-step protocol was developed by Ara et al. (1998) for in vitro rooting in aseptically grown micro-shoots from the nucellar somatic embryos. The micro-shoots were treated with IBA (5.0 mg l^{-1}) for 24 hours responded most effectively in terms of proper rooting and growth. Small internodes from greenhouse-grown mango trees were used as explants in already known culture media and conditions but did not successfully induce shoot regeneration (Reuveni and Golubowicz, 1996). The process of mango somatic embryogenesis is shown in

FIGURE 5.1 Somatic embryogenesis in mango.

Figure 5.1. Somatic embryogenesis was performed in various mango cultivars using different explants by several research team, details was given in Table 5.2.

Besides somatic embryogenesis, several groups attempted shoot bud culture in mango without any success. Yang and Lüdders (1992), Chandra et al. (2004), Thomas and Ravindra (1997), and many others tried nodal shoots from juvenile and mature plants but did not succeed in micropropagation.

5.3.2 Guava (*Psidium guajava* L.)

Guava is a highly remunerative crops for farmers. Several new varieties with skin and pulp color have been developed. However, wedge grafting is the only commercially exploited method for its multiplication. Micropropagation offers immense opportunity for bulking up precious planting material of new varieties for different agro-ecological regions.

Organogenesis

The tissue culture of nodal shoots of guava faces two major problems. First is in-borne contamination and oxidative browning. Meghwal et al. (2010) suggested etiolation of shoots of guava plants by wrapping black polythesene on the shoots, which reduces the oxidative browning significantly. Recently, Sharma et al. (2018) successfully micropropagated a wilt-resistant rootstock of guava. A combination of 0.1% carbendazim and 0.05% metalaxyl for prewashing followed by surface sterilization with 0.1% mercuric chloride for 7 minutes was found to be effective in controlling contamination. In order to control oxidative browning, applying wax to the cut ends of the explants yielded better results than the use of antioxidants. Out of all the different media preparations used, MS + 0.1mg/L IAA + 4 mg/L BAP + 30 g sucrose gave maximum proliferability (2.6 shoots/explant) and maximum leaf number/shoot (5.6) in the shortest duration *viz.* 7 days. When tissues from in vitro–grown seedlings or greenhouse-grown seedlings with a controlled environment of low light and low temperatures were used as explants, it resulted in little contamination and low phenols due to less phenolic compound synthesis in young juvenile plants (Mishra et al., 2005; Usman et al., 2013). It was suggested that field-grown plants could also be used for tissue culture; suitable fungicides were sprayed on the plants 3–4 days earlier before transportation of the plants to the greenhouse, which also reduced the contaminations via different pests and diseases in the explants. If plant parts collected from full-grown plants in natural conditions were used as explants, it required extra care before putting them on the media to reduce contamination. Several researchers standardized the methods by treating the explants with 5–10% sodium hypochlorite solution followed by 0.5–1.0% mercuric chloride solution (Ali et al., 2007; Usman et al., 2013). Various factors influence the ultimate regeneration process of plantlets like the age of the mother plant, the age of the explant, the position of the explants/shoots on the donor plant, and the environmental condition when explants were collected. Explants collected from juvenile seedlings showed a better response during establishment than those from adult woody plants (Shah et al., 2008). Organogenesis was a successful event for explants collected from in vitro (Usman et al., 2013; Mishra et al., 2014) and in vivo grown in juvenile seedlings (Shah et al., 2008; Meghwal et al., 2010; Liu and Yang, 2011; Ali et al., 2007). Organogenesis was successfully developed for Banarasi, Allahabad Safeda, Beaumont, and Safeda in India, Iran, Malaysia, and Pakistan. MS media (Murashige and Skoog, 1962) supplemented with BAP was very effective for shoot regeneration in guava (Shah et al., 2008; Liu and Yang, 2011; Usman et al., 2013; Mishra et al., 2014), which showed a higher metabolic requirement of BAP with respect to other cytokinins in guava tissues (Malik et al., 2005; Rai et al., 2008b; Rai et al., 2010). Zeatin and kinetin were also combined with BAP (Shah et al., 2008; Liu and Yang, 2011). Other additives like sucrose and adenine sulfate enhanced the bud induction and increased the growth rate, while the rooting frequency was improved by using activated carbon (Thomas, 2008; Usman et al., 2013). Media supplemented with IBA and NAA was utilized for root induction in cut-out shoots as explants from environmentally grown adult plants.

Embryogenesis

Somatic embryogenesis could be gainfully used for in vitro selection against Fusarium wilt disease in guava. Kamle et al. (2014) developed a high-frequency embryogenesis protocol for cultivars Allahabad Safeda and Banarasi by utilizing zygotic embryos as explants. Internodes, leaf disks, petals, and mesocarp were used as explants for embryogenesis; nevertheless, only mesocarp could develop embryos (Butt et al., 2013). Immature 8- to 10-week-old zygotic embryos were used as explants, and embryogenesis was found in the cultivar Banarasi (Rai et al., 2010). Embryogenesis was successfully induced in the MS media add-on with 2.4-D (Rai et al., 2008c; Kamle et al., 2014) compared to some other auxins or cytokinins. Media supplemented with nitrogen sources like ascorbic acid and L-glutamine efficaciously elicited somatic embryogenesis, whereas PEG and L-proline enhanced the maturation (Rai et al., 2009). Among the various carbon sources, sucrose (5–6%) was best for induction as well as maturation events in somatic embryos; on the other hand, the use of glucose, fructose, mannitol, maltose, and sorbitol in the media displayed repressive effects (Rai et al., 2008a, 2008c). Using 0.7–0.8% agar, the solidified media showed improved embryo induction and liquid media. The embryo germination was improved by modifying the media with half MS supplemented with sucrose (3%) (Rai et al., 2007). Somatic embryo–derived plantlets in guava were successfully established under field conditions (Kamle et al., 2014). Micropropagation of *Psidium guajava* through shoot bud culture is shown in Figure 5.2.

5.3.3 Anola (*Emblica officinalis* Gaertn)

Indian gooseberry is a cross-pollinated plant having a prolonged juvenile phase. Erect trees with multiple branches far from the ground make it difficult to approach in terms of grafting. The occurrence of successful vegetative propagation varies from genotype to genotype, and it ranges from 25% to 80% (Ram, 1982). Micropropagation was not fully utilized for the fast production of clones

FIGURE 5.2 Micropropagation of *Psidium guajava* through shoot bud culture.

for elite cultivars and rootstocks. Micropropagation of Indian gooseberry was first demonstrated by Mishra et al. (1999) under in vitro conditions, and two major problems, oxidative browning and higher contamination, were observed. Further, they improved various factors like explant waxing and treating the explants with chloramphenicol (0.1%) and bavestin (1.0%) for approximately 4–4.5 hours, followed by $HgCl_2$ treatment of the explant for 8 minutes, and all these steps cumulatively increased the success rate up to 80% (Mishra, 1997). The highest bud induction (76.4%) was observed from August to November and April to June. Moderately complex nodal segments with a light green color (10–15 nodes) were the best explants with higher bud induction (Mishra and Pathak, 2001). The highest shoot proliferation was recorded in a modified MS media add-on with kinetin (0.4 mg l^{-1}) and GA3 (1.0 mg l^{-1}). Still, root induction was not successful in in vitro–grown micro-shoots (Mishra et al., 1999). There is a demand to acquire absolute standard protocols for micropropagation of Indian gooseberry. Verma and Kant (1999) regenerated the plantlet from callus cultures with higher frequency in *E. officinalis.* Different explants like cotyledons, cotyledonary nodes, hypocotyl, shoot apices, and roots were used for morphogenesis to shoot via callus cultures. The highest shoot buds were recorded from shoot apices and cotyledonary nodes. Regenerative callus was developed in an MS media add-on with ascorbic acid (25.0 mg l^{-1}) and 2,4-D (2.0 mg l^{-1}). Shoot bud formation from the callus was achieved by an MS media add-on with kinetin (2.0 mg l^{-1}),

BA (2.0 mg l^{-1}), ascorbic acid (25.0 mg l^{-1}), citric acid (25.0 mg l^{-1}), and PVP (25.0 mg l^{-1}). The produced shoots lengthened on media without hormone, and rooting was done on half MS media supplemented with sucrose (1.5%) and IBA (2.0 mg l^{-1}). The right choice of explant is crucial; various factors like age of the explant, stage and tissue, ability to regenerate, and pest- and disease-free materials are prerequisites (Mukherjee et al., 2019). Shoots and nodal segments are generally utilized for direct organogenesis (Verma and Kant, 1999; Mishra and Pathak, 2001; Mishra et al., 2006; Goyal and Bhadauria, 2008; Patidar et al., 2010). Callus formation was achieved from leaf, cotyledons, epicotyls, hypocotyls, and embryos. Further, the calli are used for indirect organogenesis (Sehgal and Khurana, 1985; Gupta et al., 1994; Verma and Kant, 1999; Al-Sabah et al., 2012; Priyanka et al., 2014; Thilaga et al., 2013; Priyanka and Singh, 2015). Zygotic embryos, hypocotyls, and epicotyls gave a more profound response than other tissues. Other root-shoot nodes were also used for indirect somatic embryogenesis (Gour and Kant, 2009).

One of the critical steps for contamination-free culture is proper disinfection of the explants of aonla. First, the explants were washed with Tween-20 mixed water (2–3 drops in 1 liter) for 5–10 minutes; later on, successive washing with 0.1% (w/v) mercuric chloride ($HgCl_2$) for 5 minutes was carried out. Aonla shoot culture displayed higher fungal contamination than bacteria (Goyal and Bhadauria, 2008), Verma and Kant (1999) used 2% (v/v) extra, and Nayak et al. (2010) used 5% labolene as an alternative to Tween 20. Some researchers used 0.1–1.0% (w/v) bavistin to reduce the contaminations in the explants either earlier or later than Tween 20 treatment. Freshly prepared chlorine water is also used as a disinfectant of endosperm before inoculation to media (Sehgal and Khurana, 1985). The phenolic compound exclusion was controlled by treating the explants with antioxidants like citric acid, ascorbic acid, and PVP (Goyal and Bhadauria, 2008). Almost all the researchers have used MS (Murashige and Skoog, 1962) media for micropropagation of aonla (Sehgal and Khurana, 1985; Gupta et al., 1994; Verma and Kant, 1999; Mishra and Pathak, 2001; Mishra et al., 2006; Goyal and Bhadauria, 2008; Gour and Kant, 2009; Patidar et al., 2010; Nayak et al., 2010; Al-Sabah et al., 2012; Madharia and Dutta, 2016; Priyanka et al., 2014; Thilaga et al., 2013; Priyanka and Singh, 2015). Many researchers observed that half-strength MS media was efficient for root induction and proliferation of aonla (Verma and Kant, 1999; Mishra et al., 2006; Gour and Kant, 2009; Nayak et al., 2010; Al-Sabah et al., 2012). Some alternative media like B5 media and woody plant media (WPM) were also utilized for rooting by Goyal and Bhadauria (2008), but the success rate was far lower than with MS media. BAP was the most commonly used cytokinin for shoot regeneration and provided good regeneration when used exclusively or with auxin. Usually 1–4 mg l^{-1} BAP has given a good proliferation response (Verma and Kant, 1999; Gour and Kant, 2009; Madharia and Dutta, 2016). BAP along with NAA has also been tried for proliferation (Verma and Kant, 1999; Patidar et al., 2010). Combined use of kinetin and GA_3 was reported (Mishra and Pathak, 2001; Mishra et al., 2006) for shoot regeneration. The in vitro shoots are too small and require elongation. Therefore GA_3 was recommended. Different forms of auxin like NAA and IBA were used for root regeneration, but most researchers prefer to use IBA with MS media for root regeneration and proliferation (Verma and Kant, 1999; Gour and Kant, 2009; Nayak et al., 2010; Al-Sabah et al., 2012). The assemblage of IBA with NAA or only of NAA was used for rooting in a few cases (Sehgal and Khurana, 1985; Mishra et al., 2006). Micropropagation of *Emblica officinalis* Gaertn is shown in Figure 5.3. Various additional antioxidants like citric acid, ascorbic acid, and PVP were also used with the media to control the phenolic compounds' leaching (Gupta et al., 1994; Verma and Kant, 1999). A mixture of autoclaved sand, soil, and farmyard manure in a 1:1:1 (v/v) ratio or 6:2:1 was used for acclimation of micropropagated plantlets (Mishra et al., 2006; Thilaga et al., 2013). A variety of sand, peat moss, and humus in a 1:1:1 balance with bavistin was effective (Al-Sabah et al., 2012). Madharia and Dutta (2016) used sterile cocopeat mix with bavistin solution. Verma and Kant (1999) used a mixture of autoclaved soil and Soil Rite in a ratio of 3:1 with higher survival rates.

FIGURE 5.3 Micropropagation of *Emblica officinalis* Gaertn.

Callus Induction and Somatic Embryogenesis

Callus culture and somatic embryogenesis pathways of tissue culture in fruit crops proved to beneficial for genetic improvement. In vitro propagation of aonla was successfully achieved using different explants (hypocotyls, embryo, endosperm, epicotyls, and leaf) on MS media enriched with 2,4-D (Gupta et al., 1994; Verma and Kant, 1999; Priyanka et al., 2014; Priyanka and Singh, 2015). Callus induction was achieved using leaf and zygotic embryos as explants on MS media enriched with 0.45 µM 2,4-D and 22 µM BAP (Thilaga et al., 2013). MS media supplemented with 1 mg/l picloram was used as an alternative to 2,4-D. This media was utilized for efficient induction of calli and further embryogenic proliferation (Madharia and Dutta, 2016). Likewise, callus was formed from endosperm on MS media supplemented with 0.1 mg l^{-1} IAA and 0.2 mg l^{-1} BAP (Sehgal and Khurana, 1985). Callus was subcultured on MS media supplemented with BAP or kinetin or a combination of both, which resulted inshoot buds from the callus. MS + 0.2 mg l^{-1} BAP + 0.1 mg l^{-1} IAA was beneficial for shoot regeneration. In vitro rhizogenesis was achieved by culturing the shoots on liquid MS media add-on with 0.002 mg l^{-1} NAA (Sehgal and Khurana, 1985) or MS media supplemented with 2 mg/l IBA resulted in 60–80% of rooting (Verma and Kant, 1999). The increased amount of BAP (1–3 mg l^{-1} BAP) was also adequate for shoot regeneration from the callus (Verma and Kant, 1999; Madharia and Dutta, 2016). MS media with 3 mg l^{-1} BAP, 2 mg l^{-1} kinetin, and 0.1 mg l^{-1} 2,4-D was more efficient for regeneration of the shoot from the callus (Gupta et al., 1994). Somatic embryogenesis was induced to explants on MS media supplemented with 0.1 mg l^{-1} kinetin and 1 mg l^{-1} 2,4-D (Al-Sabah et al., 2012). However, somatic embryogenesis was obtained with a success rate of 33.8% on MS media supplemented with 11 µM BAP and 0.45 µM 2,4-D (Thilaga et al., 2013). IAA- and IBA-enriched MS media was utilized for direct somatic embryogenesis, and 2,4-D– and kinetin-enhanced MS media was used for indirect somatic embryogenesis via the callus (Priyanka and Singh, 2015). Somatic embryos were utilized for plantlet regeneration on MS media devoid of hormones (Al-Sabah et al., 2012; Priyanka and Singh, 2015), but Thilaga et al. (2013) reported that media supplemented with 0.46 µM kinetin or 3.78 µM ABA was used for regeneration of plantlets from somatic embryos.

5.3.4 Bael (*Aegle marmelos* Correa)

Bael is catching up fast in several parts of the country. With the advent of new bael cultivars and availability of genuine planting material, commercial orchards are coming up in a big way owing to its importance in the pharmaceutical industry. Bael is conventionally propagated through wedge grafting techniques. However, due to its importance in pharma industries, a huge surge of interest has been generated for development of micropropagation system in bael. In vitro propagation techniques of bael from in vitro–grown seedling parts, i.e., from roots (Bhati et al., 1992), hypocotyl (Hossain et al., 1995), cotyledonary node (Arumugam and Rao, 1996), excised leaf explants (Islam et al., 1993), nodal explants through axillary bud multiplication and axillary bud and shoot tips (Pati et al., 2008; Gandhi et al., 2018), have been reported. Like other hardwood perennials, bael also has phenolic compound issues; the growth and development of explants were hampered in tissue culture media by self-leached phenols, and consecutively browning media and explants retarded the in vitro induction, proliferation, and morphogenesis events (George and Sherrington, 1984; Elmore et al., 1990). Antioxidants like citric acid and ascorbic acid, with or without PVP, were utilized to control the phenolic leaching and browning before or during the culture condition (Ajithkumar and Seeni, 1998). Among all the explants, nodal segment response is best in media, and the rate of micropropagation is also high. The surface sterilization of explants was done using mercuric chloride (Tisserat, 1985; Dhar and Upreti, 1999; Bhatt and Dhar, 2004). Shoot induction and multiplication of explants in in vitro conditions were obtained by culturing them on MS media supplemented with BAP and IAA. (Dewan et al., 1992). Root induction of regenerated shoots was acquired on the half MS media add-on with IBA and IAA. Fast wilting of the in vitro–grown plants was observed when they moved to ex vitro conditions. This problem might be reduced by transferring them to a high-humidity chamber for primary hardening and then moving them to the environment (Ajithkumar and Seeni, 1998; Pati et al., 2008; Fila et al., 1998). Pati et al. (2008) reported that cocopeat supplemented with MS salt was a suitable mixture for hardening. Clonally micropropagated plants of bael were successfully established in the field. A high-frequency shoot proliferation was achieved on the same medium fortified with 3.0 mg/l BAP. The micro shoots were rooted on MS medium supplemented with 25 mg/l IBA. The rooted plants were hardened on autoclaved coconut husk containing half MS nutrient solution and 1 mg l^{-1} paclobutrazol under the mist house. The hardened plants were established in natural field conditions (Pati et al., 2008). Micropropagation of *Aegle marmelos* Corr. is shown in Figure 5.4. Different molecular markers

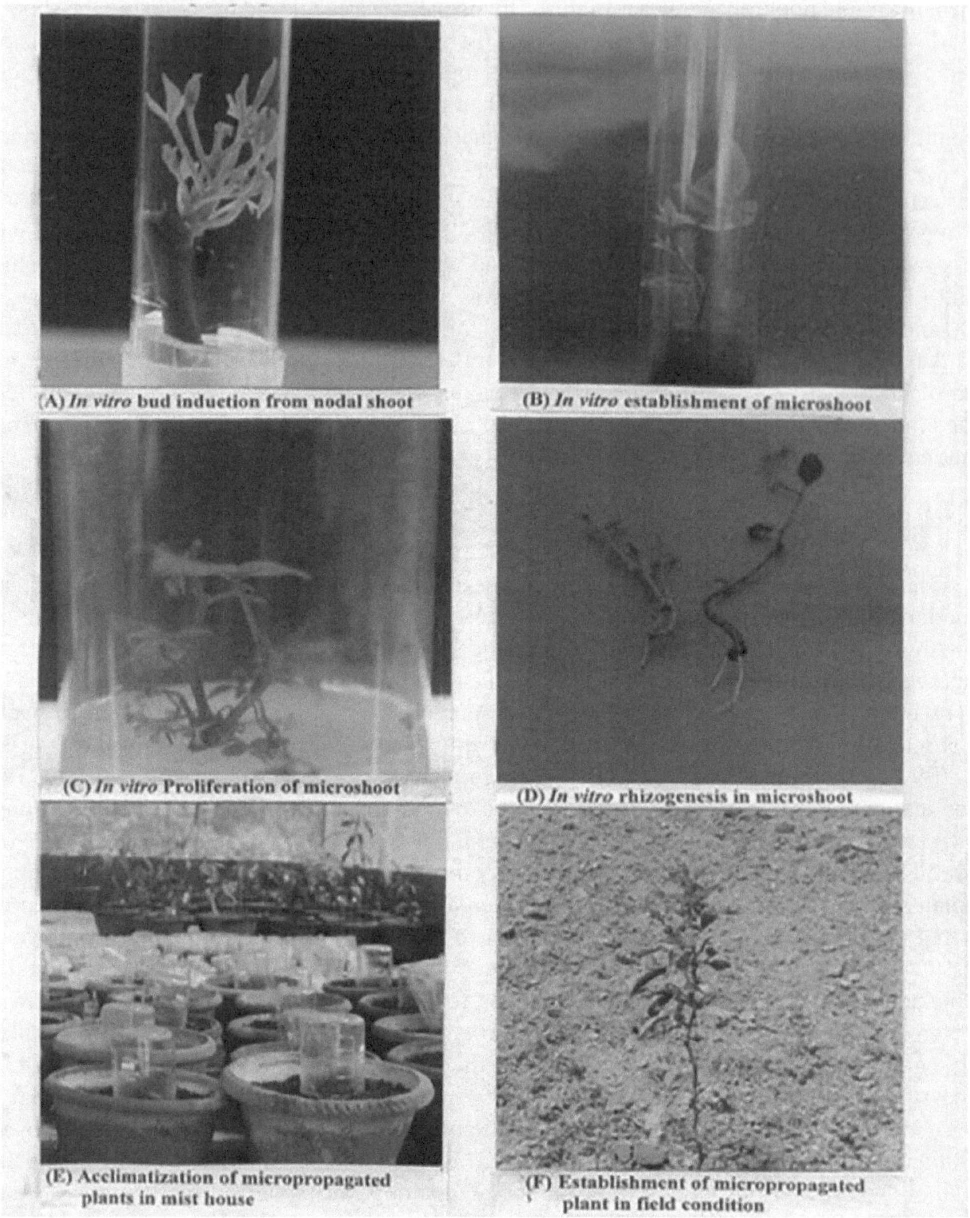

FIGURE 5.4 Micropropagation of *Aegle marmelos* Corr.

were utilized to screen the variability of tissue culture–raised micropropagated plantlets, *viz.* randome amplified polymorphic DNA (RAPD), ISSR, simple sequence repeat (SSR), etc. Pati et al. (2008) used two micro-satellite, two mini-satellites, and 12 RAPD markers to identify the variability among the in vitro–grown plantlets and donor mother plant, and no significant difference was observed; all of them showed similar banding patterns, and no polymorphic band was recorded among them, which proved the true-to-type nature of the micropropagated plants. The in vitro clonally grown micropropagated plantlets and the mother plants hold the exact genetic imprint in their body (Pati et al., 2008).

5.3.5 Jamun (*Syzizium cumini* Skeel)

Jamun, or black plum, is an essential but underexploited indigenous fruit tree of commercial value. Jamun is being propagated through grafting techniques. Owing to development of new high-yielding varieties (CISH-Jamwant, Thar Kranti), it's pertinent to develop a mass multiplication system in jamun. Micropropagation is a valuable technique for clonal mass multiplication of desired cultivars. A micropropagation protocol for jamun (*Syzygium cumini* L. Skeels.) has been worked out utilizing juvenile tissue (Roy et al., 1996) in which the explants are excised from coppiced branches. Various explants like buds, nodal segments, young shoots, and seedlings were utilized for jamun micropropagation by culturing them on MS media add-on with multiple concentrations of BA and kinetin (Yadav et al., 1990; Jain and Babbar, 2000, 2003). Initial decontamination was performed by treating them with NaOH, ascorbic acid, and citric acid. A successful micropropagation protocol of *Syzygium cumini* L. Skeels has also been developed at ICAR-CISH, Lucknow, using nodal explants from the mature tree (Mishra, 2022, Unpublished data). Nodal segments and shoot tips were used for multiple shoot induction in jamun by placing them after the disinfection process on MS media supplemented with BA, NAA, or IBA (Yadav et al., 1990). Regenerated shoots were settled on MS media add-on with NAA or IBA for root induction and finally transported to MS media devoid of any hormones for complete plantlet development (Ramashree et al., 2007). The work carried out on micropropagation on jamun is scanty and requires substantial research to develop a variety-specific protocol using mature explants.

5.3.6 Banana (*Musa* spp.)

Edible bananas usually are multiplied via suckers. Banana plants are always vegetatively propagated due to the high degree of sterility and polyploidy, resulting in a low rate of multiplication and a pathogenic effect. However, micropropagation via shoot tips as explants consistently reduced the pathogen load to the culture, and floral auspices were also used for micropropagation. The usefulness of all these explants increases the mass multiplication of bananas and avoids constraints forced by some pathogens. An in vitro micropropagation protocol was standardized and utilized for several banana cultivars (Cronauer and Krikorian, 1984; Madhulatha et al., 2004). Among different tissue culture techniques, shoot tip culture was the most popular and common for propagation purposes. Various factors determine the multiplication rate and its success up to plantlet developments. The elements are regeneration efficiency of explants, induction for the proliferation of explants, and finally, regeneration of roots; these are also influenced by cultural media and conditions along with nature and age of explants (Vuylsteke, 1998). A huge amount of research reports are available on micropropagation of banana. Shoot bud multiplication utilizes a high concentration of cytokinin in the media. Use of high levels of growth regulators during micropropagation results in undesirable clonal variability in important commercial crops of banana. The present study investigated the effects of high levels of cytokinins on micropropagation in banana (genotype AAB), and the genetic stability of plantlets was assessed using RAPD and ISSR markers. Cytokinins such as BA and kinetin were added to the routine shoot multiplication medium at concentrations up to 10 mg l^{-1}. After 12 weeks of culture involving three subcultures, the maximum number of shoot buds were produced in cultures receiving either 5 mg l^{-1} BA (80 shoot buds) or 4 mg l^{-1} kinetin (62 shoot buds). Certain morphological abnormalities observed during proliferation of shoot buds in vitro were not observed during acclimatization ex vitro. To check the genetic stability, RAPD and ISSR profiles of micropropagated plantlets obtained from different cytokinin treatments were compared with control microplants maintained on MS medium as well as the field-grown mother plant. A total of 50 RAPD and 12 ISSR primers resulted in 625 distinct and reproducible bands. Thus a total of 17,400 bands were generated showing homogeneous RAPD and ISSR patterns. The band intensity histogram of each gel confirmed their monomorphic nature with no genetic variation in all the plantlets analyzed. Based on these results, a protocol for high-rate shoot multiplication was worked out leading to uniform shoot production (Lakshman et al., 2007).

Somatic embryogenesis is a well-defined method in banana micropropagation. Nevertheless, somatic embryogenesis should be standardized for other unexplored banana cultivars, which will commercially play a significant role in their mass multiplication (Strosse et al., 2003). Somatic embryogenesis techniques aim to develop a high-performance micropropagation and regeneration valuable system for genetic improvement. Embryogenesis is considered a model for testing the totipotency of crop tissues (Komamine et al., 1992). Embryogenesis in bananas is still restricted for various reasons like genotype dependency, long period, and overall uncertainty. Based on the concern mentioned earlier, Remakanthan et al. (2014) developed a better, faster, more consistent, and more accessible protocol for banana cultivar 'Grand Maine' for somatic embryogenesis. For the study, explants were collected from 4-week-old in vitro–grown shoots, and the shoot tips were used as explants; induction of direct somatic without callus formation was performed via culturing the explants in MS media supplemented with picloram (0–8.28 μM and 0.22–4.44 μM) for 15 days. Among different combinations, 0.22 μM 6-benzylaminopurine and 4.14 μM picloram were recorded as best, as they were 100% successful for embryo induction. Transformation to plantlets (2–3%) from the somatic embryos happens periodically utilizing the MS media supplemented with NAA (0.53–2.68 μM) collectively with thidiazuron (4.54 μM) with the addition of glutamine (200 mg l-1) or BA (2.22–44.39 μM). In liquid suspension media, one-on-one embryos were isolated easily, and within 10 days, many secondary sources were generated. The embryos were shifted to filter paper shocked with MS liquid media with 4.44 μM BA in the transition into plantlets (3%) (Remakanthan et al., 2014). The effect of male blossom position and amino acid on somatic embryogenesis was reported by Nandhakumar et al. (2018). The buds, which are located at the sixth to eighth bract, gave good caulogenesis. Glutamin supplementation at 400 mg/l in medium was found to increase embryogenesis 10-fold (Nandhakumar et al., 2018).

5.3.7 Papaya (*Carica papaya* L.)

Papaya is one of the well-researched crops in terms of tissue culture (Tables 5.3 and 5.4). In vitro shoot bud culture was found optimum for cloning of female papaya plants. In vitro regeneration of shoot tips collected from greenhouse-grown plants, as well as field-grown plants, also responded (Ali and Hogan, 1976; Rajeevan and Pandey, 1986; Drew, 1988, 1992). Several researchers like Arora and Singh (1978) and Litz et al. (1983) also reported callus culture from the explants of papaya. The main difficulties in papaya micropropagation are the lower rate of culture establishment, a low proliferation rate, and high mortality during acclimatization. Aseptically grown epicotyl segments were utilized as explants for in vitro micropropagation via direct organogenesis, and petiole was used for indirect organogenesis in papaya (Hossain et al., 1993). Callus induction and proliferation were achieved on MS media supplemented with NAA and BA. Further, hard, nodular, and green callus was developed on MS media with a low cytokinin and auxin ratio. Regeneration of shoots was obtained on MS media fortified with casein hydrolysates, NAA, and MBA. Shoots were elongated on MS media devoid of hormones and root formation on half MS with NAA and GA_3 (Hossain et al., 1993). Cell culture systems were found suitable compared to the conventional method, as papaya cells can propagate at higher growth rates and are free of microorganisms and insects (Vanisree et al., 2004). Also,

TABLE 5.3
Micropropagation in Papaya

Final Outcome	Explant Source	Reference
Micropropagation and few roots	Seedling apices	Drew (1988)
	Stem section and shoot tips	Al-Mehdi and Ali Ahmed (1976)
Micropropagation and rooted plants to field	Six-month-old axillary, rooted cuttings or buds	Drew (1988)
	Six-month-old shoots	Miller and Drew (1990)
Micropropagation	Lateral shoot or main stem bud	Reuveni et al. (1990)
	Shoot apices, mature field grown	Litz and Conover (1978a)
	Shoot apices	Litz and Conover (1978b)
	Shoot apices, field/nursery plants	De Winnar (1988)
	Seedling stem, petiole shoot	Rajeevan and Pandey (1986)
	Field-grown lateral shoots	Rajeevan and Pandey (1986)
	Shoot buds	Mondal et al. (1990)
	Petioles	Yang and Ye (1992)
		Hossain et al. (1993)

TABLE 5.4
Somatic Embryogenesis in Papaya

Final Outcome	Explant Source	Reference
Embryogenic callus	Petiole	DeBruijne et al. (1974)
Somatic embryos and plantlets	Seedling stem internodes	Yie and Liaw (1977)
Somatic embryos and plantlets	Seedling shoot tips, mature shoots, internodes, seedling internodes	Mehdi and Hogan (1979)
Somatic embryos and plantlets from root	Seedling stem, cotyledon, root, leaf, shoot tip	Chen et al. (1987); Chen (1988)
Callus		
Callus	Seedling pieces	Yamamoto et al. (1986)
Somatic embryos	Seedling pieces	Yamamoto and Tabata (1989)
Somatic embryos and plantlets	Immature zygotic embryos	Fitch and Manshardt (1990).

the cell cultures are independent of climatic conditions and their changes. Modified De Fossard media supplemented with 0.5 μM BA and 0.5 μM NAA was used for papaya shoot induction via micropropagation from single-node explants. Root induction and proliferation were governed by media supplemented with 10 μM IBA and subsequently transported to Drew-Smith media devoid of any hormone (Drew, 1992). Axillary buds of dioecious papaya were used for in vitro micropropagation on MS media add-on with BA and NAA, and further moved for elongation to MS media supplemented with kinetin and NAA, and finally rooting took place in half MS media supplemented with IBA (Reuveni et al., 1990). Micropropagation in papaya from different explants was performed. All the detail outcomes were given in Table 5.3.

Fitch (1993) studied the potentiality of hypocotyl as an explant for somatic embryogenesis, and it was observed this indirect embryogenesis is a valuable protocol for somatic embryogenesis. Further, Fitch and Manshardt (1990) standardized the protocol for somatic embryogenesis using immature

FIGURE 5.5 Micropropagation of *Carica papaya* L.

zygotic embryos and were successful in regenerating complete plantlets. Immature zygotic embryos excised from 100-day-old fruit from the mid-portion of the bunch gave rise to high somatic embryogenesis when cultured on half MS supplemented with 10 mg l^{-1} 2,4-D, 400 mg l^{-1} glutamine, and 60 gm l^{-1} sucrose (Mishra et al., 2007). Shoot bud propagation in papaya is very difficult due to the presence of latex in the shoots and in-borne contamination. Therefore, somatic embryogenesis has been utilized for mass multiplication of elite cultivars in papaya (Anandan et al., 2012; Fitch, 1993; Litz and Conover, 1977; Kavitha et al., 2008; Paull et al., 2008). Somatic embryogensis in papaya utilizing different explants was performed and details were given in Table 5.4. Leaf explants from hermaphrodite mother plants were used for somatic embryogenesis in papaya via culturing on media supplemented with 2,4-D, and over time, friable callus was converted to globular embryos (Koehler et al., 2013). Micropropagation of *Carica papaya* L is shown in Figure 5.5.

5.5 CONCLUSION AND FUTURE PERSPECTIVE

The production of horticultural crops has continuously accelerated as this sector plays an essential role in nutritional security, poverty alleviation, and employment generation worldwide. In all the years, genuine planting material of fruit crops has played a decisive role in boosting production; however, the majority of planting material is produced through macropropagation methods except banana. Banana is one such crop where micropropagation technology has been exploited commercially. In India, around 100 small and big tissue culture companies are micropropagating bananas at a commercial scale. Lately, specific laboratories specialize in the clonal propagation of certain fruit species; *viz*., pineapples, palms, papaya, and pomegranate have opened up new vistas for commercial exploitation of these crops. The major bottlenecks in micropropagation of hardy woody perennial fruit crops are in vitro oxidative browning, in-borne microbial contamination, need for a variety-specific protocol, low rate of multiplication, and high mortality during acclimatization. Therefore, it is imperative that new tools and techniques should be utilized for the production of micropropagated plants. Recently, it has been demonstrated in a number of crops that bioreactors are proving to be commercially successful. Hence, bioreactor technology, particularly the temporary immersion system of a bioreactor, is proving to be handy for mass multiplication under in vitro conditions with reduced input costs. Biotization is another important technique which can enhance survival of plants during acclimatization. It involves the use of microorganisms with the potting mixture. Several such patented potting mixes are available for acclimatization. Application of robotics and artificial intelligence (AI) in tissue culture can reduce the cost of production significantly with high output.

BIBLIOGRAPHY

Aiyer, A.Y.N. (1956). The antiquity of some field and forest flora of India. Bangalore Priniting and Publishing Company.

Ajithkumar, D., & Seeni, S. (1998). Rapid clonal multiplication through *In vitro* axillary shoot proliferation of *Aegle marmelos* (L.) Corr., a medicinal tree. Plant Cell Reports, 17(5), 422–426.

Alam, M.R., Rahman, A.B., Moniruzzaman, M., Kadir, M.F., Haque, M.A., Alvi, M.R.U.H., & Ratan, M.D. (2012). Evaluation of antidiabetic phytochemicals in *Syzygium cumini* (L.) Skeels (Family: Myrtaceae). J. App. Pharm. Sci, 2(10), 94.

Ali, A.M., & Hogan, L. (1976). Tissue culture of *Carica papaya*. Hort. Sci. 11, 27.

Ali, N., Mulwa, R.M.S., Norton, M.A., & Skirvin, R.M. (2007). Radical disinfestation protocol eliminates *In vitro* contamination in Guava (*Psidium guajava* L.) seeds. Plant Cell, Tissue and Organ Culture, 91(3), 295–298.

Al-Mehdi & Ali Ahmed. (1976). Tissue culture of papaya: *Carica papaya* var. Solo. M. S. Thesis, The University of Arizona.

Al-Sabah, L., Sudhersan, C., & Jibi-Manuel, S. (2012). Somatic embryogenesis and plantlet regeneration in Amla. Am. Eurasian J. Sustain. Agric., 6, 417–421.

Anandan, R., Sudhakar, D., Balasubramanian, P., & Gutiérrez-Mora, A. (2012). *In vitro* somatic embryogenesis from suspension cultures of *Carica papaya* L. Scientia horticulturae, 136, 43–49.

Ara, H., Jaiswal, U., & Jaiswal, V.S. (1998). Rooting of microshoots of Mangifera indica L. cv. Amrapali. Curr. Sci., 240–242.

Ara, H., Jaiswal, U., & Jaiswal, V.S. (1999). Germination and plantlet regeneration from encapsulated somatic embryos of mango (*Mangifera indica* L.). Plant Cell Reports, 19(2), 166–170.

Ara, H., Jaiswal, U., & Jaiswal, V.S. (2000a). Plant regeneration from protoplasts of mango (*Mangifera indica* L.) through somatic embryogenesis. Plant Cell Reports, 19(6), 622–627.

Ara, H., Jaiswal, U., & Jaiswal, V.S. (2000b). Somatic embryogenesis and plantlet regeneration in Amrapali and Chausa cultivars of mango (*Mangifera indica* L.). Current Science, 164–169.

Ara, H., Jaiswal, U., & Jaiswal, V.S. (2004). An improved method of proliferation of proembryogenic calli of Mangifera indica L. var. Amrapali for scale-up of somatic embryo production. Indian Journal of Biotechnology, 3, 229–234.

Arora, I.K., & Singh, R.N. (1978). In vitro plant regeneration in Papaya. Current Science, 47, 839–883.

Arumugam, A., & Rao, M.V. (1996). In vitro production of plantlets from cotyledonary node cultures of *Aegle marmelos* L. Corr. Advances in Plant Sciences, 9, 181–186.

Ayyanar, M., & Babu, S.P. (2012). *Syzygium cumini* (L.) Skeels: A review of its phytochemical constituents and traditional uses. Asian Pacific Journal of Tropical Biomedicine, 2(3), 240–246.

Bhati, R., Sekhwatt, N.S., & Arya, H.C. (1992). *In vitro* regeneration of Plant lets from root segments of *Aegle marmelos*. Indian Journal of Experimental Biology, 30, 844–845.

Bhatt, I.D., & Dhar, U. (2004). Factors controlling micropropagation of *Myrica esculenta* buch.—Ham. ex D. Don: A high value wild edible of Kumaun Himalaya. African J. Biotechnology, 3(10), 534–540.

Butt, M., Usman, M., & Fatima, B. (2013). Enhanced seed germination and callogenesis under long days using leaf disc as explant in guava cultivars. Biologia (Pakistan), 59(2), 293–298.

Chadha, K.L., Ravindran, P.N., & Sahijram, L. (Eds.). (2000). Biotechnology in horticultural and plantation crops, Malhotra Publishing House, New Delhi, India.

Chandra, R., Padaria, J.C., & Srivastava, S. (2004). Factors influencing *In vitro* establishment of mango shoot buds. Indian Journal of Plant Physiology, 9, 136–144.

Chaturvedi, H.C., Agnihotri, S., Sharma, M., Sharma, A.K., Jain, M., Gupta, P., & Kidwai, N.R. (2004a). Induced nucellar embryogenesis *In vitro* for clonal multiplication of *Mangifera indica* L. var. Ambalavi: A dwarfing rootstock. Indian J. Biotechnology, 3, 22–228.

Chaturvedi, H.C., Sharma, M., Agnihotri, S., Sharma, A.K., Jain, M., & Chourasia, A. (2004b). *In vitro* development and convertibility of nucellar embryos of *Mangifera indica* L. Bappakai, a dwarfing rootstock. Plant Cell Biotechnol Mol Biol., 5(1/2), 7–14.

Chen, M.H. (1988). Tissue culture and ringspot resistant breeding in papaya. In: MA S.-S., Shii C.T., Chang W.C., Chang T.L. (eds.) Proc. Symp. Tissue culture of horticultural crops, National Taiwan University, Taipei, March 8–9.

Chen, M.H., Wang, P.J., & Maeda, E. (1987). Somatic embryogenesis and plant regeneration in *Carica papaya* L. tissue culture derived from root explants. Plant Cell Reports, 6(5), 348–351.

Chi, G.L., Pua, E.C., & Goh, C.J. (1991). Role of ethylene on de novo shoot regeneration from cotyledonary explants of *Brassica campestris* ssp. pekinensis (Lour) Olsson *In vitro*. Plant Physiology, 96(1), 178–183.

Cronauer, S.S., & Krikorian, A.D. (1984). Multiplication of Musa from excised stem tips. Annals of Botany, 53, 321–328.

De Andrade, S.R.M., Faleiro, F.G., Teixeira, J.B., de Pinto, A.C.Q., Radel, G., Cordeiro, M.R., & Ramo, V.V. (2009). Uses of alternative antibiotics to control bacterium growth in mango explants. Acta Hortic, 820, 541–546.

De Winnar, W. (1988). Clonal propagation of papaya in vitro. Plant Cell, Tissue and Organ Culture, 12(3), 305–310.

DeBruijne, E., DeLanghe, E., & Van, R.R. (1974). Int. Symp. Fytofarm. Fytiat., 26, 637–645.

Deore, A.C., Chavan, S.S., Ranade, S.S., Deshpande, R.S., & Dhonukshe, B.L. (2000). Factors affecting *in vitro* somatic embryogenesis from nucellar tissues of Alphonso mango. Annals of Plant Physiology, 14(2), 182–186.

Dewald, S.G., Litz, R.E., & Moore, G.A. (1989a). Maturation and germination of mango somatic embryos. J. Am. Soc. Hortic. Sci, 114, 837–841.

Dewald, S.G., Litz, R.E., & Moore, G.A. (1989b). Optimizing somatic embryo production in mango. J. Am. Soc. Hortic. Sci, 114(4), 712–716.

Dewan, A., Nanda, K., & Gupta, S.C. (1992). *In vitro* micropropagation of *Acacia nilotica* subsp. *indica* Brenan via cotyledonary nodes. Plant Cell Reports, 12, 18–21.

Dhar, U., & Upreti, J. (1999). *In vitro* regeneration of a mature leguminous liana (*Bauhinia vahlii* Wight & Arnott). Plant Cell Reports, 18(7), 664–669.

Drew, R.A. (1988). Rapid clonal propagation of papaya *In vitro* from mature field grown trees. HortScience, 23(3), 609–611.

Drew, R.A. (1992). Improved techniques for *in vitro* propagation and germplasm storage of papaya. Hort Science, 27(10), 1122–1124.

Eichholtz, D., Robitaille, H.A., & Hasegawa, P.M. (1979). Adventive embryony in apple. Hort Sci, 14, 699–700.

Elmore, H.W., Samples, B., Sharma, S., & Harrison, M. (1990). Influence of cultural and physiochemical factors on ascorbate stability in plant tissue culture media. Plant Cell, Tissue and Organ Culture, 20(2), 131–135.

Fila, G., Ghashghaie, J., Hoarau, J., & Cornic, G. (1998). Photosynthesis, leaf conductance and water relations of *In vitro* cultured grapevine rootstock in relation to acclimatisation. Physiologia Plantarum, 102(3), 411–418.

Fitch, M.M. (1993). High frequency somatic embryogenesis and plant regeneration from papaya hypocotyl callus. Plant Cell, Tissue and Organ Culture, 32(2), 205–212.

Fitch, M.M., & Manshardt, R.M. (1990) Somatic embryogenesis and plant regeneration from immature zygotic embryos of Papaya (*Carica papaya*). Plant Cell Reports, 9, 320–324.

Gandhi, K., Rajesh, E., Saravanan, S., & Elamvaluthi, M. (2018). Micropropagation of *Aegle marmelos* L.-A medicinal tree through culture of axillary bud and shoot tip explants from in vitro Germinated seedlings. Int. J. Sci. Res. in Biological Sciences, 5, 3.

George, E.F., & Sherrington, P.D. (1984). Plant propagation by tissue culture, handbook and directory of commercial laboratoried. Exegetics Ltd., England.

Gour, V.S., & Kant, T. (2009). *In vitro* regeneration in Emblica officinalis from juvenile root-derived callus. The Journal of Indian Botanical Society, 89(1&2), 34–36.

Goyal, D., & Bhadauria, S. (2008). *In vitro* shoot proliferation in *Emblica officinalis* var. Balwant from nodal explants. Indian Journal of Biotechnology, 7, 394–397.

Gupta, P., Patni, Y., & Kant, U. (1994). *In vitro* shoot differentiation in *Emblica officinalis* Gaertn. J Phytol Res, 7, 171–172.

Hossain, M., Islam, R., Islam, A., & Joarder, O.I. (1995). Direct organogenesis in cultured hypocotyl explants of *Aegle marmelos* Corr. Plant Tissue Culture, 5, 21–25.

Hossain, M., Rahman, S.M., Islam, R., & Joarder, O.I. (1993). High efficiency plant regeneration from petiole explants of *Carica papaya* L. through organogenesis. Plant Cell Reports, 13(2), 99–102.

ICAR-CISH Annual Report (2021). ICAR-central institute for subtropical horticulture. Lucknow, UP, India.

ICAR-CISH-Vision (2050). ICAR-Central Institute for Subtropical Horticulture, Lucknow, U.P., India.

Islam, R., Hossain, M., Joarder, O.I., & Karim, M.R. (1993). Adventitious shoot formation on excised leaf explants of *in vitro* grown seedlings of *Aegle marmelos* Corr. Journal of Horticultural Science, 68(4), 495–498.

Jain, N., & Babbar, S.B. (2000). Recurrent production of plants of black plum, *Syzygium cumini* (L.) Skeels, a myrtaceous fruit tree, from *in vitro* cultured seedling explants. Plant Cell Reports, 19(5), 519–524.

Jain, N., & Babbar, S.B. (2003). Regeneration ofjuvenile'plants of black plum, *Syzygium cumini* Skeels, from nodal exp of mature trees. Plant Cell, Tissue and Organ Culture, 73(3), 257–263.

Jana, M.M., Nadgauda, R.S., Rajmohan, K., & Mascarenhas, A.F. (1994). Rapid somatic embryogenesis from the nucelli of monoembryonic mango varieties. *In vitro* Cellular & Developmental Biology-Plant, 30(1), 55–57.

Kamle, M., Kumar, P., Bajpai, A., Kalim, S., & Chandra, R. (2014). Assessment of genetic fidelity of somatic embryogenesis regenerated guava (*Psidium guajava* L.) plants using DNA-based markers. New Zealand Journal of Crop and Horticultural Science, 42(1), 1–9.

Kavitha, P.S., Chezhiyan, N., & Kumar, N. (2008). Regeneration of papaya embryos under *in vitro* condition. In II International Symposium on Papaya 851 (pp. 213–218).

Koehler, A.D., Carvalho, C.R., Abreu, I.S., & Clarindo, W.R. (2013). Somatic embryogenesis from leaf explants of hermaphrodite *Carica papaya*: A new approach for clonal propagation. African Journal of Biotechnology, 12(18).

Komamine, A., Kawahara, R., Matsumoto, M., Sunabori, S., Toya, T., Fujiwara, A., et al. (1992). Mechanisms of somatic embryogenesis in cell cultures: physiology, biochemistry, and molecular biology. In Vitro Cellular & Developmental Biology-Plant, 28(1), 11–14.

Krishna, H., Sairam, R.K., Singh, S.K., Patel, V.B., Sharma, R.R., Grover, M., & Sachdev, A. (2008). Mango explant browning: Effect of ontogenic age, mycorrhization and pre-treatments. Scientia Horticulturae, 118(2), 132–138.

Lad, B.L., Jayasankar, S., Pliego-Alfaro, F., Moon, P.A., & Litz, R.E. (1997). Temporal effect of 2, 4-D on induction of embryogenic nucellar cultures and somatic embryo development of 'Carabao'mango. In vitro Cellular & Developmental Biology-Plant, 33(4), 253–257.

Lakshman, V., Venkataramareddy, S.R., & Neelwarne, B. (2007). Molecular analysis of genetic stability in long-term micropropagated shoots of banana using RAPD and ISSR markers. Electronic Journal of Biotechnology, 10, 1–8.

Laxmi, D.V., Sharma, H.C., Kirti, P.B., & Mohan, M.L. (1999). Somatic embryogenesis in mango (*Mangifera indica* L.) cv. Amrapali. Current Science, 1355–1358.

Litz, R.E. (1984). *In vitro* somatic embryogenesis from nucellar callus of monoembryonic *Mangifera indica* L. Hort Science, 19, 715–717.

Litz, R.E. (1985). Somatic embryogenesis in tropical fruit trees. In: Henke R.R. and Hughes K.W. (eds.), Tissue culture in Forestry and Agriculture. Constantin, M. J. and Hollanender, A., Plenem Publishing Co., New York, pp. 179–193.

Litz, R.E. (2003). *In vitro* regeneration and transformation of mango. In: Jaiwal K. and Singh R.P. (eds.), Plant genetic engineering. LLC: Sci. Tech. Publishers, Houston, USA, pp. 23–40.

Litz, R.E., & Conover, R.A. (1978a). Tissue culture propagation of papaya. In Proceedings of the Florida State Horticultural Society (Vol. 90, pp. 245–246).

Litz, R.E., & Conover, R.A. (1978b). Recent advances in papaya tissue culture. In Proceedings of the Florida State Horticultural Society (Vol. 91, pp. 180–181).

Litz, R.E., Gomez-Lim, M.A., Jayashankar, S., & de-Moura, B.L. (2000). Biotechnology for mango improvement and propagation. In: Srivastava R.P., editor. Recent advances in mango research. Prem Printing Press, Lucknow.

Litz, R.E., Hendrix, R.C., Moon, P.A., & Chavez, V.M. (1998). Induction of embryogenic mango cultures as affected by genotype, explanting, 2, 4-D and embryogenic nurse culture. Plant Cell, Tissue and Organ Culture, 53(1), 13–18.

Litz, R.E., Jarret, R.L., & Asokan, M.P. (1986). Tropical and subtropical fruits and vegetables. In: Zimmerman R.H., Griesbach R.J., Hammerschlag F.A. and Lawson R.H. (eds.), Tissue culture as a plant production system for horticultural crops. Martinus-Nijhoff Publishers, Dordrecht, pp. 237–251.

Litz, R.E., Knight Jr, R.J., & Gazit, S. (1984). *In vitro* somatic embryogenesis from *Mangifera indica* L. callus. Scientia Horticulturae, 22(3), 233–240.

Litz, R.E., Knight, R.L., & Gazit, S. (1982). Somatic embryos from culture ovules of polyembryonic *Mangifera indica* L. Plant Cell Reports, 1(6), 264–266.

Litz, R.E., Mathews, V.H., Moon, P.A., Pliego-Alfaro, F., Yurgalevitch, C., & Dewald, S.G. (1993). Somatic embryos of mango (*Mangifera indica* L.). In: Redenbaugh K. (ed.), Synseeds applications of synthetic seeds to crop improvement. CRC Press, Boca Raton, pp. 409–425.

Litz, R.E., Stephen, K.O., & Conover, R.A. (1983). In vitro growth of *Carica papaya* L. cotyledons. Scientia Horticulturae, 19(3–4), 287–293.

Liu, X., & Yang, G. (2011). Clonal propagation of guava (Psidium guajava L.) on nodal explants of mature elite cultivar. International Journal of Plant Biology, 2(1), e2.

Madharia, P., & Dutta, A.K. (2016). Indirect somatic embryogenesis in Emblica: A preliminary study in our research lab. Biosciences Biotechnology Research Asia, 9(1), 305–308.

Madhulatha, P., Anbalagan, M., Jayachandran, S., & Sakthivel, N. (2004). Influence of liquid pulse treatment with growth regulators on In vitro propagation of banana (*Musa spp.* AAA). Plant Cell, Tissue and Organ Culture, 76(2), 189–192.

Malik, S.K., Chaudhury, R., & Kalia, R.K. (2005). Rapid In vitro multiplication and conservation of Garcinia indica: A tropical medicinal tree species. Scientia Horticulturae, 106(4), 539–553.

Manzanilla Ramiriez, M.A., Robles Gonzales, M.M., & Med ina Urrutia, V.M. (2000). Advances sobre embriogenesis somatica en mango Ataulfo Haden. In: Medina Urrutia VM, Perez Barraz H, Chavez Contreras X (eds.), Control de Floracion Y Mejoramiento Genetico. INIFAP, Tecoman, Mexico, pp. 47–56.

Mathews, H., Litz, R.E., Wilde, H.D., Merkle, S.A., & Wetzstein, H.Y. (1992). Stable integration and expression of β-glucuronidase and NPT II genes in mango somatic embryos. In vitro—Plant, 28(4), 172–178.

Mathews, H., Litz, R.E., Wilde, D.H., & Wetzstein, H.Y. (1993). Genetic transformation of mango. Acta Horticulture, 341, 93–97.

Meghwal, P.R., Sharma, H.C., & Singh, S.K. (2010). Micropropagation studies on guava. Indian Journal of Horticulture, 67(4), 55–58.

Mehdi, A.A., & Hogan, L. (1979). In vitro growth and development of papaya (Carica papaya L.) and date palm (Phoenix dactylifera L.). HortScience, 14, 46.

Miller, R.M., & Drew, R.A. (1990). Effect of explant type on proliferation of *Carica papaya* L. In vitro. Plant Cell, Tissue and Organ Culture, 21(1), 39–44.

Mishra, M., Chandra, R., Pati, R., Kamle, M., Shukla, N., Seal, S., et al. (2010). Micropropagation of *Mangifera indica* L. cv. Kurukkan through somatic embryogenesis. Indian Journal of Genetics and Plant Breeding, 70(1), 85–90.

Mishra, M., Chandra, R., Pati, R., & Bajpai, A. (2005). Micropropagation of guava (*Psidium guajava* L.). In I International Guava Symposium 735, Lucknow, India (pp. 155–158).

Mishra, M., Jalil, S. U., Sharma, N., & Hudedamani, U. (2014). An agrobacterium mediated transformation system of guava (*Psidium guajava* L.) with endochitinase gene. Crop Breeding and Applied Biotechnology, 14(4), 232–237.

Mishra, M., & Pathak, R.K. (2001). Effect of nodal position and season on In vitro shoot proliferation in aonla (*Emblica officinalis* Gaertn.). Journal of Applied Horticulture, 3(2), 103–104.

Mishra, M., Pati, R., Chandra, R., & Kole, C. (2006). Clonal micropropagation of Indian gooseberry (*Emblica officinalis* Gaertn.). Indian Journal of Genetics and Plant Breeding, 66(4), 361–362.

Mishra, M., Shukla, N., & Chandra, R. (2007). Micropropagation of papaya (Carica papaya L.). In Protocols for micropropagation of woody trees and fruits. Springer, Dordrecht, pp. 437–441.

Mishra, M., Srivastawa, A.K., & Pathak, R.K. (1999). Micropropagational studies in Aonla (*Emblica officinalis* Gaertn). Annals of Plant Physiology, 13(2), 111–117.

Mishra, M.S. (1997). Micropropagational studies in anola (*E. officinalis* Gaertn). Ph D. Thesis. N.D. University of Agr. & Tech., Kumarganj, Faizabad, India.

Mondal, M., Gupta, S., & Mukherjee, B.B. (1990). *In vitro* propagation of shoot buds of *Carica papaya* L.(Caricaceae) var. Honey Dew. Plant Cell Reports, 8(10), 609–612.

Monsalud, M.J., Mathews, H., Litz, R.E., & Gray, D.J. (1995). Control of hyperhydricity of mango somatic embryos. Plant Cell, Tissue and Organ Culture, 42(2), 195–206.

Mukherjee, E., Gantait, S., Kundu, S., Sarkar, S., & Bhattacharyya, S. (2019). Biotechnological interventions on the genus Rauvolfia: Recent trends and imminent prospects. Applied Microbiology and Biotechnology, 103(18), 7325–7354.

Mullins, M.G., & Srinivasan, C. (1976). Somatic embryos and plantlets from an ancient clone of the grapevine (cv. Cabernet-Sauvignon) by apomixis *in vitro*. Journal of Experimental Botany, 27(5), 1022–1030.

Murashige, T., & Skoog, F. (1962). A revised medium for rapid growth and bio assays with tobacco tissue cultures. Physiologia Plantarum, 15(3), 473–497.

Nandhakumar, N., Kumar, K., Sudhakar, D., & Soorianathasundaram, K. (2018). Plant regeneration, developmental pattern and genetic fidelity of somatic embryogenesis derived *Musa* spp. Journal of Genetic Engineering and Biotechnology, 16(2), 587–598.

Nayak, P., Behera, P.R., Thirunavoukkarasu, M., & Chand, P.K. (2010). High frequency plant regeneration through adventitious multiple shoot organogenesis in epicotyl explants of Indian gooseberry (*Emblica officinalis* Gaertn). Scientia horticulturae, 123(4), 473–478.

Patena, L.F., Carlos-Refuerzo, L.R., & Barba, R.C. (2002). Somatic embryogenesis and plantlet regeneration in mango (*Mangifera indica* L.). *In vitro* Cellular & Developmental Biology-Plant, 38(2), 173–177.

Pati, R., Chandra, R., Chauhan, U.K., Mishra, M., & Srivastava, N. (2008). In vitro clonal propagation of bael (Aegle marmelos Corr.) CV. CISH-B1 through enhanced axillary branching. Physiology and Molecular Biology of Plants, 14(4), 337–346.

Patidar, D.K., Tripathi, M.K., Tiwari, R., Baghel, B.S., & Tiwari, S. (2010). In vitro propagation of *Emblica officinalis* from nodal segment culture. J Agri Tech, 6(2), 245–256.

Paull, R.E., Irikura, B., Wu, P., Turano, H., Chen, N.J., Blas, A., Fellman, J.K., Gschwend, A.R., Wai, C.M., & Yu, Q. (2008). Fruit development, ripening and quality related genes in the papaya genome. Trop Plant Biol, 1(3–4), 246–277.

Pliego-Alfaro, F., Litz, R.E., Moon, P.A., & Gray, D.J. (1996). Effect of abscisic acid, osmolarity and temperature on *in vitro* development of recalcitrant mango nucellar embryos. Plant Cell, Tissue and Organ Culture, 44(1), 53–61.

Priyadarshi, A., & Ram, B. (2018). A review on pharmacognosy, phytochemistry, and pharmacological activity of Carica papaya (Linn.) leaf. International Journal of Pharmaceutical Sciences and Research, 9(10), 4071–4078.

Priyanka, K.U., Rather, M.I., & Deepti Vimala, Y. (2014). *In vitro* salt stress induced enhancement of ascorbic acid in *Emblica ofcinalis*. Ann Plant Sci, 3, 588–593.

Priyanka, K.U., & Singh, K.P. (2015) Indirect, direct and secondary somatic embryogenesis in *Emblica ofcinalis*. Global J Res Anal, 4, 1–4.

Raghuvanshi, S.S., & Srivastava, A. (1995). Plant regeneration of *Mangifera indica* using liquid shaker culture to reduce phenolic exudation. Plant Cell, Tissue and Organ Culture, 41(1), 83–85.

Rai, M.K., Akhtar, N., & Jaiswal, V.S. (2007). Somatic embryogenesis and plant regeneration in *Psidium guajava* L. cv. Banarasi local. Scientia Horticulturae, 113(2), 129–133.

Rai, M.K., Asthana, P., Jaiswal, V.S., & Jaiswal, U. (2010). Biotechnological advances in guava (*Psidium guajava* L.): Recent developments and prospects for further research. Trees, London, 24(1), 1–12.

Rai, M.K., Jaiswal, V.S., & Jaiswal, U. (2008a). Alginate-encapsulation of nodal segments of guava (*Psidium guajava* L.) for germplasm exchange and distribution. The Journal of Horticultural Science and Biotechnology, 83(5), 569–573.

Rai, M.K., Jaiswal, V.S., & Jaiswal, U. (2008b). Encapsulation of shoot tips of guava (*Psidium guajava* L.) for short-term storage and germplasm exchange. Scientia Horticulturae, 118(1), 33–38.

Rai, M.K., Jaiswal, V.S., & Jaiswal, U. (2008c). Effect of ABA and sucrose on germination of encapsulated somatic embryos of guava (*Psidium guajava* L.). Scientia Horticulturae, 117(3), 302–305.

Rai, M.K., Jaiswal, V.S., & Jaiswal, U. (2009). Effect of selected amino acids and polyethylene glycol on maturation and germination of somatic embryos of guava (*Psidium guajava* L.). Scientia horticulturae, 121(2), 233–236.

Rajeevan, M.S., & Pandey, R.M. (1986). Lateral bud culture of papaya (*Carica papaya* L.) for clonal propagation. Plant Cell, Tissue and Organ Culture, 6(2), 181–188.

Raju, R., & Divya, C. (2020). Micropropagation of *Syzygium densiflorum* Wall. ex Wight & arn.: An endemic and endangered semi-evergreen tree species of the Western Ghats, India. Trees, Forests and People, 2, 100037.

Ram, S. (1982). Aonla (*E. officinalis*)—Uses, botany and culture. Directorate of Expt. Station, G.B. Pant Univ. Ag. & Tech., Pantnagar, India.

Ramashree, A.B., Varghese, T.T., Raghu, A.V., & Nabeese, N.N. (2007). Micropropagation of *Syzygium cumini* skeels. A multipurpose tree. Research Journal of Botany, 2(4), 208–213.

Rao, A.N., Sin Y.M., Kathagoda, N., & Hutchinson, J.F. (1981). Cotyledon tissue culture of some tropical trees. In: Rao, A.N. (ed.), Publisher: COSTED and ANBS. Singapore, pp. 124–137.

Remakanthan, A., Menon, T., & Soniya, G. (2014). Somatic embryogenesis in banana (*Musa acuminate* AAA cv. Grande naine): Effect of explant and culture conditions. *In vitro* Cell Dev. Biol.-Plant, 50(1), 127–136.

Reuveni, O., & Golubowicz, S. (1996). Trials of using *in vitro* techniques for vegetative propagation of mangoes. In V International Mango Symposium 455 (pp. 496–504).

Reuveni, O., Shlesinger, D.R., & Lavi, U. (1990). *In vitro* clonal propagation of dioecious *Carica papaya*. Plant Cell, Tissue and Organ Culture, 20(1), 41–46.

Rivera-Dominguez, M., Manzanilla-Ramirez, M.A., Robles-Gonzalez, M., & Gomez-Lim, M.A. (2004). Induction of somatic embryogenesis and plant regeneration of 'Ataulfo' mango (*Mangifera indica*). Plant Cell, Tissue and Organ Culture, 79(1), 101–104.

Roy, P.K., Rehman, M.M., & Roy, S.K. (1996). *In vitro* propagation of *Syzygium cumini* L. from selected elite trees. Acta Hort, 429, 489–495.

Sehgal, C.B., & Khurana, S. (1985). Morphogenesis and plant regeneration from cultured endosperm of *Emblica officinalis* Gaertn. Plant Cell Reports, 4(5), 263–266.

Shah, S.T., Zamir, R., Ahmad, J., Ali, H., & Lutfullah, G. (2008). *In vitro* regeneration of plantlets from seedlings explants of guava (*Psidium guajava* L.) cv. Safeda. Pak J Bot, 40(3), 1195–1200.

Sharma, M.R., Kumari, S., & Mishra, M. (2018). Micropropagation of wilt resistant, inter-specific (*Psidium molle* x *Psidium guajava*) rootstock of guava. Plant Archives, 18(1), 1170–1174.

Singh, S.K., Sharma, H.C., Goswami, A.M., & Singh, S.P. (2001). Abscisic acid aided normal somatic embryogenesis from nucellar tissue in mango cv. Amrapali. In: Rai M., Kumar S., Singh R. and Prasad V.S.R.K. (eds.), Recent trends in horticultural research. pp. 161–164.

Soni, S.K., Mishra, M.K., Mishra, M., Kumari, S., Saxena, S., Shukla, V., Tiwari, S., & Shirke, P. (2022). Papaya Leaf Curl Virus (PaLCuV) infection on papaya (*Carica papaya* L.) plants alters anatomical and physiological properties and reduces bioactive components. *Plants*, 11, 579. https://doi.org/10.3390/plants11050579

Strosse, H., Domergue, R., Panis, B., Escalant, J.V., & Cote, F.X. (2003). INIBAP technical guidelines 8: Banana and plantain embryogenic cell suspensions. Edited by A. Vezina and C. Picq. INIBAP, Montpellier, France.

Thilaga, S., Largia, M.J.V., Parameswari, A., Nair, R.R., & Ganesh, D. (2013). High frequency somatic embryogenesis from leaf tissue of *'Emblica officinalis'* Gaertn.-A high valued tree for non-timber forest products. Australian Journal of Crop Science, 7(10), 1480–1487.

Thomas, P. (1999). Somatic embryogenesis and plantlet regeneration from nucellar tissue of monoembryonic mango. The Journal of Horticultural Science and Biotechnology, 74(1), 135–139.

Thomas, P., & Ravindra, M.B. (1997). Shoot tip culture in mango: Influence of medium, genotype, explant factors, season and decontamination treatments on phenolic exudation, explant survival and axenic culture establishment. Journal of Horticultural Science, 72(5), 713–722.

Thomas, T.D. (2008). The role of activated charcoal in plant tissue culture. Biotechnology Advances, 26(6), 618–631.

Tisserat, B. (1985). Embryogenesis, organogenesis and plant regeneration. In: Dixon R.A. (ed.), Plant cell culture, a practical approach. IRI, Oxford, pp. 79–105.

Usman, M., Jaskani, M.J., Rehman, M.Y., Fatima, B., & Khan, I.A. (2005). *In vitro* regeneration of monoembryonic mango cultivars. Int J Biol Biotechnol, 2(1), 23–7.

Usman, M., Samad, W.A., Fatima, B., & Shah, M.H. (2013). Pollen parent enhances fruit size and quality in intervarietal crosses in guava (*Psidium guajava*). International Journal of Agriculture and Biology, 15(1), 125–129.

Vanisree, M., Lee, C.Y., Lo, S.F., Nalawade, S.M., Lin, C.Y., & Tsay, H.S. (2004). Studies on the production of some important secondary metabolites from medicinal plants by plant tissue cultures. Bot. Bull. Acad. Sin, 45(1), 1–22.

Verma, B., & Kant, U. (1999). Propagation of *Emblica ofcinalis* Gaertn. through tissue culture. Adv Plant Sci, 12, 21–25.

Vuylsteke, D.R. (1998). Field performance of banana micropropagules and somaclones. In Somaclonal variation and induced mutations in crop improvement. In: Jain, S.M., Brar, D.S. and Ahloowalia, B.S. (eds.), Springer, Dordrecht, pp. 219–231.

Xiao, J.N., Huang, X.L., Wu, Y.J., Li, X.J., Zhou, M.D., & Engelmann, F. (2004). Direct somatic embryogenesis induced from cotyledons of mango immature zygotic embryos. In vitro Cellular & Developmental Biology Plant, 40(2), 196–199.

Yadav, U., Lal, M., & Jaiswal, V.S. (1990). In vitro micropropagation of the tropical fruit tree Syzygium cuminii L. Plant Cell, Tissue and Organ Culture, 21(1), 87–92.

Yamamoto, H., Nakagawa, K., Fukui, H., & Tabata, M. (1986). Cytological changes associated with alkaloid production in cultured cells of *Coptis japonica* and *Thalictrum minus*. Plant Cell Reports, 5(1), 65–68.

Yamamoto, H., & Tabata, M. (1989). Correlation of papain-like enzyme production with laticifer formation in somatic embryos of papaya. Plant Cell Reports, 8(4), 251–254.

Yang, J.S., & Ye, C.A. (1992). Plant regeneration from petioles of *in vitro* regenerated papaya (*Carica papaya* L.) shoots. Bot. Bull. Acad. Sin, 33, 375–381.

Yang, Z., & Lüdders, P. (1992). Effect of growth regulator and media on *in vitro* shoot tip culture of different cultivars of mango (*Mangifera indica* L.) rootstocks. In IV International Mango Symposium 341 (pp. 240–247).

Yie, S.T., & Liaw, S.I. (1977). Plant regeneration from shoot tips and callus of papaya. In Vitro, 13(9), 564–568.

Zatyko, J.M., Simon, I., & Szabo, C.S. (1975). Induction of polyembryony in cultivated ovules of red currant. Plant Science Letters, 4(5), 281–283.

Zimmerman, J.L. (1993). Somatic embryogenesis: A model for early development in higher plants. The Plant Cell, 5(10), 1411.

6 Exploring the Diversity of Plant Alkaloids through a Multiomics Approach

Megha Rai and Mami Yamazaki

CONTENTS

6.1 Introduction ... 109
6.2 Occurrence, Diversity, and Function of Plant Alkaloids ... 110
6.3 Alkaloids' Biosynthetic Pathways: The Discovery in the Progress ... 112
  6.3.1 Benzylisoquinoline Alkaloids (BIAs) ... 112
  6.3.2 Monoterpene Indole Alkaloids (MIAs) ... 115
  6.3.3 Tropane Alkaloids (TAs) ... 117
  6.3.4 Purine Alkaloids ... 118
6.4 Omics Revolution: Catalyzing the Pathway Elucidation of Alkaloids ... 119
  6.4.1 Genomics ... 119
  6.4.2 Transcriptomics ... 120
  6.4.3 Proteomics ... 121
  6.4.4 Metabolomics ... 122
6.5 Multiomics Analysis: A Powerful Tool to Decrypt Biology From Huge Datasets ... 123
  6.5.1 Element-Based Approach ... 125
  6.5.2 Pathway-Based Approach ... 125
  6.5.3 Mathematical Model-Based Approach ... 126
6.6 Conclusions and Future Perspectives ... 127
Bibiliography ... 127

6.1 INTRODUCTION

Alkaloids are a diverse group of natural products, with approximately 22,000 structures reported from higher plants to date (Goyal, 2013). Before the advent of the 19th century, the purified plant-derived compounds possessed acidic properties until the extraction of alkaline potash from burnt wood. Therefore, W. Meisner, in 1818, coined the term "alkaloid" to define all the plant-derived "alkali-like" organic compounds (O'Connor, 2010). Subsequently, alkaloids with a wide array of structures and properties were discovered. In order to encompass the vast range of unrelated structures, a broad definition for alkaloids was suggested, implying that alkaloids are nitrogen-containing compounds derived from specialized metabolism (Kukula-Koch & Widelski, 2017).

Initial isolations of alkaloids were reportedly from the higher plants. The first semi-purified alkaloid "*principium somniferum opii*" was isolated by Friedrich Wilhelm Sertürner from the opium poppy in 1805 (Sertùrner, 1805). Subsequent research in the 20th century revealed the presence of alkaloids from different terrestrial and marine sources, including insects, bacteria, and fungi. Nevertheless, plants are still recognized as the primary natural source for these diverse alkaloids (Cordell & Choi, 2011; Kukula-Koch & Widelski, 2017; O'Connor, 2010). Alkaloids are distributed in 186 plant families, with the most number of alkaloids isolated from 20 families, including

DOI: 10.1201/b22917-6

Amaryllidaceae, Annonaceae, Apocynaceae, Asteraceae, Berberidaceae, Boraginaceae, Buxaceae, Celastraceae, Fabaceae, Lauraceae, Liliaceae, Loganiaceae, Menispermaceae, Papaveraceae, Piperaceae, Poaceae, Ranunculaceae, Rubiaceae, Rutaceae, and Solanaceae (Chung et al., 2016; Cordell & Choi, 2011).

Alkaloids have been recognized as a metabolite class possessing diverse biological, physiological, chemical, and pharmacological activities. They are widely used as anti-inflammatory, antipyretic, cardioprotective, stimulant, and anesthetic agents. Moreover, the isolation of alkaloids had a spectacular achievement in treating several diseases, including malaria, cancer, Parkinson's disease, diabetes, and epilepsy, among others (Seneca, 2007). The biosynthesis of these therapeutic alkaloids has challenged researchers for ages, initially to determine their structure and eventually to understand the molecular components involved in its biosynthesis and regulation in plants. In recent years, high-throughput -omics technologies have significantly improved our ability to explore and elucidate novel unknown alkaloids' biosynthetic pathways. It is anticipated that the development of multiomics technologies will yield tremendous genomics, transcriptomics, metabolomics, and proteomics datasets, which will further help to establish synthetic routes for the sustainable production of these lifesaving specialized metabolites. Here we have summarized major plant alkaloid classes and the current state-of-the-art technologies used to elucidate the biosynthesis pathways of alkaloids using multiomics approaches.

6.2 OCCURRENCE, DIVERSITY, AND FUNCTION OF PLANT ALKALOIDS

From the onset of human civilizations, alkaloids have been extensively used as an ingredient for potions to cure diseases, as well as poisons in weapons for hunting. Opium, the dried latex of opium poppy plants, is a complex mixture containing several alkaloids and is widely used as an analgesic, antidiarrheal, and antitussive agent. Globally, approximately 1100–1500 tons of opium are consumed annually worldwide (Kunnumpurath et al., 2018). The extraction and purification of morphine, representing the most abundant alkaloid in opium, led the way for the development of modern pharmaceuticals (O'Connor, 2010). The global demand for morphine-derived opiates has increased over the past 20 years, with codeine and hydrocodone representing the most consumed opiates (Kunnumpurath et al., 2018). Alkaloids continue to attract human interest owing to their tremendous diversity and health benefits. Currently, there are 60 plant-derived alkaloids approved as drugs in several countries (Cordell et al., 2001). The presence of alkaloids has been reported for over 25% of higher plants, comprising 20% of the known secondary metabolites found in plants (Facchini, 2001; Goyal, 2013; Kaur & Arora, 2015; Yang & Stockigt, 2010). Several of the alkaloid-producing plant families are known to synthesize structurally diverse classes of alkaloids. For instance, Apocynaceae, Rutaceae, and Fabaceae are known to produce 2664, 1730, and 1453 different types of alkaloids, respectively (Cordell et al., 2001). The combination and arrangement of functional groups around the central skeletal structure, together with the action of modifying enzymes on the parent structure, lead to a diverse range of alkaloids with distinct properties in plants (Bhambhani et al., 2021). It has been postulated that the evolution of plants to synthesize a myriad of natural products offers them a wide range of defense strategies that they would potentially encounter, thereby giving them an upper edge for their survival and interaction with the environment (O'Connor, 2010).

Alkaloids are generally classified into three types based on the type of nitrogen and the precursor molecule that derives the characteristic nitrogen atom to its structure: true alkaloids, protoalkaloids, and pseudoalkaloids. Both true alkaloids and protoalkaloids are derived from amino acid. However, pseudoalkaloids are derived from non-amino acid precursors. True alkaloids have the nitrogen atom in the heterocyclic ring, whereas the nitrogen atom in protoalkaloids is not a part of the heterocyclic ring. The major classes of true alkaloids, including tetrahydroisoquinoline, indole, tropane, quinolizidine, quinazoline, and pyridine, are derived from tyrosine, tryptophan, ornithine, lysine, anthranilic acid, and nicotinic acid, respectively (Figure 6.1). Tryptophan and tyrosine are

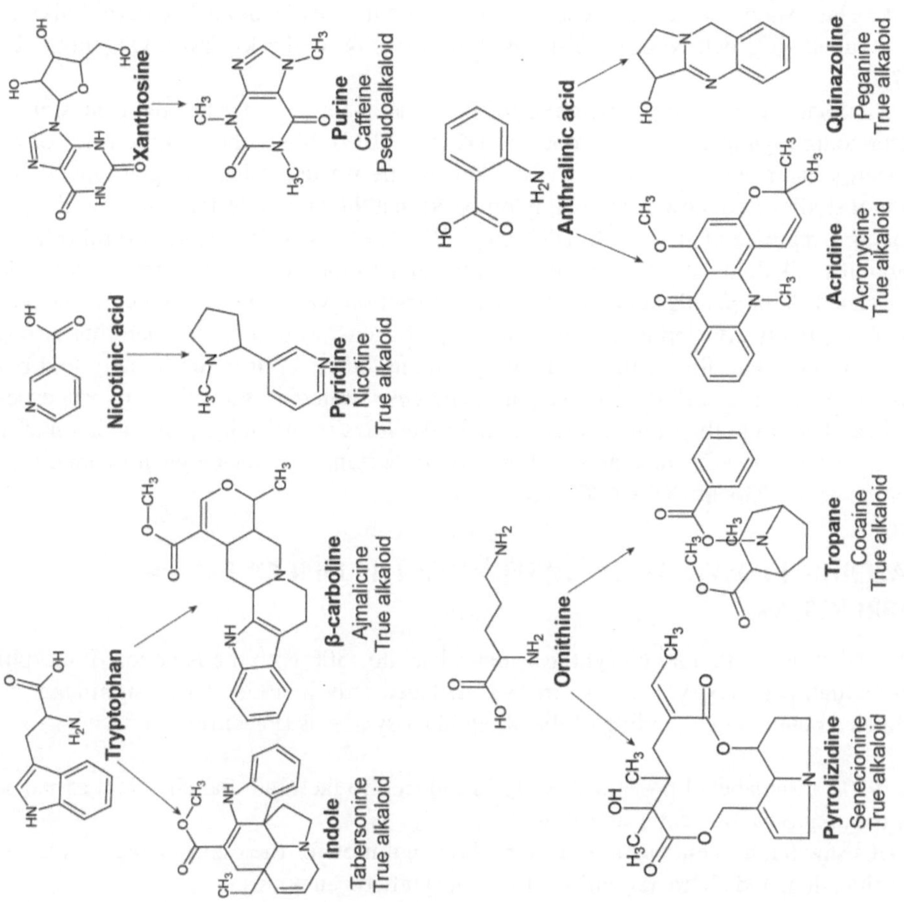

FIGURE 6.1 Major structural classes of alkaloids and their representative members. The precursor amino acids and the alkaloid skeleton are represented in bold font.

the main precursors for protoalkaloids involved in the biosynthesis of yohimbine, mescaline, and hordenine. While the pseudoalkaloids are not derived from amino acids, their nitrogen atom forms the heterocyclic ring. Major classes of pseudoalkaloids include purine alkaloids, steroid alkaloids, and ephedra alkaloids (Cordell & Choi, 2011; Kukula-Koch & Widelski, 2017; O'Connor, 2010; Seneca, 2007).

Although alkaloids are extensively explored for their therapeutic properties, their physiological role in the producing organisms remains obscure (Kurek, 2019). Since the definite roles of alkaloids are unclear, it had been earlier argued that alkaloids are waste products of plant metabolism. However, several studies have now increasingly reported that the alkaloids act to protect the plants from predators and regulate their growth (O'Connor, 2010; Seneca, 2007). Like several other secondary metabolites, alkaloids are accumulated in response to the biotic and abiotic stress conditions and are related to the plant's survival and their interaction with the environment. The plants' alkaloid content is greatly affected by the presence of pathogens, wounding, temperature, drought, and changes associated with flowering and fruiting, among others. While the toxicity and bitterness of the alkaloids, such as caffeine, are part of the defense strategies, the stimulant properties of some of them can help in pollination. Therefore, the diversities of alkaloids play an essential role in the increased survival rate of the plants and provide important pharmacological metabolites for enhanced human health (Goyal, 2013; O'Connor, 2010).

6.3 ALKALOIDS' BIOSYNTHETIC PATHWAYS: THE DISCOVERY IN THE PROGRESS

The biochemistry of plant alkaloid biosynthesis dates back to 1806 with the isolation of morphine, yet owing to its high complexity, its structure was deduced only in 1952. Three significant milestones facilitated a better understanding of the alkaloid biosynthesis (Facchini, 2001):

1. In the 1950s, radiolabeled precursors were introduced to facilitate the chemical explanation of the alkaloids' biosynthesis pathways.
2. The 1970s saw the tremendous use of plant tissue cultures for the rapid increase in isolation, purification, and characterization of the biosynthetic enzymes.
3. In the 1990s, several studies were conducted to understand the molecular mechanisms for the biosynthesis of benzylisoquinoline alkaloids, monoterpene indole alkaloids, tropane alkaloids, and purine alkaloids.

This section will focus on the present knowledge for the four classes of alkaloids, focusing primarily on the biosynthetic enzymes known so far.

6.3.1 BENZYLISOQUINOLINE ALKALOIDS (BIAS)

There are more than 2500 structurally elucidated BIAs, the majority of which are reported in basal angiosperms, including Ranunculaceae, Papaveraceae, and Magnoliaceae, among others. BIAs have a wide range of pharmacological properties, with morphine, codeine used as an analgesic, sanguinarine, berberine used as an antibiotic, noscapine used as an anticancer drug, and papaverine and tubocurarine used as a muscle relaxant (Hagel & Facchini, 2013). The current knowledge of the BIA biosynthesis defines *S*-norcoclaurine as the common intermediate formed from the condensation of dopamine and 4-hydroxyphenylacetaldehyde in the presence of norcoclaurine synthase. Further, *S*-norcoclaurine is converted to *S*-reticuline in the presence of enzymes 6-*O*-methyltransferase, coclaurine-*N*-methyltransferase, cytochrome P450 hydroxylase, and a 4′*O*-methyltransferase.

S-reticuline then diverges to produce different classes of BIAs (Figure 6.2) (Chung et al., 2016; Facchini, 2001; Hagel & Facchini, 2013; O'Connor, 2010).

In the first branch-point pathway, *S*-reticuline is methylated in the presence of a berberine bridge enzyme to form *S*-scoulerine (Figure 6.2). *S*-scoulerine then undergoes methylation in the presence of scoulerine-9-*O*-methyltransferase to form *S*-tetrahydrocolumbamine. Further, the enzyme canadine synthase catalyzes the oxidation of *S*-tetrahydrocolumbamine to form *S*-canadine with a methylenedioxy bridge. While *S*-canadine in a single oxidation reaction catalyzed by *S*-tetrahydroprotoberberine oxidase forms berberine, it undergoes multiple reactions to form noscapine. Identification of the 10-member metabolic gene cluster and its subsequent characterization revealed the entire noscapine biosynthetic pathway (Winzer et al., 2012). Catalyzed by the enzyme *S*-canadine-*N*-methyltransferase, *S*-canadine forms *S*-*N*-methylcanadine, which in the presence of *N*-methylcanadine-1-hydroxylase (CYP82Y1) forms *S*-1-hydroxyl-*N*-methylcanadine. Subsequently, through the action of enzymes CYP82X2 and CYP82X1, acetyltransferase (AT1), carboxylesterase (CXE1), and noscapine synthase (NS) form noscapine (Chen & Facchini, 2014; Dang et al., 2015; Dang & Facchini, 2012, 2014; Li et al., 2018b; Li & Smolke, 2016; Winzer et al., 2012).

The second branch-point pathway represents the biosynthesis of benzophenanthridine alkaloid sanguinarine, produced across a wide lineage of plants. The first step toward sanguinarine biosynthesis involves the formation of *S*-cheilanthifoline from *S*-scoulerine catalyzed by a cytochrome P450 enzyme, cheilanthifoline synthase. *S*-cheilanthifoline, then, in the presence of stylopine synthase, forms stylopine, which undergoes *N*-methylation by *S*-tetrahydroprotoberberine cis-*N*-methyltransferase to form *S*-cis-*N*-methylstylopine. *S*-cis-*N*-methylstylopine undergoes two sets of hydroxylation reactions catalyzed by two cytochrome P450 enzymes, *S*-cis-*N*-methylstylopine and protopine-6-hydroxylase, to form dihydrosanguinarine. In the final step, dihydrosanguinarine is converted into sanguinarine through the activity of dihydrobenzophenanthridine oxidase enzyme (Arakawa et al., 1992; Beaudoin & Facchini, 2013; Diaz Chavez et al., 2011; Ikezawa et al., 2007; Liscombe & Facchini, 2007; Takemura et al., 2013; Winkler et al., 2009).

The third branch-point pathway leading to the biosynthesis of codeine and morphine has been extensively studied in *Papaver somniferum*. In the codeine and morphine pathway, *S*-reticuline is converted to *R*-reticuline in the presence of the enzyme reticuline epimerase. *R*-reticuline, then, through a multiple-step reaction, catalyzed by the enzymes salutaridine synthase, salutaridine reductase, salutaridinol acetyltransferase, thebaine synthase, thebaine-6-*O*-demethylase, neopinone isomerase, and codeinone reductase, forms codeine. Codeine then undergoes demethylation catalyzed by codeine-*O*-demethylase to yield morphine (Chen et al., 2018; Farrow et al., 2015; Gesell et al., 2009; Grothe et al., 2001; Hagel & Facchini, 2010; Luo et al., 2006; Unterlinner et al., 1999; Winzer et al., 2015; Ziegler et al., 2006).

The extensive studies of the benzylisoquinoline biosynthetic pathway have helped the plant biologist decipher the molecular components and intermediates associated with the biosynthesis of the BIA backbone and its major branch pathways. Plants produce these therapeutic BIAs in minute quantities, and hence plant-based extraction is not sufficient for drug demand. One of the promising alternative platforms for scalable production is to reconstruct the biosynthetic pathway in the microbial system. This approach has been used for a few BIAs, including sanguinarine (Fossati et al., 2014), opiates (Galanie et al., 2015), and noscapine (Li et al., 2018b). Nevertheless, the diversity of the BIAs is enormous and still poorly understood. To metabolically engineer and synthetically produce these BIAs, a better understanding of their biosynthesis, transport, and regulation is required. Indeed, conducting the -omics studies more comprehensively and systematically will provide more insights for the same.

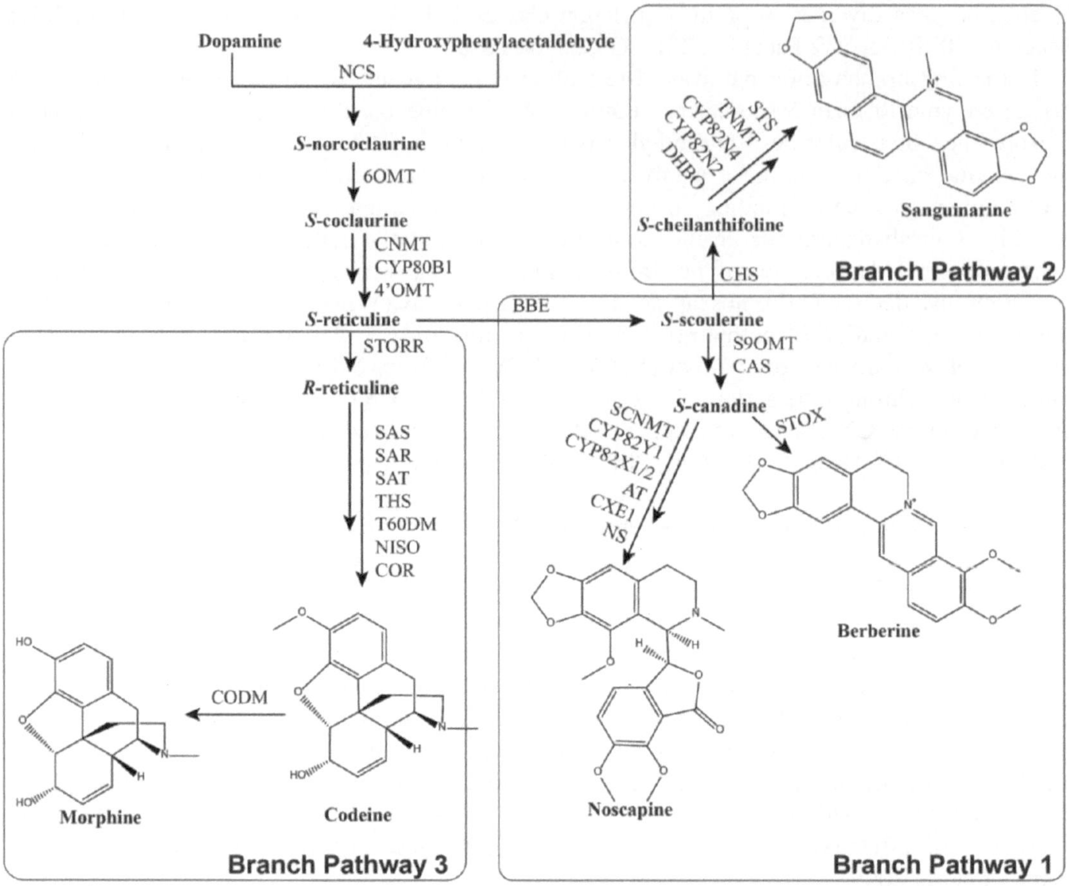

FIGURE 6.2 The biosynthesis pathway of plants' benzylisoquinoline alkaloids. The gene products and gene names are shown in bold and normal font, respectively. A solid arrow represents known biosynthetic steps, and a double arrow represents multiple-step reactions.

Abbreviations: NCS, norcoclaurine synthase; 6OMT, norcoclaurine-6-*O*-methyltransferase; CNMT, coclaurine *N*-methyltransferase; CYP80B1, *N*-methylcoclaurine-3′-monooxygenase; 4′OMT, 3′-hydroxy-*N*-methyl-(*S*)-coclaurine-4′-O-methyltransferase; BBE, berberine bridge enzyme; CHS, cheilanthifoline synthase; STS, stylopine synthase; TNMT, *S*-tetrahydroprotoberberine *cis-N*-methyltransferase; CYP82N4, *(S)-cis-N*-methylstylopine-14-hydroxylase/(*S*)-*N*-methylcanadine-14-hydroxylase; CYP82N2, protopine-6-hydroxylase; DHBO, dihydrobenzophenanthridine oxidase; S9OMT, scoulerine-9-O-methyltransferase; CAS, canadine synthase; STOX, *(S)*-tetrahydroprotoberberine-*N*-methyltransferase; SCNMT, *S*-canadine-*N*-methyltransferase; CYP82Y1, *N*-methylcanadine-1-hydroxylase; CYP82X1, (13*S*,14*R*)-13-*O*-acetyl-1-hydroxy-*N*-methylcanadine-8-hydroxylase; CYP82X2, (*S*)-1-hydroxy-*N*-methylcanadine-13-hydroxylase; AT, (13*S*,14*R*)-1,13-dihydroxy-*N*-methylcanadine 13-*O*-acetyltransferase; CXE1, 3-*O*-aceylpapaveroxine carboxylesterase; NS, noscapine synthase; STORR, reticuline epimerase; SAS, salutaridine synthase; SAR, salutaridine reductase; SAT, salutaridinol acetyltransferase; THS, thebaine synthase; T60DM, thebaine-6-*O*-demethylase; NISO, neopinone isomerase; COR, codeinone reductase; CODM, codeine-*O*-demethylase.

6.3.2 Monoterpene Indole Alkaloids (MIAs)

MIAs are a family of around 3000 compounds that consist of indole and terpenoid moieties, including quinine, vinblastine, and camptothecin, and are mainly found in Loganiaceae, Rubiaceae, and Apocynaceae plant families. The indole moiety of MIAs is derived from tryptamine, a decarboxylated product of tryptophan, and the terpenoid moiety is provided by secologanin derived from geraniol through the secoiridoid pathway (Chung et al., 2016; Facchini, 2001; O'Connor, 2010) (Figure 6.3). Further, the enzyme strictosidine synthase catalyzes the condensation reaction of tryptamine and secologanin to form strictosidine. All the identified MIAs are known to utilize strictosidine as the common intermediate.

The biosynthesis of catharanthine is initiated by the deglycosylation of strictosidine catalyzed by strictosidine-β-glucosidase to form a reactive hemiacetal intermediate that rearranges into cathenamine, which remains in equilibrium with 4,21-dehydrogeissoschizine and epi-cathenamine. Cathenamine produced undergoes reduction catalyzed by heteroyohimbine synthase to form ajmalicine, while it yields tetrahydroalstonine in the presence of the enzyme tetrahydroalstonine synthase. 4,21-dehydrogeissoschizine, in the presence of the enzymes geissoschizine synthase, geissoschizine oxidase, Redox1, and Redox2, yields stemmadenine. The acetylated form of stemmadenine, *O*-stemmadenine, catalyzed by the enzyme stemmadenine-*O*-acetyltransferase, then undergoes oxidation and subsequent reduction to form the highly labile intermediate dihydroprecondylocarpine acetate. In the final steps, dihydroprecondylocarpine acetate in the presence of the enzyme catharanthine synthase or tabersonine synthase forms catharanthine or tabersonine, respectively (Caputi et al., 2018; Geerlings et al., 2000; Qu et al., 2018).

The biosynthesis of vindoline, an intermediate in the biosynthesis of the potent anticancer drug vinblastine, uses tabersonine as the precursor. Tabersonine requires concerted action by a set of seven enzymes, including tabersonine-16-hydroxylase, 16-hydroxytabersonine-*O*-methyltransferase, tabersonine-3-oxygenase/tabersonine-3-reductase, 3-hydroxy-16-methoxy-2,3-dihydrotabersonine-*N*-methyltransferase, desacetoxyvindoline-4-hydroxylase, and deacetylvindoline-4-O-acetyltransferase, to form vindoline. In the final step for the biosynthesis of bisindole alkaloids, monomeric alkaloids, vindoline, and catharanthine, in the presence of a class III peroxidase enzyme, anhydrovinblastine synthase, undergo a dimerization reaction to form α-3′,4′-anhydrovinblastine, which further yields vinblastine and vincristine (De Carolis & De Luca, 1993; Levac et al., 2008; Power et al., 1990; Qu et al., 2015; Schroder et al., 1999; Sottomayor et al., 1998; Sottomayor & Ros Barcelo, 2003; St-Pierre & De Luca, 1995; St-Pierre et al., 1998; Vazquez-Flota et al., 1997).

The biosynthetic pathway of camptothecin from strictosidine is complex and is not fully elucidated. For the biosynthesis of camptothecin, the strictosidine is not deglycosylated, but strictosamide is produced by the formation of a lactam ring between the amine of strictosidine and secologanin derived methyl ester. While the involved gene is unknown, the study using labeled strictosamide corroborated the hypothesis (Hutchinson et al., 1974). Further, strictosamide undergoes a series of transformations in the presence of unknown enzymes to form camptothecin via the intermediates pumiloside and deoxypumiloside (Rai et al., 2021a).

While the biosynthetic pathways for some monoterpene indole alkaloid classes, including catharanthine, and tabersonine, have been well studied with the corresponding enzymes being cloned, the biosynthetic pathways and the corresponding enzymes for several MIAs remain elusive to date and are proposed based on the results obtained from feeding the isotope-labeled substrates. Therefore, extensive system-based studies to reveal the metabolic diversity of these pharmacologically active monoterpene indole alkaloids and their biosynthesis pathways are highly desired.

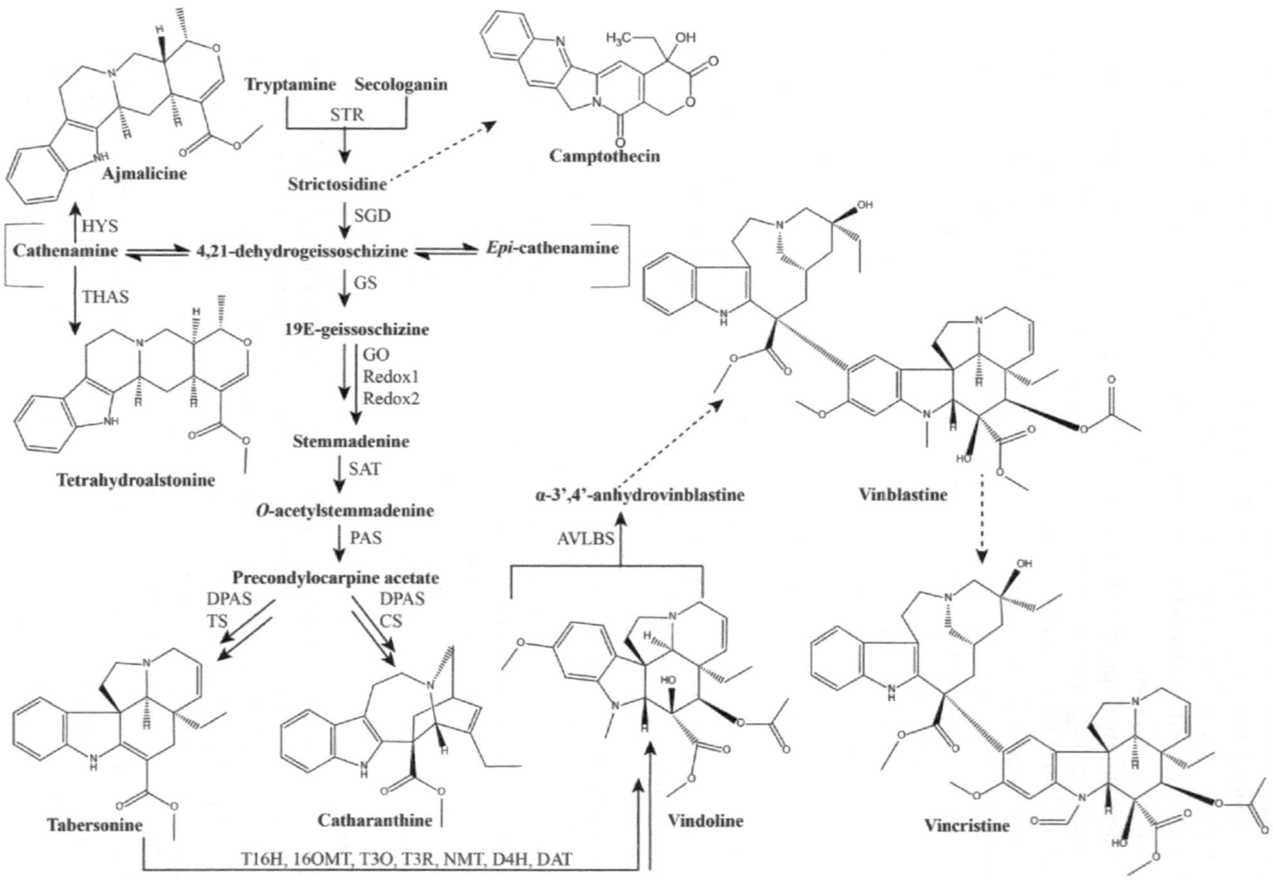

FIGURE 6.3 The biosynthesis pathway of plants' terpenoid indole alkaloids. A solid arrow represents known biosynthetic steps, a double arrow represents multiple-step reactions, and a dotted arrow represents unknown reaction steps.

Abbreviations: STR, strictosidine synthase; SGD, strictosidine-β-glucosidase; HYS, heteroyohimbine synthase; THAS, tetrahydroalstonine synthase; GS, geissoschizine synthase; GO, geissoschizine oxidase; SAT, stemmadenine-*O*-acetyltransferase; PAS, precondylocarpine synthase; DPAS, dihydroprecondylocarpine synthase; CS, catharanthine synthase; TS, tabersonine synthase; T16H, tabersonine-16-hydroxylase; 16OMT, 16-hydroxytabersonine-*O*-methyltransferase; T3O, tabersonine-3-oxygenase; T3R, tabersonine-3-reductase; NMT, 3-hydroxy-16-methoxy-2,3-dihydrotabersonine-N-methyltransferase; D4H, desacetoxyvindoline-4-hydroxylase; DAT, deacetylvindoline-4-*O*-acetyltransferase; AVLBS, anhydrovinblastine synthase.

6.3.3 Tropane Alkaloids (TAs)

TAs, characterized by a bicyclic tropane ring, are secondary metabolites reported in Solanaceae, Convolvulaceae, Proteaceae, Rhizophoraceae, Erythroxylaceae, Apocynaceae, Brassicaceae, Olacaceae, Phyllanthaceae, Moraceae, and Fabaceae plant families. All the 300 TAs known to date share the same basic skeleton but are highly diverse in their chemical, biological, and pharmacological properties. On the basis of biosynthetic pathway and structure type, TAs are classified into six major classes: tropine and its derivatives, prototropine and its derivatives, calystegines, ecgonine and its derivatives, 4α-benzyltropane and its derivatives, and pyranotropane and its derivatives. The tropine and its derivative class of TAs are the largest group, comprising 220 TAs, and representative members include hyoscyamine, scopolamine, and anisodamine. The pseudotropine and its derivative class include 21 TAs. The calystegine class of TAs, containing 14 TAs, were first reported from *Calystegia sepium*, *Convolvulus arvensis*, and *Atropa belladonna* (Tepfer et al., 1988). The ecgonine and its derivatives class of TAs, including cocaine, comprise 11 known TAs reported from plants of the Erythroxylaceae family. The 4α-benzyltropane and its derivatives and pyranotropane and its derivative group include 12 TAs each and are known to be present in plant species from the Proteaceae family (Chung et al., 2016; Facchini, 2001; Huang et al., 2021; Kohnen-Johannsen & Kayser, 2019; O'Connor, 2010).

The biosynthetic pathway of TAs utilizes a common precursor, putrescine, which is biosynthesized from L-arginine and L-ornithine (Figure 6.4). While L-ornithine in a single-step reaction undergoes decarboxylation by ornithine decarboxylase to form putrescine (Michael et al., 1996), L-arginine undergoes a three-step reaction to synthesize putrescine, catalyzed by the enzymes arginine decarboxylase, agmatine deiminase, and *N*-carbamoylputrescine amidase (Janowitz et al., 2003; Kohnen-Johannsen & Kayser, 2019; Piotrowski et al., 2003). In the next step, putrescine is methylated by putrescine-*N*-methyltransferase to form *N*-methylputrescine, representing the committed step in the biosynthesis of TAs (Hibi et al., 1994; O'Connor, 2010). *N*-methylputrescine further undergoes an oxidative deamination reaction catalyzed by *N*-methylputrescine oxidase to form 4-methylaminobutanal, which undergoes spontaneous cyclization to form *N*-methyl-Δ^1-pyrrolinium (Mizusaki et al., 1972). The *N*-methyl-Δ^1-pyrrolinium represents the branch point for the biosynthesis of TAs and nicotine, wherein by the concerted action of two enzymes, polyketide synthase and cytochrome P450, it yields tropinone, or it can condense with nicotinic acid to form nicotine (Figure 6.4) (Bedewitz et al., 2018; Kohnen-Johannsen & Kayser, 2019). The tropinone skeleton undergoes a reduction in the presence of tropinone reductase I to produce tropine, while in the presence of tropinone reductase II forms pseudotropine (Hashimoto et al., 1992). Pseudotropine is the first and only specific intermediate reported for the biosynthesis of the calystegine class of TAs (Kohnen-Johannsen & Kayser, 2019). Catalyzed by the enzyme littorine synthase, tropine condenses with phenyllactyl-CoA to form littorine. In the presence of littorine mutase, littorine rearranges to form hyoscyamine aldehyde, which is reduced by the action of hyoscyamine dehydrogenase to form hyoscyamine (Kohnen-Johannsen & Kayser, 2019; Li et al., 2006; Qiu et al., 2021). Hyoscyamine, via a two-step reaction catalyzed by hyoscyamine-6β-hydroxylase, forms scopolamine (Hashimoto et al., 1993).

Similar to the calystegine biosynthetic pathway, not much is known about the biosynthesis of cocaine, representative of the coca class of TAs. The proposed biosynthetic pathway of cocaine suggests that the *N*-methyl-Δ^1-pyrrolinium forms an oxobutanoic acid intermediate by the addition of two acetyl units (Kohnen-Johannsen & Kayser, 2019). The oxobutanoic acid intermediate then cyclizes to form methylecgonone (Jirschitzka et al., 2012), which is reduced to methylecgonine in the presence of enzyme methylecgonone reductase. In the final step, methylecgonine is converted to cocaine catalyzed by the enzyme cocaine synthase, a BAHD acyltransferase (Schmidt et al., 2015). The biosynthesis for several classes of TAs, including calystegines and ecgonine alkaloids, ise not

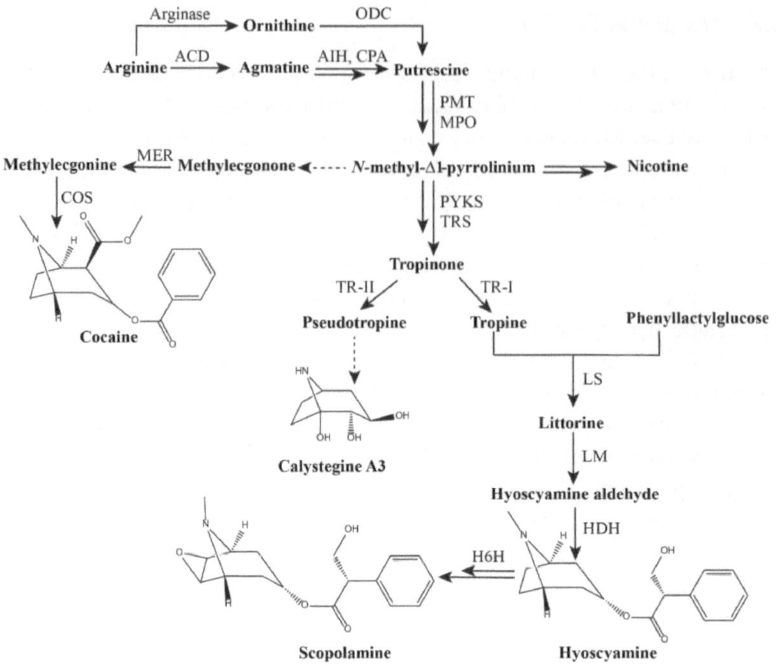

FIGURE 6.4 The biosynthesis pathway of plants' tropane alkaloids. A solid arrow represents known biosynthetic steps, a double arrow represents multiple-step reactions, and a dotted arrow represents unknown reaction steps.

Abbreviations: ODC, ornithine decarboxylase; ACD, arginine decarboxylase; AIH, agmatine deiminase; CPA, *N*-carbamoylputrescine amidase; PMT, putrescine-*N*-methyltransferase; MPO, *N*-methylputrescine oxidase; MER, methylecgonone reductase; COS, cocaine synthase; PYKS, polyketide synthase; TRS, tropinone synthase; TR-I/II, tropinone reductase I/II; LS, littorine synthase; LM, littorine mutase; HDH, hyoscyamine dehydrogenase; H6H, hyoscyamine-6β-hydroxylase.

yet characterized and remains uncertain, which will be possibly unraveled by the extensive application of -omics–based strategies.

6.3.4 Purine Alkaloids

Purine alkaloids, including methylxanthines and methyluric acids containing xanthine and a uric acid skeleton, are derived from purine nucleotides (Ashihara et al., 2013). They are reported to occur primarily in plant species widely used in the food and beverage industries. Caffeine is the most widely reported methylxanthine present in the plant species, which accumulates in the seeds and leaves of *Camellia sinensis* and *Coffea arabica*, among others, and theobromine is known to accumulate in the seeds of *Theobroma cacao*. Theacrine, a type of methyluric acid, is reported to accumulate in *Camellia sinensis* var. *kucha*. The methylxanthine and methyluric skeletons are derived from purine nucleotides, which then undergo methylation by the transfer of methyl groups from *S*-adenosyl-L-methionine (SAM) (Ashihara et al., 2008, 2013; Facchini, 2001; O'Connor, 2010).

Caffeine was first isolated from coffee and tea in 1820 and 1838, respectively. However, the biosynthetic pathway was not fully decoded until the isolation of purified caffeine synthase, and cloning of its gene sequence was accomplished in the year 2000, approximately 180 years later (Ashihara et al., 2013). The biosynthetic pathway of caffeine is now fully elucidated, with most of the studies being focused on tea and coffee plants. It consists of a four-step reaction, including three methylation steps and one nucleosidase step (Figure 6.5). Xanthosine, derived from purine

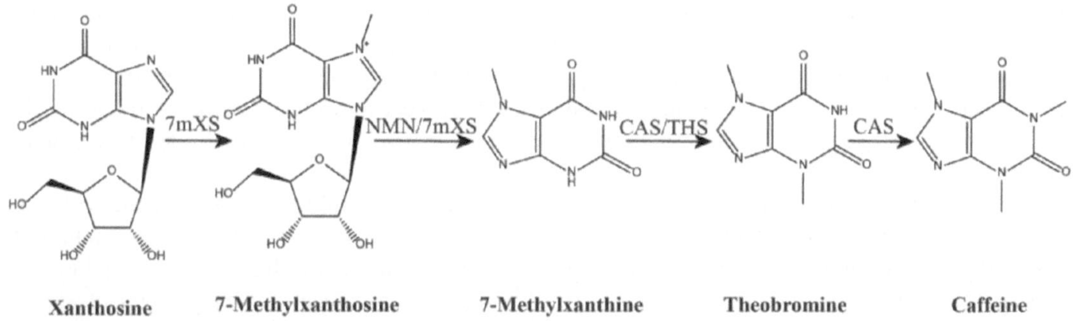

FIGURE 6.5 The biosynthesis pathway of caffeine, a representative of plants' purine alkaloids. A solid arrow represents known biosynthetic steps.

Abbreviations: 7mXS, 7-methylxanthosine synthase; NMN, *N*-methylnucleosidase; CAS, caffeine synthase; THS, theobromine synthase.

biosynthesis and an adenylate pool, guanylate pool, and SAM cofactor, acts as the substrate, which undergoes methylation in the presence of 7-methylxanthosine synthase to form *N*-methylxanthosine. Further, *N*-methylxanthosine forms the final product, caffeine, through the reactions catalyzed by *N*-methylnucleosidase, theobromine synthase, and caffeine synthase (Ashihara et al., 2020; Facchini, 2001; O'Connor, 2010). While the specific details about caffeine biosynthesis have been revealed, enormous details of the alluring purine alkaloid metabolism and biology remain elusive.

6.4 OMICS REVOLUTION: CATALYZING THE PATHWAY ELUCIDATION OF ALKALOIDS

To date, the information about alkaloids' biosynthetic and regulatory mechanisms is limited. Insufficient information about the genetic makeup, metabolite variants, and metabolism has been a major roadblock in elucidating the biosynthesis pathway of alkaloids in plants. Nevertheless, with the advances in technology and low cost for high-throughput sequencing, genomic and transcriptomic resources for diverse plant species have now been established that have fueled the field of phytochemistry in recent years. Similarly, remarkable advances in metabolomic techniques, sensitivity, computational tools, and databases have facilitated the characterization of the complete metabolome for the medicinal plants, the majority of which were previously unknown. The development of computational resources at the same pace has facilitated the processing of large datasets to unearth relevant biological interpretations. The four substantial dimensions of -omics research, including genomics, transcriptomics, metabolomics, and proteomics to measure the levels of the gene, transcripts, proteins, and metabolites, respectively, open the avenues to elucidate the biosynthesis and regulation of specialized metabolites. With the increasing number of studies reporting -omics datasets, their application in isolation of candidate genes involved in the biosynthesis and regulation of alkaloids will be exemplary.

6.4.1 Genomics

Plant genomes exhibit an impressive diversity in their complexity, size, and architecture and though poorly understood, are often linked with driving the chemodiversity. Since the first report on the genomic content of Arabidopsis in the year 2000, the number of plants being sequenced has increased rapidly in the last two decades. Therefore, the available genomic data resources have provided an unprecedented advantage to understand the biology of specialized metabolism in plants through comparative and phylogenomics approaches. Tohge et al. in 2014 used comparative genomics approaches and analyzed the genome sequences for 25 plant species to identify orthologous genes associated with the metabolism of alkaloids, terpenoids, and glucosinolates (Tohge et al., 2014). The authors suggested that comparative genomics can predict the metabo-type in a plant species.

Further, genome mining for an established genome helps identify genes associated with the biosynthetic machinery of important plant-derived natural products. Nowadays, plant biologists have shifted their attention from characterizing individual steps to validating the entire biosynthetic pathway based on genome prediction. Until recently, gene clusters were thought of as the specific characteristic of the prokaryotic genome, but genome mining studies have reported the presence of gene clusters related to the biosynthesis of specialized metabolites in diverse plants (Polturak & Osbourn, 2021; Rai, Saito et al., 2017). Tools including PhytoClust (Topfer et al., 2017), PlantClusterFinder (Schlapfer et al., 2017), and plantiSMASH (Kautsar et al., 2017) have recently been developed to predict the gene clusters based on gene length segment, co-expression pattern, and number of tandem repeats, among others. Over 30 secondary metabolite-associated gene clusters involved in the biosynthesis of specialized metabolites, including a thalianol gene cluster and α-solanine gene cluster, with a noscapine gene cluster from *Papaver somniferum* as the most prominent alkaloid gene cluster, have been reported in plant species (Polturak & Osbourn, 2021; Rai, Saito et al., 2017). These findings therefore highlight the importance of genomic resources and genome mining in the prediction and characterization of the biosynthetic pathway.

Genome-wide association studies (GWASs), a genome-facilitated approach, also represent a promising approach for understanding the pathway of interest. GWAS explores the variation in the natural populations and associates genetic variance with crucial traits (Rai et al., 2019; Tsugawa et al., 2021). This method is widely used to identify subsequent functional characterization of critical genes for the desired traits. So far, these studies have been primarily used for economically valuable crop plants, including rice (Huang et al., 2010; Li et al., 2020; Yano et al., 2016), maize (Hufford et al., 2012; Jiao et al., 2012; Lin et al., 2020), brassica (Lu et al., 2019), sesame (Dossa et al., 2019; He et al., 2019; Li et al., 2018a; Wei et al., 2015), and others (Berhe et al., 2021).

Several plant biologists are now increasingly focusing on achieving high-quality genomes for plants of commercial and medicinal value. Additionally, several consortia, for instance, the Earth Biogenome Project (Lewin et al., 2018) and the Darwin Tree of Life project (www.darwintreeoflife.org/), have attempted to sequence the whole genome of plants of diverse lineages. An exponential increase in whole-genome sequencing of the diverse plant species will establish a data-mining framework for elucidation of the alkaloids' biosynthetic pathways and undeniably open the way for novel discoveries.

6.4.2 Transcriptomics

Transcriptomics, qualitative and quantitative measurement of transcript expression, offers valuable insights into the active processes in a biological system at any given condition and/or time. Genes that are differentially expressed under different conditions and/or treatments often provide clues to understand the biosynthesis and regulation of specialized metabolism in plants. The first high-throughput quantification of transcripts was achieved through microarray technologies. Microarray-based profiling was extensively used primarily in plants with established genetic resources to link the transcripts and metabolic processes, leading to the functional characterization of several identified genes (Aharoni & Vorst, 2002; Parkinson et al., 2005; Rai, Saito, et al., 2017).

However, the inception of RNA sequencing catalyzed the transcriptome profiling of several non-model plant species. RNA sequencing does not require prior genetic information, is cost-effective, and the reads can be assembled without a reference genome. Therefore, it has been extensively used to establish genetic resources for plants of ecological and pharmacological importance. Initiatives, including the 1KP project (http://onekp.com/) (Matasci et al., 2014), which aims to establish transcriptome resources for over 1000 plants from diverse lineages; PhytoMetaSyn (Xiao et al., 2013); and the Medicinal Plant Transcriptome Project (http://apps.pharmacy.uic.edu/depts/pcrps/MedTranscriptomePlants/), among others, have performed sequencing for multiple tissues of several plants. These studies have established critical genetic resources, thus providing the framework for understanding the specialized metabolites' biosynthesis in plants.

Several studies have used transcriptome profiling to identify the biosynthetic genes involved in the specialized metabolism (Rai, Kamochi, et al., 2017; Rai, Nakamura et al., 2016; Rai et al., 2018; Rai, Yamazaki et al., 2016; Rai et al., 2017b; Sun et al., 2018). The RNA-sequencing based transcriptome assemblies for two *Podophyllum* species, *P. hexandrum* and *P. peltatum*, proposed the candidate transcripts involved in the biosynthesis of aporphine alkaloids, which were later confirmed through metabolome analysis (Marques et al., 2014, 2013). Similarly, a homology-based search of *Rauvolfia serpentina*'s root transcriptome data against the known γ-tocopherol–like *N*-methyltransferase facilitated the identification of two candidate genes involved in the indole and side-chain *N*-methylation of a monoterpenoid indole alkaloid, ajmaline (Cazares-Flores et al., 2016). In another study, RNA-sequencing data from *Catharanthus roseus* was used to identify and characterize the four critical enzymes involved in the biosynthesis of monoterpene indole alkaloids (Miettinen et al., 2014). Several other studies have reported the application of RNA-sequencing transcriptome and co-expression analysis to identify genes involved in the biosynthesis of steroid alkaloids and monoterpene indole alkaloids in *Fritillaria imperiali* (Eshaghi et al., 2019) and *Vinca minor* (Stander et al., 2020), respectively. These transcriptome-based studies have resulted in tremendous data being deposited to public databases, including National Center for Biotechnology Information (NCBI) and DNA Data Bank of Japan (DDBJ), facilitating comparative analyses between different plant species producing similar classes of specialized metabolites. A study reported by Pathak et al. (2013) used comparative transcriptome analyses between high-papaverine mutants of *Papaver somniferum* and identified candidate genes involved in the biosynthesis of papaverine. With an increasing number of transcriptome studies being undertaken for nonmodel plants, the comparative transcriptome analyses will drive our understanding of the biosynthesis and evolution of plant alkaloids.

6.4.3 Proteomics

The proteomics approach is based on high-throughput separation and identification of the proteins. The biosynthesis, modification, and regulation of protein maintain cellular homeostasis in any biological system. Knowledge about protein content, localization, protein-protein/RNA interaction, and post-translational modification facilitates understanding of cellular function and regulation. While proteomics has been widely applied to understand the plants' response under different stress conditions, it has not been used to a similar extent in understanding the specialized metabolism in nonmodel plants.

Recently, a surge in proteomics applications allowed the identification and quantification of almost all of the expressed protein due to improvements in the instrumentation, resolution, sample preparation, and bioinformatics approaches. Moreover, the wealth of genomics/transcriptomics data generated for increasing plants of diverse lineages has immensely contributed to developing a database for proteome analysis of nonmodel plants (Ahmad et al., 2016; Gong et al., 2015). Identifying proteins specific to the site of biosynthesis of specialized metabolites may provide important clues related to its biosynthesis and regulation.

Several studies have reported the successful application of proteomics to understand the biosynthesis of alkaloids. A combination of transcriptome and proteome analyses was used to propose components involved in alkaloid metabolism in opium poppy cell cultures (Desgagne-Penix et al.,

2010). Similarly, Oldham et al. (2010) conducted a shotgun proteomics approach on yeast elicited *Eschscholzia californica* suspension cultures, resulting in the identification of genes involved in the biosynthesis of benzophenanthridine alkaloids. In another study using the shotgun proteomics approach, Onoyovwe et al. (2013) revealed that the final three steps in morphine biosynthesis occur in the laticifers. While Jacobs et al. (2005), through proteome analyses of *C. roseus* culture cells, identified two isoforms of strictosidine synthases and a tryptophan synthase, representing key enzymes of monoterpene indole alkaloid biosynthesis, Champagne et al. (2012) identified 22 enzymes related to MIA biosynthesis. Decker et al. (2000), using proteome analysis of exuded latex of opium poppy, revealed several candidate genes involved in alkaloid biosynthesis, two of which were further functionally characterized by Ounaroon et al. (2003).

Previously, the lack of genetic resources hindered the proteome-based studies of plants' secondary metabolites. However, with a boom in technological advances and studies reporting the genome/transcriptome resources, proteomics represents a promising approach to identifying genes involved in alkaloids' biosynthetic machinery in the near future. Recently, a global platform to connect plant proteomics researchers has been established, the International Plant Proteomics Organization (IPPO, www.inppo.com), to help the proteomics field, now in the infant stage, prosper. Henceforward, we anticipate that several proteomics-based studies will be undertaken together with transcriptomes and/or metabolomes to understand plant alkaloids' biosynthesis.

6.4.4 Metabolomics

Over the last 10 years, metabolomics has been a principal component in functional genomics (Baran et al., 2009; Bino et al., 2004; Hall et al., 2002). Technological advances in the mass spectrometry methodology coupled with an impressive wealth of bioinformatics tools and available databases have been instrumental in our current understanding of plant chemo-diversity. The metabolomics-based approach generally is used in a top-down approach for natural products discovery, where the elicitor-mediated perturbation is introduced into the biological system to increase the production of specialized metabolites (Nakabayashi et al., 2015). Further, metabolite profiling of the samples gives important clues about these bioactive compounds' distribution and accumulation patterns. When coupled with the basics of chemistry, the results help us understand the putative intermediates in the biosynthetic pathway (Saito, 2013). Therefore, metabolome studies have a tremendous potential to reveal unmapped diversity present in plant alkaloids. The identification of metabolites, including paclitaxel, artemisinin, morphine, camptothecin, and vinblastine, among others, are a few prominent examples highlighting the importance of metabolomics studies. Metabolome profiling for diverse plant families has revealed the presence of about 2500–3500 diverse classes of metabolites, including alkaloids, iridoids, and secoiridoids (Rai, Saito et al., 2017). Hagel et al. (2015) established an important metabolic resource for studying BIA biosynthesis in plants by performing metabolome analysis of 20 taxonomically related BIA-producing plant species. Moreover, comparing metabolite profiles for loss- or gain-of-function of target genes with wild-type plants helps identify gene-specific metabolites, which helps predict the gene's function in the biosynthetic network (Tohge & Fernie, 2010). For instance, metabolome analysis coupled with a virus-induced gene silencing approach have been widely employed to study the functional characterization of several key enzymes involved in the biosynthesis of benzylisoquinoline alkaloids (Allen et al., 2008; Dang et al., 2012; Desgagne-Penix & Facchini, 2012; Farrow & Facchini, 2013; Hagel & Facchini, 2013).

The isotope labeling approach in metabolomics has been widely used to provide important cues of the precursor molecule and the intermediates, thus facilitating the elucidation of the biosynthetic pathway for the metabolite of interest. Yamazaki et al. (2004) investigated the biosynthesis of camptothecin in the hairy roots of *Oamia pumila* using ^{13}C-labeled glucose. The metabolome analysis revealed that camptothecin was derived from the 2-C-methyl-D-erythritol-4-phosphate (MEP) and shikimate pathways. Similarly, the metabolite labeling strategy across the RNAi and wild-type lines showed an alternative secoiridoids biosynthesis pathway in *Camptotheca acuminata*, where strictosidinic acid is formed instead of strictosidine (Sadre et al., 2016). In another study, Bunsupa et al. (2014), using ^{15}N-labeled lysine, revealed important insights into the anabasine biosynthetic pathway in tobacco. Tsugawa et al. (2019) used complete ^{13}C-based labeling in 12 plant species and reported 1092 structures and 344 formulae to 3604 carbon-containing metabolite ions, 67 of which were unreported earlier. Similarly, a metabolome profiling-based study by Rai et al. (2021a) using the complete ^{15}N-based stable isotope-labeled hairy roots in *O. pumila* reported 273 nitrogen-containing metabolites, including several monoterpene indole alkaloids. Several comprehensive metabolite profiling studies using multiple tissues and plant species have resulted in developing extensive metabolome datasets and continue to grow further.

With the increase in genomic sequences, recent years have seen the application of metabolic genome-wide association studies (mGWASs), which combine metabolomics datasets with the comparative genomics approaches to uncover genetic determinants associated with plant diversity. Initially, mGWAS was applied to the model species *Arabidopsis* to understand the metabolic diversity caused by genetic variations across accessions and was further extended to crop plants (Luo, 2015; Rai et al., 2019). Sauvage et al. (2014), using 163 tomato accessions, identified 44 loci associated with 19 fruit-associated metabolic traits. In another study, Wu et al. (2018) used 309 accessions of *Arabidopsis* grown under control and stress conditions and performed untargeted metabolomics-based GWAS to identify 70 critical associations between genes and metabolites, of which 8 novel associations were experimentally validated. mGWAS has been used to associate genes with both primary and secondary metabolites.

Recently, imaging mass spectrometry (IMS) has gained much attention, as it allows spatial mapping of metabolites within plant tissues and identifying the region for high and low accumulation of metabolites of interest (Yamamoto et al., 2019, 2016). When coupled with transcriptome datasets, these results will allow for identifying region-specific gene-metabolite associations, thus providing essential clues to the biosynthetic pathway. Matrix-associated laser desorption ionization (MALDI) and secondary ion mass spectroscopy (SIMS) imaging techniques, with a spatial resolution of 5–50 μm and 50 nm, respectively, are the most common types of IMS used (Buchberger et al., 2018; Nakabayashi et al., 2021). While IMS is not yet extensively applied to plants, it holds a promising approach to expanding our understanding of plants' specialized metabolism.

Several tools are increasingly available to analyze the comprehensive metabolome datasets generated by high-throughput metabolite profiling. Open-source metabolomics tools, including MS-DIAL (Tsugawa et al., 2015), MS-FINDER (Tsugawa et al., 2016), MZmine3 (Pluskal et al., 2010), and XCMS (Mahieu et al., 2016; Smith et al., 2006; Tautenhahn et al., 2012), have allowed automatic algorithms for extraction of mass features, peak alignment across different replicates and/or samples, metabolite annotation, and its quantification. Together with the expanding databases, these tools have stimulated the field of functional genomics, where novel metabolic networks can be inferred from the association between genes and bioactive metabolites.

6.5 MULTIOMICS ANALYSIS: A POWERFUL TOOL TO DECRYPT BIOLOGY FROM HUGE DATASETS

The four major dimensions of -omics research—genomics, transcriptomics, metabolomics, and proteomics—are interrelated. The biosynthesis of proteins occurs in two steps: the genetic information is transferred to messenger RNA (mRNA), which then is translated to proteins. The proteins participate in the metabolite biosynthetic pathways, leading to huge metabolite diversity. In this perspective, high-throughput -omics data and its analysis provide us with essential, but incomplete, details. Therefore, integrating several individual -omics datasets offers a promising approach to get a holistic view of the biosynthesis and regulation of the compound of interest (Figure 6.6).

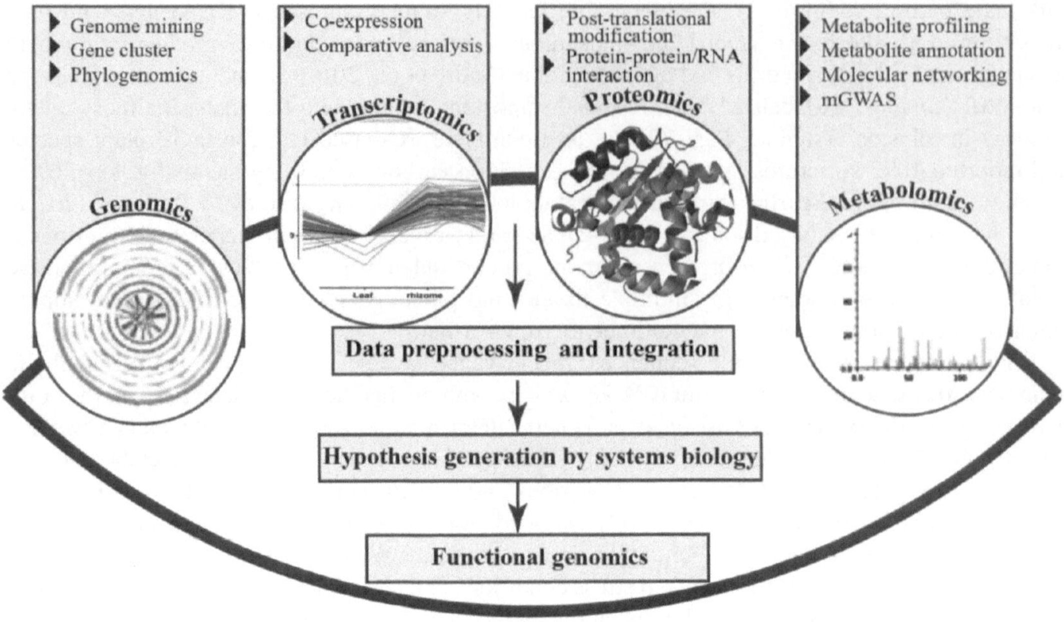

FIGURE 6.6 Integrative multiomics workflow for functional genomics.

A few of the initial attempts to integrate the -omics datasets were reported by Urbanczyk-Wochniak et al. (2003) and Hirai et al. (2004) in potato and *Arabidopsis*, respectively. Through the integrated analysis of transcriptome and metabolome data, the authors proposed the use of integrative -omics as an effective tool for hypothesis generation and subsequent functional genomics. Since then, several studies have used multiomics strategies to prioritize genes for further functional characterization. Liscombe et al. (2009) used a combination of targeted alkaloids' metabolite profiles and expressed sequence tag (EST) libraries to identify and functionally characterize *N*-methyltransferase, the first-ever enzyme discovered from the pavine alkaloid biosynthetic pathway. Similarly, the comprehensive transcriptome and metabolome profiling and its integration resulted in the functional characterization of several enzymes from the morphine biosynthetic pathway (Hagel & Facchini, 2013). To understand the biosynthesis of MIAs and anthraquinones in *O. pumila*, a complete transcriptome and metabolome analysis of *O. pumila* hairy roots led to identifying several candidate genes and transcription factors involved in the biosynthesis and regulation of camptothecin, an anticancer MIA alkaloid (Rohani et al., 2016; Udomsom et al., 2016; Yamazaki et al., 2013). Moreover, a combination of the transcriptome and metabolome datasets has resulted in the identification of genes encoding several enzymes, including salutaridine synthase (Gesell et al., 2009), norreticuline 7-*O*-methyltransferase (Pienkny et al., 2009), and noscapine synthase (Chen & Facchini, 2014) from the BIA biosynthesis pathway.

While the integration of multiomics datasets has been widely studied in animals, humans, and microbes, its use to study plants' biosynthetic pathways is relatively challenging due to diverse metabolic content, nonavailability of genomic resources, and poorly annotated genomes. However, with the rapid increase in the number of high-throughput studies being undertaken, the next step is to select a suitable methodology to integrate these datasets for accurately inferring the biological significance. There are three levels of integrating the multiomics data: element-based approach, pathway-based approach, and mathematical model-based approach.

6.5.1 Element-Based Approach

The element-based approach is among the most popular means used for integrative -omics. It relies on associations between elements of individual -omics datasets, thus offering an unbiased way forward for functional genomics. This approach can be further subclassified into correlation, clustering, and multivariate analyses (Jamil et al., 2020).

Correlation analysis represents one of the most popular methods of multiomics integration because of its simplicity. It involves estimating the correlative relationship between different -omics datasets using Pearson or Spearman correlation coefficients that measure linear or ranked associations. The positive or negative correlation coefficient value represents a strong or weak relationship between data. Several studies have reported the use of correlation analysis to propose the biosynthetic genes for several classes of secondary metabolites (Higashi & Saito, 2013; Rai et al., 2020). The study by Rischer et al. (2006) represents one of the earliest attempts to use correlation analysis between the expression data of gene tags and the accumulation profile of metabolite peaks to understand the biosynthesis of terpenoid indole alkaloids. This analysis method is relatively straightforward and assumes that differential expression of transcripts will also be observed at their protein and/or metabolite levels, negating the time lapse between transcript expression, protein modification, and thereby changes in the associated metabolites. Therefore, the correlation observed between different datasets could sometimes be ambiguous in this context.

Clustering analysis relies on concatenating -omics datasets with similar features into a single matrix. Further, machine learning algorithms, including *k*-means clustering and random forest, are used to identify the relationship between biological entities (Jamil et al., 2020). While *k*-means clustering groups the dataset into a distinct cluster based on the different trends observed in the data, a random forest defines a group of either genes, proteins, or metabolites based on previous training datasets to associate them with a specific characteristic (Ma et al., 2014; Silva et al., 2019). This analysis has been used successfully in several studies focusing on integrative -omics in plants (Acharjee et al., 2016; Keller et al., 2018; Wang et al., 2016, 2017). However, the significant difference in the dimensions of different -omics datasets and different pre-processing requirements for each of them are a few of the challenges faced by the clustering analysis (Rai, Saito et al., 2017).

The multivariate analysis represents the most appropriate method for the element-based integration approach since it allows more metadata analysis and experimental design flexibility (Rai, Saito et al., 2017). Using this approach, the researchers can predict the trends of datasets together with their variance or covariance associations (Meng et al., 2014). Moreover, the topological network showcasing the dynamic relationship between the transcript, protein, and metabolite can be elucidated (Weckwerth, 2019). Among several techniques of multivariate analysis, principal component analysis (PCA), partial least squares (PLS), and orthogonal projection to latent structures discriminant analysis (OPLS-DA) are the most common ones (Rai, Saito et al., 2017). Several studies to date have reported the use of multivariate data analysis in plant integrative -omics (De Abreu et al., 2018; Mesnage et al., 2016; Uarrota et al., 2019).

6.5.2 Pathway-Based Approach

The pathway-based approach represents the knowledge-based approach and is further subclassified into pathway mapping and co-expression analysis (Jamil et al., 2020).

Pathway mapping helps map different -omics datasets to the existing metabolic database containing important information regarding pathway annotation, including Kyoto Encyclopedia of Genes and Genomes (KEGG, www.genome.jp/kegg/), AraCyc (www.arabidopsis.org/biocyc/), CitrusCyc (www.citrusgenomedb.org/node/1136703), SolCyc (https://solgenomics.net/tools/solcyc/index.pl), and others. Software tools, including MapMan (Thimm et al., 2004) and PathVisio (van Iersel et al., 2008), have successfully been used to integrate plant-derived -omics datasets. While MapMan was used by López-Hidalgo et al. (2018) to integrate the transcriptome, proteome, and metabolome data from *Quercus ilex*, Bjornson et al. used PathVisio to study the *Arabidopsis* signaling mutant plants (2017). Although time-consuming, several studies have reported the manual reconstruction of canonical pathways from the KEGG database for soybean (Kang et al., 2018; Zhu et al., 2018) and sandalwood (Mahesh et al., 2018), based on the enzyme and/or metabolite annotation. These studies show the importance of pathway mapping in deciphering the inherent modulation of the plant biosynthetic pathways. However, the pathway mapping approach for the nonmodel plant species may fall off owing to the deficit in the annotated pathway information, thus jeopardizing the identification of associations between different biological entities.

Co-expression analysis depends on the correlations between different -omics datasets described earlier, which are then further reorganized into a weighted network. These weighted networks could be visualized using tools like Weighted Gene Co-expression Network Analysis (WGCNA) and Cytoscape, thus revealing significant clusters, hubs, and modules for subsequent biological insights into the biosynthesis pathways. In a recent study, Rai et al. (2021b) used the WGCNA approach to propose genes associated with rutin biosynthesis and associated metabolites in *Mallotus japonicus*. Another type of co-expression analysis using its integration with the pathway database has been reported for *Citrus sinensis* (He et al., 2018). In this study, He et al. used the CitrusCyc database to define pathways that showed the highest correlation between the transcript and metabolite data. Networks were drawn using the upregulated and downregulated metabolites and transcripts between a mutant variety of "Newhall" orange resistant to various fungi and a wild-type variety. Results suggested that the fungi tolerance was observed due to compositional changes in the fatty acid, thereby inducing the jasmonic acid-mediated defense response. These studies highlight the importance of integrating co-expression analysis with pathway databases to organize the -omics datasets into distinct hubs/modules, which could be pursued for further functional characterization. Therefore, a pathway-based approach could undeniably be used to integrate the plant -omics datasets to find the associations between the biosynthesis and regulation of metabolites.

6.5.3 Mathematical Model-Based Approach

The mathematical model–based approach is the most complex method of data integration. It requires thorough coverage of -omics datasets and well-characterized plants to develop a well-defined metabolic model. This approach is further divided into two subclasses: differential analysis and genome-scale analysis (Jamil et al., 2020; Rai, Saito et al., 2017)

Differential analysis to date has been reported in several fruit-related studies in plants and is divided into targeted or nontargeted pathway studies (Koc et al., 2018; Wang et al., 2018). A recently reported differential equation estimating the protein density during the ripening of tomatos is one of the examples of nontargeted pathway analysis (Belouah et al., 2019). In this study, the authors integrated transcriptome and proteome datasets obtained from nine developmental stages of tomato to acquire the rate of translation and degradation. Results suggested that the equation could predict the expression of about 50% of transcript-protein pairs. In the targeted pathway analysis approach, differential analysis is used to model a specific pathway. This approach has been successfully used in several studies. For instance, the biosynthesis of anthocyanin in *Vitis vinifera* was successfully modeled using flux-balance analysis through the metabolomics and proteomics datasets (Soubeyrand et al., 2018). Several resources, including E-flux (https://omictools.com/e-flux-tool) and Metabolic Adjustment by Differential Expression (https://omictools.com/made-tool), are available for the metabolic and flux analyses.

While differential analysis is a top-down approach, genome-scale analysis is a bottom-up approach, in which a genome-scale model is built using thorough curation prior to experimental validation (Thiele & Palsson, 2010; Voit, 2017). In this approach, each of the reactions is considered for a holistic view of mathematical evaluation (Gomes de Oliveira Dal'Molin & Nielsen, 2018). The genome-scale analysis is complicated in eukaryotes due to polyploidy, compartmentalization, and size. The PlantSEED database (https://modelseed.org/), a streamlined database, has been established to reconstruct plant metabolic networks to annotate reactions in the new plants based on the 10 well-annotated plants (Seaver et al., 2018). PlantSEED allows plant biologists to investigate metabolic processes, as was used by Scheunemann et al. (2018) to predict the metabolic flux of indole-3-acetic acid. Genome-scale analysis has also been reported in *Glycine max* (Li et al., 2015) and *Brassica napus* (Hay et al., 2014) using transcriptomics and metabolomics datasets. With a rapid increase in -omics–based studies and its integration, the metabolic model construction and/or reconstruction for plants producing an important class of secondary metabolites could be seen soon.

6.6 CONCLUSIONS AND FUTURE PERSPECTIVES

Alkaloids represent a diverse range of plants' secondary metabolites that possess various biological and pharmacological properties. While the ability of plants to synthesize such structurally diverse metabolites has been studied for more than a century, relatively little is known about the molecular components involved in their biosynthesis. There is a huge demand for plant-derived alkaloids because of their immense potential in the food and pharmacological industry. Since plants produce these alkaloids at very low concentrations, the increased demand has led to the worldwide unregulated cultivation and extensive resource exploitation, which pose serious threats to the existent ecological balance. The impact of global climate change and specific growth condition requirements of the plants make the traditional approach of acquiring plant-derived alkaloids unsustainable. Modern biotechnological advancements, including in-plant targeted genetic manipulation and synthetic biology approaches, therefore offer a great alternative for sustainable production of these alkaloids. However, these approaches require a complete understanding of the alkaloid's biosynthesis and regulation. The current understanding of major classes of alkaloid biosynthesis is limited primarily to the upstream pathway, including the biosynthesis of the skeletal structure.

To date, the synthetic biology and metabolic engineering studies are not extensive, owing to the limited understanding of their biosynthetic network. Nevertheless, the wide application of the -omics platform and the integrative -omics approach being undertaken will be useful to elucidate the elucidation of alkaloids' biosynthetic pathways. The -omics analysis and its integration will enhance our ability to identify numerous genes and/or transcription factors involved in the biosynthesis of several specialized metabolites instead of a labor-extensive method of relying on the identification of a single gene. These approaches have already been successfully used for elucidating the biosynthesis of several plant-derived metabolites, although their application to alkaloid metabolism is quite limited. The rapid advances in the generation of high-throughput -omics technologies, low data acquisition cost, and bioinformatics and computational tools for data handling and integration of -omics datasets offer a promising approach for the elucidation and metabolic engineering of the alkaloids' biosynthesis pathways. With an explosion in the number of plant biologists working on the -omics technologies, a structured deposition of their experimental datasets is the need of the hour. This will allow the reusability of the data to uncover the inherent novelty of the biosynthetic pathways and the construction of a metabolic model.

BIBILIOGRAPHY

Acharjee, A., Kloosterman, B., Visser, R. G., & Maliepaard, C. (2016). Integration of multi-omics data for prediction of phenotypic traits using random forest. *BMC Bioinformatics*, *17*(Suppl 5), 180. https://doi.org/10.1186/s12859-016-1043-4

Aharoni, A., & Vorst, O. (2002). DNA microarrays for functional plant genomics. *Plant Mol Biol*, *48*(1–2), 99–118. https://doi.org/10.1023/a:1013734019946

Ahmad, P., Abdel Latef, A. A., Rasool, S., Akram, N. A., Ashraf, M., & Gucel, S. (2016). Role of proteomics in crop stress tolerance. *Front Plant Sci*, *7*, 1336. https://doi.org/10.3389/fpls.2016.01336

Allen, R. S., Miller, J. A., Chitty, J. A., Fist, A. J., Gerlach, W. L., & Larkin, P. J. (2008). Metabolic engineering of morphinan alkaloids by over-expression and RNAi suppression of salutaridinol 7-O-acetyltransferase in opium poppy. *Plant Biotechnol J*, *6*(1), 22–30. https://doi.org/10.1111/j.1467-7652.2007.00293.x

Arakawa, H., Clark, W. G., Psenak, M., & Coscia, C. J. (1992). Purification and characterization of dihydrobenzophenanthridine oxidase from elicited Sanguinaria canadensis cell cultures. *Arch Biochem Biophys*, *299*(1), 1–7. https://doi.org/10.1016/0003-9861(92)90236-p

Ashihara, H., Ludwig, I. A., & Crozier, A. (2020). Biosynthesis of purine alkaloids. In *Plant Nucleotide Metabolism - Biosynthesis, Degradation, and Alkaloid Formation* (pp. 231–258). John Wiley & Sons Ltd. https://doi.org/10.1002/9781119476139.ch15

Ashihara, H., Sano, H., & Crozier, A. (2008). Caffeine and related purine alkaloids: Biosynthesis, catabolism, function and genetic engineering. *Phytochemistry*, *69*(4), 841–856. https://doi.org/10.1016/j.phytochem.2007.10.029

Ashihara, H., Yokota, T., & Crozier, A. (2013). Chapter four—biosynthesis and catabolism of purine alkaloids. In N. Giglioli-Guivarc'h (Ed.), *Advances in Botanical Research* (Vol. 68, pp. 111–138). Academic Press. https://doi.org/https://doi.org/10.1016/B978-0-12-408061-4.00004-3

Baran, R., Reindl, W., & Northen, T. R. (2009). Mass spectrometry based metabolomics and enzymatic assays for functional genomics. *Curr Opin Microbiol*, *12*(5), 547–552. https://doi.org/10.1016/j.mib.2009.07.004

Beaudoin, G. A., & Facchini, P. J. (2013). Isolation and characterization of a cDNA encoding (S)-cis-N-methylstylopine 14-hydroxylase from opium poppy, a key enzyme in sanguinarine biosynthesis. *Biochem Biophys Res Commun*, *431*(3), 597–603. https://doi.org/10.1016/j.bbrc.2012.12.129

Bedewitz, M. A., Jones, A. D., D'Auria, J. C., & Barry, C. S. (2018). Tropinone synthesis via an atypical polyketide synthase and P450-mediated cyclization. *Nat Commun*, *9*(1), 5281. https://doi.org/10.1038/s41467-018-07671-3

Belouah, I., Nazaret, C., Petriacq, P., Prigent, S., Benard, C., Mengin, V., Blein-Nicolas, M., Denton, A. K., Balliau, T., Auge, S., Bouchez, O., Mazat, J. P., Stitt, M., Usadel, B., Zivy, M., Beauvoit, B., Gibon, Y., & Colombie, S. (2019). Modeling protein destiny in developing fruit. *Plant Physiol*, *180*(3), 1709–1724. https://doi.org/10.1104/pp.19.00086

Berhe, M., Dossa, K., You, J., Mboup, P. A., Diallo, I. N., Diouf, D., Zhang, X., & Wang, L. (2021). Genome-wide association study and its applications in the non-model crop Sesamum indicum. *BMC Plant Biol*, *21*(1), 283. https://doi.org/10.1186/s12870-021-03046-x

Bhambhani, S., Kondhare, K. R., & Giri, A. P. (2021). Diversity in chemical structures and biological properties of plant alkaloids. *Molecules*, *26*(11). https://doi.org/10.3390/molecules26113374

Bino, R. J., Hall, R. D., Fiehn, O., Kopka, J., Saito, K., Draper, J., Nikolau, B. J., Mendes, P., Roessner-Tunali, U., Beale, M. H., Trethewey, R. N., Lange, B. M., Wurtele, E. S., & Sumner, L. W. (2004). Potential of metabolomics as a functional genomics tool. *Trends Plant Sci*, *9*(9), 418–425. https://doi.org/10.1016/j.tplants.2004.07.004

Bjornson, M., Balcke, G. U., Xiao, Y., de Souza, A., Wang, J. Z., Zhabinskaya, D., Tagkopoulos, I., Tissier, A., & Dehesh, K. (2017). Integrated omics analyses of retrograde signaling mutant delineate interrelated stress-response strata. *Plant J*, *91*(1), 70–84. https://doi.org/10.1111/tpj.13547

Buchberger, A. R., DeLaney, K., Johnson, J., & Li, L. (2018). Mass spectrometry imaging: A review of emerging advancements and future insights. *Anal Chem*, *90*(1), 240–265. https://doi.org/10.1021/acs.analchem.7b04733

Bunsupa, S., Komastsu, K., Nakabayashi, R., Saito, K., & Yamazaki, M. (2014). Revisiting anabasine biosynthesis in tobacco hairy roots expressing plant lysine decarboxylase gene by using ^{15}N-labeled lysine. *Plant Biotechnology*, *31*(5), 511–518. https://doi.org/10.5511/plantbiotechnology.14.1008a

Caputi, L., Franke, J., Farrow, S. C., Chung, K., Payne, R. M. E., Nguyen, T. D., Dang, T. T., Soares Teto Carqueijeiro, I., Koudounas, K., Duge de Bernonville, T., Ameyaw, B., Jones, D. M., Vieira, I. J. C., Courdavault, V., & O'Connor, S. E. (2018). Missing enzymes in the biosynthesis of the anticancer drug vinblastine in Madagascar periwinkle. *Science*, *360*(6394), 1235–1239. https://doi.org/10.1126/science.aat4100

Cazares-Flores, P., Levac, D., & De Luca, V. (2016). Rauvolfia serpentina N-methyltransferases involved in ajmaline and Nbeta-methylajmaline biosynthesis belong to a gene family derived from gamma-tocopherol C-methyltransferase. *Plant J*, *87*(4), 335–342. https://doi.org/10.1111/tpj.13186

Champagne, A., Rischer, H., Oksman-Caldentey, K. M., & Boutry, M. (2012). In-depth proteome mining of cultured catharanthus roseus cells identifies candidate proteins involved in the synthesis and transport of secondary metabolites. *Proteomics*, *12*(23–24), 3536–3547. https://doi.org/10.1002/pmic.201200218

Chen, X., & Facchini, P. J. (2014). Short-chain dehydrogenase/reductase catalyzing the final step of noscapine biosynthesis is localized to laticifers in opium poppy. *Plant J*, *77*(2), 173–184. https://doi.org/10.1111/tpj.12379

Chen, X., Hagel, J. M., Chang, L., Tucker, J. E., Shiigi, S. A., Yelpaala, Y., Chen, H. Y., Estrada, R., Colbeck, J., Enquist-Newman, M., Ibanez, A. B., Cottarel, G., Vidanes, G. M., & Facchini, P. J. (2018). A pathogenesis-related 10 protein catalyzes the final step in thebaine biosynthesis. *Nat Chem Biol*, *14*(7), 738–743. https://doi.org/10.1038/s41589-018-0059-7

Chung, H.-H., Sung, Y.-C., & Shyur, L.-F. (2016). Deciphering the biosynthetic pathways of bioactive compounds in planta using omics approaches. In H.-S. Tsay, L.-F. Shyur, D. C. Agrawal, Y.-C. Wu, & S.-Y. Wang (Eds.), *Medicinal Plants—Recent Advances in Research and Development* (pp. 129–165). Springer Singapore. https://doi.org/10.1007/978-981-10-1085-9_5

Cordell, G. A., & Choi, T. (2011). Alkaloids and their biosynthesis. *Phytochemistry and Pharmacognosy section, Chemical Sciences, Engineering and Technology Resources, Encyclopedia of Life Support Systems (EOLSS), Developed under the Auspices of the UNESCO.*

Cordell, G. A., Quinn-Beattie, M. L., & Farnsworth, N. R. (2001). The potential of alkaloids in drug discovery. *Phytother Res*, *15*(3), 183–205. https://doi.org/10.1002/ptr.890

Dang, T. T., Chen, X., & Facchini, P. J. (2015). Acetylation serves as a protective group in noscapine biosynthesis in opium poppy. *Nat Chem Biol*, *11*(2), 104–106. https://doi.org/10.1038/nchembio.1717

Dang, T. T., & Facchini, P. J. (2012). Characterization of three O-methyltransferases involved in noscapine biosynthesis in opium poppy. *Plant Physiol*, *159*(2), 618–631. https://doi.org/10.1104/pp.112.194886

Dang, T. T., & Facchini, P. J. (2014). CYP82Y1 is N-methylcanadine 1-hydroxylase, a key noscapine biosynthetic enzyme in opium poppy. *J Biol Chem*, *289*(4), 2013–2026. https://doi.org/10.1074/jbc.M113.505099

Dang, T. T., Onoyovwi, A., Farrow, S. C., & Facchini, P. J. (2012). Biochemical genomics for gene discovery in benzylisoquinoline alkaloid biosynthesis in opium poppy and related species. *Methods Enzymol*, *515*, 231–266. https://doi.org/10.1016/B978-0-12-394290-6.00011-2

De Abreu, E. L. F., Li, K., Wen, W., Yan, J., Nikoloski, Z., Willmitzer, L., & Brotman, Y. (2018). Unraveling lipid metabolism in maize with time-resolved multi-omics data. *Plant J*, *93*(6), 1102–1115. https://doi.org/10.1111/tpj.13833

De Carolis, E., & De Luca, V. (1993). Purification, characterization, and kinetic analysis of a 2-oxoglutarate-dependent dioxygenase involved in vindoline biosynthesis from Catharanthus roseus. *J Biol Chem*, *268*(8), 5504–5511. www.ncbi.nlm.nih.gov/pubmed/8449913

Decker, G., Wanner, G., Zenk, M. H., & Lottspeich, F. (2000). Characterization of proteins in latex of the opium poppy (Papaver somniferum) using two-dimensional gel electrophoresis and microsequencing. *Electrophoresis*, *21*(16), 3500–3516. https://doi.org/10.1002/1522-2683(20001001)21:16<3500::AID-ELPS3500>3.0.CO;2-O

Desgagne-Penix, I., & Facchini, P. J. (2012). Systematic silencing of benzylisoquinoline alkaloid biosynthetic genes reveals the major route to papaverine in opium poppy. *Plant J*, *72*(2), 331–344. https://doi.org/10.1111/j.1365-313X.2012.05084.x

Desgagne-Penix, I., Khan, M. F., Schriemer, D. C., Cram, D., Nowak, J., & Facchini, P. J. (2010). Integration of deep transcriptome and proteome analyses reveals the components of alkaloid metabolism in opium poppy cell cultures. *BMC Plant Biol*, *10*, 252. https://doi.org/10.1186/1471-2229-10-252

Diaz Chavez, M. L., Rolf, M., Gesell, A., & Kutchan, T. M. (2011). Characterization of two methylenedioxy bridge-forming cytochrome P450-dependent enzymes of alkaloid formation in the Mexican prickly poppy Argemone mexicana. *Arch Biochem Biophys*, *507*(1), 186–193. https://doi.org/10.1016/j.abb.2010.11.016

Dossa, K., Li, D., Zhou, R., Yu, J., Wang, L., Zhang, Y., You, J., Liu, A., Mmadi, M. A., Fonceka, D., Diouf, D., Cisse, N., Wei, X., & Zhang, X. (2019). The genetic basis of drought tolerance in the high oil crop Sesamum indicum. *Plant Biotechnol J*, *17*(9), 1788–1803. https://doi.org/10.1111/pbi.13100

Eshaghi, M., Shiran, B., Fallahi, H., Ravash, R., & Deri, B. B. (2019). Identification of genes involved in steroid alkaloid biosynthesis in Fritillaria imperialis via de novo transcriptomics. *Genomics*, *111*(6), 1360–1372. https://doi.org/10.1016/j.ygeno.2018.09.008

Facchini, P. J. (2001). Alkaloid biosynthesis in plants: Biochemistry, cell biology, molecular regulation, and metabolic engineering applications. *Annu Rev Plant Physiol Plant Mol Biol*, *52*, 29–66. https://doi.org/10.1146/annurev.arplant.52.1.29

Farrow, S. C., & Facchini, P. J. (2013). Dioxygenases catalyze O-demethylation and O, O-demethylenation with widespread roles in benzylisoquinoline alkaloid metabolism in opium poppy. *J Biol Chem*, *288*(40), 28997–29012. https://doi.org/10.1074/jbc.M113.488585

Farrow, S. C., Hagel, J. M., Beaudoin, G. A., Burns, D. C., & Facchini, P. J. (2015). Stereochemical inversion of (S)-reticuline by a cytochrome P450 fusion in opium poppy. *Nat Chem Biol*, *11*(9), 728–732. https://doi.org/10.1038/nchembio.1879

Fossati, E., Ekins, A., Narcross, L., Zhu, Y., Falgueyret, J. P., Beaudoin, G. A., Facchini, P. J., & Martin, V. J. (2014). Reconstitution of a 10-gene pathway for synthesis of the plant alkaloid dihydrosanguinarine in Saccharomyces cerevisiae. *Nat Commun*, *5*, 3283. https://doi.org/10.1038/ncomms4283

Galanie, S., Thodey, K., Trenchard, I. J., Filsinger Interrante, M., & Smolke, C. D. (2015). Complete biosynthesis of opioids in yeast. *Science*, *349*(6252), 1095–1100. https://doi.org/10.1126/science.aac9373

Geerlings, A., Ibanez, M. M., Memelink, J., van Der Heijden, R., & Verpoorte, R. (2000). Molecular cloning and analysis of strictosidine beta-D-glucosidase, an enzyme in terpenoid indole alkaloid biosynthesis in Catharanthus roseus. *J Biol Chem*, *275*(5), 3051–3056. https://doi.org/10.1074/jbc.275.5.3051

Gesell, A., Rolf, M., Ziegler, J., Diaz Chavez, M. L., Huang, F. C., & Kutchan, T. M. (2009). CYP719B1 is salutaridine synthase, the C-C phenol-coupling enzyme of morphine biosynthesis in opium poppy. *J Biol Chem*, *284*(36), 24432–24442. https://doi.org/10.1074/jbc.M109.033373

Gomes de Oliveira Dal'Molin, C., & Nielsen, L. K. (2018). Plant genome-scale reconstruction: From single cell to multi-tissue modelling and omics analyses. *Curr Opin Biotechnol*, *49*, 42–48. https://doi.org/10.1016/j.copbio.2017.07.009

Gong, F., Hu, X., & Wang, W. (2015). Proteomic analysis of crop plants under abiotic stress conditions: Where to focus our research? *Front Plant Sci*, *6*, 418. https://doi.org/10.3389/fpls.2015.00418

Goyal, S. (2013). Ecological role of alkaloids. In K. G. Ramawat & J.-M. Mérillon (Eds.), *Natural Products: Phytochemistry, Botany and Metabolism of Alkaloids, Phenolics and Terpenes* (pp. 149–171). Springer Berlin Heidelberg. https://doi.org/10.1007/978-3-642-22144-6_98

Grothe, T., Lenz, R., & Kutchan, T. M. (2001). Molecular characterization of the salutaridinol 7-O-acetyltransferase involved in morphine biosynthesis in opium poppy Papaver somniferum. *J Biol Chem*, *276*(33), 30717–30723. https://doi.org/10.1074/jbc.M102688200

Hagel, J. M., & Facchini, P. J. (2010). Dioxygenases catalyze the O-demethylation steps of morphine biosynthesis in opium poppy. *Nat Chem Biol*, *6*(4), 273–275. https://doi.org/10.1038/nchembio.317

Hagel, J. M., & Facchini, P. J. (2013). Benzylisoquinoline alkaloid metabolism: A century of discovery and a brave new world. *Plant Cell Physiol*, *54*(5), 647–672. https://doi.org/10.1093/pcp/pct020

Hagel, J. M., Mandal, R., Han, B., Han, J., Dinsmore, D. R., Borchers, C. H., Wishart, D. S., & Facchini, P. J. (2015). Metabolome analysis of 20 taxonomically related benzylisoquinoline alkaloid-producing plants. *BMC Plant Biol*, *15*, 220. https://doi.org/10.1186/s12870-015-0594-2

Hall, R., Beale, M., Fiehn, O., Hardy, N., Sumner, L., & Bino, R. (2002). Plant metabolomics: The missing link in functional genomics strategies. *Plant Cell*, *14*(7), 1437–1440. https://doi.org/10.1105/tpc.140720

Hashimoto, T., Matsuda, J., & Yamada, Y. (1993). Two-step epoxidation of hyoscyamine to scopolamine is catalyzed by bifunctional hyoscyamine 6 beta-hydroxylase. *FEBS Lett*, *329*(1–2), 35–39. https://doi.org/10.1016/0014-5793(93)80187-y

Hashimoto, T., Nakajima, K., Ongena, G., & Yamada, Y. (1992). Two tropinone reductases with distinct stereospecificities from cultured roots of hyoscyamus niger. *Plant Physiol*, *100*(2), 836–845. https://doi.org/10.1104/pp.100.2.836

Hay, J. O., Shi, H., Heinzel, N., Hebbelmann, I., Rolletschek, H., & Schwender, J. (2014). Integration of a constraint-based metabolic model of Brassica napus developing seeds with (13)C-metabolic flux analysis. *Front Plant Sci*, *5*, 724. https://doi.org/10.3389/fpls.2014.00724

He, Q., Xu, F., Min, M. H., Chu, S. H., Kim, K. W., & Park, Y. J. (2019). Genome-wide association study of vitamin E using genotyping by sequencing in sesame (Sesamum indicum). *Genes Genomics*, *41*(9), 1085–1093. https://doi.org/10.1007/s13258-019-00837-3

He, Y., Han, J., Liu, R., Ding, Y., Wang, J., Sun, L., Yang, X., Zeng, Y., Wen, W., Xu, J., Zhang, H., Yan, X., Chen, Z., Gu, Z., Chen, H., Tang, H., Deng, X., & Cheng, Y. (2018). Integrated transcriptomic and metabolomic analyses of a wax deficient citrus mutant exhibiting jasmonic acid-mediated defense against fungal pathogens. *Hortic Res*, *5*, 43. https://doi.org/10.1038/s41438-018-0051-0

Hibi, N., Higashiguchi, S., Hashimoto, T., & Yamada, Y. (1994). Gene expression in tobacco low-nicotine mutants. *Plant Cell*, *6*(5), 723–735. https://doi.org/10.1105/tpc.6.5.723

Higashi, Y., & Saito, K. (2013). Network analysis for gene discovery in plant-specialized metabolism. *Plant Cell Environ*, *36*(9), 1597–1606. https://doi.org/10.1111/pce.12069

Hirai, M. Y., Yano, M., Goodenowe, D. B., Kanaya, S., Kimura, T., Awazuhara, M., Arita, M., Fujiwara, T., & Saito, K. (2004). Integration of transcriptomics and metabolomics for understanding of global responses to nutritional stresses in Arabidopsis thaliana. *Proc Natl Acad Sci U S A*, *101*(27), 10205–10210. https://doi.org/10.1073/pnas.0403218101

Huang, J. P., Wang, Y. J., Tian, T., Wang, L., Yan, Y., & Huang, S. X. (2021). Tropane alkaloid biosynthesis: A centennial review. *Nat Prod Rep*, *38*(9), 1634–1658. https://doi.org/10.1039/d0np00076k

Huang, X., Wei, X., Sang, T., Zhao, Q., Feng, Q., Zhao, Y., Li, C., Zhu, C., Lu, T., Zhang, Z., Li, M., Fan, D., Guo, Y., Wang, A., Wang, L., Deng, L., Li, W., Lu, Y., Weng, Q., Liu, K., Huang, T., Zhou, T., Jing, Y., Li, W., Lin, Z., Buckler, E. S., Qian, Q., Zhang, Q. F., Li, J., & Han, B. (2010). Genome-wide association studies of 14 agronomic traits in rice landraces. *Nat Genet*, *42*(11), 961–967. https://doi.org/10.1038/ng.695

Hufford, M. B., Xu, X., van Heerwaarden, J., Pyhajarvi, T., Chia, J. M., Cartwright, R. A., Elshire, R. J., Glaubitz, J. C., Guill, K. E., Kaeppler, S. M., Lai, J., Morrell, P. L., Shannon, L. M., Song, C., Springer, N. M., Swanson-Wagner, R. A., Tiffin, P., Wang, J., Zhang, G., Doebley, J., McMullen, M. D., Ware, D., Buckler, E. S., Yang, S., & Ross-Ibarra, J. (2012). Comparative population genomics of maize domestication and improvement. *Nat Genet*, *44*(7), 808–811. https://doi.org/10.1038/ng.2309

Hutchinson, C. R., Heckendorf, A. H., Daddona, P. E., Hagaman, E., & Wenkert, E. (1974). Biosynthesis of camptothecin. I. Definition of the overall pathway assisted by carbon-13 nuclear magnetic resonance analysis. *J Am Chem Soc*, *96*(17), 5609–5611. https://doi.org/10.1021/ja00824a070

Ikezawa, N., Iwasa, K., & Sato, F. (2007). Molecular cloning and characterization of methylenedioxy bridge-forming enzymes involved in stylopine biosynthesis in Eschscholzia californica. *FEBS J*, *274*(4), 1019–1035. https://doi.org/10.1111/j.1742-4658.2007.05652.x

Jacobs, D. I., Gaspari, M., van der Greef, J., van der Heijden, R., & Verpoorte, R. (2005). Proteome analysis of the medicinal plant Catharanthus roseus. *Planta*, *221*(5), 690–704. https://doi.org/10.1007/s00425-004-1474-4

Jamil, I. N., Remali, J., Azizan, K. A., Nor Muhammad, N. A., Arita, M., Goh, H. H., & Aizat, W. M. (2020). Systematic Multi-Omics Integration (MOI) approach in plant systems biology. *Front Plant Sci*, *11*, 944. https://doi.org/10.3389/fpls.2020.00944

Janowitz, T., Kneifel, H., & Piotrowski, M. (2003). Identification and characterization of plant agmatine iminohydrolase, the last missing link in polyamine biosynthesis of plants. *FEBS Lett*, *544*(1–3), 258–261. https://doi.org/10.1016/s0014-5793(03)00515-5

Jiao, Y., Zhao, H., Ren, L., Song, W., Zeng, B., Guo, J., Wang, B., Liu, Z., Chen, J., Li, W., Zhang, M., Xie, S., & Lai, J. (2012). Genome-wide genetic changes during modern breeding of maize. *Nat Genet*, *44*(7), 812–815. https://doi.org/10.1038/ng.2312

Jirschitzka, J., Schmidt, G. W., Reichelt, M., Schneider, B., Gershenzon, J., & D'Auria, J. C. (2012). Plant tropane alkaloid biosynthesis evolved independently in the Solanaceae and Erythroxylaceae. *Proc Natl Acad Sci U S A*, *109*(26), 10304–10309. https://doi.org/10.1073/pnas.1200473109

Kang, W., Zhu, X., Wang, Y., Chen, L., & Duan, Y. (2018). Transcriptomic and metabolomic analyses reveal that bacteria promote plant defense during infection of soybean cyst nematode in soybean. *BMC Plant Biol*, *18*(1), 86. https://doi.org/10.1186/s12870-018-1302-9

Kaur, R., & Arora, S. (2015). Alkaloids-important therapeutic secondary metabolites of plant origin. *J Crit Rev*, *2*(3), 1–8.

Kautsar, S. A., Suarez Duran, H. G., Blin, K., Osbourn, A., & Medema, M. H. (2017). plantiSMASH: Automated identification, annotation and expression analysis of plant biosynthetic gene clusters. *Nucleic Acids Res*, *45*(W1), W55–W63. https://doi.org/10.1093/nar/gkx305

Keller, M., Consortium, S.-I., & Simm, S. (2018). The coupling of transcriptome and proteome adaptation during development and heat stress response of tomato pollen. *BMC Genomics*, *19*(1), 447. https://doi.org/10.1186/s12864-018-4824-5

Koc, I., Yuksel, I., & Caetano-Anolles, G. (2018). Metabolite-centric reporter pathway and tripartite network analysis of arabidopsis under cold stress. *Front Bioeng Biotechnol*, *6*, 121. https://doi.org/10.3389/fbioe.2018.00121

Kohnen-Johannsen, K. L., & Kayser, O. (2019). Tropane alkaloids: Chemistry, pharmacology, biosynthesis and production. *Molecules*, *24*(4). https://doi.org/10.3390/molecules24040796

Kukula-Koch, W. A., & Widelski, J. (2017). Chapter 9—alkaloids. In S. Badal & R. Delgoda (Eds.), *Pharmacognosy* (pp. 163–198). Academic Press. https://doi.org/10.1016/B978-0-12-802104-0.00009-3

Kunnumpurath, S., Julien, N., Kodumudi, G., Kunnumpurath, A., Kodumudi, V., & Vadivelu, N. (2018). Global supply and demand of opioids for pain management. *Curr Pain Headache Rep*, *22*(5), 34. https://doi.org/10.1007/s11916-018-0689-1

Kurek, J. (2019). *Alkaloids: Their Importance in Nature and Human Life*. BoD—Books on Demand.

Levac, D., Murata, J., Kim, W. S., & De Luca, V. (2008). Application of carborundum abrasion for investigating the leaf epidermis: Molecular cloning of Catharanthus roseus 16-hydroxytabersonine-16-O-methyltransferase. *Plant J*, *53*(2), 225–236. https://doi.org/10.1111/j.1365-313X.2007.03337.x

Lewin, H. A., Robinson, G. E., Kress, W. J., Baker, W. J., Coddington, J., Crandall, K. A., Durbin, R., Edwards, S. V., Forest, F., Gilbert, M. T. P., Goldstein, M. M., Grigoriev, I. V., Hackett, K. J., Haussler, D., Jarvis, E. D., Johnson, W. E., Patrinos, A., Richards, S., Castilla-Rubio, J. C., van Sluys, M. A., Soltis, P. S., Xu, X., Yang, H., & Zhang, G. (2018). Earth BioGenome project: Sequencing life for the future of life. *Proc Natl Acad Sci U S A*, *115*(17), 4325–4333. https://doi.org/10.1073/pnas.1720115115

Li, D., Dossa, K., Zhang, Y., Wei, X., Wang, L., Zhang, Y., Liu, A., Zhou, R., & Zhang, X. (2018a). GWAS uncovers differential genetic bases for drought and salt tolerances in sesame at the germination stage. *Genes (Basel)*, *9*(2). https://doi.org/10.3390/genes9020087

Li, L., Hur, M., Lee, J. Y., Zhou, W., Song, Z., Ransom, N., Demirkale, C. Y., Nettleton, D., Westgate, M., Arendsee, Z., Iyer, V., Shanks, J., Nikolau, B., & Wurtele, E. S. (2015). A systems biology approach toward understanding seed composition in soybean. *BMC Genomics*, *16*(Suppl 3), S9. https://doi.org/10.1186/1471-2164-16-S3-S9

Li, P., Zhou, H., Yang, H., Xia, D., Liu, R., Sun, P., Wang, Q., Gao, G., Zhang, Q., Wang, G., & He, Y. (2020). Genome-wide association studies reveal the genetic basis of fertility restoration of CMS-WA and CMS-HL in xian/indica and aus accessions of rice (Oryza sativa L.). *Rice (N Y)*, *13*(1), 11. https://doi.org/10.1186/s12284-020-0372-0

Li, R., Reed, D. W., Liu, E., Nowak, J., Pelcher, L. E., Page, J. E., & Covello, P. S. (2006). Functional genomic analysis of alkaloid biosynthesis in Hyoscyamus niger reveals a cytochrome P450 involved in littorine rearrangement. *Chem Biol*, *13*(5), 513–520. https://doi.org/10.1016/j.chembiol.2006.03.005

Li, Y., Li, S., Thodey, K., Trenchard, I., Cravens, A., & Smolke, C. D. (2018b). Complete biosynthesis of noscapine and halogenated alkaloids in yeast. *Proc Natl Acad Sci U S A*, *115*(17), E3922–E3931. https://doi.org/10.1073/pnas.1721469115

Li, Y., & Smolke, C. D. (2016). Engineering biosynthesis of the anticancer alkaloid noscapine in yeast. *Nat Commun*, *7*, 12137. https://doi.org/10.1038/ncomms12137

Lin, M., Matschi, S., Vasquez, M., Chamness, J., Kaczmar, N., Baseggio, M., Miller, M., Stewart, E. L., Qiao, P., Scanlon, M. J., Molina, I., Smith, L. G., & Gore, M. A. (2020). Genome-wide association study for maize leaf cuticular conductance identifies candidate genes involved in the regulation of cuticle development. *G3 (Bethesda)*, *10*(5), 1671–1683. https://doi.org/10.1534/g3.119.400884

Liscombe, D. K., & Facchini, P. J. (2007). Molecular cloning and characterization of tetrahydroprotoberberine cis-N-methyltransferase, an enzyme involved in alkaloid biosynthesis in opium poppy. *J Biol Chem*, *282*(20), 14741–14751. https://doi.org/10.1074/jbc.M611908200

Liscombe, D. K., Ziegler, J., Schmidt, J., Ammer, C., & Facchini, P. J. (2009). Targeted metabolite and transcript profiling for elucidating enzyme function: Isolation of novel N-methyltransferases from three benzylisoquinoline alkaloid-producing species. *Plant J*, *60*(4), 729–743. https://doi.org/10.1111/j.1365-313X.2009.03980.x

López-Hidalgo, C., Guerrero-Sanchez, V. M., Gomez-Galvez, I., Sanchez-Lucas, R., Castillejo-Sanchez, M. A., Maldonado-Alconada, A. M., Valledor, L., & Jorrin-Novo, J. V. (2018). A multi-omics analysis pipeline for the metabolic pathway reconstruction in the orphan species quercus ilex. *Front Plant Sci*, *9*, 935. https://doi.org/10.3389/fpls.2018.00935

Lu, K., Wei, L., Li, X., Wang, Y., Wu, J., Liu, M., Zhang, C., Chen, Z., Xiao, Z., Jian, H., Cheng, F., Zhang, K., Du, H., Cheng, X., Qu, C., Qian, W., Liu, L., Wang, R., Zou, Q., Ying, J., Xu, X., Mei, J., Liang, Y., Chai, Y. R., Tang, Z., Wan, H., Ni, Y., He, Y., Lin, N., Fan, Y., Sun, W., Li, N. N., Zhou, G., Zheng, H., Wang, X., Paterson, A. H., & Li, J. (2019). Whole-genome resequencing reveals Brassica napus origin and genetic loci involved in its improvement. *Nat Commun*, *10*(1), 1154. https://doi.org/10.1038/s41467-019-09134-9

Luo, J. (2015). Metabolite-based genome-wide association studies in plants. *Curr Opin Plant Biol*, *24*, 31–38. https://doi.org/10.1016/j.pbi.2015.01.006

Luo, L., Pappalardi, M. B., Tummino, P. J., Copeland, R. A., Fraser, M. E., Grzyska, P. K., & Hausinger, R. P. (2006). An assay for Fe(II)/2-oxoglutarate-dependent dioxygenases by enzyme-coupled detection of succinate formation. *Anal Biochem*, *353*(1), 69–74. https://doi.org/10.1016/j.ab.2006.03.033

Ma, C., Zhang, H. H., & Wang, X. (2014). Machine learning for Big Data analytics in plants. *Trends Plant Sci*, *19*(12), 798–808. https://doi.org/10.1016/j.tplants.2014.08.004

Mahesh, H. B., Subba, P., Advani, J., Shirke, M. D., Loganathan, R. M., Chandana, S. L., Shilpa, S., Chatterjee, O., Pinto, S. M., Prasad, T. S. K., & Gowda, M. (2018). Multi-omics driven assembly and annotation of the sandalwood (Santalum album) genome. *Plant Physiol*, *176*(4), 2772–2788. https://doi.org/10.1104/pp.17.01764

Mahieu, N. G., Genenbacher, J. L., & Patti, G. J. (2016). A roadmap for the XCMS family of software solutions in metabolomics. *Curr Opin Chem Biol*, *30*, 87–93. https://doi.org/10.1016/j.cbpa.2015.11.009

Marques, J. V., Dalisay, D. S., Yang, H., Lee, C., Davin, L. B., & Lewis, N. G. (2014). A multi-omics strategy resolves the elusive nature of alkaloids in Podophyllum species. *Mol Biosyst*, *10*(11), 2838–2849. https://doi.org/10.1039/c4mb00403e

Marques, J. V., Kim, K. W., Lee, C., Costa, M. A., May, G. D., Crow, J. A., Davin, L. B., & Lewis, N. G. (2013). Next generation sequencing in predicting gene function in podophyllotoxin biosynthesis. *J Biol Chem*, *288*(1), 466–479. https://doi.org/10.1074/jbc.M112.400689

Matasci, N., Hung, L. H., Yan, Z., Carpenter, E. J., Wickett, N. J., Mirarab, S., Nguyen, N., Warnow, T., Ayyampalayam, S., Barker, M., Burleigh, J. G., Gitzendanner, M. A., Wafula, E., Der, J. P., dePamphilis, C. W., Roure, B., Philippe, H., Ruhfel, B. R., Miles, N. W., Graham, S. W., Mathews, S., Surek, B., Melkonian, M., Soltis, D. E., Soltis, P. S., Rothfels, C., Pokorny, L., Shaw, J. A., DeGironimo, L., Stevenson, D. W., Villarreal, J. C., Chen, T., Kutchan, T. M., Rolf, M., Baucom, R. S., Deyholos, M. K., Samudrala, R., Tian, Z., Wu, X., Sun, X., Zhang, Y., Wang, J., Leebens-Mack, J., & Wong, G. K. (2014). Data access for the 1,000 Plants (1KP) project. *Gigascience*, *3*, 17. https://doi.org/10.1186/2047-217X-3-17

Meng, C., Kuster, B., Culhane, A. C., & Gholami, A. M. (2014). A multivariate approach to the integration of multi-omics datasets. *BMC Bioinformatics*, *15*, 162. https://doi.org/10.1186/1471-2105-15-162

Mesnage, R., Agapito-Tenfen, S. Z., Vilperte, V., Renney, G., Ward, M., Seralini, G. E., Nodari, R. O., & Antoniou, M. N. (2016). An integrated multi-omics analysis of the NK603 Roundup-tolerant GM maize reveals metabolism disturbances caused by the transformation process. *Sci Rep*, *6*, 37855. https://doi.org/10.1038/srep37855

Michael, A. J., Furze, J. M., Rhodes, M. J., & Burtin, D. (1996). Molecular cloning and functional identification of a plant ornithine decarboxylase cDNA. *Biochem J*, *314*(Pt 1), 241–248. https://doi.org/10.1042/bj3140241

Miettinen, K., Dong, L., Navrot, N., Schneider, T., Burlat, V., Pollier, J., Woittiez, L., van der Krol, S., Lugan, R., Ilc, T., Verpoorte, R., Oksman-Caldentey, K. M., Martinoia, E., Bouwmeester, H., Goossens, A., Memelink, J., & Werck-Reichhart, D. (2014). The seco-iridoid pathway from Catharanthus roseus. *Nat Commun*, *5*, 3606. https://doi.org/10.1038/ncomms4606

Mizusaki, S., Tanabe, Y., Noguchi, M., & Tamaki, E. (1972). N-methylputrescine oxidase from tobacco roots. *Phytochemistry*, *11*(9), 2757–2762. https://doi.org/https://doi.org/10.1016/S0031-9422(00)86509-7

Nakabayashi, R., Hashimoto, K., Mori, T., Toyooka, K., Sudo, H., & Saito, K. (2021). Spatial metabolomics using imaging mass spectrometry to identify the localization of asparaptine A in Asparagus officinalis. *Plant Biotechnol (Tokyo)*, *38*(3), 311–315. https://doi.org/10.5511/plantbiotechnology.21.0504b

Nakabayashi, R., Yang, Z., Nishizawa, T., Mori, T., & Saito, K. (2015). Top-down targeted metabolomics reveals a sulfur-containing metabolite with inhibitory activity against angiotensin-converting enzyme in asparagus officinalis. *J Nat Prod*, *78*(5), 1179–1183. https://doi.org/10.1021/acs.jnatprod.5b00092

O'Connor, S. E. (2010). 1.25—alkaloids. In H.-W. Liu & L. Mander (Eds.), *Comprehensive Natural Products II* (pp. 977–1007). Elsevier. https://doi.org/https://doi.org/10.1016/B978-008045382-8.00013-7

Oldham, J. T., Hincapie, M., Rejtar, T., Wall, P. K., Carlson, J. E., & Lee-Parsons, C. W. (2010). Shotgun proteomic analysis of yeast-elicited California poppy (Eschscholzia californica) suspension cultures producing enhanced levels of benzophenanthridine alkaloids. *J Proteome Res*, *9*(9), 4337–4345. https://doi.org/10.1021/pr1000412

Onoyovwe, A., Hagel, J. M., Chen, X., Khan, M. F., Schriemer, D. C., & Facchini, P. J. (2013). Morphine biosynthesis in opium poppy involves two cell types: Sieve elements and laticifers. *Plant Cell*, *25*(10), 4110–4122. https://doi.org/10.1105/tpc.113.115113

Ounaroon, A., Decker, G., Schmidt, J., Lottspeich, F., & Kutchan, T. M. (2003). (R,S)-Reticuline 7-O-methyltransferase and (R,S)-norcoclaurine 6-O-methyltransferase of Papaver somniferum—cDNA cloning and characterization of methyl transfer enzymes of alkaloid biosynthesis in opium poppy. *Plant J*, *36*(6), 808–819. https://doi.org/10.1046/j.1365-313x.2003.01928.x

Parkinson, H., Sarkans, U., Shojatalab, M., Abeygunawardena, N., Contrino, S., Coulson, R., Farne, A., Lara, G. G., Holloway, E., Kapushesky, M., Lilja, P., Mukherjee, G., Oezcimen, A., Rayner, T., Rocca-Serra, P., Sharma, A., Sansone, S., & Brazma, A. (2005). ArrayExpress—a public repository for microarray gene expression data at the EBI. *Nucleic Acids Res*, *33*(Database issue), D553–555. https://doi.org/10.1093/nar/gki056

Pathak, S., Lakhwani, D., Gupta, P., Mishra, B. K., Shukla, S., Asif, M. H., & Trivedi, P. K. (2013). Comparative transcriptome analysis using high papaverine mutant of Papaver somniferum reveals pathway and uncharacterized steps of papaverine biosynthesis. *PLoS One*, *8*(5), e65622. https://doi.org/10.1371/journal.pone.0065622

Pienkny, S., Brandt, W., Schmidt, J., Kramell, R., & Ziegler, J. (2009). Functional characterization of a novel benzylisoquinoline O-methyltransferase suggests its involvement in papaverine biosynthesis in opium poppy (Papaver somniferum L). *Plant J*, *60*(1), 56–67. https://doi.org/10.1111/j.1365-313X.2009.03937.x

Piotrowski, M., Janowitz, T., & Kneifel, H. (2003). Plant C-N hydrolases and the identification of a plant N-carbamoylputrescine amidohydrolase involved in polyamine biosynthesis. *J Biol Chem*, *278*(3), 1708–1712. https://doi.org/10.1074/jbc.M205699200

Pluskal, T., Castillo, S., Villar-Briones, A., & Oresic, M. (2010). MZmine 2: Modular framework for processing, visualizing, and analyzing mass spectrometry-based molecular profile data. *BMC Bioinformatics*, *11*, 395. https://doi.org/10.1186/1471-2105-11-395

Polturak, G., & Osbourn, A. (2021). The emerging role of biosynthetic gene clusters in plant defense and plant interactions. *PLoS Pathog*, *17*(7), e1009698. https://doi.org/10.1371/journal.ppat.1009698

Power, R., Kurz, W. G., & De Luca, V. (1990). Purification and characterization of acetylcoenzyme A: Deacetylvindoline 4-O-acetyltransferase from Catharanthus roseus. *Arch Biochem Biophys*, *279*(2), 370–376. https://doi.org/10.1016/0003-9861(90)90504-r

Qiu, F., Yan, Y., Zeng, J., Huang, J.-P., Zeng, L., Zhong, W., Lan, X., Chen, M., Huang, S.-X., & Liao, Z. (2021). Biochemical and metabolic insights into hyoscyamine dehydrogenase. *ACS Catalysis*, *11*(5), 2912–2924. https://doi.org/10.1021/acscatal.0c04667

Qu, Y., Easson, M. L., Froese, J., Simionescu, R., Hudlicky, T., & De Luca, V. (2015). Completion of the seven-step pathway from tabersonine to the anticancer drug precursor vindoline and its assembly in yeast. *Proc Natl Acad Sci U S A*, *112*(19), 6224–6229. https://doi.org/10.1073/pnas.1501821112

Qu, Y., Easson, M. L., Simionescu, R., Hajicek, J., Thamm, A. M. K., Salim, V., & De Luca, V. (2018). Solution of the multistep pathway for assembly of corynanthean, strychnos, iboga, and aspidosperma monoterpenoid indole alkaloids from 19E-geissoschizine. *Proc Natl Acad Sci U S A*, *115*(12), 3180–3185. https://doi.org/10.1073/pnas.1719979115

Rai, A., Hirakawa, H., Nakabayashi, R., Kikuchi, S., Hayashi, K., Rai, M., Tsugawa, H., Nakaya, T., Mori, T., Nagasaki, H., Fukushi, R., Kusuya, Y., Takahashi, H., Uchiyama, H., Toyoda, A., Hikosaka, S., Goto, E., Saito, K., & Yamazaki, M. (2021a). Chromosome-level genome assembly of Ophiorrhiza pumila reveals the evolution of camptothecin biosynthesis. *Nat Commun*, *12*(1), 405. https://doi.org/10.1038/s41467-020-20508-2

Rai, A., Kamochi, H., Suzuki, H., Nakamura, M., Takahashi, H., Hatada, T., Saito, K., & Yamazaki, M. (2017). De novo transcriptome assembly and characterization of nine tissues of Lonicera japonica to identify potential candidate genes involved in chlorogenic acid, luteolosides, and secoiridoid biosynthesis pathways. *J Nat Med*, *71*(1), 1–15. https://doi.org/10.1007/s11418-016-1041-x

Rai, A., Nakamura, M., Takahashi, H., Suzuki, H., Saito, K., & Yamazaki, M. (2016). High-throughput sequencing and de novo transcriptome assembly of Swertia japonica to identify genes involved in the biosynthesis of therapeutic metabolites. *Plant Cell Rep*, *35*(10), 2091–2111. https://doi.org/10.1007/s00299-016-2021-z

Rai, A., Nakaya, T., Shimizu, Y., Rai, M., Nakamura, M., Suzuki, H., Saito, K., & Yamazaki, M. (2018). De Novo transcriptome assembly and characterization of lithospermum officinale to discover putative genes involved in specialized metabolites biosynthesis. *Planta Med*, *84*(12–13), 920–934. https://doi.org/10.1055/a-0630-5925

Rai, A., Rai, M., Kamochi, H., Mori, T., Nakabayashi, R., Nakamura, M., Suzuki, H., Saito, K., & Yamazaki, M. (2020). Multiomics-based characterization of specialized metabolites biosynthesis in Cornus Officinalis. *DNA Res*, *27*(2). https://doi.org/10.1093/dnares/dsaa009

Rai, A., Saito, K., & Yamazaki, M. (2017). Integrated omics analysis of specialized metabolism in medicinal plants. *Plant J*, *90*(4), 764–787. https://doi.org/10.1111/tpj.13485

Rai, A., Yamazaki, M., & Saito, K. (2019). A new era in plant functional genomics. *Current Opinion in Systems Biology*, *15*, 58–67. https://doi.org/10.1016/j.coisb.2019.03.005

Rai, A., Yamazaki, M., Takahashi, H., Nakamura, M., Kojoma, M., Suzuki, H., & Saito, K. (2016). RNA-seq transcriptome analysis of Panax japonicus, and its comparison with other panax species to identify potential genes involved in the saponins biosynthesis. *Front Plant Sci*, *7*, 481. https://doi.org/10.3389/fpls.2016.00481

Rai, M., Rai, A., Kawano, N., Yoshimatsu, K., Takahashi, H., Suzuki, H., Kawahara, N., Saito, K., & Yamazaki, M. (2017b). De Novo RNA sequencing and expression analysis of aconitum carmichaelii to analyze key genes involved in the biosynthesis of diterpene alkaloids. *Molecules*, *22*(12). https://doi.org/10.3390/molecules22122155

Rai, M., Rai, A., Mori, T., Nakabayashi, R., Yamamoto, M., Nakamura, M., Suzuki, H., Saito, K., & Yamazaki, M. (2021b). Gene-metabolite network analysis revealed tissue-specific accumulation of therapeutic metabolites in mallotus japonicus. *Int J Mol Sci*, *22*(16). https://doi.org/10.3390/ijms22168835

Rischer, H., Oresic, M., Seppanen-Laakso, T., Katajamaa, M., Lammertyn, F., Ardiles-Diaz, W., Van Montagu, M. C., Inze, D., Oksman-Caldentey, K. M., & Goossens, A. (2006). Gene-to-metabolite networks for terpenoid indole alkaloid biosynthesis in Catharanthus roseus cells. *Proc Natl Acad Sci U S A*, *103*(14), 5614–5619. https://doi.org/10.1073/pnas.0601027103

Rohani, E. R., Chiba, M., Kawaharada, M., Asano, T., Oshima, Y., Mitsuda, N., Ohme-Takagi, M., Fukushima, A., Rai, A., Saito, K., & Yamazaki, M. (2016). An MYB transcription factor regulating specialized metabolisms in *Ophiorrhiza pumila*. *Plant Biotechnology*, *33*(1), 1–9. https://doi.org/10.5511/plantbiotechnology.15.1117a

Sadre, R., Magallanes-Lundback, M., Pradhan, S., Salim, V., Mesberg, A., Jones, A. D., & DellaPenna, D. (2016). Metabolite diversity in alkaloid biosynthesis: A multilane (Diastereomer) highway for camptothecin synthesis in Camptotheca acuminata. *Plant Cell*, *28*(8), 1926–1944. https://doi.org/10.1105/tpc.16.00193

Saito, K. (2013). Phytochemical genomics—a new trend. *Curr Opin Plant Biol*, *16*(3), 373–380. https://doi.org/10.1016/j.pbi.2013.04.001

Sauvage, C., Segura, V., Bauchet, G., Stevens, R., Do, P. T., Nikoloski, Z., Fernie, A. R., & Causse, M. (2014). Genome-wide association in tomato reveals 44 candidate Loci for fruit metabolic traits. *Plant Physiol*, *165*(3), 1120–1132. https://doi.org/10.1104/pp.114.241521

Scheunemann, M., Brady, S. M., & Nikoloski, Z. (2018). Integration of large-scale data for extraction of integrated Arabidopsis root cell-type specific models. *Sci Rep*, *8*(1), 7919. https://doi.org/10.1038/s41598-018-26232-8

Schlapfer, P., Zhang, P., Wang, C., Kim, T., Banf, M., Chae, L., Dreher, K., Chavali, A. K., Nilo-Poyanco, R., Bernard, T., Kahn, D., & Rhee, S. Y. (2017). Genome-wide prediction of metabolic enzymes, pathways, and gene clusters in plants. *Plant Physiol*, *173*(4), 2041–2059. https://doi.org/10.1104/pp.16.01942

Schmidt, G. W., Jirschitzka, J., Porta, T., Reichelt, M., Luck, K., Torre, J. C., Dolke, F., Varesio, E., Hopfgartner, G., Gershenzon, J., & D'Auria, J. C. (2015). The last step in cocaine biosynthesis is catalyzed by a BAHD acyltransferase. *Plant Physiol*, *167*(1), 89–101. https://doi.org/10.1104/pp.114.248187

Schroder, G., Unterbusch, E., Kaltenbach, M., Schmidt, J., Strack, D., De Luca, V., & Schroder, J. (1999). Light-induced cytochrome P450-dependent enzyme in indole alkaloid biosynthesis: Tabersonine 16-hydroxylase. *FEBS Lett*, *458*(2), 97–102. https://doi.org/10.1016/s0014-5793(99)01138-2

Seaver, S. M. D., Lerma-Ortiz, C., Conrad, N., Mikaili, A., Sreedasyam, A., Hanson, A. D., & Henry, C. S. (2018). PlantSEED enables automated annotation and reconstruction of plant primary metabolism with improved compartmentalization and comparative consistency. *Plant J*, *95*(6), 1102–1113. https://doi.org/10.1111/tpj.14003

Seneca. (2007). CHAPTER 1—Definition, typology and occurrence of alkaloids. In T. Aniszewski (Ed.), *Alkaloids—Secrets of Life* (pp. 1–59). Elsevier. https://doi.org/https://doi.org/10.1016/B978-044452736-3/50003-9

Sertùrner, F. W. (1805). Saure im opium. *Trommsdorffs Journal der Pharmazie*, *13*, 349.

Silva, J. C. F., Teixeira, R. M., Silva, F. F., Brommonschenkel, S. H., & Fontes, E. P. B. (2019). Machine learning approaches and their current application in plant molecular biology: A systematic review. *Plant Sci*, *284*, 37–47. https://doi.org/10.1016/j.plantsci.2019.03.020

Smith, C. A., Want, E. J., O'Maille, G., Abagyan, R., & Siuzdak, G. (2006). XCMS: Processing mass spectrometry data for metabolite profiling using nonlinear peak alignment, matching, and identification. *Anal Chem*, *78*(3), 779–787. https://doi.org/10.1021/ac051437y

Sottomayor, M., Lopez-Serrano, M., DiCosmo, F., & Ros Barcelo, A. (1998). Purification and characterization of alpha-3',4'-anhydrovinblastine synthase (peroxidase-like) from Catharanthus roseus (L.) G. Don. *FEBS Lett*, *428*(3), 299–303. https://doi.org/10.1016/s0014-5793(98)00551-1

Sottomayor, M., & Ros Barcelo, A. (2003). Peroxidase from Catharanthus roseus (L.) G. Don and the biosynthesis of alpha-3',4'-anhydrovinblastine: A specific role for a multifunctional enzyme. *Protoplasma*, *222*(1–2), 97–105. https://doi.org/10.1007/s00709-003-0003-9

Soubeyrand, E., Colombie, S., Beauvoit, B., Dai, Z., Cluzet, S., Hilbert, G., Renaud, C., Maneta-Peyret, L., Dieuaide-Noubhani, M., Merillon, J. M., Gibon, Y., Delrot, S., & Gomes, E. (2018). Constraint-based modeling highlights cell energy, redox status and alpha-ketoglutarate availability as metabolic drivers for anthocyanin accumulation in grape cells under nitrogen limitation. *Front Plant Sci*, *9*, 421. https://doi.org/10.3389/fpls.2018.00421

Stander, E. A., Sepulveda, L. J., Duge de Bernonville, T., Carqueijeiro, I., Koudounas, K., Lemos Cruz, P., Besseau, S., Lanoue, A., Papon, N., Giglioli-Guivarc'h, N., Dirks, R., O'Connor, S. E., Atehortua, L., Oudin, A., & Courdavault, V. (2020). Identifying genes involved in alkaloid biosynthesis in vinca minor through transcriptomics and gene co-expression analysis. *Biomolecules*, *10*(12). https://doi.org/10.3390/biom10121595

St-Pierre, B., & De Luca, V. (1995). A Cytochrome P-450 Monooxygenase catalyzes the first step in the conversion of tabersonine to vindoline in catharanthus roseus. *Plant Physiol*, *109*(1), 131–139. https://doi.org/10.1104/pp.109.1.131

St-Pierre, B., Laflamme, P., Alarco, A. M., & De Luca, V. (1998). The terminal O-acetyltransferase involved in vindoline biosynthesis defines a new class of proteins responsible for coenzyme A-dependent acyl transfer. *Plant J*, *14*(6), 703–713. https://doi.org/10.1046/j.1365-313x.1998.00174.x

Sun, L., Rai, A., Rai, M., Nakamura, M., Kawano, N., Yoshimatsu, K., Suzuki, H., Kawahara, N., Saito, K., & Yamazaki, M. (2018). Comparative transcriptome analyses of three medicinal Forsythia species and prediction of candidate genes involved in secondary metabolisms. *J Nat Med*, *72*(4), 867–881. https://doi.org/10.1007/s11418-018-1218-6

Takemura, T., Ikezawa, N., Iwasa, K., & Sato, F. (2013). Molecular cloning and characterization of a cytochrome P450 in sanguinarine biosynthesis from Eschscholzia californica cells. *Phytochemistry*, *91*, 100–108. https://doi.org/10.1016/j.phytochem.2012.02.013

Tautenhahn, R., Patti, G. J., Rinehart, D., & Siuzdak, G. (2012). XCMS Online: A web-based platform to process untargeted metabolomic data. *Anal Chem*, *84*(11), 5035–5039. https://doi.org/10.1021/ac300698c

Tepfer, D., Goldmann, A., Pamboukdjian, N., Maille, M., Lepingle, A., Chevalier, D., Denarie, J., & Rosenberg, C. (1988). A plasmid of Rhizobium meliloti 41 encodes catabolism of two compounds from root exudate of Calystegium sepium. *J Bacteriol*, *170*(3), 1153–1161. https://doi.org/10.1128/jb.170.3.1153-1161.1988

Thiele, I., & Palsson, B. O. (2010). A protocol for generating a high-quality genome-scale metabolic reconstruction. *Nat Protoc*, *5*(1), 93–121. https://doi.org/10.1038/nprot.2009.203

Thimm, O., Blasing, O., Gibon, Y., Nagel, A., Meyer, S., Kruger, P., Selbig, J., Muller, L. A., Rhee, S. Y., & Stitt, M. (2004). MAPMAN: A user-driven tool to display genomics data sets onto diagrams of metabolic pathways and other biological processes. *Plant J*, *37*(6), 914–939. https://doi.org/10.1111/j.1365-313x.2004.02016.x

Tohge, T., de Souza, L. P., & Fernie, A. R. (2014). Genome-enabled plant metabolomics. *J Chromatogr B Analyt Technol Biomed Life Sci*, *966*, 7–20. https://doi.org/10.1016/j.jchromb.2014.04.003

Tohge, T., & Fernie, A. R. (2010). Combining genetic diversity, informatics and metabolomics to facilitate annotation of plant gene function. *Nat Protoc*, *5*(6), 1210–1227. https://doi.org/10.1038/nprot.2010.82

Topfer, N., Fuchs, L. M., & Aharoni, A. (2017). The PhytoClust tool for metabolic gene clusters discovery in plant genomes. *Nucleic Acids Res*, *45*(12), 7049–7063. https://doi.org/10.1093/nar/gkx404

Tsugawa, H., Cajka, T., Kind, T., Ma, Y., Higgins, B., Ikeda, K., Kanazawa, M., VanderGheynst, J., Fiehn, O., & Arita, M. (2015). MS-DIAL: Data-independent MS/MS deconvolution for comprehensive metabolome analysis. *Nat Methods*, *12*(6), 523–526. https://doi.org/10.1038/nmeth.3393

Tsugawa, H., Kind, T., Nakabayashi, R., Yukihira, D., Tanaka, W., Cajka, T., Saito, K., Fiehn, O., & Arita, M. (2016). Hydrogen rearrangement rules: Computational MS/MS fragmentation and structure elucidation using MS-FINDER software. *Anal Chem*, *88*(16), 7946–7958. https://doi.org/10.1021/acs.analchem.6b00770

Tsugawa, H., Nakabayashi, R., Mori, T., Yamada, Y., Takahashi, M., Rai, A., Sugiyama, R., Yamamoto, H., Nakaya, T., Yamazaki, M., Kooke, R., Bac-Molenaar, J. A., Oztolan-Erol, N., Keurentjes, J. J. B., Arita, M., & Saito, K. (2019). A cheminformatics approach to characterize metabolomes in stable-isotope-labeled organisms. *Nat Methods*, *16*(4), 295–298. https://doi.org/10.1038/s41592-019-0358-2

Tsugawa, H., Rai, A., Saito, K., & Nakabayashi, R. (2021). Metabolomics and complementary techniques to investigate the plant phytochemical cosmos. *Nat Prod Rep*, *38*(10), 1729–1759. https://doi.org/10.1039/d1np00014d

Uarrota, V. G., Fuentealba, C., Hernandez, I., Defilippi-Bruzzone, B., Meneses, C., Campos-Vargas, R., Lurie, S., Hertog, M., Carpentier, S., Poblete-Echeverria, C., & Pedreschi, R. (2019). Integration of proteomics and metabolomics data of early and middle season Hass avocados under heat treatment. *Food Chem*, *289*, 512–521. https://doi.org/10.1016/j.foodchem.2019.03.090

Udomsom, N., Rai, A., Suzuki, H., Okuyama, J., Imai, R., Mori, T., Nakabayashi, R., Saito, K., & Yamazaki, M. (2016). Function of AP2/ERF transcription factors involved in the regulation of specialized metabolism in ophiorrhiza pumila revealed by transcriptomics and metabolomics. *Front Plant Sci*, *7*, 1861. https://doi.org/10.3389/fpls.2016.01861

Unterlinner, B., Lenz, R., & Kutchan, T. M. (1999). Molecular cloning and functional expression of codeinone reductase: The penultimate enzyme in morphine biosynthesis in the opium poppy papaver somniferum. *Plant J*, *18*(5), 465–475. https://doi.org/10.1046/j.1365-313x.1999.00470.x

Urbanczyk-Wochniak, E., Luedemann, A., Kopka, J., Selbig, J., Roessner-Tunali, U., Willmitzer, L., & Fernie, A. R. (2003). Parallel analysis of transcript and metabolic profiles: a new approach in systems biology. *EMBO Rep*, *4*(10), 989–993. https://doi.org/10.1038/sj.embor.embor944

van Iersel, M. P., Kelder, T., Pico, A. R., Hanspers, K., Coort, S., Conklin, B. R., & Evelo, C. (2008). Presenting and exploring biological pathways with PathVisio. *BMC Bioinformatics*, *9*, 399. https://doi.org/10.1186/1471-2105-9-399

Vazquez-Flota, F., De Carolis, E., Alarco, A. M., & De Luca, V. (1997). Molecular cloning and characterization of desacetoxyvindoline-4-hydroxylase, a 2-oxoglutarate dependent-dioxygenase involved in the biosynthesis of vindoline in Catharanthus roseus (L.) G. Don. *Plant Mol Biol*, *34*(6), 935–948. https://doi.org/10.1023/a:1005894001516

Voit, E. O. (2017). The best models of metabolism. *Wiley Interdiscip Rev Syst Biol Med*, *9*(6). https://doi.org/10.1002/wsbm.1391

Wang, J. P., Matthews, M. L., Williams, C. M., Shi, R., Yang, C., Tunlaya-Anukit, S., Chen, H. C., Li, Q., Liu, J., Lin, C. Y., Naik, P., Sun, Y. H., Loziuk, P. L., Yeh, T. F., Kim, H., Gjersing, E., Shollenberger, T., Shuford, C. M., Song, J., Miller, Z., Huang, Y. Y., Edmunds, C. W., Liu, B., Sun, Y., Lin, Y. J., Li, W., Chen, H., Peszlen, I., Ducoste, J. J., Ralph, J., Chang, H. M., Muddiman, D. C., Davis, M. F., Smith, C., Isik, F., Sederoff, R., & Chiang, V. L. (2018). Improving wood properties for wood utilization through multi-omics integration in lignin biosynthesis. *Nat Commun*, *9*(1), 1579. https://doi.org/10.1038/s41467-018-03863-z

Wang, L., Nagele, T., Doerfler, H., Fragner, L., Chaturvedi, P., Nukarinen, E., Bellaire, A., Huber, W., Weiszmann, J., Engelmeier, D., Ramsak, Z., Gruden, K., & Weckwerth, W. (2016). System level analysis of cacao seed ripening reveals a sequential interplay of primary and secondary metabolism leading to polyphenol accumulation and preparation of stress resistance. *Plant J*, *87*(3), 318–332. https://doi.org/10.1111/tpj.13201

Wang, L., Sun, X., Weiszmann, J., & Weckwerth, W. (2017). System-level and granger network analysis of integrated proteomic and metabolomic dynamics identifies key points of grape berry development at the interface of primary and secondary metabolism. *Front Plant Sci*, *8*, 1066. https://doi.org/10.3389/fpls.2017.01066

Weckwerth, W. (2019). Toward a unification of system-theoretical principles in biology and ecology—the stochastic lyapunov matrix equation and its inverse application [hypothesis and theory]. *Frontiers in Applied Mathematics and Statistics*, *5*. https://doi.org/10.3389/fams.2019.00029

Wei, X., Liu, K., Zhang, Y., Feng, Q., Wang, L., Zhao, Y., Li, D., Zhao, Q., Zhu, X., Zhu, X., Li, W., Fan, D., Gao, Y., Lu, Y., Zhang, X., Tang, X., Zhou, C., Zhu, C., Liu, L., Zhong, R., Tian, Q., Wen, Z., Weng, Q., Han, B., Huang, X., & Zhang, X. (2015). Genetic discovery for oil production and quality in sesame. *Nat Commun*, *6*, 8609. https://doi.org/10.1038/ncomms9609

Winkler, A., Puhl, M., Weber, H., Kutchan, T. M., Gruber, K., & Macheroux, P. (2009). Berberine bridge enzyme catalyzes the six electron oxidation of (S)-reticuline to dehydroscoulerine. *Phytochemistry*, *70*(9), 1092–1097. https://doi.org/10.1016/j.phytochem.2009.06.005

Winzer, T., Gazda, V., He, Z., Kaminski, F., Kern, M., Larson, T. R., Li, Y., Meade, F., Teodor, R., Vaistij, F. E., Walker, C., Bowser, T. A., & Graham, I. A. (2012). A papaver somniferum 10-gene cluster for synthesis of the anticancer alkaloid noscapine. *Science*, *336*(6089), 1704–1708. https://doi.org/10.1126/science.1220757

Winzer, T., Kern, M., King, A. J., Larson, T. R., Teodor, R. I., Donninger, S. L., Li, Y., Dowle, A. A., Cartwright, J., Bates, R., Ashford, D., Thomas, J., Walker, C., Bowser, T. A., & Graham, I. A. (2015). Plant science. Morphinan biosynthesis in opium poppy requires a P450-oxidoreductase fusion protein. *Science*, *349*(6245), 309–312. https://doi.org/10.1126/science.aab1852

Wu, S., Tohge, T., Cuadros-Inostroza, A., Tong, H., Tenenboim, H., Kooke, R., Meret, M., Keurentjes, J. B., Nikoloski, Z., Fernie, A. R., Willmitzer, L., & Brotman, Y. (2018). Mapping the arabidopsis metabolic landscape by untargeted metabolomics at different environmental conditions. *Mol Plant*, *11*(1), 118–134. https://doi.org/10.1016/j.molp.2017.08.012

Xiao, M., Zhang, Y., Chen, X., Lee, E. J., Barber, C. J., Chakrabarty, R., Desgagne-Penix, I., Haslam, T. M., Kim, Y. B., Liu, E., MacNevin, G., Masada-Atsumi, S., Reed, D. W., Stout, J. M., Zerbe, P., Zhang, Y., Bohlmann, J., Covello, P. S., De Luca, V., Page, J. E., Ro, D. K., Martin, V. J., Facchini, P. J., & Sensen, C. W. (2013). Transcriptome analysis based on next-generation sequencing of non-model plants producing specialized metabolites of biotechnological interest. *J Biotechnol*, *166*(3), 122–134. https://doi.org/10.1016/j.jbiotec.2013.04.004

Yamamoto, K., Takahashi, K., Caputi, L., Mizuno, H., Rodriguez-Lopez, C. E., Iwasaki, T., Ishizaki, K., Fukaki, H., Ohnishi, M., Yamazaki, M., Masujima, T., O'Connor, S. E., & Mimura, T. (2019). The complexity of intercellular localisation of alkaloids revealed by single-cell metabolomics. *New Phytol*, *224*(2), 848–859. https://doi.org/10.1111/nph.16138

Yamamoto, K., Takahashi, K., Mizuno, H., Anegawa, A., Ishizaki, K., Fukaki, H., Ohnishi, M., Yamazaki, M., Masujima, T., & Mimura, T. (2016). Cell-specific localization of alkaloids in Catharanthus roseus stem tissue measured with Imaging MS and Single-cell MS. *Proc Natl Acad Sci U S A*, *113*(14), 3891–3896. https://doi.org/10.1073/pnas.1521959113

Yamazaki, M., Mochida, K., Asano, T., Nakabayashi, R., Chiba, M., Udomson, N., Yamazaki, Y., Goodenowe, D. B., Sankawa, U., Yoshida, T., Toyoda, A., Totoki, Y., Sakaki, Y., Gongora-Castillo, E., Buell, C. R., Sakurai, T., & Saito, K. (2013). Coupling deep transcriptome analysis with untargeted metabolic profiling in Ophiorrhiza pumila to further the understanding of the biosynthesis of the anti-cancer alkaloid camptothecin and anthraquinones. *Plant Cell Physiol*, *54*(5), 686–696. https://doi.org/10.1093/pcp/pct040

Yamazaki, Y., Kitajima, M., Arita, M., Takayama, H., Sudo, H., Yamazaki, M., Aimi, N., & Saito, K. (2004). Biosynthesis of camptothecin. In silico and in vivo tracer study from [1–13C] glucose. *Plant Physiol*, *134*(1), 161–170. https://doi.org/10.1104/pp.103.029389

Yang, L., & Stockigt, J. (2010). Trends for diverse production strategies of plant medicinal alkaloids. *Nat Prod Rep*, *27*(10), 1469–1479. https://doi.org/10.1039/c005378c

Yano, K., Yamamoto, E., Aya, K., Takeuchi, H., Lo, P. C., Hu, L., Yamasaki, M., Yoshida, S., Kitano, H., Hirano, K., & Matsuoka, M. (2016). Genome-wide association study using whole-genome sequencing rapidly identifies new genes influencing agronomic traits in rice. *Nat Genet*, *48*(8), 927–934. https://doi.org/10.1038/ng.3596

Zhu, L., Zhou, Y., Li, X., Zhao, J., Guo, N., & Xing, H. (2018). Metabolomics analysis of soybean hypocotyls in response to phytophthora sojae infection. *Front Plant Sci*, *9*, 1530. https://doi.org/10.3389/fpls.2018.01530

Ziegler, J., Voigtlander, S., Schmidt, J., Kramell, R., Miersch, O., Ammer, C., Gesell, A., & Kutchan, T. M. (2006). Comparative transcript and alkaloid profiling in Papaver species identifies a short chain dehydrogenase/reductase involved in morphine biosynthesis. *Plant J*, *48*(2), 177–192. https://doi.org/10.1111/j.1365-313X.2006.02860.x

7 Recent Investigations on Protective Mechanisms of Flavonoids against Sarcopenia

Ishita Pandey, Srishti Vashishtha, and Sonam Chawla

CONTENTS

7.1 Clinical Definition 141
7.2 Cellular Derangements and Etiology 141
7.3 Current Pharmacological and Nonpharmacological Interventions for Sarcopenia 143
7.4 Natural Molecules as Potential Sarcopenia Management Interventions 145
7.5 Protective Roles of Flavonoids in Muscle Mass Preservation and Regeneration 145
7.5.1 Flavonol 145
7.5.2 Flavones 147
7.5.3 Flavanone 149
7.5.4 Isoflavone 149
7.5.5 Anthocyanidin 150
7.6 Conclusion 150
Bibliography 151

7.1 CLINICAL DEFINITION

Sarcopenia, or the age-related decline of skeletal muscle mass and function, is gaining medical traction. Rosenberg coined the term "sarcopenia" in 1989 to characterize this age-related decline in muscle mass. It is a term derived from the Greek phrase for "flesh poverty" (*sarx* or flesh + *penia* or loss) (Cruz-Jentoft et al., 2019). This debilitating condition is characterized by a progressive and generalized loss of muscle mass and strength, with the potential for detrimental consequences such as physical impairment, reduced quality of life (QOL), and mortality (Cruz-Jentoft et al., 2019). Primary sarcopenia refers to age-related muscle wasting, whereas secondary sarcopenia refers to muscular wasting caused by a sedentary lifestyle, chronic disease, or neoplasia (Shimokata et al., 2018). Often, sarcopenia is interlinked in terms of manifested symptoms with frailty, a geriatric syndrome comprising of suboptimal homeostatic responses, increasing the incidence of negative health-related events such as falls and hospitalizations events, disability, and mortality. Figure 7.1 indicated the points of commonality between sarcopenia and frailty (Cesari et al., 2014).

7.2 CELLULAR DERANGEMENTS AND ETIOLOGY

A bird's-eye view of the molecular aspects of sarcopenia etiology implicates loss of proteostasis, mitochondrial dysfunction, senescence, deregulated nutrient sensing, and altered cellular communication via hormonal imbalance. Interestingly, these molecular aspects of sarcopenia are classically

DOI: 10.1201/b22917-7

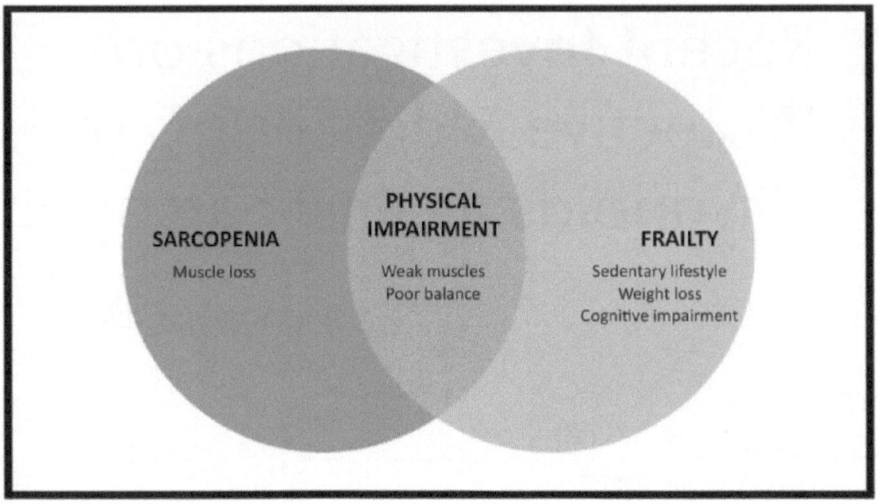

FIGURE 7.1 The link between sarcopenia, frailty, and physical impairment.
Source: Cesari et al., 2014

described as "hallmarks of ageing" (López-Otín et al., 2013; Picca et al., 2021). Derangements in distinct levels of muscle physiology are discussed in brief:

- **Motor neuron and muscle fiber deterioration:** Electrophysiological and histological evaluations in muscles indicate a decline in the number of motor units coincident with progressive aging (Hu et al., 2021). Numerically, motor units are compromised as a consequence of denervation due to loss of motor neurons, followed by apoptosis of the muscle fibers (Larsson et al., 2019).
- **Satellite cell differentiation and proliferation potential:** Satellite cells (SCs) are compact, multipotent cells in adult muscle, found between the sarcolemma and the muscle fiber's basement membrane (Laumonier and Menetrey, 2016). SCs are quiescent skeletal muscle cell precursors, and on exposure to stress (e.g., exercise mediated wear and tear), activate, proliferate, and differentiate to regenerate new myocytes/myofibers (Chen et al., 2020a, 2020b; Kaczmarek et al., 2021). Hepatocyte growth factor (HGF) is the only known growth factor activating dormant SCs and is released post-exercise to repair any wear and tear. Therefore, a lack of physical exercise in the elderly, further complicated by consumption of an imbalanced diet (fatty diet leading to obesity)/nutritional deficiency are significant variables modifying the SC's response to HGF (Pascual-Fernández et al., 2020).
- **Mitochondrial dysfunction:** Age-related decline in the skeletal muscle mitochondrial capacity/mitochondrial mass is reported to correlate with pathophysiological features in the elderly—reduced walking speed, extreme fatigue, and sarcopenia (Santanasto et al., 2016; Gonzalez-Freire et al., 2018). Pre-clinical investigations in aged skeletal muscle cells indicate a coupling of diminished mitochondrial mass, leading to lower adenosine triphosphate (ATP) generation with a decline in skeletal muscle mass (Boengler et al., 2017). Further, the molecular milieu in the aged mitochondria favors pro-apoptotic pathway activation, fission and other morphological deviations, and importantly, higher production of reactive oxygen species (ROS). Excessive mitochondrial ROS damages the cellular biomolecules, disrupts the signaling pathways activating translation, and also exaggerates the activation of proteolytic pathways—proteasomal as well as autophagy—further compromising the skeletal muscle quality (Coen et al., 2019). McKenzie and co-workers, in their landmark review article in 2002, also highlighted the role of excessive ROS in negatively influencing the quantity and quality of mitochondrial DNA in aged skeletal muscles. The pro-oxidant

locale in the aged mitochondria can lead to single-nucleotide mutations, large-scale deletions, and rearrangements in mitochondrial DNA, leading to loss of coding sequences for the electron transport chain (McKenzie et al., 2002; Herbst et al., 2021).

- **Inflammation:** The state of chronic low-grade inflammation in the elderly, diagnosed by the presence of higher levels of pro-inflammatory cytokines such as tumor necrosis factor alpha (TNF-α), C-reactive protein (CRP), interleukin (IL)-6, and IL-1, is a widely reported cause of sarcopenia and several other age-related disorders (Zembron-Lacny et al., 2019; Bano et al., 2017). This build-up of inflammatory in the aged skeletal muscle is attributed to the senescence-associated secretory phenotype of the senescent cells, deviations in the normal autophagic and mitophagy processes allowing accumulation of redundant proteins and organelles, and lastly an exaggerated mitochondrial ROS generation (Daussin et al., 2021; Wiedmer et al., 2021).
- **Hormonal alterations:** Aging-related endocrine deviations (also called "somatopause") directly correlate with the incidence of sarcopenia and are clinically indicated by low blood concentrations of growth hormone (GH), insulin-like growth factor-I (IGF-I), testosterone, and mechanical growth factor (MGF) (Bian et al., 2020). The importance of maintaining circulatory GH levels is also backed in animal studies—for instance, rodents (20 months old), when treated with subcutaneous injections of GH for 8 weeks, showed boosted antioxidant defenses coincident with lower oxidative damage to the tissues; increased protein translation and mitochondrial biogenesis; and importantly, a hightened expression of molecules responsible for muscle regeneration: myostatin, phosphorylated p38, p21 and MAFbx, and MuRF-1 (Brioche et al., 2014).

7.3 CURRENT PHARMACOLOGICAL AND NONPHARMACOLOGICAL INTERVENTIONS FOR SARCOPENIA

To date, there is no clinical recommendation for the treatment of sarcopenia. The expanding global population over the age of 60 years emphasizes the need for developing pharmacological interventions for the management of sarcopenia. The current clinical management practices are centered around protein and vitamin supplements and/or resistance training and are supported by numerous clinical studies (Cruz-Jentoft et al., 2019). Several meta-analysis studies report that sarcopenia is significantly linked with higher rates of falling, hospitalization, and mortality, besides incidence with grave degenerative disorders such as diabetes, osteoporosis, and carcinomas (Beaudart et al., 2017; Zhang et al., 2018).

Largely, the approaches for evaluating and establishing pharmacological interventions for sarcopenia can be categorized into top-down and bottom-up, as depicted in Figure 7.2. The top-down strategies include undertaking observational studies establishing the relations between functional targets and the investigational drug. Some examples of the top-down approach are:

- **Angiotensin-converting enzyme (ACE) inhibitors** have been associated with a delayed decrease in muscle mass and muscular endurance in top-down strategy studies. An understanding of angiotensin's cellular interaction in muscle by inhibition of IGF-1 activity, as well as stimulation of nuclear factor kappa light-chain enhancer of activated B cells (NF-kβ) and TNF-α has since offered support for this connection (Kilsby et al., 2017; Chen et al., 2020a, 2020b).
- **Antioxidant supplementation** leading to raised systemic antioxidant defense has improved physical function. The finding is consistent with basic findings that elevated levels of ROS and pro-inflammatory cytokines are found in many conditions associated with low muscle mass and are thought to activate NF-kβ or the ubiquitin-proteasome system via muscle RING-finger protein (MuRF)–1, resulting in protein degradation (Kilsby et al., 2017; Leitner et al., 2017). Further emphasizing the utility of antioxidants, allopurinol, a xanthine oxidase inhibitor, has been linked to improved functional outcomes in patients undergoing rehabilitation and resistance training, implying that agents that reduce oxidative stress are worth testing in sarcopenic patients (Battelli et al., 2020).

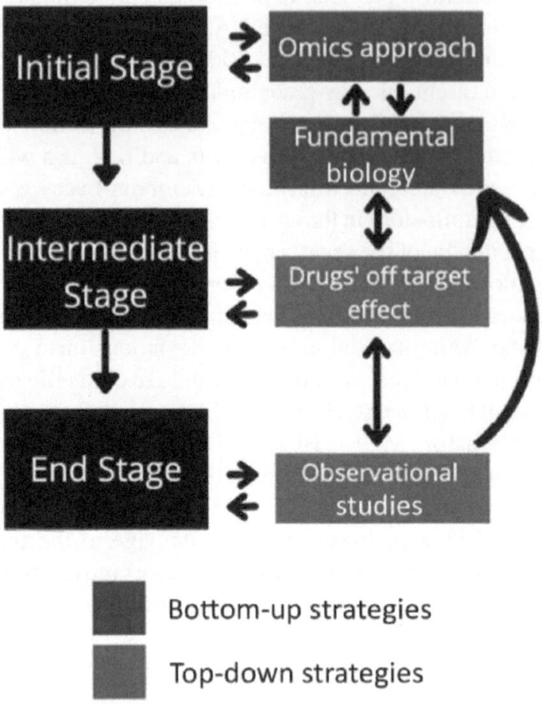

FIGURE 7.2 Strategies for selecting pharmacological interventions.
Source: Kilsby et al., 2017

The bottom-up strategy commences with identifying the key molecular pathways and regulatory junctions involved in muscular atrophy and sarcopenia and gradually moving up to establishing the most efficacious drug target with the least off-target effects. For example:

- **Myostatin**, also known as growth differentiation factor-8, is an emerging drug target against sarcopenia, atrophy, and cachexia. Myostatin expression is restricted to skeletal muscles, and its interaction with activin-A-activin type II receptors negatively regulates the phosphoinositide 3 kinase/protein kinase B (PI3K/AKT) pathway and decreases levels of myogenic regulatory factors. This molecular sequence severely impacts skeletal muscle mass. Myostatin-blocking antibodies, small interfering RNA, and the like are being explored and have been reviewed by Kwak and co-workers recently (Kwak et al., 2019).
- Besides myostatin, activin receptor, exercise mimetics (AMPK activators, PPARβ/δ agonists), and androgen receptor modulators, activation of the IGF-1/Akt/mTOR and PGC-1α/Sirt1 signaling pathway are also under various stages of pre-clinical and clinical investigation against sarcopenia (Kwak et al., 2019; Mankhong et al., 2020).

Theoretically, as with other conventional drug development programs, the pharmacological intervention should be efficacious, economical, administered orally at a suitable frequency to facilitate adherence, and have minimum off-target effects. Bioprospection of traditional medicine interventions can be a rich source of potential molecules effective in the management of sarcopenia (Kilsby et al., 2017).

7.4 NATURAL MOLECULES AS POTENTIAL SARCOPENIA MANAGEMENT INTERVENTIONS

Natural molecules have been key contributors to the modern-day medicine and drug development programs. They offer structural diversity and hence a broader functional outcome for the consumer. Moreover, their presence in the traditional medicinal formulations also endorses their efficacy and safety (Atanasov et al., 2021). Importantly, an elderly sarcopenic individual will also exhibit age associated with a slower metabolism and compromised immune system. Moreover, the subjects may also exhibit coincidences of diseases in cardiovascular system, skeletal system, etc., and are on a clinical regimen of medicines. Natural bioactive molecules for sarcopenia shall have a higher compliance and minimize the concerns associated with conventional medicine such as adverse drug reactions in the elderly or drug-drug interactions. Apart from this, natural molecules also offer the advantages of accessibility and cost. Pharmacological drugs are usually expensive and have low accessibility, especially in third world countries where natural products are not just more accessible but more economical as well. Traditional medicine formulations are well investigated and established in age-related diseases and disorders such as cognitive decline, cardiovascular derangements, etc. (Gupta et al., 2014; Varteresian et al., 2014). A recent meta-analysis study on the efficacy of various dietary phytochemicals on clinically validated diagnostic markers of sarcopenia (hand grip strength, physical performance, 30s chair stand test, and 6-min walk test) across 14 studies by Jeon and co-workers indicates a clear improvement in muscle health, although no increase in muscle mass was reported (Jeong et al., 2021). A molecular introspection of the sarcopenic muscles indicates that redox balance, inflammation control, mitochondrial functioning, and balanced proteostasis are the critical cellular disturbances for the incidence of sarcopenia (reviewed earlier, see Section 1.2). Plant-origin flavonoids are potent secondary metabolites with reported properties to offer protection in the previously listed cellular disturbances and potentially offer benefits in sarcopenic subjects.

7.5 PROTECTIVE ROLES OF FLAVONOIDS IN MUSCLE MASS PRESERVATION AND REGENERATION

Flavonoids are produced via the phenylpropanoid pathway (Kumar and Pandey, 2013; Le Marchand, 2002). They comprise a polyphenolic structure, composed of two benzene rings A and B (C6 each) connected via a C3 chain, which may (forming third fused ring C) or may not be cyclized (Kawser Hossain et al., 2016). Flavonoids are classified as flavan, flavanol, flavanone, flavanonol, flavone, flavonol, isoflavan, isoflavanone, anthocyanidin, and chalcone based on chemical modification in the described basic chemical scaffold. The classification of flavonoids has been summarized in Figure 7.3 (Kumar and Pandey, 2013). Various in vitro and in vivo studies implicate protective benefits of flavonoids in muscle mass decline related tissue level and cellular and molecular aspects. We describe reported findings along with potential underlying mechanisms thereof.

7.5.1 FLAVONOL

Flavonoids with the 3-hydroxyflavone backbone are known as flavonols (Mahalanobish et al., 2019) (IUPAC name: 3-hydroxy-2-phenyl chromen-4-one). Flavonols can be found in a large number of fruits and vegetables. Certain principal flavanols exhibiting muscle mass protective action are:

- **Quercetin**: Quercetin is the most familiar flavonoid for laymen. It is well characterized, with multiple protective health benefits in elderly and young alike, essentially owing to its powerful antioxidant and anti-inflammatory potency. It is an established over-the-counter nutraceutical supplement (Askari et al., 2013). One of the earliest studies reporting

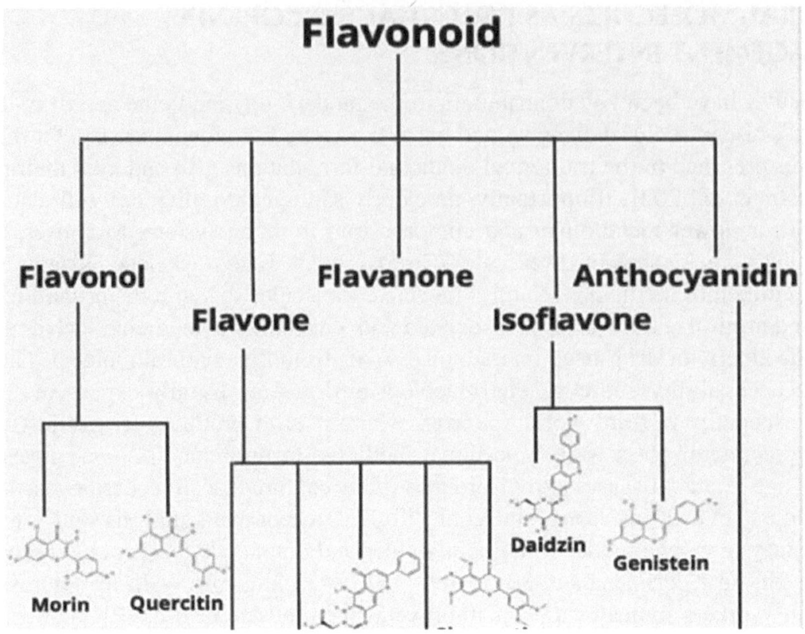

FIGURE 7.3 Classification of flavonoids.

Source: Kumar and Pandey, 2013

pre-clinical evidence for quercetin's beneficial impact on skeletal muscles showed that even short-duration consumption of quercetin by mice positively enhanced the mitochondrial biogenesis markers such as PGC-1α, cytochrome c, and mtDNA and SIRT-1 activation (Davis et al., 2009). Interestingly, Abeer and co-workers demonstrated the effect of 2 weeks of quercetin consumption on rats from progressively increasing ages (maximum up to 15 months): the antioxidant and anti-inflammatory defenses improved significantly with quercetin administration, and the age-associated shift to the pro-apoptotic milieu in the gastrocnemius tissue was controlled as indicated by upregulation of the *Bcl2* gene and downregulation of NF-κB and Bax (Abeer et al., 2021). Besides the benefit of quercetin supplementation in normal aging, there exists a volume of literature on facilitating recovery from muscle damage based on quercetin's ability to modulate proteostasis via suppression of the ubiquitin ligase atrogin-1 and muscle RING finger 1 (MuRF1), favoring muscle mass protection as shown by preserved gastrocnemius muscle weight in mice subjected to disuse-induced muscle atrophy achieved via tail suspension for 10 days (Mukai et al., 2010). In an extension of their work, Mukai and co-workers reported dietary quercetin-mediated Akt phosphorylation responsible for suppressing proteolysis in gastrocnemius muscles and also a decrease in mitochondrial peroxide generation via suppression of PGC-1α and NADH dehydrogenase expression (Mukai et al., 2016). An insightful review by Mukai and Terao on quercetin's anti-atrophic effects on skeletal muscles based on its antioxidant activities reiterates its efficacy in muscle damage models—obesity, muscle disuse, cachexia, etc. (Mukai and Terao, 2013). Nevertheless, there are several instances of clinical evidence where quercetin supplementation (short-duration/long-term consumption) did not offer any advantage in terms of physical performance improvement in soldiers, trained men and women, etc. (Bigelman et al., 2010; Sharp et al., 2012). Notable here is that there exists contrasting literature on the clinical efficacy of quercetin in facilitating recovery from muscle damage originating from various causes (exercise, inflammation, atrophy) (Bazzucchi et al., 2019; Askari et al., 2013).

- **Morin**: Almond hulls, seaweed, and guava are just a few examples of plants that contain this compound of the flavonol. Significant benefits of morin in skeletal muscles are attributed to its potent antioxidant action: in peroxide-induced oxidative damage and apoptosis in murine myoblast culture of C2C12 cells, morin protected the cells via activation of heme oxygenase-1/nuclear factor erythroid 2–related factor 2, phosphorylating and activating the Erk-mediated protection (Lee et al., 2017). Glucocorticoid-induced muscle atrophy can also benefit from morin supplementation, as demonstrated in a dexamethasone-induced atrophy in C2C12 muscles—morin-treated cells exhibited controlled ROS production and Foxo3a expression and balanced cellular proteostasis due to suppressed expression of ubiquitin ligases culminating into thicker myotubes (Ulla et al., 2021). An interesting observation was made by Issac and co-workers working with a rodent myoblast cell line—L6, wherein in addition to potent antioxidant action of morin in cells affirmed by upregulated antioxidant enzymes such as superoxide dismutase, catalase, and enhanced NO scavenging activity, the insulin signaling pathway mediators—insulin receptor tyrosinase kinase, insulin receptor substrate 1, phosphoinositide 3-kinase, glycogen synthase kinase 3-β, glycogen synthase, glucose transporter type 4—were also upregulated. These researchers contemplate an alleviation in the age-related insulin resistance due to morin treatment (Issac et al., 2021). An extremely important utility of morin supplementation in recuperating was demonstrated in case of Lewis lung carcinoma (LLC) murine models of cancer cachexia—morin supplementation controlled the reduction in muscle mass and myofiber size, in addition to controlling the tumor size. The anticachexia and antitumor benefits stemmed from direct action of morin on ribosomal protein S10 (RPS10) (Yoshimura et al., 2018).

7.5.2 Flavones

Flavones have a backbone made up of 2-phenyl-chromen-4-one. Flavones can be found in a variety of foods, primarily spices and various yellow and orange fruits and vegetables (Verma et al., 2020). Various flavones studied for their impact on aged and atrophied muscles are:

- **5,7-Dimethoxyflavone**: Black ginger rhizome (*Kaempferia parviflora*) contains 5,7-dimethoxyflavone (Lee et al., 2018). The muscle mass–enhancing effect of hydroxymethyl flavone (HMF) derivatives was demonstrated by Ono and co-workers, wherein dietary supplementation of senescence-accelerated mouse-prone 1 mice had higher soleus muscle mass and size. Additionally, the mechanistic analysis in C2C12 myotubes with the HMF derivatives indicated the crucial role of intracellular calcium flux in facilitating a muscle mass preservative effect (Ono et al., 2018). Direct benefit of 5,7-dimethoxyflavone examined in aged muscles in 18-month-old mice were examined and reported recently—oral consumption of 5,7-dimethoxyflavone enhanced exercise endurance and grip strength due to the preserved muscle mass and volume in aged mice. The said benefit was attributed to stabilizing proteostasis by decreasing the expression of ubiquitin ligases and autophagy-related genes. The master switch for this cellular mechanism is determined to be the stimulation of the phosphatidylinositol 3-kinase-Akt–mammalian target of rapamycin pathway. Additional molecular benefits observed by HMF supplementation were upregulated expression of peroxisome proliferator–activated receptor–gamma coactivator 1 alpha, nuclear respiratory factor 1, and mitochondrial transcription factor A and alleviation of inflammatory responses in the aged muscles via diminished systemic and expression of the pro-inflammatory cytokines TNK-α and IL-6 (Kim and Hwang, 2020).

- **Apigenin**: Apigenin is a 5,7,4′-trihydroxyflavone found in parsley, celery, and grapefruit (Jang et al., 2017). Pre-clinical evidence supports apigenin-mediated recovery in obesity and denervation-related muscle atrophy via downregulating the inflammatory responses and the muscle-specific ubiquitin ligases, besides enhancing the mitochondrial biogenesis mediators such as AMPK and PGC-1α (Choi et al., 2017, 2018). However, as direct evidence for the benefit of apigenin supplementation in sarcopenia, Wang and co-workers demonstrated the importance of antioxidant action in alleviating age-associated redox imbalance and benefiting aging skeletal muscles. Post 9 months of apigenin intragastric injections, the aged mice (16 months onwards) showed improved grip strength and distance covered on the treadmill. At the tissue level the morphological character and myofiber size improved in the age-matched apigenin-treated mice in the tibialis anterior, soleus muscles, and extensor digitorum longus. The antioxidant enzyme—glutathione peroxidase and superoxide dismutase—activities were boosted along with total antioxidative capability leading to controlled peroxide radicals, lipid peroxidation, and protein carbonylation (Wang et al., 2020).
- **Baicalin**: Baicalin, a glycosidic flavone, is originally sourced from *Scutellaria baicalensis* (Chinese skullcap) and has potent anti-inflammatory and anticancer activity (Emanuele et al., 2016). In a direct investigation in sarcopenic ovariectomized female rats, subcutaneous baicalin injections for 4 weeks improved the capillarization per mm^2 area of soleus, gastrocnemius, and longissimus—a key demonstrated benefit of baicalin, as the capillarization decrement per unit muscle area is a hallmark of sarcopenic muscles. Further, baicalin treatment enhanced the muscle mass of the studied distinct muscle types and also alleviated the exaggerated serum creatine kinase levels correlating clinically with increased muscle wasting. Although mechanistic evaluation of antisarcopenic activity of baicalin was not performed in this study, the authors have implicated its potential to inhibit lipoxygenases and cyclooxygenase-1, responsible for its antioxidant/anti-inflammatory action and stabilizing the proteolytic degradation due to inflammation (Saul et al., 2016). Interestingly, the stabilizing impact of baicalin on skeletal muscle proteostasis was demonstrated earlier in a mouse model of cancer-induced cachexia, wherein the muscle-specific ubiquitin ligases MuRF1 and atrogin-1 protein were downregulated in gastrocnemius muscle tissue by inactivating the NF-kβ pathway and lowering blood TNF-α and IL-6 levels (Li et al., 2014). Interestingly, the benefit of cancer-induced cachexia was replicated clinically, as well as an outcome of baicalin supplementation in subjects with head and neck cancer. It was reported that supplementation with 50 mg of baicalin every day for 3 months increased lean body mass while lowering blood levels of NF-kβ and activin A, two factors that negatively affect muscle mass (Emanuele et al., 2016).
- **Luteolin**: Luteolin is a tetra-hydroxyflavone that can be found in a range of fruits and vegetables (Hawila et al., 2019). Oral supplementation with varying doses of luteolin in rats for 14 days alleviated the pathological consequences of dexamethasone-induced muscle atrophy—the luteolin-treated rats showed improved gastrocnemius muscle mass and myofiber cross-sectional area reflecting in increased grip strength. The serum creatine kinase activity was controlled in luteolin-treated rats, further endorsing the protection conferred to skeletal muscles. Also the enzymatic antioxidant defenses and anti-apoptotic mediators (caspase-3) were boosted and controlled the damage in muscle tissue (Hawila et al., 2019). An interesting potential nutraceutical based on a luteolin/flavonoid-rich fraction of olive leaf extract was indicated to exhibit efficacy in alleviating age-related atrophy and insulin insensitivity in adipose and muscle tissues. Twenty-four-month-old rats were treated with olive leaf fraction rich in various flavanols and simple phenolic compounds (including luteolin) for 21 days, showing an age-related decrement in gastrocnemius muscle mass and protein content. The molecular mediator responsible for the benefit of olive leaf extract was

underpinned as peroxisome proliferator-activated receptor-α, leading to downregulation of sarcopenic markers such as myostatin (González-Hedström et al., 2021). Luteolin supplementation has indeed shown beneficial implications in controlling the cancer-associated inflammation-induced cachexia and muscle atrophy via controlling the activity of ubiquitin ligase muscle RING finger-containing protein 1 (MuRF1) and atrogin-1 (Chen et al., 2018). However, despite the reported benefits of luteolin treatment in muscle atrophy models, the direct utility in sarcopenia is questionable in light of a recent in vitro investigation in mice and C2C12 cells, reporting luteolin-mediated changes in the composition of muscle fiber types—oral administration of luteolin enhanced the proportion of type 1/slow-twitch muscle fibers in the gastrocnemius muscle, along with enhancing the exercise endurance and whole-body aerobic oxidation ability. The molecular pathway responsible for this significant finding was the LCN-AMPK-PGC-1α signaling axis, validated in animal muscles as well as C2C12 cells (Li et al., 2022). Age-related sarcopenia is largely characterized by a fast-twitch fiber type to slow-twitch fiber type (Dao et al., 2020). Nonetheless, the molecular shuffling due to luteolin treatment has a positive impact as evidenced by several studies; however, insights from a direct study in aged animals shall shed more light on prospective benefits.

- **Sinensetin**: Citrus species are rich sources of sinensetin, principally an anti-inflammatory flavone (Akowuah et al., 2005; Ooghe et al., 1994; Kim et al., 2019a). In an interesting investigation by Kim and co-workers, calf and thigh muscles were extracted from young (6 weeks) and old (12 months) rats, and the satellite cells were extracted from the muscles and a primary culture was established and treated with sinensetin. The primary satellite cell culture from old rats' muscles treated with sinensetin had a higher propensity to differentiate and were morphologically closer to the younger rats culture as compared to untreated primary culture from old rats. Specific markers such as myoD and myogenin were also upregulated in sinensetin-treated old rat satellite cell cultures (Kim et al., 2019b). The sinensetin-mediated antisarcopenic efficacy in the study can mechanistically be linked to an earlier study by the same group where they have reported the anti-inflammatory action of sinensetin in lipopolysaccharide-treated inflamed L6 rodent myocyte culture via suppression of NF-κB expression (Kim et al., 2019a).

7.5.3 Flavanone

Flavanones, a subclass of flavonoids, are fragrant, colorless ketones generated from flavones that are commonly found in plants as glycosides. A rich source are hops (*Humulus lupulus*) (Mukai et al., 2012, 2016). Mukai and co-workers reported 8-prenylnaringenin–mediated preservation of muscle mass in denervation-induced disuse muscle atrophy in mice. The oral supplementation of 8-prenylnaringenin raised Akt phosphorylation and lowered atrogin-1 expression (Mukai et al., 2012).

7.5.4 Isoflavone

Isoflavones are phytoestrogens present in legumes such as soybeans, peanuts, green peas, chickpeas, and alfalfa, as well as other agriculturally important legumes. Thus legumes can be of direct benefit to sarcopenic individuals as a protein source, as well bioactive isoflavones. With an average level of 1–2 mg/gram, soybean is a highly rich source of dietary isoflavones (Darbre et al., 2015). Important isoflavones that have been investigated for positively influencing skeletal muscles/aged skeletal muscles are:

- **Daidzein and genistein**: Daidzein and genistein resemble estrogen in terms of structure (Ogawa et al., 2017). Soya extract containing daidzein, genistein, and glycitein was shown to stimulate the anabolic pathways in differentiated C2C12 myotubes and increase the number and diameter of the formed myotubes. The enhanced pro-differentiation outcome was attributed to the ability of the isoflavone mixture to stimulate the estrogen receptor pathway, leading to increased expression of insulin growth factor-1 and the cognate receptor and myosin heavy chain-I expression (Zheng et al., 2018). Several other research groups evaluated the two isoflavones alone or in combination—Hirasaka and co-workers reported the TNF-α (inflammation)–induced upregulation of MuRF ubiquitin ligases leading to atrophy like the milieu in C2C12 and recovery thereof from the pre-treatment with the isoflavone mix. The isoflavone mixture seemed to work via AMPK activation and induction of SIRT-1 expression (Hirasaka et al., 2013). A more interesting observation was reported recently by Aliabadi and co-workers: metformin (a routine clinical prescription) in combination with genistein was shown to shift the cytokine buildup in the obese mice in muscles toward an anti-inflammatory composition—suppression of pro-inflammatory cytokines (TNF-α, IL-1β, MCP-1, and IL-6) vs. induction of IL-10 (Aliabadi et al., 2021). This observation can be of direct consequence in sarcopenic aged individuals on a metformin regimen for metabolic syndrome.

7.5.5 Anthocyanidin

Anthocyanidins are sugar-free versions of anthocyanins, which are the pigments that give plants their colors. In living species, at least 31 monomeric anthocyanidins have been identified, most of which are anthocyanin core components (Khoo et al., 2017). The form of anthocyanidin that has been studied for its antisarcopenic effect is delphindin:

- **Delphinidin**: Delphinidin is a blue-red pigment that can be found in a variety of fruits and vegetables. It has been shown to prevent muscle atrophy (Murata et al., 2016, 2017). Murata and co-workers investigated the delphidin treatment regimen in C2C12 cells and in muscle disuse–induced atrophy in mice. They reported that delphinidin significantly reduced MuRF1 expression by upregulating the nuclear factor of activated T cells 3 (NFATc3) and miR-23a. However, conventional pathways—Akt/p-FoxO3a—remained unaltered. These results suggest that the NFATc3/miR-23a pathway, rather than the Akt/FoxO pathway, is involved in the reduction of MuRF1 expression in response to delphinidin treatment (Murata et al., 2016, 2017).

7.6 CONCLUSION

Sarcopenia is an escalating and pervasive skeletal muscle syndrome marked by rapid muscle loss and functioning and related to adverse health implications. The objective of experimental medicine is to transform the understanding of sarcopenia pathogenesis into diagnostic, pharmacological, and therapeutic advancements. Even though extensive efforts have been made to develop treatments to alleviate muscle atrophy, sarcopenia-specific investigations are limited. Here, we emphasize investigating the therapeutic mechanisms and efficacy of natural bioactive agents as nutraceutical/combination treatment for addressing age-related sarcopenia, due to higher compliance and the multiple molecular and cellular benefits the natural molecules offer in modulating muscle biology. Specifically, flavonoids are diverse polyphenol bioactive chemicals found in food and plants and are classified into numerous groupings based on their chemical properties. The flavonoids have been evaluated in multiple settings—in vitro (L6 rodent myoblast cell line, differentiating C2C12 murine myoblast cell line forming myotubes) and in vivo models (aged mice rats, denervation/disuse/dexamethasone/inflammation induced atrophy)—and have shown significant efficacy in positively modulating the molecular and cellular hallmarks of sarcopenia. Primarily, the mechanisms cluster around antioxidant action, controlling inflammatory mediators, balancing the proteostasis, and mimicking key hormone signals for muscle mass maintenance. This makes their intervention in sarcopenia efficacious, not only devoid of the adverse reactions of therapeutic medications but also economical and accessible to the masses.

BIBLIOGRAPHY

Abeer AE, Khamis AN, Randa E, Braz A. Quercetin modulates age-induced changes in the transcript levels of some apoptosis related genes in the skeletal muscles of male rats. Pharm. Sci. 2020;56: e18861

Akowuah G, Ismail Z, Norhayati I, Sadikun A. The effects of different extraction solvents of varying polarities on polyphenols of Orthosiphon stamineus and evaluation of the free radical-scavenging activity. Food Chem. 2005;93:311–317.

Aliabadi M, Zamani-Garmsiri F, Panahi G, Tehrani SS, Meshkani R. Metformin in combination with genistein ameliorates skeletal muscle inflammation in high-fat diet fed c57BL/6 mice. Cytokine. 2021 Oct;146:155638. doi: 10.1016/j.cyto.2021.155638. Epub 2021 Jul 6. PMID: 34242900.

Alm-Eldeen A, et al. Quercetin modulates age-induced changes in the transcript levels of some apoptosis related genes in the skeletal muscles of male rats. Brazilian Journal of Pharmaceutical Sciences [online]. 2020;56:e18861. Epub 26 Apr 2021. ISSN 2175–9790. https://doi.org/10.1590/s2175-97902020000318086l.

Askari G, Ghiasvand R, Paknahad Z, et al. The effects of quercetin supplementation on body composition, exercise performance and muscle damage indices in athletes. Int J Prev Med. 2013;4(1):21–26.

Atanasov AG, Zotchev SB, Dirsch VM. International natural product sciences taskforce, Supuran CT. Natural products in drug discovery: Advances and opportunities. Nat Rev Drug Discov. 2021 Mar;20(3):200–216. doi: 10.1038/s41573-020-00114-z. Epub 2021 Jan 28. PMID: 33510482; PMCID: PMC7841765.

Bano G, Trevisan C, Carraro S, et al. Inflammation and sarcopenia: A systematic review and meta-analysis. *Maturitas*. 2017;96:10–15. doi:10.1016/j.maturitas.2016.11.006

Battelli MG, Bortolotti M, Bolognesi A, Polito L. Pro-aging effects of Xanthine Oxidoreductase products. Antioxidants. 2020;9(9), 839. doi:10.3390/antiox9090839

Bazzucchi I, Patrizio F, Ceci R, Duranti G, Sgrò P, Sabatini S, Di Luigi L, Sacchetti M, Felici F. The effects of quercetin supplementation on eccentric exercise-induced muscle damage. Nutrients. 2019 Jan 21;11(1):205. doi:10.3390/nu11010205. PMID: 30669587; PMCID: PMC6356612.

Beaudart C, Zaaria M, Pasleau F, Reginster JY, Bruyère O. Health outcomes of Sarcopenia: A systematic review and meta-analysis. PLoS One. 2017 Jan 17;12(1):e0169548. doi: 10.1371/journal.pone.0169548. PMID: 28095426; PMCID: PMC5240970.

Bian A, Ma Y, Zhou X, Guo Y, Wang W, Zhang Y, Wang X. Association between sarcopenia and levels of growth hormone and insulin-like growth factor-1 in the elderly. BMC Musculoskelet Disord. 2020 Apr 7;21(1):214. doi: 10.1186/s12891-020-03236-y. PMID: 32264885; PMCID: PMC7140321.

Bigelman KA, Fan EH, Chapman DP, Freese EC, Trilk JL, Cureton KJ. Effects of six weeks of quercetin supplementation on physical performance in ROTC cadets. Mil Med. 2010 Oct;175(10):791–798. doi: 10.7205/milmed-d-09-00088. PMID: 20968271.

Boengler K, Kosiol M, Mayr M, Schulz R, Rohrbach S. Mitochondria and ageing: Role in heart, skeletal muscle and adipose tissue. Journal of Cachexia, Sarcopenia and Muscle. 2017;8(3):349–369. https://doi.org/10.1002/jcsm.12178

Brioche T, Kireev RA, Cuesta S, Gratas-Delamarche A, Tresguerres JA, Gomez-Cabrera MC, Viña J. Growth hormone replacement therapy prevents sarcopenia by a dual mechanism: Improvement of protein balance and of antioxidant defenses. J Gerontol A Biol Sci Med Sci. 2014 Oct;69(10):1186–1198. doi:10.1093/gerona/glt187. Epub 2013 Dec 3. PMID: 24300031.

Cesari M, Landi F, Vellas B, Bernabei R, Marzetti E. Sarcopenia and physical frailty: Two sides of the same coin. Frontiers in Aging Neuroscience. 2014;6. doi:10.3389/fnagi.2014.00192

Chen L, Zhang H, Chi M, Yang Q, Guo C. Drugs for the treatment of muscle atrophy. In (Ed.), Background and Management of Muscular Atrophy. IntechOpen. 2020a. https://doi.org/10.5772/intechopen.93503

Chen T, Li B, Xu Y, Meng S, Wang Y, Jiang Y. Luteolin reduces cancer-induced skeletal and cardiac muscle atrophy in a Lewis lung cancer mouse model. Oncol Rep. 2018 Aug;40(2):1129–1137. doi: 10.3892/or.2018.6453. Epub 2018 May 21. PMID: 29845270.

Chen W, Datzkiw D, Rudnicki MA. Satellite cells in ageing: use it or lose it. Open Biology. 2020b;10(5):200048. doi:10.1098/rsob.200048

Choi WH, Jang YJ, Son HJ, Ahn J, Jung CH, Ha TY. Apigenin inhibits sciatic nerve denervation-induced muscle atrophy. Muscle Nerve. 2018;58:314–318.

Choi WH, Son HJ, Jang YJ, Ahn J, Jung CH, Ha TY. Apigenin ameliorates the obesity-induced skeletal muscle atrophy by attenuating mitochondrial dysfunction in the muscle of obese mice. Mol. Nutr. Food Res. 2017;61:1700218.

Coen PM, Musci RV, Hinkley JM, Miller BF. Mitochondria as a target for mitigating sarcopenia. Frontiers in Physiology. 2019;9:1883. https://doi.org/10.3389/fphys.2018.01883.

Cruz-Jentoft AJ, Bahat G, Bauer J, Boirie Y, Bruyère O, Cederholm T, Cooper C, Landi F, Rolland Y, Sayer AA, Schneider SM, Sieber CC, Topinkova E, Vandewoude M, Visser M, Zamboni M. Writing group for the European Working Group on Sarcopenia in Older People 2 (EWGSOP2), and the extended group for EWGSOP2. Sarcopenia: Revised European consensus on definition and diagnosis. Age Ageing. 2019 Jan 1;48(1):16–31. doi: 10.1093/ageing/afy169

Dao T, Green AE, Kim YA, et al. Sarcopenia and muscle aging: A brief overview. Endocrinol Metab (Seoul). 2020;35(4):716–732. doi: 10.3803/EnM.2020.405

Darbre PD. What are endocrine disruptors and where are they found? Endocrine Disruption and Human Health. 2015:3–26. doi: 10.1016/b978-0-12-801139-3.00001-6

Daussin FN, Boulanger E, Lancel S. From mitochondria to sarcopenia: Role of inflammaging and RAGE-ligand axis implication. Exp Gerontol. 2021;146:111247. doi: 10.1016/j.exger.2021.111247

Davis JM, Murphy EA, Carmichael MD, Davis B. Quercetin increases brain and muscle mitochondrial biogenesis and exercise tolerance. Am. J. Physiol. Regul. Integr. Comp. Physiol. 2009;296:R1071–R1077.

Emanuele E, Bertona M, Pareja-Galeano H, Fiuza-Luces C, Morales JS, Sanchis-Gomar F, Lucia A. Baicalin supplementation reduces serum biomarkers of skeletal muscle wasting and may protect against lean body mass reduction in cancer patients: Results from a pilot open-label study. Neuro Endocrinol Lett. 2016 Jul;37(3):213–216.

Gonzalez-Freire M, Scalzo P, D'Agostino J, Moore ZA, Diaz-Ruiz A, Fabbri E, Zane A, Chen B, Becker KG, Lehrmann E, Zukley L, Chia CW, Tanaka T, Coen PM, Bernier M, de Cabo R, Ferrucci L. Skeletal muscle ex vivo mitochondrial respiration parallels decline in vivo oxidative capacity, cardiorespiratory fitness, and muscle strength: The Baltimore Longitudinal Study of Aging. Aging Cell. 2018 Apr;17(2):e12725. doi:10.1111/acel.12725.

González-Hedström D, Priego T, Amor S, et al. Olive leaf extract supplementation to old wistar rats attenuates aging-induced sarcopenia and increases insulin sensitivity in adipose tissue and skeletal muscle. Antioxidants (Basel). 2021;10(5):737. Published 2021 May 7. doi:10.3390/antiox10050737

Gupta C, Prakash D. Nutraceuticals for geriatrics. J Tradit Complement Med. 2014;5(1):5–14. Published 2014 Dec 17. doi:10.1016/j.jtcme.2014.10.004

Hawila N, Hedya S, Maaly AE, Abdin A. Luteolin attenuates dexamethasone-induced skeletal muscle atrophy in male albino rats. Med. J. Cairo Univ. 2019;87:3365–3374.

Herbst A, Lee CC, Vandiver AR, Aiken JM, McKenzie D, Hoang A, Allison D, Liu N, Wanagat J. Mitochondrial DNA deletion mutations increase exponentially with age in human skeletal muscle. Aging Clin Exp Res. 2021 Jul;33(7):1811–1820. doi: 10.1007/s40520-020-01698-7. Epub 2020 Sep 23. PMID: 32965609; PMCID: PMC7985047.

Hirasaka K, Maeda T, Ikeda C, Haruna M, Kohno S, Abe T, Ochi A, Mukai R, Oarada M, Eshima-kondo S. Ohno A, Okumura Y, Terao J, Nikawa T. Isoflavones derived from soy beans prevent MuRF1-mediated muscle atrophy in C2C12 myotubes through SIRT1 activation. J. Nutr. Sci. Vitaminol. 2013;59:317–324.

Hu Yang C-C, Tu SJ, Huang IJ, Ganbat D, Guo L-Y. Characteristics of the electrophysiological properties of neuromuscular motor units and its adaptive strategy response in lower extremity muscles for seniors with pre-sarcopenia: A preliminary study. International Journal of Environmental Research and Public Health. 2021;18(6):3063. https://doi.org/10.3390/ijerp

Issac PK, Karan R, Guru A, Pachaiappan R, Arasu MV, Al-Dhabi NA, Choi KC, Harikrishnan R, Raj JA. Insulin signaling pathway assessment by enhancing antioxidant activity due to morin using in vitro rat skeletal muscle L6 myotubes cells. Mol Biol Rep. 2021 Aug;48(8):5857–5872. doi: 10.1007/s11033-021-06580-x. Epub 2021 Jul 23. PMID: 34302266.

Jang YJ, Son HJ, Choi YM, Ahn J, Jung CH, Ha TY. Apigenin enhances skeletal muscle hypertrophy and myoblast differentiation by regulating Prmt7. Oncotarget. 2017;8:78300–78311.

Jeong HY, Kwon O. Dietary phytochemicals as a promising nutritional strategy for sarcopenia: A systematic review and meta-analysis of randomized controlled trials. Appl Biol Chem. 2021;64:60. https://doi.org/10.1186/s13765-021-00633-2

Kaczmarek A, Kaczmarek M, Ciałowicz M, Clemente FM, Wolański P, Badicu G, Murawska-Ciałowicz E. The role of satellite cells in skeletal muscle regeneration: The effect of exercise and age. Biology. 2021;10:1056. https://doi.org/10.3390/biology10101056

Kawser Hossain M, Abdal Dayem A, Han J, Yin Y, Kim K, Kumar Saha S, Yang GM, Choi HY, Cho SG. Molecular mechanisms of the anti-obesity and anti-diabetic properties of flavonoids. Int. J. Mol. Sci. 2016;17:569.

Khoo HE, Azlan A, Tang ST, Lim SM. Anthocyanidins and anthocyanins: Colored pigments as food, pharmaceutical ingredients, and the potential health benefits. Food & Nutrition Research. 2017;61(1):1361779. https://doi.org/10.1080/16546628.2017.1361779

Kilsby AJ, Sayer AA, Witham MD. Selecting potential pharmacological interventions in sarcopenia. Drugs & Aging. 2017;34(4):233–240. doi:10.1007/s40266-017-0444-z

Kim C, Hwang JK. The 5,7-dimethoxyflavone suppresses srcopenia by rgulating protein turnover and mitochondria biogenesis-related pathways. Nutrients. 2020;12:1079.

Kim JA, Ha SE, Kim SM, Vetrivel P, Kim GS, Kim EH. Anti-inflammatory activities of sinensetin on LPS-stimulated L6 skeletal muscle cells. J. Biomed. Transl. Res. 2019a;20:1–7.

Kim JA, Kim SM, Ha SE, Vetrivel P, Saralamma VVG, Kim EH, Kim GS. Sinensetin regulates age-related sarcopenia in cultured primary thigh and calf muscle cells. BMC Complement. Altern. Med. 2019b;19:287.

Kumar S, Pandey AK. Chemistry and biological activities of flavonoids: An overview. Sci. World. J. 2013;2013:162750.

Kwak JY, Kwon KS. Pharmacological interventions for treatment of sarcopenia: Current status of drug development for sarcopenia. Ann Geriatr Med Res. 2019;23(3):98–104. doi:10.4235/agmr.19.0028

Larsson L, Degens H, Li M, Salviati L, Lee Y, Thompson W, . . . Sandri M. Sarcopenia: Aging-related loss of muscle mass and function. Physiological Reviews. 2019;99(1):427–511. doi: 10.1152/physrev.00061.201

Laumonier T, Menetrey J. Muscle injuries and strategies for improving their repair. Journal of Experimental Orthopaedics. 2016;3(1). doi:10.1186/s40634-016-0051-7

Lee MH, Han MH, Lee DS, Park C, Hong SH, Kim GY, Hong SH, Song KS, Choi IW, Cha HJ, Choi YH. Morin exerts cytoprotective effects against oxidative stress in C2C12 myoblasts via the upregulation of Nrf2-dependent HO-1 expression and the activation of the ERK pathway. Int J Mol Med. 2017 Feb;39(2):399–406. doi:10.3892/ijmm.2016.2837. Epub 2016 Dec 23. PMID: 28035409.

Lee S, Kim C, Kwon D, Kim MB, Hwang JK. Standardized Kaempferia parviflora Wall. ex Baker (Zingiberaceae) extract inhibits fat accumulation and muscle atrophy in ob/ob mice. Evid. Based Complement. Altern. Med. 2018;2018:8161042.

Leitner LM, Wilson RJ, Yan Z, Gödecke A. Reactive oxygen species/nitric oxide mediated inter-organ communication in skeletal muscle wasting diseases. Antioxidants & Redox Signaling. 2017;26(13):700–717. https://doi.org/10.1089/ars.2016.6942

Le Marchand L. Cancer preventive effects of flavonoids—a review. Biomed. Pharmacother. 2002;56:296–301 lean body mass reduction in cancer patients: Results from a pilot open-label study. Neuro. Endocrinol. Lett. 2016;37:213–216.

Li B, Wan L, Li Y, Yu Q, Chen P, Gan R, Yang Q, Han Y, Guo C. Baicalin, a component of Scutellaria baicalensis, alleviates anorexia and inhibits skeletal muscle atrophy in experimental cancer cachexia. Tumor Biol. 2014;35:12415–12425.

López-Otín C, Blasco MA, Partridge L, Serrano M, Kroemer G. The hallmarks of aging. Cell. 2013;153(6):1194–1217. doi: 10.1016/j.cell.2013.05.039

Mahalanobish S, Saha S, Dutta S, Ghosh S, Sil PC. Anti-inflammatory efficacy of some potentially bioactive natural products against rheumatoid arthritis. Discovery and Development of Anti-Inflammatory Agents from Natural Products. 2019:61–100. doi: 10.1016/b978-0-12-816992-6.00003-6

Mankhong S, Kim S, Moon S, Kwak HB, Park DH, Kang JH. Experimental models of sarcopenia: Bridging molecular mechanism and therapeutic strategy. Cells. 2020;9(6):1385. Published 2020 Jun 2. doi: 10.3390/cells9061385

McKenzie D, Bua E, McKiernan S, Cao Z, Aiken JM, Wanagat J. Mitochondrial DNA deletion mutations: A causal role in sarcopenia. Eur J Biochem. 2002 Apr;269(8):2010–2015. doi: 10.1046/j.1432-1033.2002.02867.x.

Mukai R, Horikawa H, Fujikura Y, Kawamura T, Nemoto H, Nikawa T, et al. Prevention of disuse muscle atrophy by dietary ingestion of 8-prenylnaringenin in denervated mice. PLoS ONE. 2012;7(9):e45048. https://doi.org/10.1371/journal.pone.0045048

Mukai R, Horikawa H, Lin PY, Tsukumo N, Nikawa T, Kawamura T, Nemoto H, Terao J. 8-Prenylnaringenin promotes recovery from immobilization-induced disuse muscle atrophy through activation of the Akt phosphorylation pathway in mice. Am J Physiol Regul Integr Comp Physiol. 2016 Dec 1;311(6):R1022–R1031. doi: 10.1152/ajpregu.00521.2015. Epub 2016 Sep 14. PMID: 27629889.

Mukai R, Nakao R, Yamamoto H, Nikawa T, Takeda E, Terao J. Quercetin prevents unloading-derived disused muscle atrophy by attenuating the induction of ubiquitin ligases in tail-suspension mice. Journal of Natural Products. 2010;73(10):1708–1710.

Mukai R, Terao J. Role of dietary flavonoids in oxidative stress and prevention of muscle atrophy. The Journal of Physical Fitness and Sports Medicine. 2013;2(4):385–392.

Murata M, Kosaka R, Kurihara K, Yamashita S, Tachibana H. Delphinidin prevents disuse muscle atrophy and reduces stress-related gene expression. Biosci. Biotechnol. Biochem. 2016;80:1636–1640.

Murata M, Nonaka H, Komatsu S, Goto M, Morozumi M, Yamada S, Lin I-C, Yamashita S, Tachibana H. Delphinidin prevents muscle atrophy and upregulates miR-23a expression. J. Agric. Food Chem. 2017;65:45–50.

Ogawa M, Kitano T, Kawata N, Sugihira T, Kitakaze T, Harada N, Yamaji R. Daidzein down-regulates ubiquitin-specific protease 19 expression through estrogen receptor b and increases skeletal muscle mass in young female mice. J. Nutr. Biochem. 2017;49:63–70.

Ono S, Yoshida N, Maekawa D, Kitakaze T, Kobayashi Y, Kitano T, Fujita T, Okuwa-Hayashi H, Harada N, Nakano Y, Yamaji R. 5-Hydroxy-7-methoxyflavone derivatives from Kaempferia parviflora induce skeletal muscle hypertrophy. Food Sci Nutr. 2018 Nov 20;7(1):312–321. doi: 10.1002/fsn3.891. PMID: 30680186; PMCID: PMC6341173.

Ooghe WC, Ooghe SJ, Detavernier ClM, Huyghebaert A. Characterization of orange juice (Citrus sinensis) by polymethoxylated flavones. J. Agric. Food Chem. 1994;42:2191–2195.

Pascual-Fernández J, Fernández-Montero A, Córdova-Martínez A, Pastor D, Martínez-Rodríguez A, Roche E. Sarcopenia: Molecular pathways and potential targets for intervention. International Journal of Molecular Sciences. 2020;21(22):8844. https://doi.org/10.3390/ijms2122884

Peiyuan Li, Ruilin Xu, Yanfei Shi, Xiaoyang Shi, Xinyuan Zhang, Jinjie Li, Guangning Kou. Luteolin increases slow muscle fibers via FLCN-AMPK-PGC-1α signaling pathway. Journal of Functional Foods. 2022;88:104876. ISSN 1756-4646. https://doi.org/10.1016/j.jff.2021.104876.

Picca A, Calvani R. Molecular mechanism and pathogenesis of sarcopenia: An overview. Int J Mol Sci. 2021 Mar 16;22(6):3032. doi: 10.3390/ijms22063032. PMID: 33809723; PMCID: PMC8002369.

Santanasto AJ, Glynn NW, Jubrias SA, et al. Skeletal Muscle Mitochondrial Function and Fatigability in Older Adults. The Journals of gerontology. Series A, Biological Sciences and Medical Sciences. 2015 Nov;70(11):1379–1385. DOI: 10.1093/gerona/glu134.

Saul D, Kling JH, Kosinsky RL, Hoffmann DB, Komrakova M, Wicke M, Menger B, Sehmisch S. Effect of the Lipoxygenase Inhibitor Baicalein on Muscles in Ovariectomized Rats. J Nutr Metab. 2016;2016:3703216. doi: 10.1155/2016/3703216. Epub 2016 Dec 5. PMID: 28050282; PMCID: PMC5165164.

Sharp MA, Hendrickson NR, Staab JS, McClung HL, Nindl BC, Michniak-Kohn BB. Effects of short-term quercetin supplementation on soldier performance. J Strength Cond Res. 2012 Jul;26(Suppl 2):S53–S60. doi: 10.1519/JSC.0b013e31825cf22d. PMID: 22614228.

Shimokata H, Shimada H, Satake S, Endo N, Shibasaki K, Ogawa S, Arai H. Chapter 2 epidemiology of sarcopenia. Geriatrics & Gerontology International. 2018;18:13–22. doi: 10.1111/ggi.13320

Toda K, Hitoe S, Takeda S, Shimoda H. Black ginger extract increases physical fitness performance and muscular endurance by improving inflammation and energy metabolism. Heliyon. 2016a;2:e00115.

Toda K, Takeda S, Hitoe S, Nakamura S, Matsuda H, Shimoda H. Enhancement of energy production by black ginger extract containing polymethoxy flavonoids in myocytes through improving glucose, lactic acid and lipid metabolism. J. Nat. Med. 2016b;70:163–172.

Ulla A, Uchida T, Miki Y, Sugiura K, Higashitani A, Kobayashi T, Ohno A, Nakao R, Hirasaka K, Sakakibara I, Nikawa T. Morin attenuates dexamethasone-mediated oxidative stress and atrophy in mouse C2C12 skeletal myotubes. Arch Biochem Biophys. 2021 Jun 15;704:108873. doi: 10.1016/j.abb.2021.108873. Epub 2021 Apr 10. PMID: 33848514.

Varteresian T, Lavretsky H. Natural products and supplements for geriatric depression and cognitive disorders: An evaluation of the research. Curr Psychiatry Rep. 2014;16(8):456. doi: 10.1007/s11920-014-0456-x
Verma ML, Sharma S, Saini R, Rani V, Kushwaha R. Bioflavonoids. Biotechnological Production of Bioactive Compounds. 2020:69–105. doi: 10.1016/b978-0-444-64323-0.00003-5
Wang D, Yang Y, Zou X, Zhang J, Zheng Z, Wang Z. Antioxidant apigenin relieves age-related muscle atrophy by inhibiting oxidative stress and hyperactive mitophagy and apoptosis in skeletal muscle of mice. J Gerontol A Biol Sci Med Sci. 2020;75(11):2081–2088. doi: 10.1093/gerona/glaa214
Wiedmer P, Jung T, Castro JP, et al. Sarcopenia—molecular mechanisms and open questions. Ageing Res Rev. 2021;65:101200. doi: 10.1016/j.arr.2020.101200.
Yoshimura T, Saitoh K, Sun L, Wang Y, Taniyama S, Yamaguchi K, Uchida T, Ohkubo T, Higashitani A, Nikawa T, Tachibana K, Hirasaka K. Morin suppresses cachexia-induced muscle wasting by binding to ribosomal protein S10 in carcinoma cells. Biochem Biophys Res Commun. 2018 Dec 2;506(4):773–779. doi: 10.1016/j.bbrc.2018.10.184. Epub 2018 Oct 31. PMID: 30389140.
Yoshino M, Naka A, Sakamoto Y, Shibasaki A, Toh M, Tsukamoto S, Kondo K, Iida K. Dietary isoflavone daidzein promotes Tfam expression that increases mitochondrial biogenesis in C2C12 muscle cells. J. Nutr. Biochem. 2015;26:1193–1199.
Zembron-Lacny A., et al. The relation of inflammaging with skeletal muscle properties in elderly men. American Journal of Men's Health. 2019;13(2):1557988319841934. doi:10.1177/1557988319841934.
Zhang X, Zhang W, Wang C, Tao W, Dou Q, Yang Y. Sarcopenia as a predictor of hospitalization among older people: A systematic review and meta-analysis. BMC Geriatr. 2018 Aug 22;18(1):188. doi: 10.1186/s12877-018-0878-0. PMID: 30134867; PMCID: PMC6103964.
Zheng W, Hemker ML, Xie M, Soukup ST, Diel P. Anabolic activity of a soy extract and three major isoflavones in C2C12 myotubes. Planta Med. 2018 Sep;84(14):1022–1029. doi: 10.1055/a-0598-4812. Epub 2018 Apr 12. PMID: 29649842.

8 Herbal Medicinal Compounds and Their Anti–Breast Cancer Actions

A Mechanistic Perspective

Daniel A. Gideon, Pushparaj Annadurai, Abhinav Parashar, Rachel Grace Murthy, Kandavel Dhandayuthapani, Sriram Srinivasan, Sivaramakrishnan R, and R. Jasmine

CONTENTS

8.1 Introduction ... 158
8.1.1 Incidence of Breast Cancer ... 158
8.1.2 Breast Cancer and Its Causes ... 158
8.1.3 Treatment Modalities for Breast Cancer and the Relevance of Phytomedicine ... 159
8.2 Breast Cancer Signaling Pathways and Phytochemicals as Pharmacological Agents ... 161
8.2.1 Molecular Mechanism of Breast Cancer and Targets of Signaling Pathways ... 161
8.2.1.1 MaSC Conversion to BCSCs ... 161
8.2.1.2 Other Signaling Pathways in Breast Cancer Cells ... 161
8.2.2 The History of Medicine ... 163
8.2.3 Relevance of Phytochemicals in Breast Cancer Therapy ... 164
8.2.3.1 Curcumin ... 169
8.2.3.2 Capsaicin ... 170
8.2.3.3 Quercetin ... 170
8.2.3.4 Resveratrol ... 170
8.2.3.5 Gallic Acid ... 171
8.2.4 How Phytochemicals Prevent/Treat Breast Cancer ... 171
8.2.4.1 Inhibition of Breast Cancer Stem Cell Proliferation ... 173
8.2.4.2 Binding to Proteins/Enzymes and Blocking Substrate/Protein Partner Binding ... 173
8.2.4.3 Antioxidant Action and Inhibition of Electron Transfer (Redox Activity) ... 174
8.3 Conclusion ... 175
Bibliography ... 175

DOI: 10.1201/b22917-8

8.1 INTRODUCTION

8.1.1 Incidence of Breast Cancer

A report shows that in the year 2018 alone, 18.1 million new cases of cancer were estimated to have occurred worldwide; these numbers are predicted to reach a colossal 23.6 million new cases annually by the year 2030 (Bray et al., 2018). Breast cancer is on the rise, with about 2.3 million reported cases every year globally. There are three major subtypes of breast cancer based on the incidence (presence/absence) of three molecular markers—human epidermal growth factor receptor 2 (ERBB2/HER2), estrogen receptor (ER) and progesterone receptor (PR/NR3C3/PgR), which in turn leads to i) ERBB2/HER2 negative/hormone receptor (HR) positive (in ~70% of all cases), ii) ERBB2 positive (15–20% of cases) and iii) triple negative breast cancer (TNBC), which lacks all three key markers. There are many classifications of breast cancer types based on location of the tumor, stage and grade of the cancer; source; and receptor status. Based on the location of cancer, the four varieties of breast cancer are i) luminal A, ii) luminal B, iii) basal-like and iv) EGFR2/HER2-enriched (ER negative) types (Bagaria et al., 2014; Harbeck et al., 2019). However, in clinical practice, there are five distinguishable types of breast cancer: i) HR positive (in which tumors express either ER or PR), ii) TNBC (in which there is no detectable ER, HER2 or PR), iii) luminal B-like HER2+, iv) luminal B-like HER2- and v) luminal A-like ER+ and PR+ (but HER2-) (Bagaria et al., 2014).

8.1.2 Breast Cancer and Its Causes

Cancer is perhaps the most challenging non-communicable ailment that humans continue to battle in recent times. The hallmarks of cancer are i) establishment of replicative immortality, ii) induction of angiogenesis, iii) resistance to apoptosis, iv) sustained proliferative signaling, v) evasion of growth suppressors, vi) establishment of tumor invasion and metastasis, vii) accomplishment of mutations and genomic instability, viii) dysregulation of cellular energetics, ix) tumor evasion of immune responses and x) establishment of inflammation leading to tumor promotion (Hanahan and Weinberg, 2011). Both genetic and environmental causes of carcinogenesis have been uncovered; despite the advances in cancer research, the etiology of cancer is still puzzling and warrants further investigation (Haile et al., 2012). In a generic sense, all cancers are caused because of inheritable and non-inheritable causes; among the non-inheritable causes, lifestyle factors (smoking, high-fat diet, toxic chemical exposure, alcohol consumption and sedentary lifestyle) are the major contributors to cancer risk (Bosetti et al., 2005). Besides lifestyle changes, environmental factors like viral infections (Epstein-Barr virus [EBV], human immunodeficiency virus [HIV], cytomegalovirus [CMV], etc.); receiving high-dose chemotherapy early on in life; and the presence of comorbidities like diabetes, obesity, acid reflux disease, stress, insomnia and chronic inflammation are conditions which promote carcinogenesis (Gideon and James, 2021). Weakened immunity and poor dietary choices caused by macronutrient/vitamin/mineral deficiencies, high vegetable oil consumption, gut dysbiosis, etc., are other potential risk factors in the complex interplay of events (and causative factors) leading to carcinogenesis. Alterations in gene expression and hallmark capabilities of breast cancer occur because of mutations in critical genes like *APC*, *p53*, *BRCA1*, *BRCA2*, *MMR*, *PALB2*, *CHEK2*, *PTEN*, *CDH1*, *NF1*, *STK11*, *PIK3CA*, *MUTYH*, *BRP1*, *PMS2*, *RAD51C*, *RAD51D*, *BARD1*, etc. (Harbeck et al., 2019). Almost all of these genes are involved in one of the following critical cellular functions which govern genomic stability and cellular fate: DNA repair (double strand break repair or mismatch repair), cell cycle checkpoint regulation, tumor suppression, and growth factor binding. Among these genes, *BRCA1* (20% of cases), *BRCA2* (20%), *CHEK2* (5%) and *TP53* (<1%) mutations are responsible for the majority of the breast cancer incidence, and the remaining 54% of mutations are linked to other candidate genes, whose roles in breast cancer incidence continue to be uncovered (Wooster and Weber, 2003). The tremendous diversity of etiological causes and genes involved in carcinogenesis also makes breast cancer a very heterogeneous disease to deal with and hence, cancer therapy aimed at breast cancer management is quite complicated, even for those who are well-versed with the etiology of the disease. The salient mechanisms involved in breast cancer pathogenesis and progression and the types as well as stages of breast cancer are summarized in Figure 8.1.

8.1.3 Treatment Modalities for Breast Cancer and the Relevance of Phytomedicine

Based on the stage of the disease and its molecular subtype, both systemic and loco-regional therapeutic approaches are used for breast cancer treatment. Among the systemic therapies, the prominent avenues for treating breast cancer are i) endocrine therapy (for HR-positive type); ii) chemotherapy, using natural as well as synthetic organic compounds (and radiation chemotherapy); iii) anti-HER2 therapy (drugs and monoclonal antibodies which target HER2 and prevent EDGF binding); iv) bone-stabilizers; v) poly(ADP-ribose) polymerase (PARP) inhibitors; and vi) immunotherapy (Fleege and Cobain, 2022; Samadder et al., 2019). Combinations of these therapeutic avenues have augmented the chances for increasing the efficacy of breast cancer treatment by upto 70–80%. However, in advanced stages of breast cancer, treatment prognosis is dismal and therefore, cancer has to be diagnosed at very early stages to increase the chance of curing the disease. Hence, governmental agencies in western countries encourage women over 60 years of age to get screened for breast cancer every year.

Enhanced understanding of the critical molecular mechanisms of cancer initiation and progression has paved the way for identification and screening of a vast number of anticancer drugs. Advanced methods like high-throughput screening have largely enhanced the pace of anticancer drug research. Almost every facet/druggable target in breast cancer has been explored, and the advent of ligand-based drug design (LBDD), structure-based drug design (SBDD), molecular docking, molecular dynamics simulations and other in silico methodologies has amplified our understanding of the structure–activity relationships of drug candidates. Based on the cancer-related phenomenon that anticancer drugs tackle, several types of breast cancer drugs such as selective anti-inflammatory drugs, monoclonal antibody therapies (for immune activation), telomerase inhibitors, cell cycle inhibitors (eg. which bind to cyclin-dependent kinases), EGFR inhibitors, aerobic glycolysis (aka the Warburg effect) inhibitors, proapoptotic BH3 mimics, PARP inhibitors and VEGF inhibitors (Hanahan and Weinberg, 2011) have become routine anticancer agents that we possess in our arsenal. Currently, antibody drug conjugates (ADCs), which specifically target HER2 and tyrosine kinase inhibitors (TKIs), and combinatorial therapies (ADCs, TKIs and chemotherapeutic agents), which synergistically block multiple breast cancer targets, have become the new norm in breast cancer therapy. Knowledge of the state of the art in breast cancer pathology and pathophysiology can improve the clinical therapeutic regimen and improve treatment outcomes. Hence, clarity in understanding the breast cancer targets and ligands that inhibit them is of the utmost importance for the discovery of novel biotherapeutics and natural agents to tackle breast cancer. Targeted breast cancer therapy involving the use of carbon nanotubes and nanoparticle-based drug delivery systems is also an emerging treatment modality (Zheng et al., 2020). Despite these phenomenal advances in breast cancer treatment avenues, only modest improvements are seen in overall survival (OS) rates, disease-free survival (DFS), progression-free survival (PFS) rates, health-related quality of life/quality of life (HR-QOL/QOL), objective response rate (ORR), complete response (CR), etc., which are the most common clinical outcomes that are monitored in cancer clinical trials. These parameters are the defining endpoints in human clinical trials, and improvements in these indices support the effectiveness of a treatment modality. Moreover, there is a significant development of chemotherapeutic drug resistance by several cancer types. Hence, the need to develop newer therapeutic strategies and novel chemopreventive agents has become pivotal in addressing breast cancer–related morbidity and mortality. While it is generally presumed that plant formulations are toxicity-free,

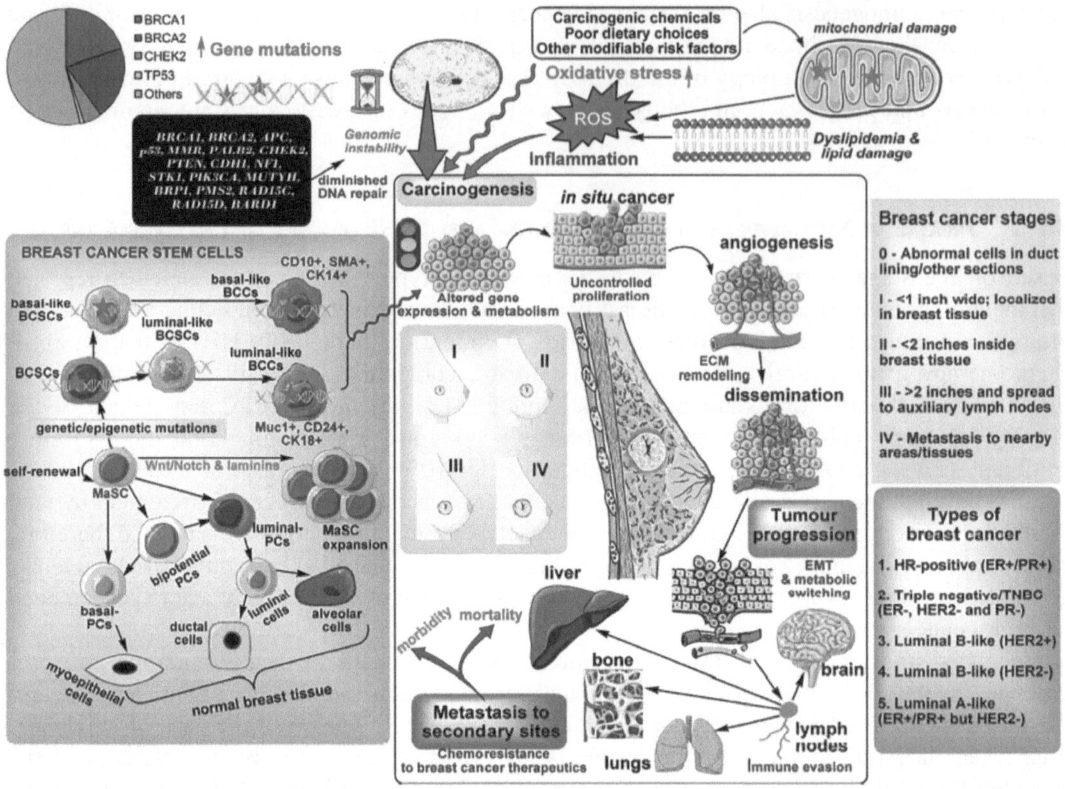

FIGURE 8.1 Breast cancer: Overview of mechanisms of carcinogenesis, stages and types. The key genetic defects in breast cancer pathogenesis are listed. *BRCA1*, *BRCA2*, *CHEK2* and *TP53* alone account for 20%, 20%, 5% and <1%, respectively, and other genes are responsible for >50% of the cases. Oxidative stress can be caused by diverse factors like carcinogenic chemical exposure, viral infections, poor dietary choices, alcohol consumption and obesity. ROS can cause damage to macromolecules like DNA, RNA and proteins and tilt the redox homeostasis in cells. The risk for breast cancer is higher in women, especially as age increases (>50–60 years). Breast cancer stem cells (BCSCs) serve as tumor initiation cells (TICs) and are formed from normal mammary stem cells (MaSCs), which gain self-renewal potential and give rise to both BCSCs and non-BCSCs. Clonal evolution gives rise to tumor heterogeneity through mutations, deletions and gene amplifications that are passed on to succeeding cellular generations of basal-like BCSCs and luminal-like BCSCs, which further differentiate into their corresponding basal-like and luminal-like breast cancer cells. Pathogenesis is a result of diminished DNA repair, because most of the gene mutations linked to breast cancer are involved either in DNA repair or cell cycle checkpoint control. Carcinogenesis further progresses to allow in situ cancer mass formation, ECM remodeling, dissemination into blood tissue, spread to nearby lymph nodes and thence, to other distant secondary sites (metastasis), causing morbidity and mortality. Dysregulation of mitochondrial metabolism, prevalence of the Warburg effect, metabolic switching (from TCA-dependent to glycolytic switch and glutamine-fueled ATP synthesis) and epithelial to mesenchymal transition (EMT) promotes cancer proliferation, metastasis and invasion. Based on the histological and molecular characteristics (presence/absence of ER, PR and HER2), there are five major types of breast cancer. The stages of breast cancer are 0–IV based on the size and the spread of the tumor from its site of origin.

it is important to understand that both natural and synthetic drug candidates can be toxic and even carcinogenic and hence, each drug/molecule has to be explored for safety and efficacy in pre-clinical studies and clinical trials before being sold on the market and administered to patients.

8.2 BREAST CANCER SIGNALING PATHWAYS AND PHYTOCHEMICALS AS PHARMACOLOGICAL AGENTS

8.2.1 Molecular Mechanism of Breast Cancer and Targets of Signaling Pathways

8.2.1.1 MaSC Conversion to BCSCs

Figure 8.2 shows the processes involved in the conversion of MaSCs into BCSCs. First, MaSCs acquire self-renewal potential. Expansion of MaSCs occurs through Wnt/Notch signaling events. When MaSCs acquire genetic and epigenetic mutations/errors, they become transformed into BCSCs and non-BCSCs. When non-BCSCs encounter hypoxia as well as pro-inflammatory signals like interleukin (IL)-6, IL-8, and transforming growth factor-beta (TGF-β), they become BCSCs, which on the other hand, in response to IL-6, Notch, Wnt and GM-CSF signaling, undergo clonal expansion to form multiple BCSCs. The BCSCs themselves divide into either basal-like or luminal-like BCSCs, and these form basal-like and luminal-like breast cancer cells (BCs) which possess varied surface markers. Basal-like BCSCs possess CD10+, SMA+ and CK14+, while luminal-like BCSCs express Muc1+, CD24+ and CK18+ markers (Shipitsin and Polyak, 2008). While BCSCs are usually $CD44^{high}$, $CD24^{low}$ and $ALDH^{high}$, non-BCSCs are $CD44^{low}$, $CD24^{high}$ and $ALDH^{low}$. Aldehyde dehydrogenase (ALDH) is a key marker in BCSCs and proffers chemoresistance to apoptosis through the Nfr2-Keap1 signaling pathway (Kamble et al., 2021). Cancer stem cell signaling pathways are discussed in detail in other excellent review articles (Butti et al., 2019; Farooqi, 2021).

8.2.1.2 Other Signaling Pathways in Breast Cancer Cells

HER2 and VEGF Signaling

HER2/ERBB2 is a key transducer of cellular signals such as EGF, HBEGF, β-cellulin, TGFα, amphiregulin, epigen and epiregulin. HER2 forms heterodimers with HER1, HER3 and HER4 and recognizes various sets of these extracellular signaling agents (Hayes, 2019). The binding of the hormonal ligand activates the intrinsic tyrosine kinase activity of the HER2 intracellular domain, which is associated with several other proteins, and this triggers key pathways like the RAS-RAF-MEK-ERK1/2 cascade, which triggers the transactivation of p38 and CREB—through which, more nuclear transcription mediators are phosphorylated (Chen et al., 2015). These events elicit cellular proliferation, cellular survival, breast tumor cell migration and invasion. Also, HER2 signaling activates the PI3K/AKT/mTOR pathway, enhancing cancer cell cycle progression and survival (Nagaria et al., 2017). VEGF binds to VEGFR2 and activates multiple signaling pathways like the Src-PI3K/AKT pathway, Cdc42-p38-MAPK route, FAK/paxilin, PLCγ-mediated RAS/RAF/MEK/ERK cascade and CALN/NFAT-mediated COX2 activity. Also, VEGF signaling activates the JAK/STAT pathway, and these events pave the way for cancer cell migration, invasion, survival, proliferation and focal adhesion (Zhu and Zhou, 2015).

ER and PR Signaling

Both estrogen and progesterone nuclear receptors are involved in breast cancer pathogenesis, invasiveness, migration and metastasis. Estrogen receptor α is responsible for driving 75% of all breast cancers; ER is responsible for initiation as well as cancer progression. Although ER-targeting pharmacological agents and phytochemicals are known, ER signaling is complex and leads to metastasis as well as incurable breast cancer. Estrogen binding to ER α and β causes its heterodimerization and transactivation into the nucleus. ER can be phosphorylated by several

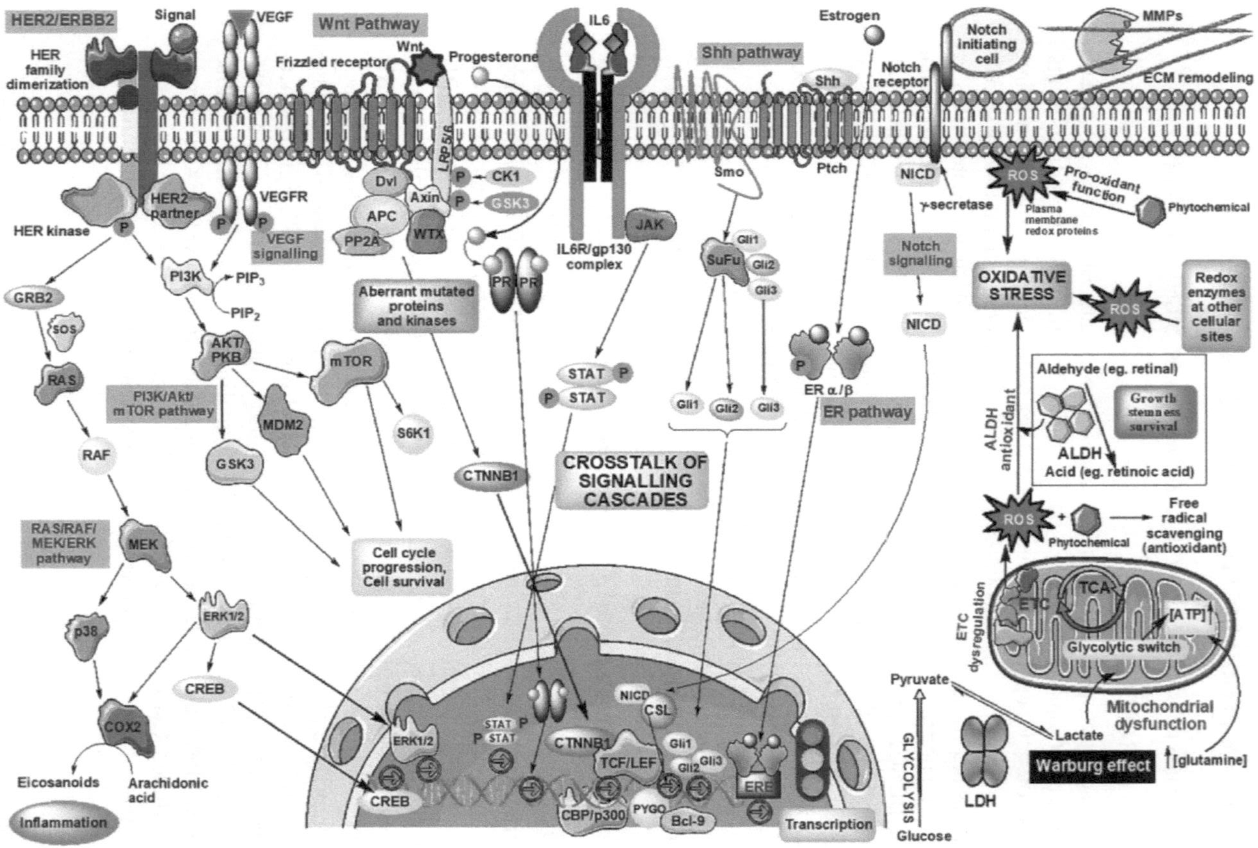

FIGURE 8.2 Signaling pathways and key pharmacological targets in breast cancer. The most prominent signaling pathways involved in carcinogenesis of breast cancer cells and BCSCs are shown. The HER-2 mediated RAS/RAF/MEK/ERK pathway, VEGF signaling, Wnt pathway, progesterone signaling, IL6-JAK/STAT pathway, Shh signaling cascade, Notch signaling pathway and estrogen receptor signaling pathway are discussed briefly in the text. The salient features of the signal pathways alone are shown for the sake of brevity. Several hallmarks of breast cancer are spurious signaling by mutated/aberrant proteins and kinases, ROS formation from multiple cellular sites (mitochondria, cell membrane, etc., causing oxidative stress), oxidative stress, signaling cross-talks, inflammation, alterations in the proteome and transcriptome owing to gene expression and epigenetic alterations, ALDH-mediated antioxidant action (and enhanced cancer cell survival), increased LDH expression, glycolytic switch, higher glutamine synthesis (and higher ATP production through glutamine metabolism), ETC dysregulation, mitochondrial dysfunction, etc., can be routinely identified in literature reports. Undoubtedly, breast cancer is a complex disease involving several significant individual-specific genetic and environmental factors, making it a very heterogeneous disease.

other kinases and hence, ER signaling involves significant crosstalk as this protein is modulated by epigenetics via signaling of multiple growth factors and cytokines (Mann et al., 2011; Renoir et al., 2013). ER transactivates to the nucleus in the presence of estrogens like estradiol and recruits transcriptional coactivators at the estrogen response element (ERE) to proffer cancer chemoresistance, proliferation and survival. ER signaling becomes altered in breast cancer; in breast tumors obtained through surgical excision, the transcriptomes of luminal mammary epithelial cells which are ER+ and breast tumor cells which are ER+ were compared and it was found that the cistromes were very different. By enriching the GRHL2 DNA binding motif in the ER cistrome specifically in breast cancer, the disparate functioning of ER in breast cancer was identified. This is a significant breakthrough that can essentially foster novel drug development by finding new targets for modulating ER function in breast cancer (Chi et al., 2019). Similarly, progesterone receptor (PR)-B is involved in transducing progesterone signaling, while PR-A is a transcriptional inhibitor of most steroid hormone receptors (Wen et al., 1994). PR signaling involves receptor dimerization, transactivation and binding to the progesterone response element; recruitment of transcriptional coactivators; and expression of PR-dependent genes, which cause a number of events through several pathways involving key proteins like c-Src, Wnt4, RANKL, EGFR, FGF, cyclin D1 and MAPK (Daniel et al., 2011).

Impaired Mitochondrial Metabolism, Glycolytic Switch and Warburg Effect

The metabolic switch from catabolism to anabolism, a key hallmark of breast cancer, is linked to mitochondrial dysregulation, which plays a major part in breast cancer energy metabolism and proliferation. In BCSCs/TICs which gain self-renewal potential, RB1 (retinoblastoma 1/transcriptional corepressor 1) loss increases glutamine production by enhancing glutamine transport into the cell via SLCA15 (glutamine transporter) expression. Additionally, multiple oncogenic alterations like mutation of PI3K, MYC amplification and loss of TP53 are known to occur. Since RB1 loss increases mitochondrial permeability transition (MPT) and increases oxidative phosphorylation, Zacksenhaus et al. (2017) scrutinized the contribution of RB1 in the metabolic switch during breast cancer. RB1 (in conjunction with other mitochondrial genes like *FER/FER-T*, *STAT3* and *CHCHD2*) is implicated in oxidative phosphorylation (OXPHOS); it hampers pyruvate usage by pyruvate dehydrogenase, and hence, there is a greater dependency on glutamine as the fuel for adenosine triphosphate (ATP) production. The Warburg effect (aerobic glycolysis) involves lactate production (higher lactate dehydrogenase [LDH] expression and activity) and low pyruvate-dependent cellular respiration, and together, these factors drive the metabolic switch (Zacksenhaus et al., 2017).

About 95 transcripts linked to mitochondrial biogenesis and mitochondrial protein synthesis have been differentially expressed in normal vs. breast cancer tissue, and about 40 of these genes coded for ribosomal proteins of mitochondria which are involved in production of mitochondrial oxidative phosphorylation (mOXPHOS) complexes. These genes and their protein products are involved in diverse cancer-associated processes like mitochondrial lipid synthesis, mitochondrial biogenesis and mitochondrial translation, which support metabolic symbiosis between glycolytic tumor stroma and the adjoining oxidative epithelial cells (Sotgia et al., 2012). This phenomenal work paves the way for the discovery of novel pharmacological targets for breast cancer management.

8.2.2 The History of Medicine

The history of medicine, as expressed by "anonymous" (Green et al., 2002), is summarized as follows: In 2000 BC, "here, eat this root" appeared to be both scientific and logical to our ancestors. In 1000 AD, the root was considered to be "heathen" and prayers were deemed to ward off diseases.

In the modern era and post-industrial revolution times, prayers were considered to be superstition and potions/concoctions were formulated. Then, pills and isolated/purified compounds were considered to be much more effective and "ancient" and "unscientific" treatments were shunned. This led to the antibiotic era, and specific inhibitors that target very critical receptors responsible for microbial virulence had been discovered and sold on the market. But due to prevalence of increased antibiotic resistance and resistance to cancer chemotherapeutics, more powerful (and, perhaps, more toxic) anticancer agents have been discovered. History always repeats itself, and change is always constant in academia and research studies. Now, mankind is going back to "the bugs have won"/"chemoresistance is high" and hence, the "let us eat this root" philosophy. This trend is very evident in the literature on breast cancer phytomedicine. The literature is currently inundated with myriad reports of purified compounds, phytochemical formulations/herbal concoctions/admixtures and phytochemical adjuvants which enhance the potency of routinely used chemotherapeutic agents (and also decrease the side effects thereof) for numerous cancer types (Fernando et al., 2019; Zhang et al., 2021).

8.2.3 Relevance of Phytochemicals in Breast Cancer Therapy

Several naturally occurring phytochemicals have varied biological activities, one of which is anticancer/antitumor potential. This term "antitumour potential" is a consequence of several other biological activities like pro-oxidant, antioxidant, antiangiogenic, antimetastatic, antiproliferative, anti-inflammatory and pro-apoptotic activities of phytochemicals. Bioactivity-guided secondary metabolite fractioning through chromatographic separation methods is often the starting point in investigation of anticancer potential of plant compounds; the fractionated/purified metabolites are then tested in vitro in cell lines and in vivo in animal cancer models for their safety and anticancer efficacy. This chapter summarizes the recent advancements in phytochemical- and phytomedicine-based treatments for breast cancer and discusses the important phytochemical compounds (and their targets). Also, this chapter draws on several published reviews and research reports to offer a summary of the mechanisms by which breast cancer begins, progresses and metastasizes to other sites in the body and delineates the mechanisms of action of natural anti-breast cancer compounds. Drug discovery and pharmacotherapy in western medicine is aimed at targeting specific cellular receptors. In order to inhibit several targets simultaneously, combinations of drugs are used and their synergistic effects are often reported. But there are some phytochemicals which possess the ability to simultaneously block several receptors (belonging to diverse disease processes and different cellular pathways). Such molecules should be exploited in future investigations when considering the formulation of nutraceuticals/novel therapeutics. The broad classes of phytochemicals are carotenoids, alkaloids, betalains, phytosterols, polyphenols/phenolics (which possess chemically diverse compounds like flavonoids and flavanols, stilbenes, phenolic acids, coumarins, tannins, lignans, catachins/apicatechins, etc.), terpenoids and organosulfur compounds (Vo et al., 2020). Interestingly, isolated/single phytochemicals may not be able to prevent cancer, as clinical trials do not prove chemoprevention by isolated compounds (Miller and Snyder, 2012). However, dietary compounds/combinations of compounds may be able to reduce cancer risk. While clinical trials are restricted to deciphering the mechanisms by which isolated phytochemicals or formulations manifest anticancer activity, concrete proof for chemopreventive effects of phytochemicals is largely elusive, and well-designed clinical trials may be required to obtain substantial proof. However, there is a plethora of research in different set-ups (in vitro, in silico, in situ and in vivo) which have proven the ability of phytochemicals to both prevent and treat breast cancer (Kapinova et al., 2018). Hence, we undertake a daunting task of succinctly summating the key phytochemicals which prevent (and treat) breast cancer, the pathways involved therein and the diverse mechanisms by which various phytochemicals act against breast cancer and briefly discuss their emerging roles as adjuvants in cancer chemotherapy clinical trials. The compounds presented in Table 8.1 are merely

TABLE 8.1
Representative Phytochemicals and Their Cytotoxic Potential (IC_{50}) in Breast Cancer Cell Lines

Phytochemicals (phytochemical class)	Cell Lines	IC50 Values	Reference	Breast Cancer Protein Pathways	Reference
Curcumin (polyphenol)	i) T47D ii) MCF-7 iii) MDA-MB-415 iv) MDA-MB-231 v) MDA-MD-468 vi) BT-20	i) 2.07 ± 0.08 μM ii) 1.32 ± 0.06 μM iii) 4.69 ± 0.06 μM iv) 11.32 ± 2.13 μM v) 18.61 ± 3.12 μM vi) 16.23 ± 2.16 μM	(Hu et al., 2018)	Up-regulating—p53 (mutated) and Bax down-regulating—MDM2 and Bcl-2	(Wang et al., 2021)
Quercetin (flavonol)	i) MCF-7	i) 118.55 ± 22.5	(Elsayed et al., 2021)	i) JAK/STAT1 ii) p38 MAPK signaling	i) (Qiu et al., 2021) ii) (Kashyap et al., 2019)
Resveratrol (phytoalexin)	i) MCF-7 ii) MDA-MD-231	i) 120 μM ii) 370 μM	(Casanova et al., 2012)	i) p53 (mutated) ii) RhoA/Lats1/YAP	i) (Shi et al., 2011) ii) (Kim et al., 2017b)
Capsaicin (alkaloid)	i) MDA-MD-231	i) 120 μM	(Ferreira et al., 2015)	i) EGFR/HER-2 ii) p53 (mutant)	i) (Thoennissen et al., 2010) ii) (Garufi et al., 2016)
Pterostilbene (stilbenes)	i) MDA-MB-231 ii) T-47D	i) 45.7 ± 0.01 μM ii) 63.1 ± 0.11 μM	(Elsherbini et al., 2021)	p53 (mutant)	(Elsherbini et al., 2021)
Gallic acid (polyphenol)	i) MCF-7 ii) MDA-MD-231	i) 80 μM ii) 80 μM	(Sales et al., 2018)	P53, Mcl-1 and P21	(Rezaei-Seresht et al., 2019)
Anthocyanins (flavonoids)	i) MCF-7	i) 34.85 ± 2.02	(Barrios et al., 2010)	p53	(Liu et al., 2021)
Piperine (alkaloids)	i) 4T1	i) 105 ± 1.08 μM	(Lai et al., 2012)	p53 Mutated	(Mitra et al., 2021)
Sulforaphane (isothiocyanates)	i) KPL-1 ii) MCF-7 iii) MDA-MB-231 iv) MDA-MB-468 v) SK-BR-3 vi) SUM-159 vii) BT549 viii) T47D	i) 19.1 ± 1.4 μM ii) 12.38 μM iii) 16.04 μM iv) 8.1 μM v) 25 μM vi) 10 μM vii) 20.47 μM viii) 9.5 μM	i) (Kanematsu et al., 2011) ii) (Kuran et al., 2020)	i) Nrf2 Mutant ii) ErbB2/ ER-PI3K-AktmTOR-S6K1	i) (Su et al., 2018) ii) (Pawlik et al., 2013)

(Continued)

TABLE 8.1
(Continued)

Phytochemicals (phytochemical class)	Cell Lines	IC50 Values	Reference	Breast Cancer Protein Pathways	Reference
Thymoquinone (1,4-benzoquinones)	i) MCF-7	i) 25 μM	(Motaghed et al., 2013)	PTEN, NF-κB, Akt	(Arafa et al., 2011)
Phytol (acyclic diterpenoids)	i) MCF-7 ii) MDA-MB-231	i) 8.97 ± 0.41 ii) 69.67 ± 2.99	(Pejin et al., 2014)	-	-
Beta-sitosterol (phytosterols)	i) MDA-MB-231	i) 63.5 μM	(Eyong et al., 2018)	MAPK and PI3K	(Grattan, 2013)
Anthraquinone (phenolic)	i) MCF-7 ii) SK-BR-3	i) 60 ?g ii) 150 ?g	(Esmat et al., 2006)	p53 and p21	(Yeap et al., 2015)
Geraniin (tannin)	MCF-7	9.94 μM	(Zhai et al., 2016)	p38 MAPK	(Zhai et al., 2016)
Squalene (triterpenoids)	i) MCF-7 ii) Hs 578T iii) SK-BR-3 iv) MDA-MB-468 v) T-47D vi) MDA-MB-231	i) 32.1 ± 3.1 ii) 52.2 ± 4.0 iii) 84.9 ± 6.8 iv) 115.4 ± 10.3 v) 140.8 ± 18.4 vi) 220.9 ± 8.8	(Brown et al., 2016)	—	—
Gingerol (Gingerol)	i) 4T1Br4 ii) L929 iii) MDA-MB-231	i) 32.2 ± 0.5 μM ii) 55.9 ± 2.3 μM iii) 60.6 ± 2.9 μM	(AC et al., 2020)	p53 /BAX/BCL-2 (Mutant)	(Sp et al., 2021)
Genistein (isoflavones)	4T1	50 μM	(Ikawati et al., 2020)	MAPK, NF-κB, PI3K/Akt, ERK ½, Wnt/β-catenin	(Tuli et al., 2019)
Fisetin (flavonols)	MCF-7	267.10 ± 48.96 μM	(Buranrat and Junking, 2022).	PTEN/Akt/GSK3β	(Li et al., 2018a)

representative/salient phytochemicals with well-known chemopreventive and therapeutic effects. Several review articles have exhaustively discussed the mechanisms and structure-function correlations of phytochemicals and their molecular targets by which they mediate (and modulate) diverse pathways (Saleh et al., 2021). Besides, the multimodal and holistic effects of these phytochemicals are apparent from their ability to bind to (and potentially inhibit) several targets at one time. In vitro studies with mammalian and, specifically, human breast cancer cell lines (and comparison of cytotoxic and other beneficial effects in normal breast cell lines like MCF-10A) provide impactful insights into the dosage, IC_{50}, receptor target binding, target expression (Western blotting) and gene expression changes associated with crude/purified phytochemical treatment of normal vs. breast cancer cell lines (MCF-7, MDA-MB-231, MDA-MB-415, MDA-MB-468, KP-L1, 4T1, SK-BR-3, PTK, T-47D and BT-20).

Since not all cell lines possess the same proteome and may have differential gene expression changes in response to phytochemical treatment, it is beneficial to explore the mechanism and efficacy of crude extracts and purified phytochemicals in multiple cell lines at one time. Through in silico techniques, the molecular interactions between phytochemical ligands and their receptor targets have been explored in sufficient detail, and the insights obtained through computer-aided drug design have far-reaching implications. These data, though very limited, can pave the way for further Quantitative Structure Activity Relationship (QSAR)–based studies/explorations to find better lead compounds, preferably compounds which can bind simultaneously to several targets. Modifications/tweaking of key functional (electron donating as well as electron withdrawing) groups and pocket binding exploration using 3D-QSAR can help in lowering the IC_{50} values (to obtain nanomolar/sub-nanomolar values) and increase phytochemical-based drug specificity and potency against breast cancer targets. Such details are apparent from IC_{50} values obtained for diverse phytochemicals (and their functional group modifications) in different breast cancer cell lines. If one compound potently inhibits multiple targets at one time, the need to use several combinations of drugs becomes unnecessary if one drug can effectively target many pathways at one time. Collateral damage caused by the use of multiple drugs for combating breast cancer and immunotherapy, as well as radiation therapy for breast cancer, is known to cause pneumonitis and pulmonary fibrosis, with patients also dying from treatment-related side effects (Omarini et al., 2019). HER2-targeting therapy often causes interstitial lung disease and pneumonitis (Hackshaw et al., 2020). In a study, locoregional therapy was found to proffer a 34% enhanced survival rate (lower hazard of death) when compared to systemic therapy in stage 4 treatment-naïve breast cancer patients (Soran et al., 2018). As the stage of cancer advances, the efficacy of breast cancer cure also diminishes. It is therefore important to identify new sources of anticancer drugs and novel drug candidates which can simultaneously target multiple receptors involved in breast cancer, which could potentially have much higher therapeutic potential and exhibit much lower toxicity to normal cells. Interestingly, a wide variety of phytochemicals seem to be able to target almost all the breast cancer signaling pathways and the proteins/enzymes which participate in those pathways; several review articles have been written on the protective and therapeutic actions of phytochemicals (Blackley and Loi, 2019). In Table 8.1, the mechanisms of action of these phytochemicals are also briefly mentioned, and the pathways targeted by anti–breast cancer phytochemicals are discussed in Table 8.2.

Since this chapter only focuses on representative phytochemicals and can perhaps never be able to fully discuss all the phytochemicals which potentially possess anti–breast cancer activity, only salient phytochemicals and their mechanisms of action are briefly discussed (Figure 8.3).

TABLE 8.2
Phytochemical Inhibitions of Breast Cancer Signal Transduction Pathways

Breast Cancer Signaling Pathways	Inhibitor Phytochemicals	Reference
Wnt	i) Curcumin	(Kakarala et al., 2010)
	ii) Piperine	
Shh	i) Curcumin	i) (Li et al., 2018b)
	ii) Resveratrol	ii)(Galdo et al., 2019)
	iii) Isoflavones	iii) (Huang et al., 2013)
	iv) Quercetin	iv) (Ezzati et al., 2020)
Notch signaling pathway	i) Resveratrol	i) (Dong et al., 2020)
	ii) Curcumin	ii) (Gallardo and Calaf, 2016)
	iii) Honokiol	iii) (Yuan et al., 2015)
	iv) Genistein	iv) (Pan et al., 2012)
	v) EGCG	v) (Farooqi et al., 2020)
	vi) Sulforaphane	vi) (Castro et al., 2019)
JAK-STAT signalling pathway	i) Resveratrol	i) (Kohandel et al., 2021)
	ii) Curcumin	ii) (Farghadani and Naidu, 2021)
	iii) Cryptotanshinone	iii) (Zhou et al., 2014)
	iv) Ursolic acid	iv) (Chen, 2017)
	v) Cucurbitacin	v) (Lan et al., 2013)
	vi) Thymoquinone	vi) (Şakalar et al., 2016)
	vii) EGCG	vii) (Suarez et al., 2021)
	viii) Capsaicin	viii) (Kaymaz et al., 2014)
	ix) Quercetin	ix) (Ezzati et al., 2020)
	x) Flavopiridol	x) (Hosford and Miller, 2014)
	xi) Celastrol	xi) (Dandawate et al., 2016)
	xii) Butein	xii) (Mendonca et al., 2019)
MAPK-ERK signaling pathway	i) Curcumin	i) (Wang et al., 2018a)
	ii) EGCG	ii) (Deguchi et al., 2002)
	iii) Sulforaphane	iii) (Nandini et al., 2020)
	iv) Genistein	iv) (Hu et al., 2020)
	v) Resveratrol	v) (Tang et al., 2008a)
	vi) Isothiocyanates	vi) (Kang and Wang, 2010)
PI3K/Akt/mTOR signaling pathway	i) EGCG	i) (Narayanankutty, 2020)
	ii) Delphinidin	ii) (Chen et al., 2018)
	iii) Resveratrol	iii) (Jiang et al., 2009)
	iv) Genistein	iv) (Kaushik et al., 2019)
	v) Indole-3-carbinol	v) (Xu et al., 2019)
	vi) Silibinin	vi) (Binienda et al., 2020)
	vii) Curcumin	vii) (Khan et al., 2020)
Micro-RNA signaling pathway	Cardamonin, curcumin and resveratrol	(Varghese et al., 2020)
VEGFR pathway	i) Quercetin	i) (Balakrishnan et al., 2016)
	ii) Curcumin	ii) (Carroll et al., 2008)
	iii) Resveratrol	iii) (Kim et al., 2017a)
LnC RNA	i) Curcumin	i) (Hu et al., 2018)
ALDH pathways	i) Quercetin	i) (Wang et al., 2018b)
	ii) Piperine and curcumin	ii) (Aumeeruddy and Mahomoodally, 2019)
LDH signaling pathway	i) Thymoquinone	i) (Sutton et al., 2014)
	ii) Curcumin	ii) (Khorsandi et al., 2019)

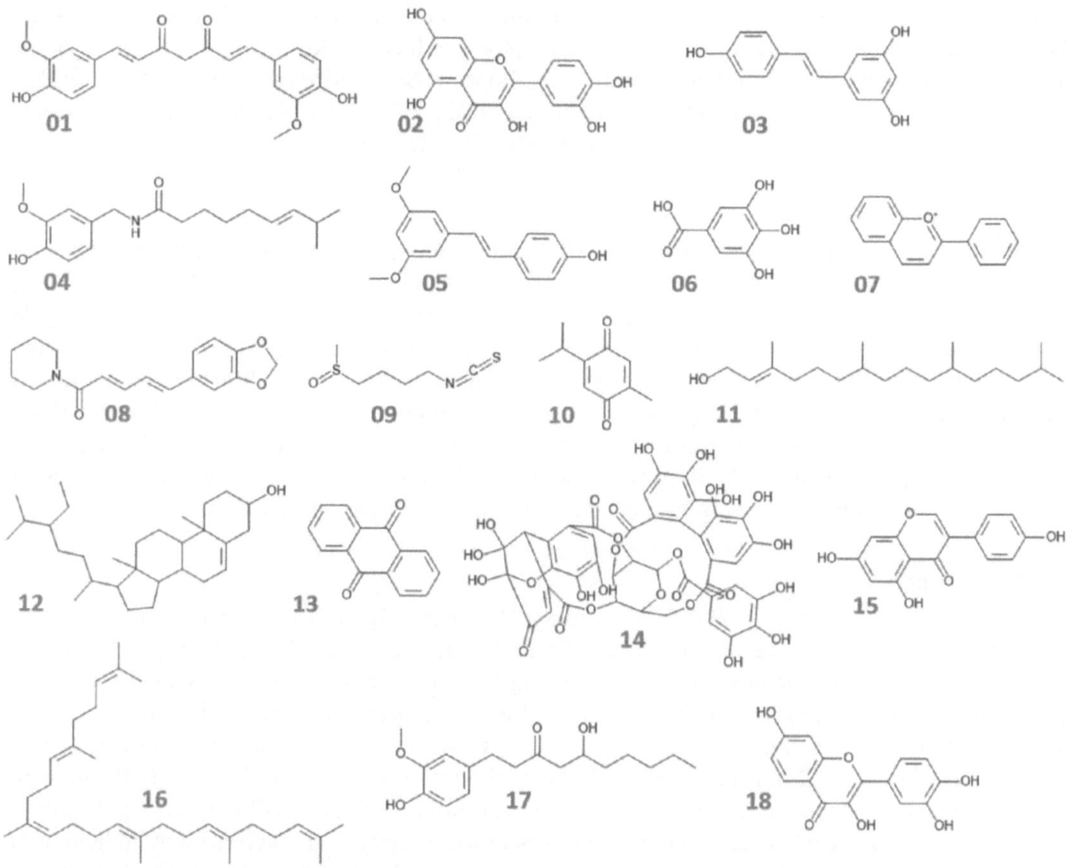

FIGURE 8.3 Structure of phytochemicals used to treat breast cancer: 1. Curcumin, 2. quercetin, 3. resveratrol, 4. capsaicin, 5. pterostilbene, 6. gallic acid, 7. anthocyanins, 8. piperine, 9. sulforaphane, 10. thymoquinone, 11. phytol, 12. beta-sitosterol, 13. anthraquinone, 14. geraniin, 15. squalene, 16. gingerol, 17. genistein, 18. fisetin.

8.2.3.1 Curcumin

The bioactive polyphenol, curcumin [(1E,6E)-1,7-bis (4-hydroxy-3-methoxyphenyl)-1,6-heptadiene-3,5-dione], is routinely isolated from *Curcuma longa* L. tubers and possesses a dark yellow color and hydrophobic nature. It is a profound antioxidant, anti-inflammatory and anticancer compound with known ability to inhibit several breast cancer–related signaling pathways responsible for cellular proliferation, enhanced cancer cell survival and angiogenesis as well as metastasis (Sajadimajd et al., 2020). It acts through inhibition of several cancer-related pathways like Wnt/β-catenin, JAK/STAT, PI3K/AKT/mTOR, MAPK, NF-kB, and p53 signaling pathways. It inhibits the expression and activity of several proteins like PI3K, β-catenin, Wnt, NFkB, RAS, p53, cyclin, CDKs, uPA, MMPs, HER2, VEGF, VEGFR and OPN (Song et al., 2019). In almost all cancer cell lines, curcumin has a low IC_{50} value of <20 μM. Moreover, it also tremendously affects the apoptotic cascade and also influences the cell cycle in breast cancer cells (Farghadani and Naidu, 2021). Curcumin analogs and derivatives synthesized and validated using 2D/3D-QSAR approaches have better anticancer activity; for example, curcumin–amino acid conjugates were synthesized, and their 2D-QSAR studies revealed that the curcumin derivatives possessed antitumor activity comparable to that of doxorubicin and 5-fluorouracil (Panda et al., 2020b). Several curcumin derivatives inhibit breast cancer cell growth via diverse mechanisms such as attenuation of ROS-YAP-JNK signaling (Wang et al., 2019), G2/M cell cycle arrest and apoptosis induction (Ali et al., 2017) and blockage of cell migration through inhibition of matrix metalloproteinases (MMPs) (Nirgude et al., 2020). Moreover, a curcumin metal complex (Mohammed et al., 2020) and curcumin nanocomposites (Tajbakhsh et al., 2018) were shown to possess good anticancer activities.

8.2.3.2 Capsaicin

The alkaloid capsaicin (*trans*-8-methyl-*N*-vanillyl-6-nonenamide) is found in chilli peppers and exhibits profound anticancer effects in ER-positive (e.g. MCF-7, T47D and BT-474) as well as ER negative (SKBR-3 and MDA-MB231) breast cancer cell lines and was found to induce cell cycle arrest, trigger apoptosis, decrease EGFR and HER2 expression, elicit G_0/G_1 cell cycle arrest and inhibit EGFR/HER2-sponsored cell migration (Thoennissen et al., 2010). It also down-regulated the FBI-1–mediated NFkB pathway and decreased the protein expression levels of Ki-67, Bcl-2 and survivin. FBI-1, when overexpressed, reversed capsaicin-mediated antiproliferation and pro-apoptotic effect, while conversely, its gene silencing was found to show the opposite effect. Besides this, it also affects cancers by inhibiting Notch signaling in BCSCs (Shim and Song, 2015).

8.2.3.3 Quercetin

Quercetin [2-(3,4-dihydroxyphenyl)-5,7-dihydroxy-4H-1-benzopyran-4-one] belongs to a hydroxylated phenolic group of flavonoids (flavanols) and possesses antioxidant, antihyperlipidemic, anti-inflammatory and anticancer properties (among various other biological activities). Quercetin therapy for breast cancer was found to elevate p53 expression levels in basal-like MDA-MB-468 cells. Quercetin exerts its anti–breast cancer activities by serving as an antagonist of ER, HER2 and PR (Ezzati et al., 2020). It modulates cellular proliferation signals by inhibiting PI3K and p38 MAPK and stabilizes telomeric DNA structures. It also inhibits metastasis by inhibiting the activity of MMPs, mTOR, c-MYC and TGFβ. Its pro-apoptotic activity is characterized by induction of BAX and cytochrome *c* release and activation of caspase-3 (Baydas et al., 2005). Furthermore, it inhibits BCSCs by inhibiting Wnt-β catenin signaling and the Notch and Shh pathways (Ezzati et al., 2020). Quercetin is also used as an adjuvant in conventional chemotherapy to enhance the killing of BCSCs and to prevent re-emergence of cancer and tumor recurrence. Another important discovery is that quercetin triggers a reactive oxygen species (ROS)–dependent apoptotic pathway in MCF-7 cells, and this shows the pro-oxidant role of quercetin in triggering apoptosis in breast cancer cells (Wu et al., 2018). Both quercetin and its water-soluble analog, quercetin-5′,8-disulfonate, were found to exert potent anti–breast cancer activity in vitro in LoVo and MCF-7 cells; the quercetin derivative was found to generate ROS such as superoxide and hydrogen peroxide and also concomitantly activate ROS-dependent pathways to inhibit growth as well as apoptosis (Zhang et al., 2012).

8.2.3.4 Resveratrol

Resveratrol [3,5,4′-trihydroxy-*trans*-stilbene], a polyphenol, possesses a wide spectrum of biological activity and is a non-flavonoid phytoestrogen/phytoalexin found abundantly in grapes, pomegranate, berries and soy beans; it is known to bind to multiple breast cancer targets at one time and also protect against several cancers. It possesses anti-inflammatory, antiproliferative and pro-apoptotic actions and lowers the resistance of breast cancer resistance protein (BRCP) (Le Corre et al., 2005). Resveratrol is known to modulate several miRNAs (miR-125b-5p, miR-200c, miR-409–3p, miR-122–5p and miR-542–3p) and regulates the action of important proteins related to apoptosis such as Bcl-2, CDKs and XIAP (Venkatadri et al., 2016). Resveratrol also inhibited the expression of MMP-2 and suppressed the PI3K-AKT signaling pathway via IGF-1 mediated MMP-2 action in MDA-MB-435, thereby attenuating cancer cell migration (Tang et al., 2008b). Additionally, resveratrol is useful as a chemotherapy adjuvant and was found to chemosensitize breast cancer cells to melphalan by inducing cell cycle arrest (Casanova et al., 2012). Molecular targets of resveratrol include i) p53 and Rb tumor suppressor genes; ii) cell cycle regulators like cyclins, CDKs, p21WAF1, p27KIP, INK, etc.; iii) ATM and ATR (checkpoint kinases); iv) diverse transcription factors like NFkB, AP-1, c-Fos and c-Jun; v) angiogenesis and metastasis factors like VEGF and MMP2/9 and inflammatory mediators like COX-2; and vi) apoptotic pathway modulators like Bax, Bak, PUMA, Noxa, TRAIL, APAF, survivin, Akt, Bcl-2 and Bcl-X_L (Athar et al., 2009). Besides causing gene expression changes and direct binding-based protein and enzyme inhibition, resveratrol and its analogues also possess both antioxidant and pro-oxidant properties (Murias et al., 2005).

8.2.3.5 Gallic Acid

Gallic acid (3,4,5-trihydroxy-benzoic acid), a polyhydroxyl phenolic compound, is found in several plants, fruits and foods. It has been called the "molecular rival of cancer" (Verma et al., 2013). While gallic acid possesses diverse biological activities, it is well-known for its anti–breast cancer activity. It was found to decrease the expression of cyclin A, CDK2, cyclin B1 and Cdc2/CDK1 and conversely, to increase the levels of cell division–negative regulatory proteins like $p27^{kip1}$ and $p27^{Cip1}$. Also, since gallic acid prevented the proteasomal degradation of $p27^{kip1}$, it was found to induce G2/M phase arrest in MCF-7 cells (Hsu et al., 2011). A combination of curcumin and gallic acid was found to induce apoptosis in MDA-MB-231 cells. This phytochemical combination was also found to enhance oxidative stress and cytotoxicity and glutathione depletion also occurred. By raising the expression levels of Bax, cleaved caspase-3 and PARP and by down-regulating Bcl-2 levels, apoptosis was triggered (Moghtaderi et al., 2018). Additionally, gallic acid was found to abrogate the EGFR/Src/AKT/ERK-mediated MMP-9 expression in MCF-7 cells and thereby decreased metastasis and invasion.

8.2.4 How Phytochemicals Prevent/Treat Breast Cancer

Breast cancers differ based on the molecular characteristics, molecular etiology and location of the tumor. Several signaling pathways are involved in the pathogenesis and progression of breast cancer. As discussed earlier, phytochemicals target almost all the breast cancer pathways involved in both initiation and progression of disease, and hence, they can serve as pharmacological agents for chemoprevention as well as treatment of breast cancer. An overview of how phytochemicals prevent and treat cancer is presented in Figure 8.4 (left panel). These diverse actions of phytochemicals may occur together in permutations/combinations. Table 8.2 shows the various pathways that phytochemicals inhibit (and merely representative examples are cited therein).

Depending on the prevailing cellular redox conditions, oxygen concentration, mitochondrial activity/metabolism, concentration of the phytochemical and the redox properties of the phytochemical, end outcomes can change. Phytochemicals can have both antioxidant and pro-oxidant actions, depending on the prevailing cellular physicochemical conditions such as pH, temperature, concentrations of reactants and the rate constants of multiple competing reactions which simultaneously occur in the cellular reaction milieu. This is why some phytochemicals can have both antioxidant and pro-oxidant actions. Also, this is the reason why phytochemicals which yield effective antitumor actions in one patient may not work in another patient at the same dosage. These discrepancies and conundrums are better explained by murburn (Mild UnRestricted BURNing/mured burning) concept, the fundamental and logical explanation for diverse cellular biochemical phenomena such as cellular respiration/oxidative phosphorylation, heme peroxidase chemistry, cytochrome P450–mediated xenobiotic metabolism, anaerobic respiration, drug biology/pharmacology, nerve conduction, molecular vision, cellular homeostasis, thermogenesis, photoreception, photosynthesis and hormesis (and idiosyncratic dose responses) (Manoj et al., 2021). Redox enzymes are known for their non-classical kinetic profiles and their substrate-binding promiscuities, and hence, redox reactions mediated by such enzymes may or may not involve active-site substrate binding (Manoj

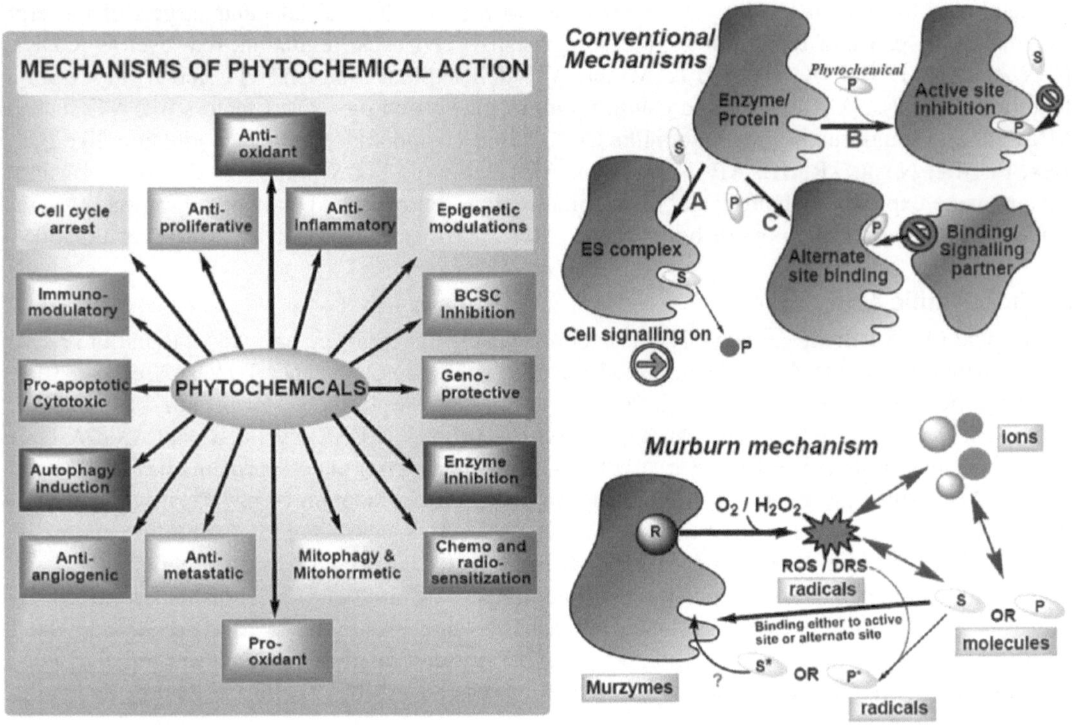

FIGURE 8.4 Overview of phytochemicals and their biological actions and conventional vs. murburn mechanisms of phytochemical action: *Left panel:* Phytochemicals possess a plethora of biological actions that are well documented in breast cancer research, which is responsible for their cancer-preventive and -ameliorative actions. *Top right panel:* In routine cellular reactions, enzymes recognize their substrates or binding partners and transduce cell signals through product formation or protein-protein complex formation (A). Phytochemicals may bind to active sites/allosteric sites and block either substrate binding (B) or partner protein binding (C) (in cell signaling cascades) and thereby inhibit cellular enzymatic reactions/cell signal cascades. *Bottom right panel:* Since most phytochemicals also have antioxidant functions and redox properties, their interactions with ROS/DRS (diffusible reduced species) can lead to formation of radicals of phytochemicals or other molecules (including substrates). Redox enzymes possessing redox cofactors/coenzymes like $NAD(P)^{+}/NAD(P)H$, $FAD^{+}/FADH_2$, heme, etc. (denoted as R in a gray sphere) are usually capable of oxygen activation and formation of ROS/DRS (including RNS, peroxide, RSS and other reactive species), which can modulate the functions of the enzymes directly or attack substrate/phytochemical molecules bound either to the classical active site or to the surface (alternative binding) and lead to formation of S*/p*, which may or may not affect cellular metabolism and signaling. The murburn concept is a logical, well-rounded cellular phenomenon that is explained by the complex interplay between enzymes (murzymes), molecules, unbound ions and radicals. Murburn reaction chemistry has tremendous implications in cellular physiology and medicine. See text for more details.

et al., 2010). Murburn chemistry involves the complex interplay of entities like murzymes, unbound ions, organic molecules and diffusible reduced oxygen species/diffusible reduced species (DROS/DRS) (see Figure 8.4, bottom right panel). The complex reaction networks occurring in the cellular milieu are governed by mass-charge principles (fundamental physical and chemical logic); hence, the spatiotemporal concentrations of a product of interest (or a biochemical reaction outcome) can be dictated by a stochastic and non-predetermined chemoenzymatic murburn redox reaction network (Manoj et al., 2016b). Phytochemicals can comprehensively target the breast cancer tumor microenvironment (Malla et al., 2020). This is why there is a need to understand phytochemical anticancer action in a newer light and to identify potential avenues to improve cancer medicine. The conventional mechanisms and the murburn explanations are discussed next.

8.2.4.1 Inhibition of Breast Cancer Stem Cell Proliferation

A report reveals that ALDH is responsible for chemotherapy and radiation resistance of BCSCs and that inhibition of this enzyme using the ALDH-specific inhibitor diethylaminobenzaldehyde or by adding all-trans retinoic acid resulted in significantly lower cancer colony formation (Croker and Allan, 2012). The presence of ALDH in $ALDH^{high}CD44^{+}$ BCSCs facilitated chemoradiation resistance owing to its potent antioxidant activity. Another reason for this could be i) conversion of reactive aldehydes to carboxylates (and thereby prevention of DNA damage and apoptosis) and ii) generation of NAD(P)H from NAD(P)+ during the aldehyde to carboxylate conversion (more reduced nicotinamide can support more cellular anabolism as well as ATP synthesis that would facilitate better cell division, EMT and metastasis). The Notch and Wnt pathways are known to have roles in embryonic development, and dysregulated signaling of Notch, Wnt/β-catenin, Shh and Hippo are responsible for initiation of several cancers and solid tumors (Saeg and Anbalagan, 2018). β-catenin knockdown affects mitochondrial biogenesis and disrupts lipid metabolism in breast cancer cells, and hence, Wnt signaling is often deemed to be a pharmacological target (Vergara et al., 2017). Since the Wnt pathway, for instance, is involved in normal tissue homeostasis, blocking the Wnt/β-catenin signaling pathway entirely may not be safe and can have dire consequences. The role of TRAF-2 and Nck-interacting kinase (TNIK) had been explored in colorectal cancer (CRC), and novel synthetic azole inhibitors which shut down Wnt-dependent gene transcription and ameliorate CRC were identified (Masuda et al., 2016). Blocking TNIK abrogates Wnt signaling and cancer cell stemness almost entirely, but there is significant doubt over whether it is safe to block this key pathway. However, phytochemicals and phytomedicinal formulations/concoctions work on several targets at once and may not possess nanomolar/picomolar K_i values for key proteins, and hence, there is a probability that milder inhibition of several receptors at once (judging from the low micromolar K_d/K_i values reported for phytochemicals) may have chemotherapeutic effects. This synergistic effect of several phytochemicals may not disadvantage a single target/pathway and may holistically combat cancer progression. Curcumin and piperine inhibit Wnt signaling (Kakarala et al., 2010), while curcumin, resveratrol, isoflavones and quercetin block the Shh signaling pathway. Resveratrol, curcumin, honokiol, genistein, sulforaphane and EGCG inhibit the Notch signaling pathway (Dandawate et al., 2016). Multiple phytochemicals block the JAK-STAT pathway, which is normally activated by IL-6 (Arumuggam et al., 2015).

8.2.4.2 Binding to Proteins/Enzymes and Blocking Substrate/Protein Partner Binding

All enzymes possess an active site where the enzymatic reaction occurs. For example, kinases have a characteristic active site where ATP binding occurs and a protein partner or small molecule substrate bound to it. Binding of a phytochemical or synthetic drug inhibits the enzyme. Since cell signaling relies on phosphorylation signals from one protein to another in a cascade, this binding-based mechanism of signal relay is prevented by inhibitor binding to the active site instead of a substrate or a protein substrate partner. Phytochemicals can bind to manifold proteins and enzymes which relay phosphorylation signals (kinases), phosphatases, aromatases, cytochrome P450s, ATP citrate lyase, fatty acid synthase, HK11, urokinase plasminogen activator, MMPs, ALDH, LDHA/LDH5, and other cell signal transduction pathway components (Baig et al., 2019). Dietary polyphenols are not just antioxidants; phenolic acids, flavonoids, lignans and stilbenes are known to bind to ER α and β and thereby block ER-mediated signaling pathways by mimicking the structure of endogenous estrogens and the consequences of ER-dependent signaling, such as survival (against apoptosis), EMT, proliferation, chemoresistance and invasion (Cipolletti et al., 2018). Our group has investigated the pharmacological potentials of crude extracts of relatively poorly explored medicinal plants such as *Plecospermum spinosum* Trecul. (Annadurai et al., 2021) and *Aganosma cymosa* (Roxb.) G. Don (Annadurai et al., 2022) in MCF-7 cells and through in silico approaches, the compounds identified therein (by GC-MS)—coprostanol, pyrocatechol, indole and 4-methylcatechol. These ligands had better binding scores (-ΔG) for PI3K-110δ, MMP9, mTOR and ER and were found to dock at the active sites of these proteins.

Kinases are often mutated and dysregulated in cancer; kinase inhibition is one of the mechanisms by which anticancer phytochemicals cause gene expression changes. There are phytochemicals which can inhibit both wild-type and mutant (and abnormally active) kinases. In a study, fistein and quercetin were found to inhibit the migration of TNBC through inhibition of kinases of the PI3K/AKT and JAK-STAT pathways and MAPK (Thakuri et al., 2020). Through binding as well as redox activity, many phytochemicals possess the ability to trigger apoptosis in cancer cells and possess little or no cytotoxic effects in normal breast cells. Through the avenue of protein-drug interactions and antioxidant as well as immunomodulatory mechanisms, some phytochemicals trigger autophagy and either prevent or treat cancer (Patra et al., 2020; Sun et al., 2019; Younas et al., 2018). It is important to identify phytochemicals which specifically bind to mutant kinases; for example, Panda et al. identified that epicatechin and epigallocatechin were able to bind and inhibit mutated CHK2 kinase, a protein involved in CRC (Panda et al., 2020a). It is also important to understand that redox mechanisms can modulate pharmacological actions of drugs; phytochemicals/synthetic drug molecules can themselves react with ROS to form toxic products, and this aspect must be addressed to improve phytochemical/phytochemical-derived drug efficacy and toxicity profiles (Gideon et al., 2022 In press).

Phytochemicals may serve as competitive inhibitors or as non-competitive/mixed inhibitors by binding to alternative sites other than the active site. This is depicted in Figure 8.4 (top right panel). The contrast between conventional mechanisms of phytochemical action and the murburn perspective is discussed in the pertinent section. We had provided a new perspective on LDH and shown how it acts as a murzyme in Cori's cycle and the Warburg effect (Manoj et al., 2022). There are several LDHA inhibitors such as EGCG, galloflavin, gossypol, sesamol, machilin A, etc., which are obtained purely from plant sources, and some of them have nanomolar IC_{50} values for purified LDHA (Deiab et al., 2013; Manoj et al., 2022). In cells, kinase-mediated phosphorylation reactions are crucial in cell signaling. Hemoglobin was recently shown to possess adenosine diphosphate (ADP) binding sites, and in red blood cells (RBCs), which lack mitochondria, ATP synthesis is most likely achieved by hemoglobin's murburn chemistry (Parashar et al., 2021). The phosphorylation activity of kinases can be modulated not just by mutations but also by oxygen concentrations and prevailing redox conditions (Griendling et al., 2000). This development is a significantly important concept that requires further experimental investigations because cellular phosphorylation reactions (and ATP synthesis) can be catalyzed by DROS/DRS (Manoj and Bazhin, 2021). In cancer cell signaling, there is overt emphasis on protein-protein interactions. Such a complex/sophisticated explanation of breast cancer signaling is revealed by the protein-protein network. The sheer complexity of the interactome and the reliance of protein-protein interactions for phosphorylation and electron transfer cannot fully explain how or why kinase signaling in cancer exhibits so much signaling crosstalk. The relevance of the murburn concept in cancer kinase signaling must be explored further because Manoj and co-workers have shown that cytochrome P450–catalyzed xenobiotic oxidation reactions and chloroplast electron transfer reactions do not involve fastidious protein-protein interactions (Gideon et al., 2021; Gideon et al., 2020; Manoj et al., 2016a).

8.2.4.3 Antioxidant Action and Inhibition of Electron Transfer (Redox Activity)

Through their antioxidant action, phytochemicals also prevent electron transfer from one protein to another by scavenging free radicals in the reaction milieu. Several phytochemicals are known to inhibit and modulate electron transfer reactions. This is because phytochemicals can donate electrons to redox enzymes like peroxidases, oxidases, oxygenases and peroxygenases via binding as well as through murburn mechanisms, and phytochemicals can modulate electron transfer reactions in mitochondrial electron transfer chain (ETC). Thus, by modulating the electron transfer reactions, phytochemicals potentially decrease oxidative stress by preventing ROS formation and not merely by scavenging ROS prevalent in cancer cellular environments. For example, xanthumol inhibits oxidative phosphorylation in mitochondria by binding to NADH dehydrogenase (complex I) and inhibiting electron transfer reactions (Zhang et al., 2015). Another report discusses the involvement of phytochemicals in inhibiting several complexes of mitochondrial ETC (Sehrawat et al., 2013). Phytochemicals (especially, redox active ones) perturb cellular redox homeostasis and introduce either reductive or oxidative stress and lead to cellular dysfunction and apoptosis (Gaikwad and Srivastava, 2021). Plant-derived molecules like anacardic acid, salicylic acid and oleic acid (considered to be mitochondrial-targeted drugs/mitocans) alter cellular bioenergetics in breast cancer cells and cause mitochondrial uncoupling and may have phenomenal therapeutic implications in breast cancer therapy (Radde et al., 2016). Free radicals are known to modulate the activity of several proteins/enzymes; for example, nitric oxide (a reactive nitrogen species or RNS) inhibits multiple heme-containing enzymes like nitric oxide synthase and mitochondrial complex IV. Hence, enzymatic inhibition is not just an active-site phenomenon but can also involve non-enzymatic reactions in the redox enzyme's vicinity. Redox additives tremendously modulate redox enzymatic reactions (Manoj et al., 2016b). One of the often-ignored aspects in pharmacology/drug biology is that drug molecules (and phytochemicals themselves) can serve as either reductants or oxidants. Hence, there is a need to go beyond the classical theoretical framework and design "redox-neutral" molecules which may not participate in redox enzymatic reactions. It is close to impossible to determine all the possible proteins/cellular macromolecules that can bind to and interact with currently available pharmaceutical compounds and phytochemicals, and hence, computer-aided drug design may help in this regard to design and synthesize target-specific molecules which will not disturb cellular redox homeostasis.

8.3 CONCLUSION

Breast cancer is an incredibly complex disease. Several phytochemical molecules belonging to different phytochemical classes have been reviewed herein and their mechanisms of action, such as antioxidant, pro-oxidant, enzyme inhibitory, immunomodulatory, antiproliferative, autophagy-inducing, and mitohormetic mechanisms have been discussed. The signaling pathways and key factors involved in breast cancer pathogenesis and pathophysiology can be targeted by bioactive plant compounds which have chemopreventive as well as breast cancer therapeutic effects. The role of redox reactions and the murburn concept in explaining the various pharmacological actions of phytochemicals has been addressed. Murburn explanations for redox enzyme mechanistic chemistry can aid in better explaining how phytochemicals function in breast cancer chemoprevention as well as pharmacological management. Although clinical trials of breast cancer with isolated phytochemicals have failed to show a chemopreventive effect, phytochemical research in breast cancer is a promising avenue for identification of novel compounds. Phytochemicals and phytochemical formulations block several pathways at one time, and surprisingly, some phytochemicals have been found to inhibit several proteins in multiple breast cancer signaling pathways. Redox reactions have fundamental implications in cancer therapy, and identification of novel phytochemical-based drugs through computer-aided drug design can provide safer alternatives to conventional chemotherapeutic agents.

BIBLIOGRAPHY

AC, B.M.M., Tomasin, R., Luna-Dulcey, L., Graminha, A., Teles, R., da Silva, V., da Silva, J., Vieira, P., Annabi, B., and Cominetti, M. (2020). [10]-Gingerol improves doxorubicin anticancer activity and decreases its side effects in triple negative breast cancer models. Cellular Oncology (Dordrecht) *43*, 915–929.

Ali, N.M., Yeap, S.K., Abu, N., Lim, K.L., Ky, H., Pauzi, A.Z.M., Ho, W.Y., Tan, S.W., Alan-Ong, H.K., and Zareen, S. (2017). Synthetic curcumin derivative DK1 possessed G2/M arrest and induced apoptosis through accumulation of intracellular ROS in MCF-7 breast cancer cells. Cancer Cell International *17*, 1–12.

Annadurai, P., Annadurai, V., Yongkun, M., Pugazhendhi, A., and Dhandayuthapani, K. (2021). Phytochemical composition, antioxidant and antimicrobial activities of Plecospermum spinosum Trecul. Process Biochemistry *100*, 107–116.

Annadurai, P., Gideon, D.A., Nirusimhan, V., Sivaramakrishnan, R., Dhandayuthapani, K., and Pugazhendhi, A. (2022). Deciphering the pharmacological potentials of Aganosma cymosa (Roxb.) G. Don using in vitro and computational methods. Process Biochemistry *114*, 119–133.

Arafa, E.-S.A., Zhu, Q., Shah, Z.I., Wani, G., Barakat, B.M., Racoma, I., El-Mahdy, M.A., and Wani, A.A. (2011). Thymoquinone up-regulates PTEN expression and induces apoptosis in doxorubicin-resistant human breast cancer cells. Mutation Research/Fundamental and Molecular Mechanisms of Mutagenesis *706*, 28–35.

Arumuggam, N., Bhowmick, N.A., and Rupasinghe, H.V. (2015). A review: Phytochemicals targeting JAK/STAT signaling and IDO expression in cancer. Phytotherapy Research *29*, 805–817.

Athar, M., Back, J.H., Kopelovich, L., Bickers, D.R., and Kim, A.L. (2009). Multiple molecular targets of resveratrol: Anti-carcinogenic mechanisms. Archives of Biochemistry and Biophysics *486*, 95–102.

Aumeeruddy, M.Z., and Mahomoodally, M.F. (2019). Combating breast cancer using combination therapy with 3 phytochemicals: Piperine, sulforaphane, and thymoquinone. Cancer *125*, 1600–1611.

Bagaria, S.P., Ray, P.S., Sim, M.-S., Ye, X., Shamonki, J.M., Cui, X., and Giuliano, A.E. (2014). Personalizing breast cancer staging by the inclusion of ER, PR, and HER2. JAMA Surgery *149*, 125–129.

Baig, M.H., Adil, M., Khan, R., Dhadi, S., Ahmad, K., Rabbani, G., Bashir, T., Imran, M.A., Husain, F.M., and Lee, E.J. (2019). Enzyme targeting strategies for prevention and treatment of cancer: Implications for cancer therapy. Seminars in Cancer Biology *56*, 1–11.

Balakrishnan, S., Bhat, F., Raja Singh, P., Mukherjee, S., Elumalai, P., Das, S., Patra, C., and Arunakaran, J. (2016). Gold nanoparticle—conjugated quercetin inhibits epithelial—mesenchymal transition, angiogenesis and invasiveness via EGFR/VEGFR-2-mediated pathway in breast cancer. Cell Proliferation *49*, 678–697.

Barrios, J., Cordero, C.P., Aristizabal, F., Heredia, F.J., Morales, A.L., and Osorio, C. (2010). Chemical analysis and screening as anticancer agent of anthocyanin-rich extract from uva caimarona (Pourouma cecropiifolia Mart.) fruit. Journal of Agricultural and Food Chemistry *58*, 2100–2110.

Baydas, G., Reiter, R., Akbulut, M., Tuzcu, M., and Tamer, S. (2005). Melatonin inhibits neural apoptosis induced by homocysteine in hippocampus of rats via inhibition of cytochrome c translocation and caspase-3 activation and by regulating pro-and anti-apoptotic protein levels. Neuroscience *135*, 879–886.

Binienda, A., Ziolkowska, S., and Pluciennik, E. (2020). The anticancer properties of silibinin: Its molecular mechanism and therapeutic effect in breast cancer. Anti-Cancer Agents in Medicinal Chemistry (Formerly Current Medicinal Chemistry-Anti-Cancer Agents) *20*, 1787–1796.

Blackley, E.F., and Loi, S. (2019). Targeting immune pathways in breast cancer: Review of the prognostic utility of TILs in early stage triple negative breast cancer (TNBC). The Breast *48*, S44–S48.

Bosetti, C., Spertini, L., Parpinel, M., Gnagnarella, P., Lagiou, P., Negri, E., Franceschi, S., Montella, M., Peterson, J., and Dwyer, J. (2005). Flavonoids and breast cancer risk in Italy. Cancer Epidemiology and Prevention Biomarkers *14*, 805–808.

Bray, F., Ferlay, J., Soerjomataram, I., Siegel, R. L., Torre, L. A., and Jemal, A. (2018). Global cancer statistics 2018: GLOBOCAN estimates of incidence and mortality worldwide for 36 cancers in 185 countries. CA: A Cancer Journal for Clinicians *68*(6), 394–424.

Brown, D.N., Caffa, I., Cirmena, G., Piras, D., Garuti, A., Gallo, M., Alberti, S., Nencioni, A., Ballestrero, A., and Zoppoli, G. (2016). Squalene epoxidase is a bona fide oncogene by amplification with clinical relevance in breast cancer. Scientific Reports *6*, 1–13.

Buranrat, B., and Junking, M. (2022). Effect of fisetin on the proliferation and migration of human breast and cervical cancer cells. Tropical Journal of Pharmaceutical Research *21*, 79–85.

Butti, R., Gunasekaran, V.P., Kumar, T.V., Banerjee, P., and Kundu, G.C. (2019). Breast cancer stem cells: Biology and therapeutic implications. The International Journal of Biochemistry & Cell Biology *107*, 38–52.

Carroll, C.E., Ellersieck, M.R., and Hyder, S.M. (2008). Curcumin inhibits MPA-induced secretion of VEGF from T47-D human breast cancer cells. Menopause *15*, 570–574.

Casanova, F., Quarti, J., da Costa, D.C.F., Ramos, C.A., da Silva, J.L., and Fialho, E. (2012). Resveratrol chemosensitizes breast cancer cells to melphalan by cell cycle arrest. Journal of Cellular Biochemistry *113*, 2586–2596.

Castro, N.P., Rangel, M.C., Merchant, A.S., MacKinnon, G., Cuttitta, F., Salomon, D.S., and Kim, Y.S. (2019). Sulforaphane suppresses the growth of triple-negative breast cancer stem-like cells in vitro and in vivo. Cancer Prevention Research *12*, 147–158.

Chen, H. (2017). Ursolic acid inhibited proliferation and invasion of MDA-MB-231 human breast cancer cells via regulating cellular signal transduction pathways. Thesis submitted to Cornell University. https://doi.org/10.7298/X49P2ZT2.

Chen, J., Zhu, Y., Zhang, W., Peng, X., Zhou, J., Li, F., Han, B., Liu, X., Ou, Y., and Yu, X. (2018). Delphinidin induced protective autophagy via mTOR pathway suppression and AMPK pathway activation in HER-2 positive breast cancer cells. BMC Cancer *18*, 1–13.

Chen, X.-Y., Zhou, J., Luo, L.-P., Han, B., Li, F., Chen, J.-Y., Zhu, Y.-F., Chen, W., and Yu, X.-P. (2015). Black rice anthocyanins suppress metastasis of breast cancer cells by targeting RAS/RAF/MAPK pathway. BioMed Research International *2015*.

Chi, D., Singhal, H., Li, L., Xiao, T., Liu, W., Pun, M., Jeselsohn, R., He, H., Lim, E., and Vadhi, R. (2019). Estrogen receptor signaling is reprogrammed during breast tumorigenesis. Proceedings of the National Academy of Sciences *116*, 11437–11443.

Cipolletti, M., Solar Fernandez, V., Montalesi, E., Marino, M., and Fiocchetti, M. (2018). Beyond the antioxidant activity of dietary polyphenols in cancer: The modulation of estrogen receptors (ers) signaling. International Journal of Molecular Sciences *19*, 2624.

Croker, A.K., and Allan, A.L. (2012). Inhibition of aldehyde dehydrogenase (ALDH) activity reduces chemotherapy and radiation resistance of stem-like ALDHhiCD44+ human breast cancer cells. Breast Cancer Research and Treatment *133*, 75–87.

Dandawate, P.R., Subramaniam, D., Jensen, R.A., and Anant, S. (2016). Targeting cancer stem cells and signaling pathways by phytochemicals: Novel approach for breast cancer therapy. Seminars in Cancer Biology *40*, 192–208.

Daniel, A.R., Hagan, C.R., and Lange, C.A. (2011). Progesterone receptor action: Defining a role in breast cancer. Expert Review of Endocrinology & Metabolism *6*, 359–369.

Deguchi, H., Fujii, T., Nakagawa, S., Koga, T., and Shirouzu, K. (2002). Analysis of cell growth inhibitory effects of catechin through MAPK in human breast cancer cell line T47D. International Journal of Oncology *21*, 1301–1305.

Deiab, S., Mazzio, E., Messeha, S., Mack, N., and Soliman, K. (2013). High-throughput screening to identify plant derived human LDH-A inhibitors. European Journal of Medicinal Plants *3*, 603.

Dong, J., Yang, W., Han, J., Cheng, R., and Li, L. (2020). Effects of Notch signaling components from breast cancer cells treated in culture with resveratrol. Research in Veterinary Science *132*, 369–378.

Elsayed, A.M., Sherif, N.M., Hassan, N.S., Althobaiti, F., Hanafy, N.A., and Sahyon, H.A. (2021). Novel quercetin encapsulated chitosan functionalized copper oxide nanoparticles as anti-breast cancer agent via regulating p53 in rat model. International Journal of Biological Macromolecules *185*, 134–152.

Elsherbini, A.M., Sheweita, S.A., and Sultan, A.S. (2021). Pterostilbene as a phytochemical compound induces signaling pathways involved in the apoptosis and death of mutant P53-breast cancer cell lines. Nutrition and Cancer *73*, 1976–1984.

Esmat, A.Y., Tomasetto, C., and Rio, M.-C. (2006). Cytotoxicity of a natural anthraquinone (Aloin) against human breast cancer cell lines with and without ErbB-2: Topoisomerase II-alpha coamplification. Cancer Biology & Therapy *5*, 97–103.

Eyong, K.O., Bairy, G., Eno, A.A., Taube, J., Hull, K.G., Folefoc, G.N., Foyet, H.S., and Romo, D. (2018). Triterpenoids from the stem bark of Vitellaria paradoxa (Sapotaceae) and derived esters exhibit cytotoxicity against a breast cancer cell line. Medicinal Chemistry Research *27*, 268–277.

Ezzati, M., Yousefi, B., Velaei, K., and Safa, A. (2020). A review on anti-cancer properties of Quercetin in breast cancer. Life Sciences *248*, 117463.

Farghadani, R., and Naidu, R. (2021). Curcumin: Modulator of key molecular signaling pathways in hormone-independent breast cancer. Cancers *13*, 3427.

Farooqi, A.A. (2021). Regulation of deregulated cell signaling pathways by pomegranate in different cancers: Re-interpretation of knowledge gaps. Seminars in Cancer Biology *73*, 294–301.

Farooqi, A.A., Pinheiro, M., Granja, A., Farabegoli, F., Reis, S., Attar, R., Sabitaliyevich, U.Y., Xu, B., and Ahmad, A. (2020). EGCG mediated targeting of deregulated signaling pathways and non-coding RNAs in different cancers: Focus on JAK/STAT, Wnt/β-Catenin, TGF/SMAD, NOTCH, SHH/GLI, and TRAIL mediated signaling pathways. Cancers *12*, 951.

Fernando, W., Rupasinghe, H.V., & Hoskin, D.W. (2019). Dietary phytochemicals with anti-oxidant and pro-oxidant activities: A double-edged sword in relation to adjuvant chemotherapy and radiotherapy? *Cancer Letters*, *452*, 168–177.

Ferreira, A.K., Tavares, M.T., Pasqualoto, K.F.M., de Azevedo, R.A., Teixeira, S.F., Ferreira-Junior, W.A., Bertin, A.M., de-Sá-Junior, P.L., Barbuto, J.A.M., and Figueiredo, C.R. (2015). RPF151, a novel capsaicin-like analogue: In vitro studies and in vivo preclinical antitumor evaluation in a breast cancer model. Tumor Biology *36*, 7251–7267.

Fleege, N.M.G., and Cobain, E.F. (2022). Breast cancer management in 2021: A primer for the OB GYN. *Best Practice & Research Clinical Obstetrics & Gynaecology 82*, 30–45.

Gaikwad, S., and Srivastava, S.K. (2021). Role of phytochemicals in perturbation of redox homeostasis in cancer. Antioxidants *10*, 83.

Galdo, R.-D., Natalia, A., Lara Montero, Á., and Wertheimer, E.V. (2019). Role of Hedgehog signaling in breast cancer: Pathogenesis and therapeutics. Cells *8*, 375.

Gallardo, M., and Calaf, G.M. (2016). Curcumin inhibits invasive capabilities through epithelial mesenchymal transition in breast cancer cell lines. International Journal of Oncology *49*, 1019–1027.

Garufi, A., Pistritto, G., Cirone, M., and D'Orazi, G. (2016). Reactivation of mutant p53 by capsaicin, the major constituent of peppers. Journal of Experimental & Clinical Cancer Research *35*, 1–9.

Gideon, D.A., Annadurai, P., Nirusimhan, V., Parashar, A., James, J., and Dhayabaran, V.V. (2022). Evaluation of the Anticancer Activities of Isatin-Based Derivatives. In: Chakraborti S. (eds.) Handbook of Oxidative Stress in Cancer: Therapeutic Aspects. Springer, Singapore. https://doi.org/10.1007/978-981-16-1247-3_51-1 (In Press).

Gideon, D.A., Nirusimhan, V., E, J.C., Sudarsha, K., and Manoj, K.M. (2021). Mechanism of electron transfers mediated by cytochromes c and b 5 in mitochondria and endoplasmic reticulum: Classical and murburn perspectives. Journal of Biomolecular Structure and Dynamics, 1–18.

Gideon, D.A., Nirusimhan, V., and Manoj, K.M. (2020). Are plastocyanin and ferredoxin specific electron carriers or generic redox capacitors? Classical and murburn perspectives on two photosynthetic proteins. Journal of Biomolecular Structure and Dynamics, 1–15.

Gideon, D.A.M., and James, J. (2021). Assessing the Contributions of Lipid Profile and Oxidative Lipid Damage to Carcinogenesis. In Handbook of Oxidative Stress in Cancer: Mechanistic Aspects. Springer, Singapore. pp. 1–15.

Grattan, B.J. (2013). Plant sterols as anticancer nutrients: Evidence for their role in breast cancer. Nutrients *5*, 359–387.

Green, A.R., Carrillo, J.E., and Betancourt, J.R. (2002). Why the disease-based model of medicine fails our patients. Western Journal of Medicine *176*, 141.

Griendling, K.K., Sorescu, D., Lassègue, B., and Ushio-Fukai, M. (2000). Modulation of protein kinase activity and gene expression by reactive oxygen species and their role in vascular physiology and pathophysiology. Arteriosclerosis, Thrombosis, and Vascular Biology *20*, 2175–2183.

Hackshaw, M.D., Danysh, H.E., Singh, J., Ritchey, M.E., Ladner, A., Taitt, C., Camidge, D.R., Iwata, H., and Powell, C.A. (2020). Incidence of pneumonitis/interstitial lung disease induced by HER2-targeting therapy for HER2-positive metastatic breast cancer. Breast Cancer Research and Treatment *183*, 23–39.

Haile, R.W., John, E.M., Levine, A.J., Cortessis, V.K., Unger, J.B., Gonzales, M., Ziv, E., Thompson, P., Spruijt-Metz, D., and Tucker, K.L. (2012). A review of cancer in US Hispanic populations. Cancer Prevention Research *5*, 150–163.

Hanahan, D., and Weinberg, R.A. (2011). Hallmarks of cancer: The next generation. Cell *144*, 646–674.

Harbeck, N., Penault-Llorca, F., Cortes, J., Gnant, M., Houssami, N., Poortmans, P., Ruddy, K., Tsang, J., and Cardoso, F. (2019). Breast cancer. Nature Reviews Disease Primers *5*, 1–31.

Hayes, D.F. (2019). HER2 and breast cancer—a phenomenal success story. New England Journal of Medicine *381*, 1284–1286.

Hosford, S.R., and Miller, T.W. (2014). Clinical potential of novel therapeutic targets in breast cancer: CDK4/6, Src, JAK/STAT, PARP, HDAC, and PI3K/AKT/mTOR pathways. Pharmacogenomics and Personalized Medicine *7*, 203.

Hsu, K.-C., Li-Chan, E.C., and Jao, C.-L. (2011). Antiproliferative activity of peptides prepared from enzymatic hydrolysates of tuna dark muscle on human breast cancer cell line MCF-7. Food Chemistry *126*, 617–622.

Hu, C., Zhou, Q., Yang, B., Xiao, W., Jiao, K., Yang, K., Zhou, M., and Li, Z. (2020). ERK-mediated phosphorylation of EZH2 regulates HER2 expression on long-term genistein-induced acquired endocrine resistance in estrogen-receptor-positive breast cancer cells. Research Square. https://doi.org/10.21203/rs.3.rs-108061/v1.

Hu, S., Xu, Y., Meng, L., Huang, L., and Sun, H. (2018). Curcumin inhibits proliferation and promotes apoptosis of breast cancer cells. Experimental and therapeutic medicine *16*, 1266–1272.

Huang, Y.-C., Chao, K., Liao, H.-F., and Chen, Y.-J. (2013). Targeting sonic hedgehog signaling by compounds and derivatives from natural products. Evidence-Based Complementary and Alternative Medicine *2013*.

Ikawati, M., Jenie, R.I., Utomo, R.Y., Amalina, N.D., Ilmawati, G.P.N., Kawaichi, M., and Meiyanto, E. (2020). Genistein enhances cytotoxic and antimigratory activities of doxorubicin on 4T1 breast cancer cells through cell cycle arrest and ROS generation. J Appl Pharm Sci *10*, 95–104.

Jiang, H., Shang, X., Wu, H., Gautam, S.C., Al-Holou, S., Li, C., Kuo, J., Zhang, L., and Chopp, M. (2009). Resveratrol downregulates PI3K/Akt/mTOR signaling pathways in human U251 glioma cells. Journal of Experimental Therapeutics & Oncology *8*, 25.

Kakarala, M., Brenner, D.E., Korkaya, H., Cheng, C., Tazi, K., Ginestier, C., Liu, S., Dontu, G., and Wicha, M.S. (2010). Targeting breast stem cells with the cancer preventive compounds curcumin and piperine. Breast Cancer Research and Treatment *122*, 777–785.

Kamble, D., Mahajan, M., Dhat, R., and Sitasawad, S. (2021). Keap1-Nrf2 pathway regulates ALDH and contributes to radioresistance in breast cancer stem cells. Cells *10*, 83.

Kanematsu, S., Yoshizawa, K., Uehara, N., Miki, H., Sasaki, T., Kuro, M., Lai, Y.-C., Kimura, A., Yuri, T., and Tsubura, A. (2011). Sulforaphane inhibits the growth of KPL-1 human breast cancer cells in vitro and suppresses the growth and metastasis of orthotopically transplanted KPL-1 cells in female athymic mice. Oncology Reports *26*, 603–608.

Kang, L., and Wang, Z.Y. (2010). Breast cancer cell growth inhibition by phenethyl isothiocyanate is associated with down-regulation of oestrogen receptor-α36. Journal of Cellular and Molecular Medicine *14*, 1485–1493.

Kapinova, A., Kubatka, P., Golubnitschaja, O., Kello, M., Zubor, P., Solar, P., and Pec, M. (2018). Dietary phytochemicals in breast cancer research: Anticancer effects and potential utility for effective chemoprevention. Environmental health and preventive medicine *23*, 1–18.

Kashyap, D., Garg, V.K., Tuli, H.S., Yerer, M.B., Sak, K., Sharma, A.K., Kumar, M., Aggarwal, V., and Sandhu, S.S. (2019). Fisetin and quercetin: promising flavonoids with chemopreventive potential. Biomolecules *9*, 174.

Kaushik, S., Shyam, H., Agarwal, S., Sharma, R., Nag, T.C., Dwivedi, A.K., and Balapure, A.K. (2019). Genistein potentiates Centchroman induced antineoplasticity in breast cancer via PI3K/Akt deactivation and ROS dependent induction of apoptosis. Life Sciences *239*, 117073.

Kaymaz, B.T., Çetintaş, V.B., Aktan, Ç., and Kosova, B. (2014). MicroRNA-520a-5p displays a therapeutic effect upon chronic myelogenous leukemia cells by targeting STAT3 and enhances the anticarcinogenic role of capsaicin. Tumor Biology *35*, 8733–8742.

Khan, K., Quispe, C., Javed, Z., Iqbal, M.J., Sadia, H., Raza, S., Irshad, A., Salehi, B., Reiner, Ž., and Sharifi-Rad, J. (2020). Resveratrol, curcumin, paclitaxel and miRNAs mediated regulation of PI3K/Akt/mTOR pathway: Go four better to treat bladder cancer. Cancer Cell International *20*, 1–19.

Khorsandi, K., Hosseinzadeh, R., and Shahidi, F.K. (2019). Photodynamic treatment with anionic nanoclays containing curcumin on human triple-negative breast cancer cells: Cellular and biochemical studies. Journal of Cellular Biochemistry *120*, 4998–5009.

Kim, D.H., Sung, B., Kim, J.-A., Kang, Y.J., Hwang, S.Y., Hwang, N.-L., Suh, H., Choi, Y.H., Im, E., and Chung, H.Y. (2017a). HS-1793, a resveratrol analogue, downregulates the expression of hypoxia-induced HIF-1 and VEGF and inhibits tumor growth of human breast cancer cells in a nude mouse xenograft model. International Journal of Oncology *51*, 715–723.

Kim, Y.N., Choe, S.R., Cho, K.H., Cho, D.Y., Kang, J., Park, C.G., and Lee, H.Y. (2017b). Resveratrol suppresses breast cancer cell invasion by inactivating a RhoA/YAP signaling axis. Experimental & Molecular Medicine *49*, e296–e296.

Kohandel, Z., Farkhondeh, T., Aschner, M., Pourbagher-Shahri, A.M., and Samarghandian, S. (2021). STAT3 pathway as a molecular target for resveratrol in breast cancer treatment. Cancer Cell International *21*, 1–9.

Kuran, D., Pogorzelska, A., and Wiktorska, K. (2020). Breast cancer prevention-is there a future for sulforaphane and its analogs? Nutrients *12*, 1559.

Lai, L.-H., Fu, Q.-H., Liu, Y., Jiang, K., Guo, Q.-M., Chen, Q.-Y., Yan, B., Wang, Q.-Q., and Shen, J.-G. (2012). Piperine suppresses tumor growth and metastasis in vitro and in vivo in a 4T1 murine breast cancer model. Acta Pharmacologica Sinica *33*, 523–530.

Lan, T., Wang, L., Xu, Q., Liu, W., Jin, H., Mao, W., Wang, X., and Wang, X. (2013). Growth inhibitory effect of Cucurbitacin E on breast cancer cells. International Journal of Clinical and Experimental Pathology *6*, 1799.

Le Corre, L., Chalabi, N., Delort, L., Bignon, Y.J., and Bernard-Gallon, D. (2005). Resveratrol and breast cancer chemoprevention: Molecular mechanisms. Molecular Nutrition & Food Research *49*, 462–471.

Li, J., Gong, X., Jiang, R., Lin, D., Zhou, T., Zhang, A., Li, H., Zhang, X., Wan, J., and Kuang, G. (2018a). Fisetin inhibited growth and metastasis of triple-negative breast cancer by reversing epithelial-to-mesenchymal transition via PTEN/Akt/GSK3β signal pathway. Frontiers in Pharmacology, 772.

Li, X., Wang, X., Xie, C., Zhu, J., Meng, Y., Chen, Y., Li, Y., Jiang, Y., Yang, X., and Wang, S. (2018b). Sonic hedgehog and Wnt/β-catenin pathways mediate curcumin inhibition of breast cancer stem cells. Anti-Cancer Drugs *29*, 208–215.

Liu, J., Zhou, H., Song, L., Yang, Z., Qiu, M., Wang, J., and Shi, S. (2021). Anthocyanins: promising natural products with diverse pharmacological activities. Molecules *26*, 3807.

Malla, R.R., Deepak, K., Merchant, N., and Dasari, V.R. (2020). Breast tumor microenvironment: Emerging target of therapeutic phytochemicals. Phytomedicine *70*, 153227.

Mann, M., Cortez, V., and Vadlamudi, R.K. (2011). Epigenetics of estrogen receptor signaling: role in hormonal cancer progression and therapy. Cancers *3*, 1691–1707.

Manoj, K.M., Baburaj, A., Ephraim, B., Pappachan, F., Maviliparambathu, P.P., Vijayan, U.K., Narayanan, S.V., Periasamy, K., George, E.A., and Mathew, L.T. (2010). Explaining the atypical reaction profiles of heme enzymes with a novel mechanistic hypothesis and kinetic treatment. PloS One *5*, e10601.

Manoj, K.M., and Bazhin, N.M. (2021). The murburn precepts for aerobic respiration and redox homeostasis. Progress in Biophysics and Molecular Biology *167*, 104–120.

Manoj, K.M., Gade, S.K., Venkatachalam, A., and Gideon, D.A. (2016a). Electron transfer amongst flavo- and hemo-proteins: Diffusible species effect the relay processes, not protein—protein binding. RSC Advances *6*, 24121–24129.

Manoj, K.M., Nirusimhan, V., Parashar, A., Edward, J., and Gideon, D.A. (2022). Murburn precepts for lactic-acidosis, Cori cycle, and Warburg effect: Interactive dynamics of dehydrogenases, protons, and oxygen. *Journal of Cellular Physiology 237*, 1902–1922.

Manoj, K.M., Parashar, A., Venkatachalam, A., Goyal, S., Singh, P.G., Gade, S.K., Periyasami, K., Jacob, R.S., Sardar, D., and Singh, S. (2016b). Atypical profiles and modulations of heme-enzymes catalyzed outcomes by low amounts of diverse additives suggest diffusible radicals' obligatory involvement in such redox reactions. Biochimie *125*, 91–111.

Masuda, M., Uno, Y., Ohbayashi, N., Ohata, H., Mimata, A., Kukimoto-Niino, M., Moriyama, H., Kashimoto, S., Inoue, T., and Goto, N. (2016). TNIK inhibition abrogates colorectal cancer stemness. Nature Communications *7*, 1–14.

Mendonca, P., Horton, A., Bauer, D., Messeha, S., and Soliman, K.F. (2019). The inhibitory effects of butein on cell proliferation and TNF-α-induced CCL2 release in racially different triple negative breast cancer cells. PLoS One *14*, e0215269.

Miller, P.E., and Snyder, D.C. (2012). Phytochemicals and cancer risk: A review of the epidemiological evidence. Nutrition in Clinical Practice *27*, 599–612.

Mitra, S., Anand, U., Jha, N.K., Shekhawat, M.S., Saha, S.C., Nongdam, P., Rengasamy, K.R., Proćków, J., and Dey, A. (2021). Anticancer applications and pharmacological properties of piperidine and piperine: A comprehensive review on molecular mechanisms and therapeutic perspectives. Frontiers in Pharmacology *12*, 772418–772418.

Moghtaderi, H., Sepehri, H., Delphi, L., and Attari, F. (2018). Gallic acid and curcumin induce cytotoxicity and apoptosis in human breast cancer cell MDA-MB-231. BioImpacts: BI *8*, 185.

Mohammed, F., Rashid-Doubell, F., Taha, S., Cassidy, S., and Fredericks, S. (2020). Effects of curcumin complexes on MDA-MB-231 breast cancer cell proliferation. International Journal of Oncology *57*, 445–455.

Motaghed, M., Al-Hassan, F.M., and Hamid, S.S. (2013). Cellular responses with thymoquinone treatment in human breast cancer cell line MCF-7. Pharmacognosy Research *5*, 200.

Murias, M., Jäger, W., Handler, N., Erker, T., Horvath, Z., Szekeres, T., Nohl, H., and Gille, L. (2005). Antioxidant, prooxidant and cytotoxic activity of hydroxylated resveratrol analogues: structure—activity relationship. Biochemical Pharmacology *69*, 903–912.

Nagaria, T.S., Shi, C., Leduc, C., Hoskin, V., Sikdar, S., Sangrar, W., and Greer, P.A. (2017). Combined targeting of Raf and Mek synergistically inhibits tumorigenesis in triple negative breast cancer model systems. Oncotarget *8*, 80804.

Nandini, D., Rao, R.S., Deepak, B., and Reddy, P.B. (2020). Sulforaphane in broccoli: The green chemoprevention!! Role in cancer prevention and therapy. Journal of Oral and Maxillofacial Pathology: JOMFP *24*, 405.

Narayanankutty, A. (2020). Phytochemicals as PI3K/Akt/mTOR inhibitors and their role in breast cancer treatment. Recent Patents on Anti-Cancer Drug Discovery *15*, 188–199.

Nirgude, S., Mahadeva, R., Koroth, J., Kumar, S., Kumar, K.S.S., Gopalakrishnan, V., S Karki, S., and Choudhary, B. (2020). ST09, a novel curcumin derivative, blocks cell migration by inhibiting matrix metalloproteases in breast cancer cells and inhibits tumor progression in EAC mouse tumor models. Molecules *25*, 4499.

Omarini, C., Bettelli, S., Caprera, C., Manfredini, S., Barbolini, M., Moscetti, L., Isca, C., Toss, A., Barbieri, E., and Cortesi, L. (2019). Differential molecular pathways expression in HER2 positive early breast cancer according to hormone receptor status. Journal of Cancer Research and Clinical Oncology *145*, 821–828.

Pan, H., Zhou, W., He, W., Liu, X., Ding, Q., Ling, L., Zha, X., and Wang, S. (2012). Genistein inhibits MDA-MB-231 triple-negative breast cancer cell growth by inhibiting NF-κB activity via the Notch-1 pathway. International Journal of Molecular Medicine *30*, 337–343.

Panda, R., Sahu, S.N., Khan, F., and Pattanayak, S.K. (2020a). Binding performance of phytochemicals with mutant threonine-protein kinase Chk2 protein: An in silico study. *AIP Conference Proceedings (AIP Publishing LLC). 2270*, 020003.

Panda, S.S., Girgis, A.S., Thomas, S.J., Capito, J.E., George, R.F., Salman, A., El-Manawaty, M.A., and Samir, A. (2020b). Synthesis, pharmacological profile and 2D-QSAR studies of curcumin-amino acid conjugates as potential drug candidates. European Journal of Medicinal Chemistry *196*, 112293.

Parashar, A., Jacob, V.D., Gideon, D.A., and Manoj, K.M. (2021). Hemoglobin catalyzes ATP-synthesis in human erythrocytes: A murburn model. Journal of Biomolecular Structure and Dynamics, 1–13.

Patra, S., Mishra, S.R., Behera, B.P., Mahapatra, K.K., Panigrahi, D.P., Bhol, C.S., Praharaj, P.P., Sethi, G., Patra, S.K., and Bhutia, S.K. (2020). Autophagy-modulating phytochemicals in cancer therapeutics: Current evidences and future perspectives. Seminars in Cancer Biology *80*, 205–217.

Pawlik, A., Wiczk, A., Kaczyńska, A., Antosiewicz, J., and Herman-Antosiewicz, A. (2013). Sulforaphane inhibits growth of phenotypically different breast cancer cells. European Journal of Nutrition *52*, 1949–1958.

Pejin, B., Kojic, V., and Bogdanovic, G. (2014). An insight into the cytotoxic activity of phytol at in vitro conditions. Natural Product Research *28*, 2053–2056.

Qiu, D., Yan, X., Xiao, X., Zhang, G., Wang, Y., Cao, J., Ma, R., Hong, S., and Ma, M. (2021). To explore immune synergistic function of Quercetin in inhibiting breast cancer cells. Cancer Cell International *21*, 1–14.

Radde, B.N., Alizadeh-Rad, N., Price, S.M., Schultz, D.J., and Klinge, C.M. (2016). Anacardic acid, salicylic acid, and oleic acid differentially alter cellular bioenergetic function in breast cancer cells. Journal of Cellular Biochemistry *117*, 2521–2532.

Renoir, J.-M., Marsaud, V., and Lazennec, G. (2013). Estrogen receptor signaling as a target for novel breast cancer therapeutics. Biochemical Pharmacology *85*, 449–465.

Rezaei-Seresht, H., Cheshomi, H., Falanji, F., Movahedi-Motlagh, F., Hashemian, M., and Mireskandari, E. (2019). Cytotoxic activity of caffeic acid and gallic acid against MCF-7 human breast cancer cells: An in silico and in vitro study. Avicenna Journal of Phytomedicine *9*, 574.

Saeg, F., and Anbalagan, M. (2018). Breast cancer stem cells and the challenges of eradication: A review of novel therapies. Stem Cell Investigation *5*.

Sajadimajd, S., Bahramsoltani, R., Iranpanah, A., Patra, J.K., Das, G., Gouda, S., Rahimi, R., Rezaeiamiri, E., Cao, H., and Giampieri, F. (2020). Advances on natural polyphenols as anticancer agents for skin cancer. Pharmacological Research *151*, 104584.

Şakalar, Ç., İzgi, K., İskender, B., Sezen, S., Aksu, H., Çakır, M., Kurt, B., Turan, A., and Canatan, H. (2016). The combination of thymoquinone and paclitaxel shows anti-tumor activity through the interplay with apoptosis network in triple-negative breast cancer. Tumor Biology *37*, 4467–4477.

Saleh, H.A., Yousef, M.H., and Abdelnaser, A. (2021). The anti-inflammatory properties of phytochemicals and their effects on epigenetic mechanisms involved in TLR4/NF-κB-mediated inflammation. Frontiers in Immunology *12*.

Sales, M.S., Roy, A., Antony, L., Banu, S.K., Jeyaraman, S., and Manikkam, R. (2018). Octyl gallate and gallic acid isolated from Terminalia bellarica regulates normal cell cycle in human breast cancer cell lines. Biomedicine & Pharmacotherapy *103*, 1577–1584.

Samadder, N.J., Giridhar, K.V., Baffy, N., Riegert-Johnson, D., and Couch, F.J. (2019). Hereditary cancer syndromes—A primer on diagnosis and management: Part 1: Breast-ovarian cancer syndromes. Mayo Clinic Proceedings *99*, 1084–1098.

Sehrawat, A., Kim, S.-H., Vogt, A., and Singh, S.V. (2013). Suppression of FOXQ1 in benzyl isothiocyanate-mediated inhibition of epithelial—mesenchymal transition in human breast cancer cells. Carcinogenesis *34*, 864–873.

Shi, Y., Yang, S., Troup, S., Lu, X., Callaghan, S., Park, D.S., Xing, Y., and Yang, X. (2011). Resveratrol induces apoptosis in breast cancer cells by E2F1-mediated up-regulation of ASPP1. Oncology Reports *25*, 1713–1719.

Shim, Y., and Song, J.M. (2015). Quantum dot nanoprobe-based high-content monitoring of notch pathway inhibition of breast cancer stem cell by capsaicin. Molecular and Cellular Probes *29*, 376–381.

Shipitsin, M., and Polyak, K. (2008). The cancer stem cell hypothesis: In search of definitions, markers, and relevance. Laboratory Investigation *88*, 459–463.

Song, X., Zhang, M., Dai, E., and Luo, Y. (2019). Molecular targets of curcumin in breast cancer. Molecular Medicine Reports *19*, 23–29.

Soran, A., Ozmen, V., Ozbas, S., Karanlik, H., Muslumanoglu, M., Igci, A., Canturk, Z., Utkan, Z., Ozaslan, C., and Evrensel, T. (2018). Randomized trial comparing resection of primary tumor with no surgery in stage IV breast cancer at presentation: protocol MF07–01. Annals of Surgical Oncology *25*, 3141–3149.

Sotgia, F., Whitaker-Menezes, D., Martinez-Outschoorn, U. E., Salem, A. F., Tsirigos, A., Lamb, R., Sneddon, S., Hulit, J., Howell, A., and Lisanti, M. P. (2012). Mitochondria "fuel" breast cancer metabolism: Fifteen markers of mitochondrial biogenesis label epithelial cancer cells, but are excluded from adjacent stromal cells. Cell Cycle, 11(23), 4390–4401.

Sp, N., Kang, D.Y., Lee, J.-M., Bae, S.W., and Jang, K.-J. (2021). Potential antitumor effects of 6-gingerol in p53-dependent mitochondrial apoptosis and inhibition of tumor sphere formation in breast cancer cells. International Journal of Molecular Sciences *22*, 4660.

Su, X., Jiang, X., Meng, L., Dong, X., Shen, Y., and Xin, Y. (2018). Anticancer activity of sulforaphane: The epigenetic mechanisms and the Nrf2 signaling pathway. Oxidative Medicine and Cellular Longevity *2018*.

Suarez, N.G., Torres, S.R., and Annabi, B. (2021). EGCG targeting of adipogenesis reveals a STAT3-mediated paracrine oncogenic control of triple-negative breast cancer cell invasive phenotype. Molecules *10*, 1506.

Sun, C.-Y., Zhang, Q.-Y., Zheng, G.-J., and Feng, B. (2019). Autophagy and its potent modulators from phytochemicals in cancer treatment. Cancer Chemotherapy and Pharmacology *83*, 17–26.

Sutton, K.M., Greenshields, A.L., and Hoskin, D.W. (2014). Thymoquinone, a bioactive component of black caraway seeds, causes G1 phase cell cycle arrest and apoptosis in triple-negative breast cancer cells with mutant p53. Nutrition and Cancer *66*, 408–418.

Tajbakhsh, A., Hasanzadeh, M., Rezaee, M., Khedri, M., Khazaei, M., ShahidSales, S., Ferns, G.A., Hassanian, S.M., and Avan, A. (2018). Therapeutic potential of novel formulated forms of curcumin in the treatment of breast cancer by the targeting of cellular and physiological dysregulated pathways. Journal of Cellular Physiology *233*, 2183–2192.

Tang, F.-Y., Chiang, E.-P.I., and Sun, Y.-C. (2008a). Resveratrol inhibits heregulin-β1-mediated matrix metalloproteinase-9 expression and cell invasion in human breast cancer cells. The Journal of Nutritional Biochemistry *19*, 287–294.

Tang, F.Y., Shih, C.J., Cheng, L.H., Ho, H.J., and Chen, H.J. (2008b). Lycopene inhibits growth of human colon cancer cells via suppression of the Akt signaling pathway. Molecular Nutrition & Food Research *52*, 646–654.

Thakuri, B.K.C., Zhang, J., Zhao, J., Nguyen, L.N., Nguyen, L.N., Schank, M., Khanal, S., Dang, X., Cao, D., and Lu, Z. (2020). HCV-associated exosomes upregulate RUNXOR and RUNX1 expressions to promote MDSC expansion and suppressive functions through STAT3—miR124 axis. Cells *9*, 2715.

Thoennissen, N., O'kelly, J., Lu, D., Iwanski, G., La, D., Abbassi, S., Leiter, A., Karlan, B., Mehta, R., and Koeffler, H. (2010). Capsaicin causes cell-cycle arrest and apoptosis in ER-positive and-negative breast cancer cells by modulating the EGFR/HER-2 pathway. Oncogene *29*, 285–296.

Tuli, H.S., Tuorkey, M.J., Thakral, F., Sak, K., Kumar, M., Sharma, A.K., Sharma, U., Jain, A., Aggarwal, V., and Bishayee, A. (2019). Molecular mechanisms of action of genistein in cancer: Recent advances. Frontiers in Pharmacology, 1336.

Varghese, E., Liskova, A., Kubatka, P., Samuel, S.M., and Büsselberg, D. (2020). Anti-angiogenic effects of phytochemicals on miRNA regulating breast cancer progression. Biomolecules *10*, 191.

Venkatadri, R., Muni, T., Iyer, A., Yakisich, J., and Azad, N. (2016). Role of apoptosis-related miRNAs in resveratrol-induced breast cancer cell death. Cell Death & Disease *7*, e2104.

Vergara, D., Stanca, E., Guerra, F., Priore, P., Gaballo, A., Franck, J., . . . Maffia, M. (2017). β-Catenin knockdown affects mitochondrial biogenesis and lipid metabolism in breast cancer cells. *Frontiers in Physiology 8*, 544.

Verma, S., Singh, A., and Mishra, A. (2013). Gallic acid: molecular rival of cancer. Environmental Toxicology and Pharmacology *35*, 473–485.

Vo, T.T.T., Chu, P.-M., Tuan, V.P., Te, J.S.-L., and Lee, I.-T. (2020). The promising role of antioxidant phytochemicals in the prevention and treatment of periodontal disease via the inhibition of oxidative stress pathways: Updated insights. Antioxidants *9*, 1211.

Wang, H., Zhang, K., Liu, J., Yang, J., Tian, Y., Yang, C., Li, Y., Shao, M., Su, W., and Song, N. (2021). Curcumin regulates cancer progression: Focus on ncRNAs and molecular signaling pathways. Frontiers in Oncology *11*, 1202.

Wang, L., Wang, C., Tao, Z., Zhao, L., Zhu, Z., Wu, W., He, Y., Chen, H., Zheng, B., and Huang, X. (2019). Curcumin derivative WZ35 inhibits tumor cell growth via ROS-YAP-JNK signaling pathway in breast cancer. Journal of Experimental & Clinical Cancer Research *38*, 1–17.

Wang, R., Li, J., Zhao, Y., Li, Y., and Yin, L. (2018a). Investigating the therapeutic potential and mechanism of curcumin in breast cancer based on RNA sequencing and bioinformatics analysis. Breast Cancer *25*, 206–212.

Wang, R., Yang, L., Li, S., Ye, D., Yang, L., Liu, Q., Zhao, Z., Cai, Q., Tan, J., and Li, X. (2018b). Quercetin inhibits breast cancer stem cells via downregulation of aldehyde dehydrogenase 1A1 (ALDH1A1), chemokine receptor type 4 (CXCR4), mucin 1 (MUC1), and epithelial cell adhesion molecule (EpCAM). Medical Science Monitor: International Medical Journal of Experimental and Clinical Research *24*, 412.

Wen, D.X., Xu, Y.-F., Mais, D.E., Goldman, M.E., and McDonnell, D.P. (1994). The A and B isoforms of the human progesterone receptor operate through distinct signaling pathways within target cells. Molecular and Cellular Biology *14*, 8356–8364.

Wooster, R., and Weber, B.L. (2003). Breast and ovarian cancer. New England Journal of Medicine *348*, 2339–2347.

Wu, F., Zhou, C., Zhou, D., Ou, S., Zhang, X., and Huang, H. (2018). Structure characterization of a novel polysaccharide from Hericium erinaceus fruiting bodies and its immunomodulatory activities. Food & Function *9*, 294–306.

Xu, X., Rajamanickam, V., Shu, S., Liu, Z., Yan, T., He, J., Liu, Z., Guo, G., Liang, G., and Wang, Y. (2019). Indole-2-carboxamide derivative LG25 inhibits triple-negative breast cancer growth by suppressing akt/mTOR/NF-κB signalling pathway. Drug Design, Development and Therapy *13*, 3539.

Yeap, S., Akhtar, M.N., Lim, K.L., Abu, N., Ho, W.Y., Zareen, S., Roohani, K., Ky, H., Tan, S.W., and Lajis, N. (2015). Synthesis of an anthraquinone derivative (DHAQC) and its effect on induction of G2/M arrest and apoptosis in breast cancer MCF-7 cell line. Drug Design, Development and Therapy *9*, 983.

Younas, M., Hano, C., Giglioli-Guivarc'h, N., and Abbasi, B.H. (2018). Mechanistic evaluation of phytochemicals in breast cancer remedy: current understanding and future perspectives. RSC Advances *8*, 29714–29744.

Yuan, X., Wu, H., Xu, H., Xiong, H., Chu, Q., Yu, S., Wu, G.S., and Wu, K. (2015). Notch signaling: An emerging therapeutic target for cancer treatment. Cancer Letters *369*, 20–27.

Zacksenhaus, E., Shrestha, M., Liu, J. C., Vorobieva, I., Chung, P. E., Ju, Y., . . . Jiang, Z. (2017). Mitochondrial OXPHOS induced by RB1 deficiency in breast cancer: implications for anabolic metabolism, stemness, and metastasis. *Trends in Cancer*, *3*(11), 768–779.

Zhai, J.-W., Gao, C., Ma, W.-D., Wang, W., Yao, L.-P., Xia, X.-X., Luo, M., Zu, Y.-G., and Fu, Y.-J. (2016). Geraniin induces apoptosis of human breast cancer cells MCF-7 via ROS-mediated stimulation of p38 MAPK. Toxicology Mechanisms and Methods *26*, 311–318.

Zhang, H., Zhang, M., Yu, L., Zhao, Y., He, N., and Yang, X. (2012). Antitumor activities of quercetin and quercetin-5′, 8-disulfonate in human colon and breast cancer cell lines. Food and Chemical Toxicology *50*, 1589–1599.

Zhang, M., Chen, X., & Radacsi, N. (2021). New tricks of old drugs: Repurposing non-chemo drugs and dietary phytochemicals as adjuvants in anti-tumor therapies. *Journal of Controlled Release*, *329*, 96–120.

Zhang, X., Yuan, X., Shi, H., Wu, L., Qian, H., and Xu, W. (2015). Exosomes in cancer: Small particle, big player. Journal of Hematology & Oncology *8*, 1–13.

Zheng, Y., Li, Z., Chen, H., and Gao, Y. (2020). Nanoparticle-based drug delivery systems for controllable photodynamic cancer therapy. European Journal of Pharmaceutical Sciences *144*, 105213.

Zhou, J., Xu, X.-Z., Hu, Y.-R., Hu, A.-R., Zhu, C.-L., and Gao, G.-S. (2014). Cryptotanshinone induces inhibition of breast tumor growth by cytotoxic CD4+ T cells through the JAK2/STAT4/perforin pathway. Asian Pacific Journal of Cancer Prevention *15*, 2439–2445.

Zhu, X., and Zhou, W. (2015). The emerging regulation of VEGFR-2 in triple-negative breast cancer. Frontiers in Endocrinology *6*, 159.

9 Aromatic Plants as a Source of Natural Antioxidants and Their Potential to Reduce Disease Risk

Indira Mikkili, Krupanidhi Sreerama, Venkateswarulu T C, Abraham Peele Karlapudi, and Vidya Prabhakar Kodali

CONTENTS

9.1 Introduction 185
9.2 Aromatic Plants 186
9.3 Natural Antioxidants From Aromatic Plants 188
9.3.1 Eugenol 188
9.3.2 Thymol 189
9.3.3 Carvacrol 189
9.3.4 Quercetin 192
9.3.5 Kaempferol 192
9.3.6 Gingerol 192
9.3.7 Luteolin 192
9.3.8 Apigenin 192
9.3.9 Rosmanol 193
9.3.10 Rosmarinic Acid 193
9.3.11 Cyanidin 193
9.3.12 Curcumin 193
9.3.13 Chrysin 193
9.3.14 Diosmetin 194
9.3.15 Carnosol 194
9.3.16 Carnosic Acid 194
9.4 Mechanism of Action of Natural Antioxidant Molecules in Averting Disease Risk 194
9.5 Conclusion and Future Aspects 195
Acknowledgments 195
Bibliography 195

9.1 INTRODUCTION

Aromatic plants are economically important plants used for their aroma and flavor (Samarth et al., 2017). Ethnobotanists mostly describe aromatic plants along with medicinal plants together, and they are categorized as medicinal and aromatic plants (MAPs) (Kandari et al., 2012). Aromatic plants are used as natural medicines, generally termed natural biochemical factories (Thomas et al., 2000). Worldwide nearly 1500 species of aromatic plants are known for their beneficial effects, out of which only 500 species have been studied in detail (Skaria, 2007). Aromatic plants are used in

DOI: 10.1201/b22917-9

food, spices, cooking, perfumes, liquor, cosmetics, fragrances, insect repellents, botanical pesticides, herbal drinks, medicines and pharmaceuticals (Irshad et al., 2020).

There is an increasing demand for aroma chemicals, flavoring agents, essential oils and pharmaceuticals in the world market (Sharmeen et al., 2021). In the world market, the pure essential oils have great demand due to their unique fragrances and natural origin (Sharmeen et al., 2021). The global essential oil market is witnessing high growth due to its natural fragrances over synthetic fragrances as result of health benefits (Irshad et al., 2020; Sharmeen et al., 2021). Though the COVID-19 pandemic had a devastating effect on several industries, the medicinal and aromatic plants market experienced positive growth in the world market (Sharmeen et al., 2021).

The aromatic compounds are present in various parts of the plant such as the seed, flower, fruit, bark, stem, root, foliage and wood (Samarth et al., 2017; Pandey et al., 2020). The pharmaceutical industries are targeting medicinal plants, aromatic herbs and phytochemical compounds for identifying potential drug molecules for treating various diseases (Boukhatem and Setzer, 2020). Herbal medicines, spices, essential oils and distilled natural products are a rich source of phytochemical compounds for the discovery and development of novel drug molecules for therapeutic purposes (Pan et al., 2013; Boukhatem and Setzer, 2020).

9.2 AROMATIC PLANTS

The classification of aromatic plants is done based on the importance, growth habit, habitat, duration of the crop, propagation and economic importance (Skaria et al., 2007; Pandey et al., 2020).

Based on the importance: The aromatic plants are classified into major aromatic plants and minor aromatic plants.

i. **Major aromatic plants:** The major aromatic plants are grown for essential oil extraction. The plants are extensively used in perfumes and cosmetics due to their aromatic principles, fragrances and essential oils.
   Examples: *Cymbopogon citratus* (lemon grass), *Chrysopogon zizanioides* (vetiver), *Matricaria chamomilla* (chamomile), *Eucalyptus globulus* (eucalyptus), *Melaleuca alternifolia* (tea tree) and *Pogostemon cablin* (patchouli).
ii. **Minor aromatic plants:** The plants that produce secondary products or by-products of natural volatile oils are considered minor aromatic plants.
   Examples: *Apium graveolens* (celery), *Cinnamomum verum* (cinnamon), *Anethum graveolens* (dill), *Tagetes erecta* (marigold) and *Abelmoschus moschatus* (ambrette).

Based on the growth habit: Based on the growth habit, the aromatic plants are classified into grasses, shrubs, herbs and trees.

i. **Aromatic grasses:** Aromatic grasses are used in aromatherapy due to the essential oils present in them.
   Examples: *Cymbopogon citratus* (lemon grass), *Cymbopogon nardus* (citronella grass) and *Cymbopogon martinii* (palmarosa grass).
ii. **Aromatic shrubs:** Shrubs differ from trees by their height and multiple stems. Some shrubs produce fragrance from foliage and scent from their flowers.
   Examples: *Pogostemon cablin* (patchouli), *Jasminum officinale* (jasmine), *Rosa indica* (rose), *Pelargonium graveolens* (geranium) and *Skimmia laureola* (skimmia)
iii. **Aromatic herbs:** Aromatic herbs have a strong pleasant smell, and the aromatic compounds are present in the surface glands of the leaf, veins of the leaf and stems. They are more perceptible by smell and better when chewed.

Examples: *Thymus vulgaris* (thyme), *Salvia rosmarinus* (rosemary), *Petroselinum crispum* (parsley), *Matricaria chamomilla* (chamomile), *Tagetes erecta* (marigold), *Trachyspermum ammi* (ajowan) and *Mentha piperita* (mint).

iv. **Aromatic trees:** Aromatic trees are popular for their fragrances and perfumes.
Examples:*Mangolia champaca* (champak), *Eucalyptus globulus* (eucalyptus), *Melaleuca alternifolia* (tea tree), *Cinnamomum camphora* (camphor) and *Cinnamomum verum* (cinnamon).

Based on the habitat: The aromatic plants are distributed in tropical, subtropical and temperate regions.

i. **Tropical:** The hot, humid environment is suitable for growth of a wide variety of aromatic plants. The commercially important plants cultivated in this region are *Cymbopogon citratus* (lemon grass), *Cymbopogon nardus* (citronella), *Cymbopogon martini* (palmarosa), *Ocimum sanctum* (ocimum), *Santalum album* (sandalwood), *Cinnamomum verum* (cinnamon) and *Pogostemon cablin* (patchouli).
ii. **Subtropical:** The aromatic plants grow in a subtropical type of climate are *Mentha piperita* (mint), *Chrysopogon zizanioides* (vetiver), *Trachyspermum ammi* (ajowan), *Eucalyptus globulus* (eucalyptus), *Cymbopogon nardus* (citronella), *Salvia rosmarinus* (rosemary), *Mentha canadensis* (Japanese mint), *Foeniculum vulgare* (fennel) and *Artemisia pallens* (davana).
iii. **Temperate:** The temperate regions have wider temperature ranges (warm, cold and moderate) compared to the tropical climate. The plants grown in this region are *Foeniculum vulgare* (fennel), *Trachyspermum ammi* (ajowan), *Mentha piperita* (peppermint), *Matricaria chamomilla* (chamomile) and *Mentha citrata* (bergamot mint).

Based on Crop Duration:

i. **Annuals:** Generally, annual plants grow only in one season and later they will die. The aromatic plants grown in one season are *Ocimum sanctum* (ocimum), *Trachyspermum ammi* (ajowan), *Artemisia pallens* (davana), *Lagoecia cuminoides* (lagoecia) and *Matricaria chamomilla* (chamomile).
ii. **Biennials:** The biennial plants grow in two growing seasons. They are *Apium graveolens* (celery) and *Chrysopogon zizanioides* (vetiver).
iii. **Perennials:** The plants regrow every spring season and have shorter blooming periods. They are *Mentha piperita* (mint), *Cymbopogon citratus* (lemon grass), *Cymbopogon martinii* (palmarosa), *Rosa indica* (rose), *Pelargonium graveolens* (geranium), *Cinnamomum verum* (cinnamon), *Melaleuca alternifolia* (tea tree) and *Cananga odorata (ylang-ylang).*

Based on Propagation:

i. **Vegetative propagation:** *Pelargonium graveolens* (geranium), *Cymbopogon nardus* (citronella), *Jasminum officinale* (jasmine), *Rosa indica* (rose), *Pogostemon cablin* (patchouli) and *Polianthes tuberosa* (tuberose).
ii. **Sexual mode of propagation:** *Cuminum cyminum* (cumin), *Artemisia pallens* (davana), *Salvia sclarea* (clary sage), *Cinnamomum camphora* (camphor), *Eucalyptus globulus* (eucalyptus), *Cymbopogon martinii* (palmarosa) and *Cananga odorata* (cananga)
iii. **Both vegetative and sexual mode:** *Tagetes erecta* (marigold), *Bursera penicillate* (linaloe), *Cymbopogon citratus* (lemon grass), *Cymbopogon martinii* (palmarosa), *Chrysopogon zizanioides* (vetiver), *Salvia rosmarinus* (rosemary) and *Thymus vulgaris* (thyme)

Based on economic importance: Various parts of the aromatic plants have high economic value in the market. Based on the economic value, the plants are classified under each category.

i. **Seeds:** *Apium graveolens* (celery), *Trachyspermum ammi* (ajowan), *Salvia sclarea* (clary sage) and *Abelmoschus moschatus* (ambrette).
ii. **Root:** *Santalum album* (sandalwood), *Chrysopogon zizanioides* (vetiver), *Cinnamomum camphora* (camphor) and *Sassafras albidum* (silk sassafras).
iii. **Wood:** *Santalum album* (sandalwood), *Cinnamomum camphora* (camphor), *Bursera penicillate* (linaloe), *Ocotea cymbarum* and *Ocotea pretiosa.*
iv. **Bark:** *Cinnamomum verum* and *Cinnamomum burmannii.*
v. **Leaf:** *Melaleuca alternifolia* (tea tree), *Laurus nobilis* (bay leaf), *Eucalyptus globulus* (eucalyptus), *Cinnamomum camphora* (camphor), *Mentha piperita* (mint), *Cymbopogon citratus* (lemon grass), *Sassafras albidum* (sassafras) and *Skimmia laureola* (skimmia).
vi. **Flower:** *Rosa indica* (rose), *Tagetes erecta* (marigold), *Jasminum officinale* (jasmine), *Mangolia champaca* (champak), *Matricaria chamomilla* (chamomile), *Polianthes tuberosa* (tuberose) and *Artabotrys hexapetalus* (tail grape).
vii. **Flowering tops:** *Thymus vulgaris* (thyme), *Artemisia pallens* (davana) and *Cymbopogon martinii* (palmarosa).
viii. **Fruit:** *Anethum graveolens* (dill), *Bursera penicillate* (linaloe) and *Tasmannia lanceolata* (mountain pepper).
ix. **Herbage:** *Cymbopogon nardus* (citronella), *Pogostemon cablin* (patchouli), *Ocimum basilicum* (sweet basil), *Salvia rosmarinus* (rosemary) and *Pelargonium graveolens* (geranium).

9.3 NATURAL ANTIOXIDANTS FROM AROMATIC PLANTS

Aromatic plants are a rich source of natural antioxidants (Figure 9.1). The antioxidant activity is due to phenolic compounds (Embuscado, 2015). The phytochemical constituents which give aroma and flavor to aromatic plants are polyphenols, phenolics, flavonoids, flavanols, terpenoids, quinines, oxygen substituted derivatives, alkaloids, glycosides, polypeptides and essential oils. They are also a rich source for vitamins, fatty acids and sugars (Christaki et al., 2012). The bioactive compounds have high redox properties, and these chemical structures are responsible for chelating metal ions, neutralizing free radicals, quenching singlet oxygen and decomposing peroxides (Christaki et al., 2012). The natural antioxidants reduce oxidative stress, which is caused by a high concentration of free radicals in the cells and tissues of the body (Nimse and Pal, 2015). The oxidative stress is due to various factors such as x-ray radiation, gamma radiation, ultraviolet (UV) radiation, polluted food, psychological stress, adverse effects of environmental conditions, drug addiction, alcoholism and smoking (Yashin et al., 2017). The oxidative stress leads to various diseases, including cardiac, inflammation, aging and cancer. Therefore, the aromatic plants and their natural antioxidants are used to prevent various diseases (Engwa, 2018; Dhalaria et al., 2020) (Table 9.1).

9.3.1 Eugenol

Eugenol is a phenyl propanoid with a spicy aroma extracted from clove buds (*Eugenia caryophyllata).* The flower buds of clove consist of 60–90% of eugenol (Nejad et al., 2017). It is also found in various aromatic plants such as nutmeg, bay leaf, basil, cinnamon, tulasi, ginger, turmeric, pepper, thyme, oregano, mace and marjoram (Khalil et al., 2017). Eugenol shows antioxidant capacity and free radical scavenging activity. Eugenol and its derivatives have been used in medicine as an

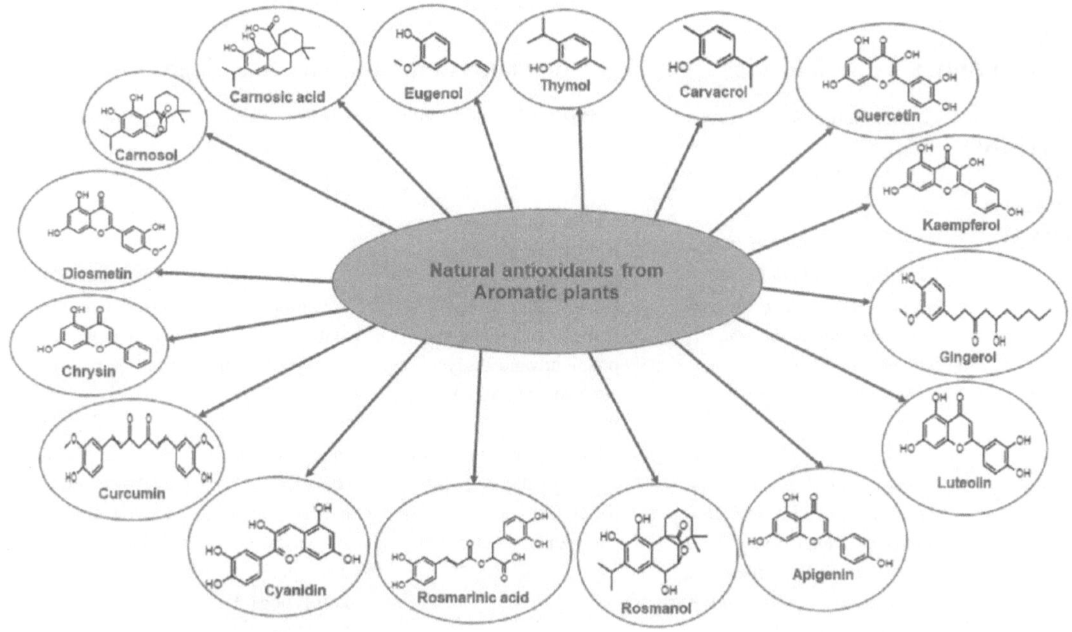

FIGURE 9.1 Natural antioxidant molecules from aromatic plants.

Source: Pubchem

anesthetic and local antiseptic (Pramod et al., 2010). The pharmacological activities of the eugenol are antioxidant, analgesic, antimicrobial, anticancer, anti-inflammatory and for gastrointestinal and respiratory complaints (Nisar et al., 2021; Zari et al., 2021).

9.3.2 Thymol

Thymol is a natural monoterpenoid derivative of p-cymene isomeric with carvacrol (Escobar et al., 2020). It is extracted from aromatic plants such as *Thymus vulgaris*, *Trachyspermum ammi*, *Thymus zygis*, *Thymus glandulosus*, *Origanum compactum*, *Illicium verum*, *Euphrasia rostkoviana*, *Lagoecia cuminoides*, *Monarda didyma*, *Monarda fistula* and *Origanum dictamnus* (Salehi et al., 2018). It occupies 10–64% of the thymol oil. Thymol has high antioxidant activity with health-promoting activities (Salehi et al., 2018). Thymol has various pharmacological activities including antioxidant, analgesic, anti-inflammatory, antiseptic, antibacterial, antispasmodic, antitumor and antifungal. Further, it also showed therapeutic actions against neurological, cardiovascular, metabolic, gastrointestinal and rheumatological diseases (Nagoor Meeran et al., 2017).

9.3.3 Carvacrol

This is a phenolic compound belonging to the monoterpenoid category found in essential oils of *Origanum vulgare*, *Lepidium flavum*, *Thymus vulgaris*, *Thymus zygis*, *Citrus aurantium*, *Satureja hortensis*, *Origanum majorana*, *Origanum dictamnus*, *Nigeria sativa* and *Satureja montana* (Sharifi-Rad et al., 2018). Carvacrol has antimicrobial, antiparasitic, antispasmodic, antitumor, anti-inflammatory, analgesic and antihepatotoxic activities (Can Baser, 2008).

TABLE 9.1
Natural Antioxidant Molecules from Aromatic Plants and Their Pharmacological Activities

Bioactive Compound and Molecular Formula	Aromatic Plant Sources	Mechanism of Action	Diseases	References
Eugenol $C_{10}H_{12}O_2$	*Eugenia caryophyllata, Cinnamomum verum, Curcuma longa, Pimpinella anisum, Ocimum basilicum*	Free radical scavenging activity; induction of DNA damage and apoptosis; potent advanced glycation end product (AGE) inhibitor	Cancer, inflammation, diabetes, neuroprotection	Nejad et al., 2017; Nisar et al., 2021; Singh et al., 2016
Thymol $C_{10}H_{14}O$	*Thymus vulgaris, Thymus zygis, Thymus glandulosus, Origanum compactum, Origanum dictamnus, Trachyspermum ammi, Illicium verum, Eupheasia rostkoviana, Lagoecia cuminoides, Monarda didyma, Monarda fistula*	DNA fragmentation, cell cycle arrest, caspase-dependent and -independent apoptosis	Inflammation, cancer, hyperlipidemia, Alzheimer's disease, gastrointestinal and metabolic disorders	Agarwal et al., 2020; Nagoor Meeran et al., 2017
Carvacrol $C_{10}H_{14}O$	*Origanum vulgare, Origanum dictamnus, Origanum majorana, Thymus vulgaris, Thymus zygis, Citrus aurantium, Satureja hortensis, Nigeria sativa, Satureja montana, Lepidium flavum*	Inhibition of cell proliferation and induction of apoptosis	Inflammation, cancer and hepatic-related disorders	Fan et al., 2015; Aristatile et al., 2013
Quercetin $C_{15}H_{10}O_7$	*Salvia officinalis, Anethum graveolens, Camellia sinensis, Trachyspermum ammi, Foeniculum vulgare, Petroselinum crispum*	Regulates antioxidant enzymes, free radical scavenging, binding to metal ions, pancreatic β-cell proliferation	Cardiac diseases, cancer, inflammation, arthritis, diabetes and cardiac diseases	Zhang et al., 2011; Xu et al., 2019; Salehi et al., 2020
Kaempferol $C_{15}H_{10}O_6$	*Camellia oleifera, Camellia sinensis, Delphinium formosum, Delphinium consolida, Delphinium hybridum, Helichrysum simillimum, Citrus medica,*	Inhibition of matrix metalloproteinases (MMP-2, MMP-3 and MMP-9); downregulation of AKT phosphorylation	Neurological disorders, cancer, inflammation, angiogenesis and cardiac diseases	Silva dos Santos et al., 2021; Wang et al., 2019
Gingerol $C_{17}H_{26}O_4$	*Zingiber officinalis,*	Inhibition of inflammatory mediators, iinhibition of cell proliferation; scavenging of hydroxyl radicals	Inflammation, angiogenesis, cancer, cardiac and neurological disorders	Tiwari and Mishra, 2016; Salehi et al., 2019
Luteolin $C_{15}H_{10}O_6$	*Apium graveolens, Piper nigrum, Mentha piperita, Salvia Rosmarinus, Origanum vulgare, Petroselinum crispum, Thymus vulgaris, Matricaria chamomilla*	Induces apoptosis, induces angiogenesis, chelation of metal ions, inhibits anti-inflammatory mediators	Cancer, inflammation	Dong et al., 2017; Wang et al., 2018

Bioactive Compound and Molecular Formula	Aromatic Plant Sources	Mechanism of Action	Diseases	References
Apigenin $C_{15}H_{10}O_5$	*Matricaria chamomilla, Apium graveolens, Pteroselinum crispum, Thymus vulgaris, Origanum vulgare, Ocimum basilicum*	Reduces reactive oxygen species and carcinogenesis, inhibits the inflammatory mediators	Diabetes, cancer, inflammation, Alzheimer's disease	Ginwala et al., 2019; Yan et al., 2017
Rosmanol $C_{20}H_{26}O_5$	*Rosmarinus officinalis*	Reduces oxidative stress, induces apoptosis	Inflammation, cancer	Nie et al., 2019; Jiang et al., 2021
Rosmarinic acid $C_{18}H_{16}O_8$	*Rosmarina officinalis, Salvia officinalis, Ocimum basilicum, Origanum vulgare, Thymus vulgaris, Mentha piperita, Melissa officinalis, Origanum majorana*	Reduces oxidative stress, reduction of reactive oxygen species, beta amyloid aggregation inhibition and cytokine inhibition	Diabetes, inflammation, cardiac diseases, neurological diseases, arthritis, autoimmune disorders	Runtuwene et al., 2016; Taguchi et al., 2017; Ghasemzadeh et al., 2020; Stansbury et al., 2012
Cyanidin $C_{15}H_{11}O_6^+$	*Rosa damascene, Rosa rugosa, Rosa hybrida*	Reduces oxidative stress, stimulation of antioxidant enzymes	Neurological diseases	Li et al., 2021; Salehi et al., 2020
Curcumin $C_{21}H_{20}O_6$	*Curcuma longa, Curcuma aromaticus, Curcuma manga, Curcuma phaeocaulis, Curcuma zedoaria, Etlingera elatior, Costus speciosus*	Inhibits accumulation of beta amyloid protein, reduces oxidative stress, inhibits inflammatory cytokines	Inflammation, cancer, thrombosis, cardiac diseases, neurological diseases and hepatotoxicity	Ramaholimihaso et al., 2020; Zheng et al., 2018; Blaslov, 2017
Chrysin $C_{15}H_{10}O_4$	*Passiflora carulea, P. incarnata, P. edulis* and *Oroxylum indicum*	Reduces oxidative stress and inhibits the inflammatory mediators	Cancer, inflammation, diabetes, neurological and hepatic diseases	Farkhondeh et al., 2019
Diosmetin $C_{16}H_{12}O_6$	*Citrus plant sps., Gallinum verum*	Activates antioxidant enzymes, inhibits cell proliferation, cell cycle arrest	Inflammation, depression, cancer and cardiac diseases	Yang et al., 2017; Garg et al., 2022
Carnosol $C_{20}H_{26}O_4$	*Rosmarina officinalis, Salvia officinalis, Thymus vulgaris, Ocimum basilicum*	Inhibition of anti-inflammatory cytokines, inhibition of NO and PGE-2 production	Cancer and inflammation	de Oliveira, 2016
Carnosic acid $C_{20}H_{28}O_4$	*Rosmarinus officinalis, Salvia officinalis, Origanum vulgare*	Inhibition of anti-inflammatory cytokines, inhibition of NO and COX-2 expression	Inflammation, cancer and metabolic disorders	de Oliveira, 2016

9.3.4 Quercetin

Quercetin is a potent flavonoid used as an antioxidant produced by sage, parsley, tea, dill, ajowan and fennel (David et al., 2016; Proestos and Varzakas, 2017). It is a strong antioxidant and scavenges the free radicals by binding to the metal ions (Kaurinovic and Vastag, 2019). Quercetin improves the heart rate and reduces the cancer risk in humans. It inhibits the oxidation of low-density lipoprotein (LDL) cholesterol. Further, it shows anti-inflammatory activity by inhibiting the lipoxygenase enzyme and inflammatory mediators (David et al., 2016).

9.3.5 Kaempferol

Kaempferol is a flavanol produced by plant genera such as citrus, camellia, delphinium and berries (Devi et al., 2015). It has wide range of pharmacological activities such as antioxidant, antimicrobial, antidiabetic, anticancer, analgesic, anti-inflammatory, neuroprotective and cardioprotective properties (Calderon-Montano et al., 2011). Kaempferol modulates the proinflammatory NF-KB activation via the NIK/IKK and MAPK signaling pathway. Thus, it regulates the anti-inflammatory activity (Ren et al., 2019).

9.3.6 Gingerol

Gingerol is a phenolic compound extracted from *Zingiber officinalis* (Mao et al., 2019). It is the main bioactive compound in ginger, and it showed high antioxidant activity by reducing the oxidative stress in the human body (Yusof and Anum, 2016). It also showed anti-inflammatory activity, anticancer activity, inhibition of angiogenesis, regulates the cell cycle and cytotoxic activities (Wang et al., 2014). Gingerol belongs to the alkaloids and is relative to the capsaicin and piperine molecules (Beltran et al., 2013). Many studies were done related to pharmacological activities, both in in vitro and in vivo, including antifungal, anticancer, anti-inflammatory, gastrointestinal and neuroprotective (Mao et al., 2019) activities. The gingerol derivative compounds are 6-gingerol, 8-gingerol, 10-gingerol and 12-gingerol (Mao et al., 2019).

9.3.7 Luteolin

Luteolin is a yellow-colored flavonoid polyphenol widely distributed in many plant families (Jakimiuk et al., 2021). It is an aglycone and in glycoside form exerts various biological activities. It is commonly found in celery, pepper, peppermint, rosemary, parsley, thyme, chamomile and oregano (Kytidou et al., 2020). The pharmacological activities are antioxidant, antimicrobial, anticancer, antidiabetic and anti-inflammatory (Bhaskarachary and Joshi, 2018). Luteolin acts as a powerful antioxidant by preventing inflammation and allergies. It also suppresses cancer-promoting protein expression, induces apoptosis and inhibits angiogenesis (Kytidou et al., 2020; Jakimiuk et al., 2021).

9.3.8 Apigenin

Apigenin is an aglycone moiety naturally occurring in glycosides belonging to the flavone class (Ali et al., 2017). The compound constitutes about 68% of total flavonoids and is particularly abundant in the flowers of the chamomile plant (Venigalla et al., 2015). It also found in celery, parsley, oregano, thyme, basil and celeriac. It is widely distributed in plants and is a well-studied phenolic compound among the flavonoid class (Shankar et al., 2017). Apigenin enhances the expression antioxidant enzymes such as superoxide dismutase, catalase and glutathione synthetase, thereby reducing oxidative stress. The pharmacological activities of apigenin are antioxidant, anti-inflammatory, antidiabetic, anticancer and cell cycle arrest (Jung, 2014; Salehi et al., 2019). Further, it improves learning ability and memory in Alzheimer's disease and provides beneficial effects in depression and insomnia (Salehi et al., 2019).

9.3.9 Rosmanol

Rosmanol is a polyphenol extracted from the aromatic plant *Rosmarinus officinalis*. It has high antioxidant activity (Nieto et al., 2018). The other phenolic compounds present in the rosemary plant are rosmadial, carnosic acid, carnosol and rosmarinic acid (Aludatt et al., 2018). Rosmanol showed inhibitory activity of LDL oxidation (Zeng et al., 2001). The rosmanol derivatives such as 7-methoxy rosmanol, 7-ethoxy rosmanol, 11-methoxy rosmanol, 11-ethoxy rosmanol, 12-methoxy rosmanol and 12-ethoxy rosmanol act as hepatic drugs by showing antilipid peroxidation activity (Marrero et al., 2002).

9.3.10 Rosmarinic Acid

Rosmarinic acid is a phenolic compound and ester of caffeic acid produced by plants such as rosemary, sage, basil, oregano, thyme, mint, lemon balm and marjoram (Ramos-Hryb et al., 2018). It is mainly found in the families of Laminaceae and Boroginaceae. It has therapeutic applications such as antioxidant, anti-inflammatory, antidiabetic, cardioprotective and neuroprotection (Petersen and Simmonds, 2003). Further, it also showed antimicrobial, antiviral and antiarthritic properties (Luo et al., 2020).

9.3.11 Cyanidin

Cyanidin is a natural plant pigment belonging to the anthocyanidin group (Dorman et al., 2021). It is mainly found in pigmented leaves, fruits and flowers (Giuliani et al., 2016). The plant sources are berries, plums, red cabbage, red onion and apples. The aromatic plants such as *Rosa damascene, Rosa rugosa* and *Rosa hybrida* contain a cyanidin molecule in their petals (Cendrowski et al., 2017). The cyanidin derivatives are cyanidin-3-O-glucoside, cyanidin-3-*O*-rutinoside, cyanidin-3-*O*-galactoside and cyanidin-3,5-diglucoside (Jaiswal et al., 2019). The cyanidin and cyanidin-3-*O* glucoside showed inhibitory activity on oxidation of LDLs. The cyanidin glucosides have neuroprotective activity by inhibiting the oxidative stress and DNA fragmentation in neuronal cells (Jaiswal et al., 2019).

9.3.12 Curcumin

Curcumin is a polyphenolic compound extracted from roots of *Curcuma longa* L (Sharifi-Rad et al., 2020). The compound is also found in *C. aromaticus, C. zedoaria, C. mangga, C. xanthorrhiza, C. phaeocaulis, Etlingera elatior* and *Costus speciosus* (Carvalho et al., 2016). Curcumin is a natural antioxidant reduces oxidative stress. The other pharmacological activities are anti-inflammatory, anticancer, analgesic, antiseptic, antithrombotic, cardioprotective, neuroprotective and hepatoprotective effects (Sharifi-Rad et al., 2020).

9.3.13 Chrysin

Chrysin is a natural flavonoid compound widely distributed in plants (Al-Oudat et al., 2019). Chemically it is a 5,7-dihydroxy flavone contained in flowers such as *Passiflora carulea*, *P. incarnata, P. edulis* and *Oroxylum indicum* (Stompor-Goracy et al., 2021). It has antioxidant, anti-inflammatory, anticancer, antidiabetic, neurodegenerative and hepatoprotective activity (Al-Oudat et al., 2019; Stompor-Goracy et al., 2021).

9.3.14 Diosmetin

Diosmetin is an aglycone part of the flavonoid glycoside widely distributed in natural plants, particularly citrus plants (Barreca et al., 2020). Chemically it is a 3′,5,7-trihydroxy-4′-methoxy flavone and also found in *Gallium verum* (Zhao et al., 2011). The pharmacological activities are antimicrobial, antioxidant, anti-inflammatory, antidepressant, anticancer, antineoplastic, cardioprotective and anticytokine (Shukla et al., 2019). The glycosylated flavones of diosmetin compounds are diosmetin 6,8-di-*C*-glucoside, diosmetin 8-*C*-glucoside, diosmetin 7-*O*-rutinoside and diosmetin 7-*O*-neohesperoside (Wang et al., 2017; Barreca et al., 2020).

9.3.15 Carnosol

Carnosol is a polyphenolic diterpene compound found in rosemary plants (Arya and Saldanha et al., 2019). It is also found in *Salvia officinalis*, *Thymus vulgaris* and *Ocimum basilicum* (Loussouarn et al., 2017). It acts as an anti-inflammatory agent by inhibiting the process of inflammation. It also prevents the activation of carcinogens, thereby acting as an anticancer agent (Johnson, 2011).

9.3.16 Carnosic Acid

Carnosic acid is a phenolic diterpenoid specifically present in the Lamiaceae family (Sarker and Nahar, 2020). It is abundantly found in the rosemary plant (*Rosmarinus officinalis*). It is also found in *Origanum vulgare* and *Salvia officinalis* (Loussouarn et al., 2017). The pharmacological activities are antioxidant, antimicrobial, anti-inflammatory, anticancer and metabolic disorders (Park et al., 2019; Goncalves et al., 2022).

9.4 MECHANISM OF ACTION OF NATURAL ANTIOXIDANT MOLECULES IN AVERTING DISEASE RISK

The natural antioxidant molecules and their various mechansims for reducing disease risk are shown in Figure 9.2. The bioactive compounds have shown free radical scavenging activity and inhibit reactive oxygen species production. Further, they inhibit proinflammatory mediators like

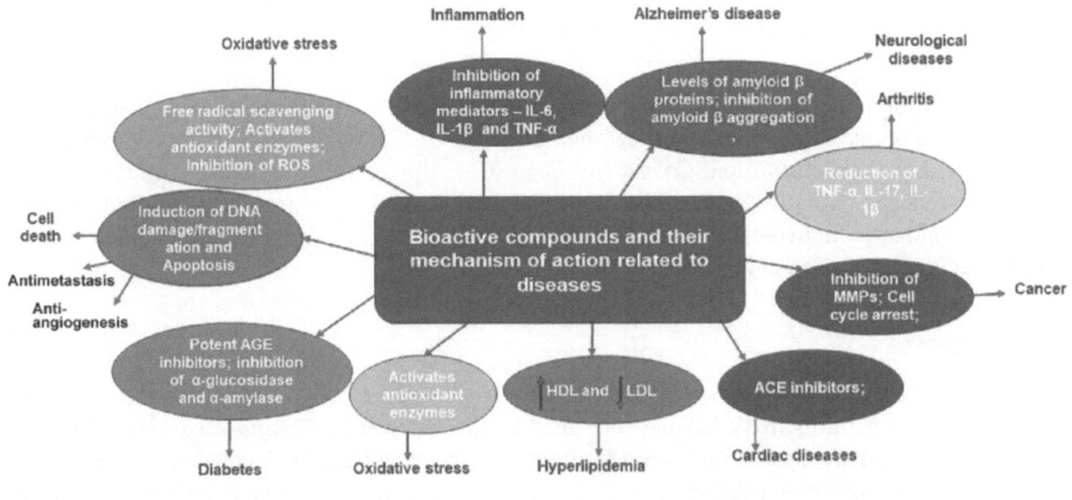

FIGURE 9.2 Overall mechanism of action of bioactive compounds in reducing disease risk.

interleukin (IL)-6, tumor necrosis factor alpha (TNF-α) and IL-1β (Nunes et al., 2020). The expression of inducible nitroc oxide (iNOS), NF-κB, prostaglandin synthesis and COX-2 is inhibited, which leads to anti-inflammatory activity (Miguel, 2010; das Chagas Pereira de Andrade and Mendes, 2020). The induction of cell cycle arrest leads to anticancer activity. In diabetes, the bioactive compounds act as potent advanced glycation end product (AGE) inhibitors and competitively block the binding of glucose molecules (Singh et al., 2016). In the case of cardiovascular diseases, the compounds attenuate inflammation of the myocardium by inhibiting the release of lysosomal enzymes. They also downregulate the expression of proinflammatory cytokines such as IL-6, IL-1β and TNF-α (Saleh et al., 2021). The increased levels of high-density lipoproteins and decreased levels of LDL result in regulation of hyperlipidemia (Bahmani et al., 2015). The neuroprotective mechanism in Alzheimer's disease is due to the increased levels of beta amyloid protein that leads to a reduction of age-related dementia (Ayaz et al., 2019; Gregory et al., 2021).

9.5 CONCLUSION AND FUTURE ASPECTS

The aromatic plants are rich in polyphenolic compounds, and several studies have showed antioxidant, antimicrobial, anti-inflammatory, antifungal, antiparasitic and antiprotozoal properties. In recent years, the demand for these plants and their derivatives have gained more attention due to their natural availability, which are generally recognized as safe. Hence, the aromatic plants and their derivatives have the potential to become a new generation of drug molecules or substances for treating human diseases. The excessive usage, less cultivation and extensive trade of aromatic plants in the world had become a major threat to the survival of the aromatic plants. For effective conservation, there is a need to identify cultivar varieties, development of various strategies, use of technological interventions and contribution of stakeholders. Further, there is a need to take efforts to protect and conserve aromatic plants by local governments, researchers, scientists, nongovernmental organizations (NGOs) and international groups. In addition to this, the bioactive compound production can be increased by using biotechnological applications for the development of efficient therapeutic molecules. Further, a good knowledge of the production of therapeutic compounds by modulating the biosynthetic pathways is essential. Hence, the commercial therapeutic molecules from natural aromatic plants could increase the quality of life and decrease the threats of various diseases.

ACKNOWLEDGMENTS

The authors acknowledge the management of VFSTR (Deemed to Be University) for providing a platform to carry out the work. This work was supported by DST FIST networking facility project LSI-576/2013 and Centre of Excellence in Biotechnology Department at VFSTR University, India.

BIBLIOGRAPHY

Agarwal, S., Tripathi, S. and Mishra, N. 2020. Pharmacological potential of thymol. In Innovations in Food Technology. Springer 489–500.

Ali, F., Rahul, Naz F., Jyoti, S. and Siddique, Y.H. 2017. Health functionality of apigenin: A review. International Journal of Food Properties 20(6): 1197–1238.

Al-Oudat, B.A., Alqudah, M.A., Audat, S.A., Al-Balas, Q.A., El-Elimat, T., Hassan, M.A., Frhat, I.N. and Azaizeh, M.M. 2019. Design, synthesis, and biologic evaluation of novel chrysin derivatives as cytotoxic agents and caspase-3/7 activators. Drug Design, Development and Therapy 13: 423–433.

Aludatt, M.H., Rababah, T., Alhamad, M.N., Gammoh, S., Al-Mahasneh, M.A., Tranchant, C.C. and Rawshdeh, M. 2018. Pharmaceutical, nutraceutical and therapeutic properties of selected wild medicinal plants: Thyme, spearmint, and rosemary. In Therapeutic, Probiotic, and Unconventional Foods. Academic Press. 275–290.

Aristatile, B., Al-Assaf, A.H. and Pugalendi, K.V. 2013. Carvacrol suppresses the expression of inflammatory marker genes in D-galactosamine-hepatotoxic rats. Asian Pacific Journal of Tropical Medicine 6(3): 205–211.

Arya, R. and Saldanha, S.N. 2019. Dietary phytochemicals, epigenetics, and colon cancer chemoprevention. In Epigenetics of Cancer Prevention. Academic Press. 205–229.

Ayaz, M., Sadiq, A., Junaid, M., Ullah, F., Ovais, M., Ullah, I., Ahmed, J. and Shahid, M. 2019. Flavonoids as prospective neuroprotectants and their therapeutic propensity in aging associated neurological disorders. Frontiers in Aging Neuroscience 11: 155.

Bahmani, M., Mirhoseini, M., Shirzad, H., Sedighi, M., Shahinfard, N. and Rafieian-Kopaei, M. 2015. A review on promising natural agents effective on hyperlipidemia. Journal of Evidence-Based Complementary & Alternative Medicine 20(3): 228–238.

Barreca, D., Mandalari, G., Calderaro, A., Smeriglio, A., Trombetta, D., Felice, M.R. and Gattuso, G. 2020. Citrus flavones: An update on sources, biological functions, and health promoting properties. Plants 9(3): 288.

Beltran, L.R., Dawid, C., Beltrán, M., Gisselmann, G., Degenhardt, K., Mathie, K., Hofmann, T. and Hatt, H. 2013. The pungent substances piperine, capsaicin, 6-gingerol and polygodial inhibit the human two-pore domain potassium channels TASK-1, TASK-3 and TRESK. Frontiers in Pharmacology 4: 141.

Bhaskarachary, K. and Joshi, A.K. 2018. Natural bioactive molecules with antidiabetic attributes: Insights into structure–activity relationships. In Studies in Natural Products Chemistry Vol. 57. Elsevier. 353–388.

Blaslov, K. 2017. Curcumin-a polyphenol with molecular targets for diabetes control. Endocr Oncol Metab 3: 43–48.

Boukhatem, M.N. and Setzer, W.N. 2020. Aromatic herbs, medicinal plant-derived essential oils, and phytochemical extracts as potential therapies for coronaviruses: Future perspectives. Plants 9(6): 800.

Calderon-Montano, M. J., Burgos-Morón, E., Pérez-Guerrero, C. and López-Lázaro, M. 2011. A review on the dietary flavonoid kaempferol. Mini Reviews in Medicinal Chemistry 11(4): 298–344.

Can Baser, K.H. 2008. Biological and pharmacological activities of carvacrol and carvacrol bearing essential oils. Current Pharmaceutical Design 14(29): 3106–3119.

Carvalho, A.C., Gomes, A.C., Pereira-Wilson, C. and Lima, C.F. 2016. Mechanisms of action of curcumin on aging: Nutritional and pharmacological applications. In Molecular Basis of Nutrition and Aging chapter 35. Elsevier. 491–511.

Cendrowski, A., Scibisz, I., Mitek, M., Kieliszek, M. and Kolniak-Ostek, J. 2017. Profile of the phenolic compounds of *Rosa rugosa* petals. Journal of Food Quality 2017: 1–10.

das Chagas Pereira de Andrade, F. and Mendes, A.N. 2020. Computational analysis of eugenol inhibitory activity in lipoxygenase and cyclooxygenase pathways. Scientific Reports 10(1): 1–14.

Christaki, E., Bonos, E., Giannenas, I. and Florou-Paneri, P. 2012. Aromatic plants as a source of bioactive compounds. Agriculture 2(3): 228–243.

David, A.V.A., Arulmoli, R. and Parasuraman, S. 2016. Overviews of biological importance of quercetin: A bioactive flavonoid. Pharmacognosy Reviews 10(20): 84–89.

de Oliveira, M.R. 2016. The dietary components carnosic acid and carnosol as neuroprotective agents: A mechanistic view. Molecular Neurobiology 53(9): 6155–6168.

Devi, K.P., Malar, D.S., Nabavi, S.F., Sureda, A., Xiao, J., Nabavi, S.M. and Daglia, M. 2015. Kaempferol and inflammation: From chemistry to medicine. Pharmacological Research 99: 1–10.

Dhalaria, R., Verma, R., Kumar, D., Puri, S., Tapwal, A., Kumar, V., Nepovimova, E. and Kuca, K. 2020. Bioactive compounds of edible fruits with their anti-aging properties: A comprehensive review to prolong human life. Antioxidants 9(11): 1123.

Dong, H., Yang, X., He, J., Cai, S., Xiao, K. and Zhu, L. 2017. Enhanced antioxidant activity, antibacterial activity and hypoglycemic effect of luteolin by complexation with manganese (II) and its inhibition kinetics on xanthine oxidase. Rsc Advances, 7(84): 53385–53395.

Dorman, G., Flachner, B., Hajdú, I. and Andras, C. 2021. Target identification and poly pharmacology of nutraceuticals. In Nutraceuticals. Elsevier. 315–343.

Embuscado, M.E. 2015. Spices and herbs: Natural sources of antioxidants—a mini review. Journal of Functional Foods 18: 811–819.

Engwa, G.A. 2018. Free radicals and the role of plant phytochemicals as antioxidants against oxidative stress-related diseases. In Phytochemicals: Source of Antioxidants and Role in Disease Prevention. BoD—Books on Demand 7. 49–74.

Escobar, A., Perez, M., Romanelli, G. and Blustein, G. 2020. Thymol bioactivity: A review focusing on practical applications. Arabian Journal of Chemistry 13(12): 9243–9269.

Fan, K., Li, X., Cao, Y., Qi, H., Li, L., Zhang, Q. and Sun, H. 2015. Carvacrol inhibits proliferation and induces apoptosis in human colon cancer cells. Anti-Cancer Drugs 26(8): 813–823.

Farkhondeh, T., Samarghandian, S. and Bafandeh, F. 2019. The cardiovascular protective effects of chrysin: A narrative review on experimental researches. Cardiovascular & Hematological Agents in Medicinal Chemistry (Formerly Current Medicinal Chemistry-Cardiovascular & Hematological Agents) 17(1): 17–27.

Garg, M., Chaudhary, S.K., Goyal, A., Sarup, P., Kumari, S., Garg, N., Vaid, L. and Shiveena, B. 2022. Comprehensive review on therapeutic and phytochemical exploration of diosmetin: A promising moiety. Phytomedicine Plus 2(1): 100179.

Ghasemzadeh Rahbardar, M. and Hosseinzadeh, H. 2020. Effects of rosmarinic acid on nervous system disorders: An updated review. Naunyn-Schmiedeberg's Archives of Pharmacology 393(10): 1779–1795.

Ginwala, R., Bhavsar, R., Chigbu, D.G.I., Jain, P. and Khan, Z.K. 2019. Potential role of flavonoids in treating chronic inflammatory diseases with a special focus on the anti-inflammatory activity of apigenin. Antioxidants 8(2): 35.

Giuliani, A., Cerretani, L. and Cichelli, A. 2016. Colors: Properties and determination of natural pigments. Encyclopedia of Food and Health 2: 273–283.

Goncalves, C., Fernandes, D., Silva, I. and Mateus, V. 2022. Potential anti-inflammatory effect of *Rosmarinus officinalis* in preclinical *In Vivo* models of inflammation. Molecules 27(3): 609.

Gregory, J., Vengalasetti, Y.V., Bredesen, D.E. and Rao, R.V. 2021. Neuroprotective herbs for the management of Alzheimer's disease. Biomolecules 11(4): 543.

Irshad, M., Subhani, M.A., Ali, S. and Hussain, A. 2020. Biological importance of essential oils. In Essential Oils-Oils of Nature. Intech Open 1–14.

Jaiswal, Y.S., Guan, Y., Moon, K.H. and Williams, L.L. 2019. Anthocyanins: Natural sources and traditional therapeutic uses. In Flavonoids-A Coloring Model for Cheering up Life. Intech Open 1–15.

Jakimiuk, K., Wink, M. and Tomczyk, M. 2021. Flavonoids of the caryophyllaceae. Phytochemistry Reviews 21: 179–218.

Jiang, D., Xu, J., Liu, S., Nasser, M.I., Wei, W., Mao, T., Liu, X., Zou, X., Li, J. and Li, X. 2021. Rosmanol induces breast cancer cells apoptosis by regulating PI3K/AKT and STAT3/JAK2 signaling pathways. Oncology Letters 22(2): 1–13.

Johnson, J.J. 2011. Carnosol: A promising anti-cancer and anti-inflammatory agent. Cancer Letters 305(1): 1–7.

Jung, W.W. 2014. Protective effect of apigenin against oxidative stress-induced damage in osteoblastic cells. International Journal of Molecular Medicine 33(5): 1327–1334.

Kandari, L.S., Phondani, P.C., Payal, K.C., Rao, K.S. and Maikhuri, R.K. 2012. Ethnobotanical study towards conservation of medicinal and aromatic plants in upper catchments of Dhauli Ganga in the central Himalaya. Journal of Mountain Science 9(2): 286–296.

Kaurinovic, B. and Vastag, D. 2019. Flavonoids and phenolic acids as potential natural antioxidants. Intech Open: 1–20.

Khalil, A.A., ur Rahman, U., Khan, M.R., Sahar, A., Mehmood, T. and Khan, M. 2017. Essential oil eugenol: Sources, extraction techniques and nutraceutical perspectives. RSC Advances 7(52): 32669–32681.

Kytidou, K., Artola, M., Overkleeft, H.S. and Aerts, J.M. 2020. Plant glycosides and glycosidases: A treasure-trove for therapeutics. Frontiers in Plant Science 11: 357.

Li, P., Feng, D., Yang, D., Li, X., Sun, J., Wang, G., Tian, L., Jiang, X. and Bai, W. 2021. Protective effects of anthocyanins on neurodegenerative diseases. Trends in Food Science & Technology 117: 205–217.

Loussouarn, M., Krieger-Liszkay, A., Svilar, L., Bily, A., Birtic, S. and Havaux, M. 2017. Carnosic acid and carnosol, two major antioxidants of rosemary, act through different mechanisms. Plant physiology 175(3): 1381–1394.

Luo, C., Zou, L., Sun, H., Peng, J., Gao, C., Bao, L., Ji, R., Jin, Y. and Sun, S. 2020. A review of the anti-inflammatory effects of rosmarinic acid on inflammatory diseases. Frontiers in Pharmacology 11: 153.

Mao, Q.Q., Xu, X.Y., Cao, S.Y., Gan, R.Y., Corke, H. and Li, H.B. 2019. Bioactive compounds and bioactivities of ginger (Zingiber officinale Roscoe). Foods 8(6): 185.

Marrero, J.G., Andrés, L.S. and Luis, J.G. 2002. Semi synthesis of rosmanol and its derivatives: Easy access to abietatriene diterpenes isolated from the genus Salvia with biological activities. Journal of Natural Products 65(7): 986–989.

Miguel, M.G. 2010. Antioxidant and anti-inflammatory activities of essential oils: A short review. Molecules 15(12): 9252–9287.

Nagoor Meeran, M.F., Javed, H., Al Taee, H., Azimullah, S. and Ojha, S.K. 2017. Pharmacological properties and molecular mechanisms of thymol: Prospects for its therapeutic potential and pharmaceutical development. Frontiers in Pharmacology 8: 380.

Nejad, S.M., Özgüneş, H. and Başaran, N. 2017. Pharmacological and toxicological properties of eugenol. Turkish Journal of Pharmaceutical Sciences 14(2): 201–206.

Nie, J.Y., Li, R., Wang, Y., Tan, J., Tang, S.H. and Jiang, Z.T. 2019. Antioxidant activity evaluation of rosemary ethanol extract and their cellular antioxidant activity toward HeLa cells. Journal of Food Biochemistry 43(7): e12851.

Nieto, G., Ros, G. and Castillo, J. 2018. Antioxidant and antimicrobial properties of rosemary (*Rosmarinus officinalis L.*): A review. Medicines 5(3): 98.

Nimse, S.B. and Pal, D. 2015. Free radicals, natural antioxidants, and their reaction mechanisms. RSC Advances 5(35): 27986–28006.

Nisar, M.F., Khadim, M., Rafiq, M., Chen, J., Yang, Y. and Wan, C.C. 2021. Pharmacological properties and health benefits of eugenol: A comprehensive review. Oxidative Medicine and Cellular Longevity 2021: 2497354.

Nunes, C.D.R., Barreto Arantes, M., Menezes de Faria Pereira, S., Leandro da Cruz, L., de Souza Passos, M., Pereira de Moraes, L., Vieira, I.J.C. and Barros de Oliveira, D. 2020. Plants as sources of anti-inflammatory agents. Molecules 25(16): 3726.

Pan, S.Y., Zhou, S.F., Gao, S.H., Yu, Z.L., Zhang, S.F., Tang, M.K., Sun, J.N., Ma, D.L., Han, Y.F., Fong, W.F. and Ko, K.M. 2013. New perspectives on how to discover drugs from herbal medicines: CAM's outstanding contribution to modern therapeutics. Evidence-Based Complementary and Alternative Medicine 2013:1–25

Pandey, A.K., Kumar, P., Saxena, M.J. and Maurya, P. 2020. Distribution of aromatic plants in the world and their properties. In Feed Additives. Academic Press 89–114.

Park, J., Rho, S.J. and Kim, Y.R. 2019. Enhancing antioxidant and antimicrobial activity of carnosic acid in rosemary (*Rosmarinus officinalis L.*) extract by complexation with cyclic glucans. Food Chemistry 299: 125119.

Petersen, M. and Simmonds, M.S. 2003. Rosmarinic acid. Phytochemistry 62(2): 121–125.

Pramod, K., Ansari, S.H. and Ali, J. 2010. Eugenol: A natural compound with versatile pharmacological actions. Natural Product Communications 5(12): 1999–2006.

Proestos, C. and Varzakas, T. 2017. Aromatic plants: Antioxidant capacity and polyphenol characterisation. Foods 6(4): 28.

Ramaholimihaso, T., Bouazzaoui, F. and Kaladjian, A. 2020. Curcumin in depression: Potential mechanisms of action and current evidence: A narrative review. Frontiers in Psychiatry 11:1–16.

Ramos-Hryb, A.B., Cunha, M.P., Kaster, M.P. and Rodrigues, A.L.S. 2018. Natural polyphenols and terpenoids for depression treatment: Current status. Studies in Natural Products Chemistry 55: 181–221.

Ren, J., Lu, Y., Qian, Y., Chen, B., Wu, T. and Ji, G. 2019. Recent progress regarding kaempferol for the treatment of various diseases. Experimental and Therapeutic Medicine 18(4): 2759–2776.

Runtuwene, J., Cheng, K.C., Asakawa, A., Amitani, H., Amitani, M., Morinaga, A., Takimoto, Y., Kairupan, B.H.R. and Inui, A. 2016. Rosmarinic acid ameliorates hyperglycemia and insulin sensitivity in diabetic rats, potentially by modulating the expression of PEPCK and GLUT4. Drug Design, Development and Therapy 10: 2193–2202.

Saleh, H.A., Yousef, M.H. and Abdelnaser, A. 2021. The anti-inflammatory properties of phytochemicals and their effects on epigenetic mechanisms involved in TLR4/NF-κB-mediated inflammation. Frontiers in Immunology 12: 1–29.

Salehi, B., Fokou, P.V.T., Yamthe, L.R.T., Tali, B.T., Adetunji, C.O., Rahavian, A., Mudau, F.N., Martorell, M., Setzer, W.N., Rodrigues, C.F. and Martins, N. 2019. Phytochemicals in prostate cancer: From bioactive molecules to upcoming therapeutic agents. Nutrients 11(7): 1483.

Salehi, B., Machin, L., Monzote, L., Sharifi-Rad, J., Ezzat, S.M., Salem, M.A., Merghany, R.M., El Mahdy, N.M., Kılıç, C.S., Sytar, O. and Sharifi-Rad, M. 2020. Therapeutic potential of quercetin: New insights and perspectives for human health. Acs Omega 5(20): 11849–11872.

Salehi, B., Mishra, A.P., Shukla, I., Sharifi-Rad, M., Contreras, M.D.M., Segura-Carretero, A., Fathi, H., Nasrabadi, N.N., Kobarfard, F. and Sharifi-Rad, J. 2018. Thymol, thyme, and other plant sources: Health and potential uses. Phytotherapy Research 32(9): 1688–1706.

Salehi, B., Sharifi-Rad, J., Cappellini, F., Reiner, Ž., Zorzan, D., Imran, M., Sener, B., Kilic, M., El-Shazly, M., Fahmy, N.M. and Al-Sayed, E. 2020. The therapeutic potential of anthocyanins: Current approaches based on their molecular mechanism of action. Frontiers in Pharmacology 11: 1300.

Salehi, B., Venditti, A., Sharifi-Rad, M., Kręgiel, D., Sharifi-Rad, J., Durazzo, A., Lucarini, M., Santini, A., Souto, E.B., Novellino, E. and Antolak, H. 2019. The therapeutic potential of apigenin. International Journal of Molecular Sciences 20(6): 1305.

Samarth, R.M., Samarth, M. and Matsumoto, Y. 2017. Medicinally important aromatic plants with radioprotective activity. Future Science OA 3(4): FSO247.

Sarker, S.D. and Nahar, L. 2020. Medicinal Natural Products: A Disease-Focused Approach. Academic Press.

Shankar, E., Goel, A., Gupta, K. and Gupta, S. 2017. Plant flavone apigenin: An emerging anticancer agent. Current Pharmacology Reports 3(6): 423–446.

Sharifi-Rad, J., Rayess, Y.E., Rizk, A.A., Sadaka, C., Zgheib, R., Zam, W., Sestito, S., Rapposelli, S., Neffe-Skocińska, K., Zielińska, D. and Salehi, B. 2020. Turmeric and its major compound curcumin on health: Bioactive effects and safety profiles for food, pharmaceutical, biotechnological and medicinal applications. Frontiers in Pharmacology 11: 1021.

Sharifi-Rad, M., Varoni, E.M., Iriti, M., Martorell, M., Setzer, W.N., del Mar Contreras, M., Salehi, B., Soltani-Nejad, A., Rajabi, S., Tajbakhsh, M. and Sharifi-Rad, J. 2018. Carvacrol and human health: A comprehensive review. Phytotherapy Research 32(9): 1675–1687.

Sharmeen, J.B., Mahomoodally, F.M., Zengin, G. and Maggi, F. 2021. Essential oils as natural sources of fragrance compounds for cosmetics and cosmeceuticals. Molecules 26(3): 666.

Shukla, R., Pandey, V., Vadnere, G.P. and Lodhi, S. 2019. Role of flavonoids in management of inflammatory disorders. In Bioactive Food as Dietary Interventions for Arthritis and Related Inflammatory Diseases. Academic Press 293–322.

Silva dos Santos, J., Goncalves Cirino, J.P., de Oliveira Carvalho, P. and Ortega, M.M. 2021. The pharmacological action of kaempferol in central nervous system diseases: A review. Frontiers in Pharmacology 11: 2143.

Singh, P., Jayaramaiah, R.H., Agawane, S.B., Vannuruswamy, G., Korwar, A.M., Anand, A., Dhaygude, V.S., Shaikh, M.L., Joshi, R.S., Boppana, R. and Kulkarni, M.J. 2016. Potential dual role of eugenol in inhibiting advanced glycation end products in diabetes: Proteomic and mechanistic insights. Scientific Reports 6(1): 1–13.

Skaria, B.P., Joy, P.P., Mathew, S., Mathew, G., Joseph, A., Joseph, R. 2007. Aromatic plants. Horticultural Sciences 7: 7–22.

Stansbury, J., Saunders, P., Winston, D. and Zampieron, E. 2012. Rosmarinic acid as a novel agent in the treatment of autoimmune disease. Journal of Restorative Medicine 1(1): 112–116.

Stompor-Goracy, M., Bajek-Bil, A. and Machaczka, M. 2021. Chrysin: Perspectives on contemporary status and future possibilities as pro-health agent. Nutrients 13(6): 2038.

Taguchi, R., Hatayama, K., Takahashi, T., Hayashi, T., Sato, Y., Sato, D., Ohta, K., Nakano, H., Seki, C., Endo, Y. and Tokuraku, K. 2017. Structure—activity relations of rosmarinic acid derivatives for the amyloid β aggregation inhibition and antioxidant properties. European Journal of Medicinal Chemistry 138: 1066–1075.

Thomas, J., Joy, P.P., Mathew, S. and Skaria, B.P. 2000. Plant sources of aroma chemicals and medicines in India. Chemical Industry Digest (Special Millennium Issue 104–108).

Tiwari, M.K. and Mishra, P.C. 2016. Anti-oxidant activity of 6-gingerol as a hydroxyl radical scavenger by hydrogen atom transfer, radical addition and electron transfer mechanisms. Journal of Chemical Sciences 128(8): 1199–1210.

Venigalla, M., Gyengesi, E. and Münch, G. 2015. Curcumin and Apigenin—novel and promising therapeutics against chronic neuroinflammation in Alzheimer's disease. Neural Regeneration Research 10(8): 1181.

Wang, S., Yang, C., Tu, H., Zhou, J., Liu, X., Cheng, Y., Luo, J., Deng, X., Zhang, H. and Xu, J. 2017. Characterization and metabolic diversity of flavonoids in citrus species. Scientific Reports 7(1): 1–10.

Wang, S., Zhang, C., Yang, G. and Yang, Y. 2014. Biological properties of 6-gingerol: A brief review. Natural Product Communications 9(7): 1027–1030.

Wang, X., Yang, Y., An, Y. and Fang, G. 2019. The mechanism of anticancer action and potential clinical use of kaempferol in the treatment of breast cancer. Biomedicine & Pharmacotherapy 117: 109086.

Wang, Y., Wang, J., Gong, X., Wen, X. and Gu, X. 2018. Luteolin: Anti-breast cancer effects and mechanisms. Journal of Exploratory Research in Pharmacology 3(3): 85–90.

Xu, D., Hu, M.J., Wang, Y.Q. and Cui, Y.L. 2019. Antioxidant activities of quercetin and its complexes for medicinal application. Molecules 24(6): 1123.

Yan, X., Qi, M., Li, P., Zhan, Y. and Shao, H. 2017. Apigenin in cancer therapy: Anti-cancer effects and mechanisms of action. Cell & Bioscience 7(1): 1–16.

Yang, Y., Gong, X.B., Huang, L.G., Wang, Z.X., Wan, R.Z., Zhang, P., Zhang, Q.Y., Chen, Z. and Zhang, B.S. 2017. Diosmetin exerts anti-oxidative, anti-inflammatory and anti-apoptotic effects to protect against endotoxin-induced acute hepatic failure in mice. Onco Target 8(19): 30723.

Yashin, A., Yashin, Y., Xia, X. and Nemzer, B. 2017. Antioxidant activity of spices and their impact on human health: A review. Antioxidants 6(3): 70.

Yusof, M. and Anum, Y. 2016. Gingerol and its role in chronic diseases. Drug Discovery from Mother Nature 177–207.

Zari, A.T., Zari, T.A. and Hakeem, K.R. 2021. Anticancer properties of Eugenol: A review. Molecules 26(23): 7407.

Zeng, H.H., Tu, P.F., Zhou, K., Wang, H., Wang, B.H. and Lu, J.F. 2001. Antioxidant properties of phenolic diterpenes from Rosmarinus officinalis. Acta Pharmacologica Sinica 22(12): 1094–1098.

Zhang, M., Swarts, S.G., Yin, L., Liu, C., Tian, Y., Cao, Y., Swarts, M., Yang, S., Zhang, S.B., Zhang, K. and Ju, S. 2011. Antioxidant properties of quercetin. In Oxygen Transport to Tissue XXXII. Springer 283–289.

Zhao, R., Chen, Z., Jia, G., Li, J., Cai, Y. and Shao, X. 2011. Protective effects of diosmetin extracted from *Galium verum L.* on the thymus of U14-bearing mice. Canadian Journal of Physiology and Pharmacology 89(9): 665–673.

Zheng, J., Cheng, J., Zheng, S., Feng, Q. and Xiao, X. 2018. Curcumin, a polyphenolic curcuminoid with its protective effects and molecular mechanisms in diabetes and diabetic cardiomyopathy. Frontiers in Pharmacology 9: 472.

10 Phytopharmaceuticals of Medicinal Plants for the Treatment of Common Illnesses

Venkata Kanaka Srivani Maddala, Asha S, and Sachidanand Singh

CONTENTS

10.1 Introduction 201
10.2 Common Diseases/Illnesses 202
10.3 Importance of Phytopharmaceuticals and Their Benefits and Applications 202
10.4 Extraction Process 202
10.4.1 Methods of Extraction 203
10.4.2 Phytopharmaceutical Drugs Manufacturing Process 203
10.5 Current Status of Phytopharmaceuticals 204
10.6 Role of Pharmaceuticals in Treating Illness 205
10.7 Significance 209
10.8 Conclusion and Future Prospects 209
Bibliography 210

10.1 INTRODUCTION

According to the World Health Organization, about 75% (three-quarters) of the world's population use medicinal plants to treat ill health. The application of herbal medicine improved historically all over the world (Chikezie 2015), and 40% of countries use herbal medicine for the treatment of diseases (Pandey et al. 2013). In developing countries, 80% of the population depends on traditional herbal medicine (Ekor 2013). Phytopharmaceuticals are the plant-based compounds that can be assessed in the form of fruits, purees, vegetables, prepacked fruits, vegetable products and supplements. They control disease pathways, disrupt the pathogenic cycle and target specific receptors. Herbal medicines have been used to treat number of diseases in both ancient times and today. Resveratrol, p-lutein, omega-3 fatty acids, phytoestrogens and terpenoids are examples. Phytopharmaceuticals may include purified and standard fractions of at least four bioactive or photochemical components of a medicinal plant extract that play a vital role in preventing diseases in both humans and animal. Herbal medicines are in great demand today, and the global market is projected to reach $111 billion by 2023, and India will occupy a strong market share. The Council of Scientific and Industrial Research (CSIR) launched its phytopharmaceutical mission in 2018, and its main aim was focused on increasing the availability of medicinal plants

DOI: 10.1201/b22917-10

availability with the highest demand. The northeastern region is rich in terms of a variety of flora and fauna, so the Department of Biotechnology launched their phytopharmaceutical mission to increase the cultivation and conservation of medicinal plants and produce phytopharmaceuticals. Medicinal plants are cultivated on the basis of phytomedicine in various countries, and it is big business today. Some side effects of phytopharmaceuticals are similar to drug interactions (Stiefelhagen 2002). Carnivora, Esberitox N Iascodar M, Pascotox forte injectopas may cause nausea, shivering and fever (Becker et al. 1988). So it is necessary to evaluate the efficacy, toxicity and safety of phytopharmaceuticals for further use.

10.2 COMMON DISEASES/ILLNESSES

Illness refers to poor health or sickness. Infections occur due to the presence of bacteria, viruses, fungi and parasites in the environment. Illness depends on the health condition of the person, pollution, stress, malnutrition and other manmade activities. Common cold, flu, diarrhea, skin rashes, allergies, conjunctivitis, depression, diabetes, cancer, arthritis, asthma, common heart and lung diseases, cough, dehydration, fever, etc., are the examples of illness. Almost all diseases lead to illness, so many drugs have been discovered to treat illnesses, but they may lead to side effects or a negative outcome. Instead of synthetic drugs suggested by the doctors, herbal medicines can be used which are eco-friendly and will have fewer side effects.

10.3 IMPORTANCE OF PHYTOPHARMACEUTICALS AND THEIR BENEFITS AND APPLICATIONS

Phytopharmaceuticals have been produced in Baden-Wurttemberg for many generations. Plants or its their flowers, seeds, fruits and stems are ground and reduced in size and consist of a number of chemical compounds, and these are the basis for the finished herbal products called phytopharmaceuticals. These are sold in the form of granules, tablets, capsules and lozenges. Baden-Wurttemberg is the company producing herbal medicines. It uses complex extraction procedures to produce herbal extracts with high concentrations of the elements sought. They produce special extracts which help to remove the unwanted substances and improve the efficacy of the herbal medicine. Phytopharmaceutical drugs are the purified and standardized fractions with at least four phytochemicals that are extracts of plant parts that can be used externally or internally to mitigate an illness. Some of the phytopharmaceuticals are flavonoids, isoflavonoids, lycopenes, carotenoids, limonoids, omega-3 fatty acids, resveratrol, terpenoids, phytosterol and phytoestrogens, which show antibacterial, anti-aging, antispasmodic, antidiabetic and antimicrobial activity. Phytopharmaceuticals are becoming hugely popular in both developed and developing countries due to their multiple treatment applications (Evans 2009). Applications of phytopharmaceuticals and their benefits presented in Figure 10.1.

10.4 EXTRACTION PROCESS

Extraction plays a vital role in the separation of active portions of plants or their parts from inactive components by using a standard solvent, and the products obtained from the plants are used externally. Standard extraction procedures help to obtain a therapeutically desired portion and remove the inert material by using a solvent called menstrum. The obtained extract can be used in the form of fluids or tinctures and further processed in any dosage form such as tablets and fractionated to produce individual chemical entities known as hyoscine and vincristine and standardized to get the final quality of the herbal product.

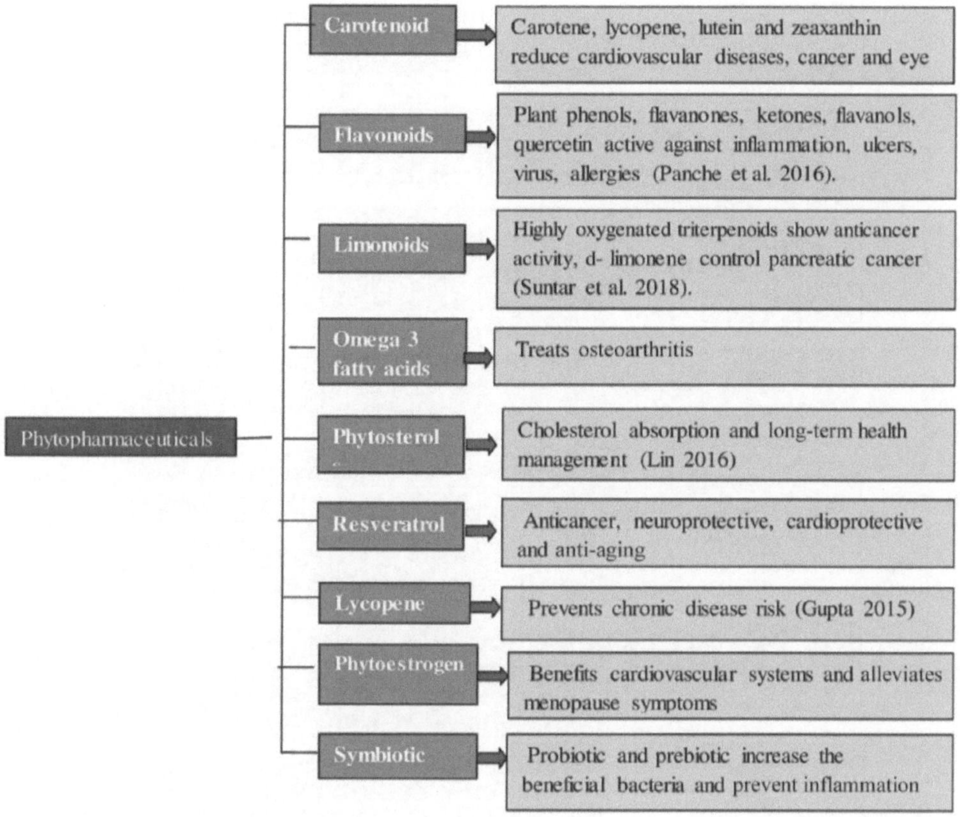

FIGURE 10.1 Applications of phytopharmaceuticals.

10.4.1 Methods of Extraction

Infusion, decoction, maceration, percolation, digestion steam distillation, successive solvent extraction and super-critical fluid extraction are the methods of extraction shown in Figure 10.2.

10.4.2 Phytopharmaceutical Drugs Manufacturing Process

Raw materials are collected such as a plant or its parts and the exudates are subjected to a stabilization and extraction process. To obtain herbal preparations, extraction, distillation, purification, concentration or fermentation is done. Stabilization is needed to maintain purity, identity and strength. Fractionating the plant material helps to provide isolated extractives, which includes pretreatment of the plant material to provide a fluidized plant material subjected to high-frequency pulses and shear force without changing the bioactive properties of bioactive components and provide the first liquid fraction. It consists of the extractives that should be isolated from the fractionated plant material (Mitchell and Brandenburg 2015). Clinical studies are conducted to predict the outcome of the patient and to find out the effect of the drug. The manufacturing process of phytopharmaceutical products is shown in Figure 10.3.

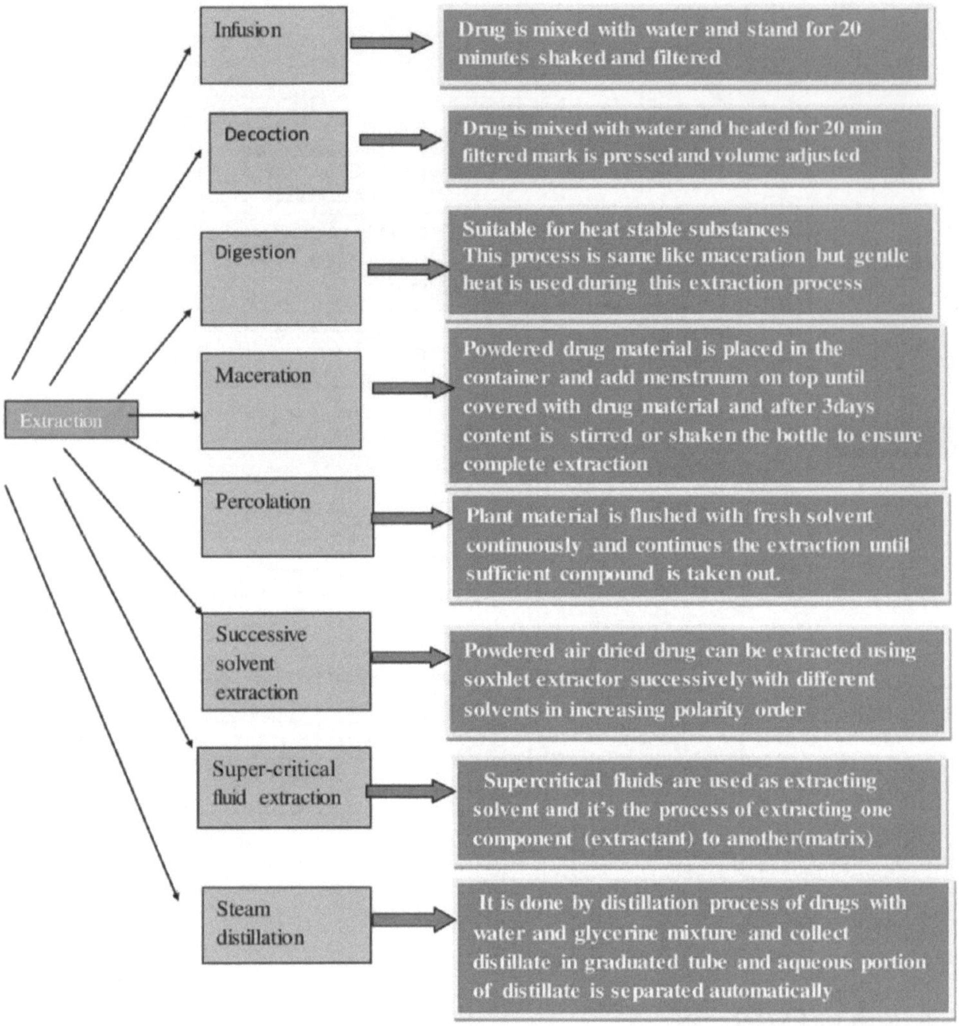

FIGURE 10.2 Methods of extraction: Pharmaceuticals.

10.5 CURRENT STATUS OF PHYTOPHARMACEUTICALS

There are many plants with medicinal properties, but only few are recommended, as they are poorly regulated and controlled by health authorities. For many drugs, they efficacy and quality are not assured. Many new drug applications are investigated, but investigational new drug applications (INDS) require information regarding the safety and proper clinical protocol.

New phytopharmaceutical regulations permit drug development based on the application of advanced techniques like extraction, fractionation and modern formulation development. The new regulations increase innovations of new medicinal drug development from plants in a scientific way that accepts herbal products.

Phytopharmaceuticals were well developed and used in a successful way for the treatment of common illnesses. The key benefits include less toxicity and minimum effects on their environment. But the therapeutic efficacy is lost in some compounds due to poor solubility and specificity at the target site (Mahomoodally et al. 2021).

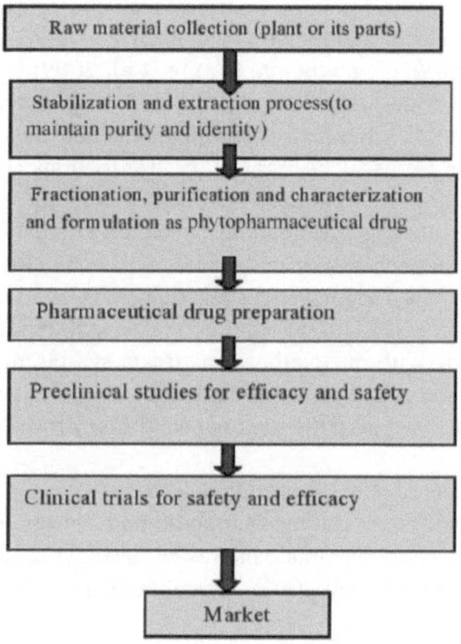

FIGURE 10.3 Phytopharmaceutical products manufacturing process.

Herbal medicines have been used for a long time to treat illnesses, and the consumption of phytomedicine is safer than synthetic medicine, leading to tremendous growth in phytopharmaceuticals. To treat minimal side effects, clinical trials and scientific approaches are required. So application of nanotechnology in herbal medicine extracts was developed to curb the side effects (Barkat et al. 2020).

10.6 ROLE OF PHARMACEUTICALS IN TREATING ILLNESS

Cultivation of medicinal plants can be done based on the presence of the phytomedicine in them. The crops that are used for the extraction and preparation of drugs, vaccines and pharmaceuticals are pharmaceutical crops. Pharmaceutical substances are used for treatment, diagnosis, restoring or correcting and modifying organic substances. There are many plant species which consist of active agents and have been used in pharmaceuticals are far back as the 18th century (Bernath Jeno 1998). Medicinal plants have active ingredients or secondary metabolites that possess biologic activities. Herbal medicines are prepared from the active ingredients of herbal plants' oil, gums and other secretions (Doughari 2012; Sasidharan et al. 2011; Rungsung et al. 2015).

Phytopharmaceuticals help in treating skin cancer and play a vital role in skin cancer management, targeting multiple molecular pathways without affecting other body parts (Iqubal et al. 2021). Death rates are increasing due to diabetes, and it is very difficult to treat, but the combination therapies of synthetic and herbal drugs will treat it. Combinations of many phytochemicals with synthetic drugs can control the effects of synthetic drug treatment (Banerjee et al. 2020). Phytopharmaceuticals also play a vital role in cancer chemoprevention and are mainly focused on the role of natural supplements in cancer prevention and therapy. The focus is also on cancer health risk factors and protective abilities of various natural products (Bagchi 2004). Many phytochemicals show anticancer activity in various stages of tumor progression (Solowey et al. 2014).

Vinblastine, vincristine from *Catharanthus roseus*, paclitaxel and docetaxel from plants of taxus or yews and podophyllotoxin from podophyllaceae are used to treat cancer (Hyder et al. 2015).

Docetaxel treats neck cancer, gastric cancer, breast cancer and prostate cancer. Paclitaxel treats breast cancer, small lung cell cancer and pancreatic cancer.

Camptothecin from *Camptothec aacuminate* (Cragg et al. 2009) is the chemotherapeutic agent used in the form of hydrochloride and treats ovarian and lung cancer. Treatment of upper and lower tract treatment infections by phytopharmaceuticals became successful in Germany. It helps to reduce the antibiotic prescriptions for respiratory tract infections (Baars et al. 2019). *Ageratina pichinchensis* extract is used for the treatment of mouth ulcers and skin diseases. The chemical compound 7-O-(β–D-glucopyranosyl) galactin was isolated and used for healing wounds, and this compound induced cell proliferation of normal human skin cells (Romero-Cerecero et al. 2015). The substances from Silexan (WS 1265, lavender oil) treats anxiety disorders, WS 5570 (hypericul extract) treats major depression and EGb 761 (ginkgo biloba leaf extract treats dementia syndrome and a combination of EGb 761 with our medications treats symptomatic dementia (Tomino et al. 2021). Ginkgo biloba leaf extract consists of standardized extracts of flavonglycosides and terpene-lactones, and more than 5 ppm of ginkgolic acid helps with brain function, tinnitus and dizziness (Kaschel 2009; Amieva 2013).

Solidago canadensis consists of active ingredients like flavonoids, saponins, hydroxycinnamates and minerals and shows anti-inflammatory, spasmolytic and diuretic properties. The polyphenols present in it show multivarious effect that made interested to study the accurate composition extracts obtained by different technologies (Apati 2003). Green tea (*Camellia sinensis*) extract can be used to treat skin itching, mild diarrhea, cracks and insect bites. *Adenanthera pavonina, Peltophorum pterocarpum* and *Pipernigrum* have antioxidant, antibacterial, anti-inflammatory and neuropharm-cological activities with hexane and ethanol extracts and β-sitosterol glucoside, bergenin and piperine. Arjunin, lactone arjunetin, phyrocatechol and tannin from the bark of *Terminlia arjuna* treat ulcers, hypertension and diabetes (Amalraj et al. 2017). Flavonoids, tannins, lignin, terpenoids and phenolic compounds extracted from the bark of *Saracaindica* and *Chebulin* and gallic acid and tannic acid from the fruits of *Terminalia chebulin* act as chemoprotective agents. The plant parts of *Sidacordifolia* L. reduce fever and nervous fits. The active agents hexane and ethanol extracts and β-sitosterol glucoside, bergenin and piperine of *Adenanthera pavonina, Peltophorumptero carpum* and *Pipernigrum* show antiinflammatory, neuroprotective, antioxidant and hepatoprotective activities (Tasleem 2016). Motions (loose stool or diarrhea) and nervous depression can be controlled by extracts of *Datura stramonium* and *Datura metel*. Revitonil R tablets from root extracts of *Echinacea puepurea* and *Glycyrrhiza* can curb infections (Wagner and Jurcic 2002).

Extracts of *Claviceps purpurea* ergocristine cure migraine, ergocriptine inhibits the action of adrenaline and ergocorminre regulates involuntary reactions to stress on multiorgan systems in the body. The roots of *Withonia sominifera* treat rheumatic pain, joint pains and nervous system disorders; the leaves prevent ulcers, skin lesions and boils; and they show antitumor, anti-arthritic, anti-inflammatory, antioxidant and antistress properties. The *Curcumalonga* L. rhizome consists of curcumin, demethoxycurcumin, and bisdemethoxycurcumin (called curcuminoids), which are used as an antiseptic and to control spleen disorders, cough and rheumatism. It also acts as and antifungal, anti-inflammatory and antibacterial agent in the case of cholera and syphilis (Krup 2013). *Digitalis lantana*, with digoxin, digitoxin and lanthanide C, controls heart diseases. Quinine from *Cinchona* is the first pharmaceutical molecule in industries to cure malaria. *Atropa belladonna*, with the active agents hyoscyamine, atropine and scopolamine, controls motion sickness and acts as a depressant of nerve endings. Cough and pain can be reduced from the papaverine, morphine, narcodine and codeine active agents of *Papaver somniferum*. Berberine from *Berberis vulgaris* controls bacillary dysentery. Phytopharmaceuticals derived from the medicinal plants also treat symptoms of COVID-19. Secondary metabolites and other chemical compounds from plants treat the symptoms by inhibiting viral cell entry or treat the underlying infections and became innovative treatments (Gyawali et al. 2020; Gurung et al. 2020). The roots of *Boerhaviad iffusa* treats gynecological problems. It also has hepatoprotective properties and controls gastrointestinal problems. Many herbal medicines and their role in treatments are provided in Table 10.1.

TABLE 10.1
Herbal Plants, Their Active Agents and Pharmaceutical Actions

S. No	Name of the plant	Scientific Name	Active Agents	Pharmaceutical Action
1	St. John's wort extract	*Hypericum perforatum* Flowers and aerial parts	Hypericiherba Naphthodianthrone hypericin Phloroglucinol derivative hyperforin	Reduce depression Antiviral effect Antibacterial effect
2	Canadian golden rod	*Solidago canadensis*	Flavonoids, saponins, hydroxycinnamates	Anti-inflammatory, reduction of excessive smooth muscle contractility Help the body to get rid of excess fluids
3	Cone flower	*Echinacea purpurea*	Extract alone or combination with other plant extracts	Treats upper respiratory tract infections (Hudson 2012)
4	Chaste tree	*Vitex agnus-castus*	Fruit extract	Reduce premenstrual syndrome (Loch et al. 2000)
5	Green tea	*Camellia sinensis*	Green tea extract	Oral use: control diarrhea, supplement of weight reduction Treats insect bites and cracks Antioxidant, antitumor, anti-inflammatory
			EFLA[R]942 (green tea extract)	Antiinflammatory activity and elastase inhibition. It inhibits neutrophil elastase (H LE) that leads to inflammatory diseases rheumatoid arthritis (Momohara et al. 1997). Prevents diarrea, suppresses necrosis, weight loss, increase energy vitality
6	Pacific yew	*Taxus brevifolia*	Paclitaxel from bark	Fight against certain types of cancer
7	Spinach	*Spinacea oleracea* L.	Rutin, quercetin, kaempferol	Anticarcinogenic, anti-bacterial, anti-asthmatic, anti-aging, antihypersensitive properties Antioxidant activity by spinach's water-soluble polyphenolic flavonoids (Ashok Kumar et al. 2013)
8	Arjuna	*Terminlia arjuna*	Arjunin, lactonearjunetin, phyrocatechol and tannin from the bark	Treats ulcers, hypertension, inflammation, diabetes, leukorrhea (Amalraj and Gopi Sreeraj 2017)
9	Asoka tree	*Saraca indica*	Bark consists of flavonoids, tannins, lignin, terpenoids, phenolic compounds Leaf juice mixed with cumin seeds	Antiulcer, anticancer and chemoprotective agents Treats stomach pain
			Dried flowers	Diabetes, hemorrhagic dysentery
10	Chebulic myrobalan	*Terminalia chebula*	Fruits consists of chebulin, gallic acid, tannic acid and resins	Anti-inflammatory, treats stomach pain, anti helimenthics. It helps in treating helminthiasis, gastropathy, anorexia (Promila and Madan 2018)

(Continued)

TABLE 10.1
(Continued)

S. No	Name of the plant	Scientific Name	Active Agents	Pharmaceutical Action
11	Flannel weed	*Sida cordifolia* L.	Chemical constituents: alkaloids, phytosterol, mucin, resin	
			Main portion of alkaloid-ephedrine and rutin	Aphrodisiac properties
			Plant parts	Control fits, fever, heart irregularity, nervous disorders, gonorrhea
			Root juice	Wound healing, facial paralysis, sciatica
			Leaves	Ophthalmic disorders, rheumatism, shivering, fits, improve sexual health
12	Red beed tree Copper pod, yellow flamboyant Black pepper	*Adenanthera pavonina* *Peltophorum pterocarpum* *Piper nigrum*	Using hexane and ethanol extracts and β-sitosterol glucoside, bergenin and piperine	Antioxidant, antibacterial, anti-inflammatory, neuropharmacological and hepatoprotective activities (Tasleem 2016)
13	Aswagandha (Indian ginseng)	*Withania somnifera*	Alkanoids, withanolides (steroidal lactones) Flavonol glycosides, sterols, phenolics, trophine, anahygrine, withasomnine, pseudotrophine	Antioxidant, antibacterial, anti-inflammatory, neuropharmacological and hepatoprotective, anticancer, anti-ulcer, antidiabetic and many other traditional benefits (Azgomi et al. 2018)
14	Gymnema/Cow plant	*Gymnema sylvestre*	Gymnemic acid from leaves	Treats glycosuria and urinary diseases, neutralizes excess sugar in diabetic patients (Pothuraju Ramesh et al. 2014), leaves control cholesterol
			Gurmarin and gymnemic acid	Control/block sweet taste (Bamoal, Verma and Nagi 2018)
15	Turmeric	*Curcuma longa* L. rhizome	Curcumin, demethoxycurcumin, bis-demethoxy curcumin (called curcuminoids)	Antiseptic, control spleen disorders, cough, rheumatism Antifungal, anti-inflammatory and antibacterial agent
16	Foxglove	*Digitalis lanata*	Lanatoside C, digoxin	Heart muscle activity, controls cardiac arrhythmias
17	Ginkgo	*Ginkgo biloba*	Ginkgolides	Memory loss and cerebral circulation
18	Nightshade	*Atropa belladona*	Atropine, scopolamine, hycoscyamine	Control motion sickness and depressant of nerve endings
19	Opium poppy	*Papaver somniferum*	Papaverine, morphine, narcodine, codeine	Reduce pain, cough Control pain in coronary and cerebral thromboses
20	Quinine	*Cinchona officianalis*	Quinine, quinidine	Cure malaria
21	Thorn apple	*Datura stramonium*, *Datura metel*	Atropine, hyaciamine, scopalamine	Control motion sickness, depressant of nerve endings
22	Periwinkle	*Catharanthus roseus*	Vinblastine, vincristine	Cure leukemia in children, Hodgkin's disease
23	Barberry	*Berberis vulgaris*	Berberine	Control bacillary dysentery

S. No	Name of the plant	Scientific Name	Active Agents	Pharmaceutical Action
24	Ergot	*Claviceps prupurea*	Ergocristine Ergocriptine Ergocorminre	Treat migraines Inhibit the action of adrenaline Regulate involuntary reactions to stress on multiorgan systems in the body
25	Milk thistle	*Silybum marianum*	Silybin, silymarin	Used as antioxidant, protection of liver
26	Cone flower and Licorice	Root extract of *Echinacea puepurea* and *Glycyrrhiza*	Revitonil R tablets	Show anti-infection activity (Wagner and Jurcic 2002)
27	Empress tree	*Paulownia tomentosa*	Tomentin A	Inhibits PL pro (papain-like protease) prevents SARS-CoV
	Indian surge tree	*Euphorbia neriifolia*	Friedelin and epiteraxerol	Antiviral activity (Das et al. 2021)
28	Punarnava	*Boerhavia diffusa*	Roots	Used as for gynecological, hepatoprotective and gastrointestinal indications

10.7 SIGNIFICANCE

Many studies have revealed that phytopharmaceuticals play a vital role in treating illness and have proved to be an effective natural remedy for skin infections, different types of cancer, fever, dysentery, malaria, cough, depressions, diabetes, allergies, skin infections and others. They show antibacterial, antiviral and antifungal activities without any side effects. Many of the research findings demonstrate the importance of phytopharmaceuticals and their role in the prevention of almost all diseases. There are many medicinal plants still unknown with active ingredients, and focus on these medicinal plants by extending the research can make innovative use of herbal medicines in the future.

10.8 CONCLUSION AND FUTURE PROSPECTS

Plant-based phytopharmaceuticals play a significant role in the current system of medicine and health care. The quality of plant materials and the manufacturing processes for phytopharmaceuticals are regulated by food laws that lack the specificity required for botanical drugs. Pharmaceutical professionals should take up the responsibility for safe maintenance and improvement of herbal medicine. The increasing importance of plant-based products like essential oils, extracts, phytomolecules and fragrance have attracted the pharmaceutical industry.

It is necessary to understand the chemical composition and quantity of pharmacologically active ingredients in the formulations. The preparation of pharmaceutical drugs and their history are not clearly defined, and hence efforts are made to standardize the drug components. Even though the modern age is dominated by synthetic drugs, phytomedicine plays a strong role in the prevention of illness. The extracts of *Commipora guggul, Ginkgo biloba, Silibum marianum* and *Gingiber officinale* are herbal medicines in India. Approximately 1300 medicinal plants have medical purposes, and almost all the plants are part of Ayurvedic medicine and help in treating liver diseases, so it is expected to herbal extracts will be produced under the umbrella of phytopharmaceuticals. In the future, super-critical extracts are used to improve bioavailability, decrease the toxic nature of drugs and improve dosage forms. The herbal industry will be dominated by nano-phytopharmaceuticals and phytopharmaceuticals in the future.

BIBLIOGRAPHY

Amalraj, A. and Gopi, S. 2017. Medicinal properties of *Terminalia arjuna* (Roxb.) Wight & Arn.: a review. *Journal of Traditional and Complementary Medicine* 7(1):65–78.

Amieva, H., Meillon, C., Helmer, C., Barberger-Gateau, P. and Dartigues, J.F., 2013. Ginkgo biloba extract and long-term cognitive decline: A 20-year follow-up population-based study. *PLoS One 8*(1):e52755.

Apati, P.G. 2003. *Antioxidant Constituents in Solidago canadensis L. and its traditional Phytopharmaceuticals*. Diss. Semmelweis University.

Ashok kumar, C.K. and Divyasree, M.S. Joshna. 2013. Review on south Indian edible leafy vegetables. *J Global trends pharm sci 4*(4):1248–1256.

Azgomi, R.N.D., Zomorrodi Nazemyieh, H., FazliJou, S.M.B., Bazargani, H.S., Nejatbaksh, F., Jazani, A.M. and AsrBadr, Y.A. 2018. Effects of *Withoniasomnifera* on reproductive system: A systematic review of the available evidence. *Biomed Res*:1–17. https://doi.org/10.1155/2018/4076430

Baars, Erik W. et al. 2019. The contribution of complementary and alternative medicine to reduce antibiotic use: A narrative review of health concepts, prevention, and treatment strategies. *Evidence-Based Complementa)y and Alternative Medicine*:5365608. http://doi.org/10.1155/2019/5365608.

Bagchi D. and Preusany. 2004. Phytochemicals, H. G. *Phytopharmaceuticals in cancer chemoprevention*. Boca Raton: CRC Press.

Bamoal, N., Verma, P. and Negi, C. 2018. Review of some traditional medicinal plants. *Int J Life Sci Scienti Res 4*(1):1550–1556.

Banerjee, M., et al. 2020. A systematic review on synthetic drugs and phytopharmaceuticals used to manage diabetes. *Current Diabetes Reviews 16*(4):340–356.

Barkat, M.A., Das, S.S., Beg, S. and Ahmad, F.J. 2020. Nanotechnology-based phytotherapeutics: Current status and challenges. *Nanophytomedicine*:1–17.

Becker, K.P., Ditter, B., Nimsky, C., Urbaschek, R. and Urbasehek, B. 1988. Endotoxin contents of phytopharmaceuticals: Correlation with clinically observed side effects. *Deutsche Medizinische Wochenschrift (1946) 113*(3):83–87.

Bernath, J. 1998. Pharmaceutical plants (plants used in pharmaceutical preparations). *Cultivated Plants, Primarily as Food Sources* 2.

Chikezie, P.C. and Ojiako, O.A. 2015. Herbal medicine yesterday today and tomorrow. *Altern Integr. Med* 4(3):195.

Cragg, Gordon M., and Newman, David J. 2009. Nature: A vital source of leads for anticancer drug development. *Phytochemistry Reviews 8*(2):313–331.

Das, A., Pandita, D., Jain, G.K., Agarwal, P., Grewal, A.S., Khar, R.K. and Lather, V. 2021. Role of phytoconstituents in the management of COVID-19. *Chemico-Biological Interactions 341*:109449.

Doughari, James Hamuel. 2012. *Phytochemicals: Extraction methods, basic structures and mode of action as potential chemotherapeutic agents*. Rijeka, Croatia: INTECH Open Access Publisher.

Ekor, M. 2013. The growing use of herbal medicines: Issues relating to adverse reactions and challenges in monitoring safety. *Front Pharmacol* 4:177.

Evans, W.C. (ed.). 2009. *Trease and evans pharmacognosy*, Sixteenth Edition. Edinburgh: Saunders-Elsevier. ISBN: 9780702029332.

Gurung, A.B., Ali, M.A., Lee, J., Farah, M.A. ad Al-Anazi, K.M. 2020. Unravelling lead antiviral phytochemicals for the inhibition of SARS-CoV-2 Mpro enzyme through in silico approach. *Life Sciences* 255:117831.

Gyawali, R., Paudel, P.N., Basyal, D., Setzer, W.N., Lamichhane, S., Paudel, M.K., Gyawali, S. and Khanal, P. 2020. A review on ayurvedic medicinal herbs as remedial perspective for COVID-19. *Journal of Karnali Academy of Health Sciences 3*:1–21. doi:10.3126/jkahs.v3i0.29116

Hudson, J.B. 2012. Applications of Phytomedicine *Echinaceapurpurea* purple coneflower in infectious diseases. *Journal of Biomedicine and Biotechnology 2012*:1–16.

Hyder, I., Yedlapudi, D., Kalivendi, S.V., Khazir, J., Ismail, T., Nalla, N., Miryala, S. and Kumar, H.M.S., 2015. Synthesis and Biological evaluation of novel 4β-[(5-substituted)-1, 2, 3, 4-tetrazolyl] podophyllotoxins as anticancer compounds. *Bioorganic & Medicinal Chemistry Letters 25*(14):2860–2863.

Iqubal, M.K., Khan, M.A., Agarwal, N.B., Ali, J. and Baboota, S. 2021. Nanoformulations-based advancement in the delivery of phytopharmaceuticals for skin cancer management. *Journal of Drug Delivery Science and Technology 66*:102912.

Kaschel, R., 2009. Ginkgo-Extrakt bei Personen mit nachlassender mentaler Leistungsfähigkeit. Studienergebnisse differenziert betrachten. *Pharmazie in unserer Zeit 38*(5):432–439.

Krup, V., Prakash, L.H. and Harini, A. 2013. Pharmacological activities of turmeric (Curcuma longa Linn): A review. *J Homeop Ayurv Med 2*(4):1–4.

Loch, Ernst-Gerhard, Selle, Hartmut and Boblitz, Normann. 2000. Treatment of premenstrual syndrome with a phytopharmaceutical formulation containing Vitex agnus castus. *Journal of Women's Health & Gender-Based Medicine 9*(3):315–320.

Mahomoodally, M.F., Sadeer, N., Edoo, M. and Venugopala, K.N. 2021. The potential application of novel drug delivery systems for phytopharmaceuticals and natural extracts—current status and future perspectives. *Mini Reviews in Medicinal Chemistry 21*(18):2731–2746.

Mitchell, M. and Brandenburg, J.E. 2015 Process for fractionation and extraction of herbal plant material to isolate extractives for pharmaceuticals and nutraceuticals. *U.S. Patent Application 14*(454):998.

Momohara, S., Kashiwazaki, S., Inoue, K., Saito, S. and Nakagawa, T. 1997. Elastase from polymorphonuclear leukocyte in articular cartilage and synovial fluids of patients with rheumatoid arthritis. *Clinical Rheumatology 16*(2):133–139.

Pandey, M. M., Rastogi, S. and Rawat, A.K.S. 2013. Indian traditional ayurvedic system of medicine and nutritional supplementation. *Evidence-Based Complementary and Alternative Medicine 2013*:article id 376327. https://doi.org/10.1155/2013/376327

Pothuraju, R., et al. 2014. A systematic review of *Gymnema sylvestre* in obesity and diabetes management. *Journal of the Science of Food and Agriculture 94*(5):834–840.

Promila, P. and Madan V.K. 2018. Therapeutic and phytochemical profiling of *Terminalia chebula* Retz A review. *J Med Plant studies 6*(2):21–31.

Romero-Cerecero, O., Zamilpa, A. and Tortoriello, J. 2015. Pilot study that evaluated the clinical effectiveness and safety of a phytopharmaceutical elaborated with an extract of Ageratinapichinchensis in patients with minor recurrent aphthous stomatitis. *Journal of Ethnopharmacology* 173:225–230.

Rungsung, W., Ratha, K.K., Dutta, S., Dixit, A.K., Hazra, J. 2015. Secondary metabolites of plants in drugs discovery. *World J Pharm Res 4*:604–613.

Sasidharan, S., Chen, Y., Saravanan, D., Sundram, K.M. and Yoga Latha, L. 2011. Extraction, isolation and characterization of bioactive compounds from plants extracts. *Afr J Tradit Complement Altern Med 8*:1–10.

Solowey, E., Lichtenstein, M., Sallon, S., Paavilainen, H., Solowey, E. and Lorberboum-Galski, H. 2014. Evaluating medicinal plants for anticancer activity. *The Scientific World Journal*:721402.

Stiefelhagen, P. 2002. Drug interactions caused by phytopharmaceuticals. Don't underestimate herbs! *MMW Fortschritte der Medizin 144*(6):6–9.

Tasleem, F. 2016. *Biomedical Analysis on Phytopharmaceuticals*. Diss. University of Karachi Karachi.

Tomino, C., et al. 2021. Mild cognitive impairment and mild dementia: The role of Ginkgo biloba (EGb 761®). *Pharmaceuticals 14*(4):305.

Wagner, H. and Jurcic, K. 2002. Immunological studies of Revitonil®, a phytopharmaceutical containing *Echinacea purpurea* and *Glycyrrhizaglabra* root tract. *Phytomedicine* 9(5):390–397.

11 A Disease-Driven Approach to Medicinal Natural Products

Jasmine R, Deena Priscilla H, and Sowmiya S

CONTENTS

11.1 Introduction 213
11.2 Conventional Drugs: Pros and Cons 214
11.3 Limitations of Conventional Drugs 214
11.4 Alternative Drugs From Natural Sources 215
11.4.1 Drugs from Plants-Phytocompounds 216
11.4.2 Commercial Drugs From Medicinal Plants 216
11.4.3 Drugs From Marine Sources 216
11.4.4 Drugs From Microbial Metabolites: Probiotics 217
11.5 Targets: Antimicrobial, Antidiabetic, Antioxidant, and Anticancer 218
11.6 Disease-Specific Regulation of Molecules and Pathways 218
11.7 Significance 220
11.8 Conclusion and Future Prospects 220
Bibliography 220

11.1 INTRODUCTION

Natural products, especially compounds obtained from varied sources such as plants, animals, marine organisms, and microbes, have been known to cure human ailments from time immemorial. Our traditional claims have recorded the use of herbs to relieve pain and wrapped plant paste at the site of wounds to bring about healing. This supports the idea that natural products have been relied upon solely to treat diseases. The advent of technology, especially in the fields of molecular biology, combinatorial chemistry, and genetic engineering, have escalated the use of natural products in drug discovery, design, and development (Hong-Fang Ji et al. 2009). Natural products tend to play a prominent role in drug discovery in the fight against cancer, multidrug-resistant infections (Atanasov et al. 2021; Harvey et al. 2015) and to combat cardiovascular diseases (Newman and Cragg 2016; Waltenberger et al. 2016). The structural diversity of chemical entities derived from natural products that could aid in the modification of altered biological function has been the bringer of hope for several diseases (Sircar 1982; Akshada Amit Koparde et al. 2019). Owing to all the support from technological advancements, ancient findings, and proof, biological diversity (biodiversity) with precise expertise based on research experience has provided a unique opportunity to conquer and reestablish medicinal natural products as a major source of drugs to prevent and cure many diseases. Nature has always astounded us with its unique features and outstanding phenomena that have helped us attain greater heights, carving a niche in the field of herbal medicine (Panda 2004). The disease-inhibiting possibility providing basic bioactive compounds that are non-toxic and highly effective strategies have made natural drug substances to be of vital and appreciable medicinal value and therapeutic importance. In the present scenario, many lifesaving drugs have been offered by natural drug substances to be

DOI: 10.1201/b22917-11

of vital and appreciable medicinal value and therapeutic importance, especially plants. These natural products could be ideally labeled as cornerstones of drug discovery and development (Akshada Amit Koparde et al. 2019).

It is time to reveal how natural products have helped to change and promote the process of modern drug discovery. At this juncture, it has now become crucial to provide disease-specific drugs based on their molecular mechanism (Sun et al. 2022). The advantages of the use of natural products when compared to traditional medicine includes a good degree of tolerance, less toxic, polypharmacological behavior, economic, reliable, and safe (Harvey et al. 2015; Newman and Cragg 2020; Zhang et al. 2021; Mok et al. 2020). The creation of local compounds from natural products have been challenging, but there has been tremendous scope pertaining to advances in screening, identification of mode of action of selected active moiety, and selection of a suitable entity specified towards a biological target of a particular disease (Shi et al. 2021; Fellmann et al. 2017; Schirle and Jenkins 2016). First, the spread of infectious diseases arising from resistance to drugs is a major cause of concern as this impacts global public health at large not only as an epidemic but also giving rise to pandemics (Burkle 2020; Noah and Fidas, 2000; Chopra 2000). The amplitude of secondary metabolites present in plants, animals, microorganisms, and marine organisms seems to be an inevitable part of modern-day drug discovery and development. Lead compounds obtained from plant-based systems have contributed enormously to the field of health care to treat diabetes, cancer, etc. Microbial metabolites as chemotherapeutics have been used as an antidote to cure infectious diseases and in some cancers too. Marine microbes and phytoplankton also have increased the variety of compounds available for the treatment of diseases. Finally, active peptide and certain endogenous neurotransmitters obtained from human beings and animals have found a place in the area of drug research (Zongre Guo 2017). In this chapter, we have focused on the entire role of natural products rich in medicinal value as potential drugs for the treatment and prevention of various diseases.

11.2 CONVENTIONAL DRUGS: PROS AND CONS

Conventional drugs hold a significant position in global health care. Conventional drugs focus on making an accurate diagnosis and treating symptoms. A wide range of tests has been developed for diagnosing and studying the progression of diseases. This also focuses on the development of antibiotics and vaccines to fight against infectious diseases. Conventional medicine also helps in treating accident victims. It focuses more on curing diseases than maintaining general health (Jamison et al. 2006). Natural products and traditional remedies have already made significant contributions to modern medicine, including the isolation of the first pharmacologically active molecule, morphine, from opium. Natural products and traditional medicines provide unequaled benefits when it comes to developing novel pharmaceuticals, such as extensive clinical experience and a unique diversity of chemical structures and biological activity (Yuan et al. 2016). The therapeutic approach to alternative traditional medicine as a possibility for a coordinated search for novel chemical entities has been prompted by the high cost of pharmaceuticals and the rise in drug resistance to prevalent ailments such as malaria, bacterium infections, and sexually transmitted diseases (Fokunang et al. 2011).

11.3 LIMITATIONS OF CONVENTIONAL DRUGS

Due to the increased multidrug resistance in bacteria, the use of antibiotics and other allopathic medicines has been reduced (Ventola 2015). Extensive and/or inappropriate use of antibiotics, inadequate hygiene, an increasing aging and immune-compromised population, and lack of rapid diagnostics have fueled this problem. It should be understood that it is inevitable to prevent resistance and that it can only be slowed down (Kumar 2018). This can be achieved only through persistent research and development of new drugs. The discovery of natural product–based antimicrobial agents has been the most important breakthrough in the history of medicine. Hence, researchers have begun attempts to modify the natural products, using various biotechnological methods, in order to improve the drug likeness and clinical potential (Wang et al. 2017).

11.4 ALTERNATIVE DRUGS FROM NATURAL SOURCES

Since ancient times, natural products have played an important role in the prevention, management, and treatment of various human ailments. These products are obtained from a wide variety of sources such as plants, marine sources, and microbes (Khalifa et al. 2019). Owing to their diversified chemical properties, natural products tend to be one of the prime resources for new drug leads covering several therapeutic areas for drug discovery and development in the future (Cragg and Newman 2013). One of the essential contributions of natural products is the development of anti-infective drugs such as antibiotics, antiviral agents, analgesic agents, antimicrobial agents, anti-inflammatory agents, antioxidant agents, etc. (Parham et al. 2020). A large number of natural products possess biological and pharmacological properties, and some have clinically or pre-clinically proven medicinal properties (Sultana et al. 2022). For the past two decades, about one-third of all the Food and Drug Administration (FDA)–approved drugs are natural products (Patridge et al. 2016). Among the modern drugs obtained from natural products, drug products based on phytobiotics obtained from higher plants have remained to be one of the major sources (Veeresham 2012). Earlier, plant extracts and oils were used in their crude forms, but due to the recent progress in scientific research, isolation, structural studies, and formulation of specialized products using the active ingredients from those plants have become possible (Pan et al. 2014). Natural products are both advantageous and disadvantageous over conventional synthetic drugs. This is due to their higher molecular weight, more sp3 carbons and oxygen atoms, fewer nitrogen and halogen atoms, higher numbers of H-bond acceptors and donors, and greater molecular rigidity when compared with synthetic compound libraries (Destro 2018). Evolutionary optimizations of the structures of natural products have made them serve specialized biological functions, which explains their relevance in medicine for the treatment of various ailments (Kulka 2013) (Figure 11.1).

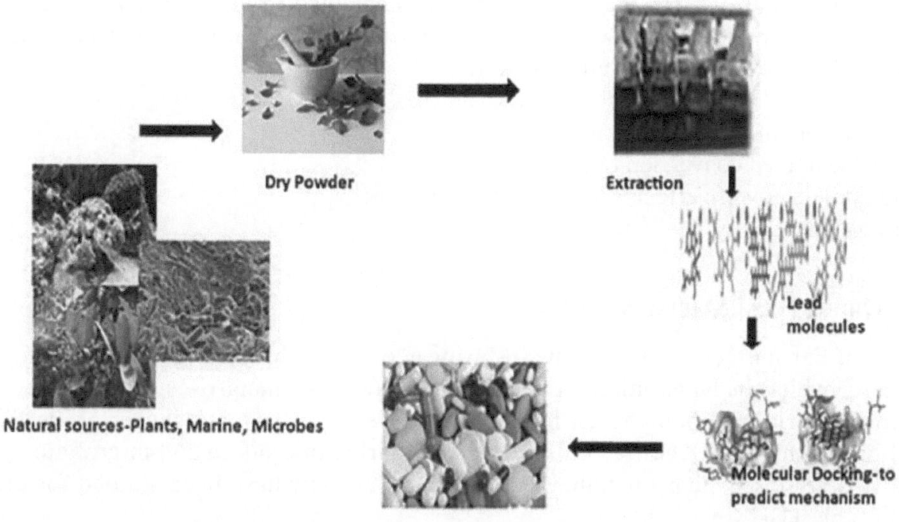

FIGURE 11.1 Flow chart depicting the process of drug development from natural sources.

11.4.1 Drugs from Plants-Phytocompounds

Plant species have been used for medicinal purposes since ancient times due to better patient tolerance and acceptance. The plant kingdom consists of 300,000–400,000 higher species. But only 5–15% of the terrestrial plant species have been pharmacologically studied. About 10,000–15,000 plant species globally have been reported as medicinal plants. About 25% of all medicines have plant origins (Gurnani et al. 2014). Secondary metabolites present in plants include tannins, terpenoids, alkaloids, and flavonoids, which have been shown to have antibacterial activities in vitro (Cowan 1999).

11.4.2 Commercial Drugs From Medicinal Plants

The term phytobiotics is applied to compounds derived from leaves, roots, tubers, or fruits of herb species, spices, and other plants. They may be used in their solid form, dried form, and ground form or as extracts. These compounds tend to have an effect on the stimulation of appetite, stimulation of endogenous secretions, stimulation of immunity and antimicrobial, coccidiostatic, antiviral, anti-inflammatory, antioxidant, and anthelmintic activities. They are also known as phytogenic feed additives (Krauze 2021). About 1000 different plant species have anticancer potential. Some important plant-based anticancer drugs include vincristine and vinblastine (isolated from *Catharanthus roseus*), etoposide (isolated from *Podophyllum peltatum*), and paclitaxel/Taxol (isolated from the bark of *Taxus brevifolia*) (Bhanot et al. 2011; Noble 1990).

Owing to its high nutritive values, *Moringa oleifera* is cultivated for a wide range of uses such as human consumption, feed for livestock, and use in cosmetics. It has a low calorific value. Hence it is recommended as a diet for the obese. Its pods are fibrous. Thus, it is used to treat digestive problems. The leaves are rich in minerals (such as calcium, magnesium, iron, potassium, zinc, and copper), vitamins (like beta-carotene from vitamin A; vitamin B such as pyridoxine, folic acid, and nicotinic acid; and vitamins C, D, and E), and other essential phytochemicals (like tannins, flavonoids, saponins, anthraquinones, alkaloids, sterols, terpenoids, and reducing sugar). Extracts from the leaves are used to treat malnutrition and augment breast milk in lactating mothers. It is used as a potential antioxidant, anti-inflammatory, antidiabetic, anticancer, and antimicrobial agent. Its seed is extensively used in water treatment, as it is regarded as a natural coagulant (Sagarika Sarkar and Sauris Panda 2021).

According to (Krauze 2021), the active compounds obtained from *M. oleifera* include chlorogenic acid, caffeic acid, ascorbic acid, flavonoids, phenolics, and carotenoids. Upon addition to poultry feed, the following benefits are observed:

- Improvement in egg production
- Decrease in feed conversion ratio (FCR)
- Reduction in adverse effects of environmental stress
- Antioxidant activity

11.4.3 Drugs From Marine Sources

About 70% of Earth is composed of water. Marine ecosystems have a high level of biodiversity. Marine flora, which includes microalgae, fungus, seaweeds, mangroves, bacteria, cyanobacteria, actinobacteria, and halophytes, has been discovered to have anticancer properties in in vitro and in vivo models (Khalifa et al. 2019). The marine microflora and microfauna produce structurally diverse secondary metabolites, of which very few have been studied for their bioactivities (Table 11.1).

TABLE 11.1
Variety of Marine Sources for Compounds Used in Clinical Trials and Drugs

S. No	Marine Sources	Compounds	Drug	Action
1.	Algae	Peptide	Cryptohycins Kahalaide	Prostate and breast cancer
2.	Tunicate, Ascidiacea	Alkaloid, peptide	ET-743 Aplidine	Antineoplastic against bone marrow cancer
3.	Sponges	Polyketide, alkaloid, peptide	E-7389, Eribuline Citarabine Cryptophycins	Breast cancer
4.	Actinomyces	Polyketide	Salinosporamide A	Multiple myeloma
5.	Mollusk	Peptide	Dolastatin-10 Khalalide F	Pancreatic, Breast, and Lung Cancer
6.	Bryoza	Polyphenol	Bryostatin 1	Anticancer and Alzheimer's disease
7.	Shark cartilage	Peptide	AE–941	Kidney cancer

The exploration of marine bioactive compounds began in the 1950s. It started with the isolation of two typical nucleotides, namely "spongothymidine" and "spongouridine," from *Crytotheca crypta*. These were observed to have antiviral and anticancer properties (Haefner 2003). The launch of self-contained underwater breathing apparatus (SCUBA) in the 1970s and remotely operated vehicles (ROVs) in the 1990s paved the way for further exploration of marine biodiversity. Polyphenols, polysaccharides, and alkaloids are among the anticancer components extracted from marine species that are highly active, physiologically powerful, and prevalent.

There are about 30,000 marine algal species distributed worldwide. Marine algae have been used by mankind for various purposes such as food, medicine, and fertilizers. Modern algal research is based on the existing knowledge and adequate availability of algal species (Dias et al. 2012). Bioactive compounds isolated from marine algae possess antioxidative, anti-inflammatory, and anticancer characteristics. For example, *Callophyllis japonica* and *Gracilaria tenuistipitata* show antioxidative properties, and the flavonoids produced by *Ulva lactuca* and *Sargassum wightii* show antioxidant properties (Lee et al. 2013).

About 45% of the total bioactive microbial secondary metabolites are obtained from Actinomycetes. They produce a wide range of antibiotics such as anthracyclines, aminoglycosides, glycopeptides, beta-lactams, macrolides, polyketides, actinomycins, and tetracyclines (Subramani and Aalbersberg 2012).

11.4.4 Drugs From Microbial Metabolites: Probiotics

Of all the natural compounds isolated to date, about 5% have microbial origins. Of these, 20–25% have been classified as biologically active compounds. Of these, about 10% have been isolated from microbes, 45% have been isolated from Actinomycetes, 38% have been isolated from fungi, and 17% have been isolated from unicellular bacteria (Demain and Sanchez 2009). Some natural bioactive compounds derived from fungi include penicillin, vancomycin, mevinolin, norsolorinic acid, sequoiatone A and B, torreyanic acid, 5-hydroxyramulosin, helvolic acid, and ergosterol peroxide. Some natural bioactive compounds derived from bacteria include actinomycin, nisin, rapamycin, and amrubicin hydrochloride (Mushtaq et al. 2018) (Figure 11.2 and Table 11.2).

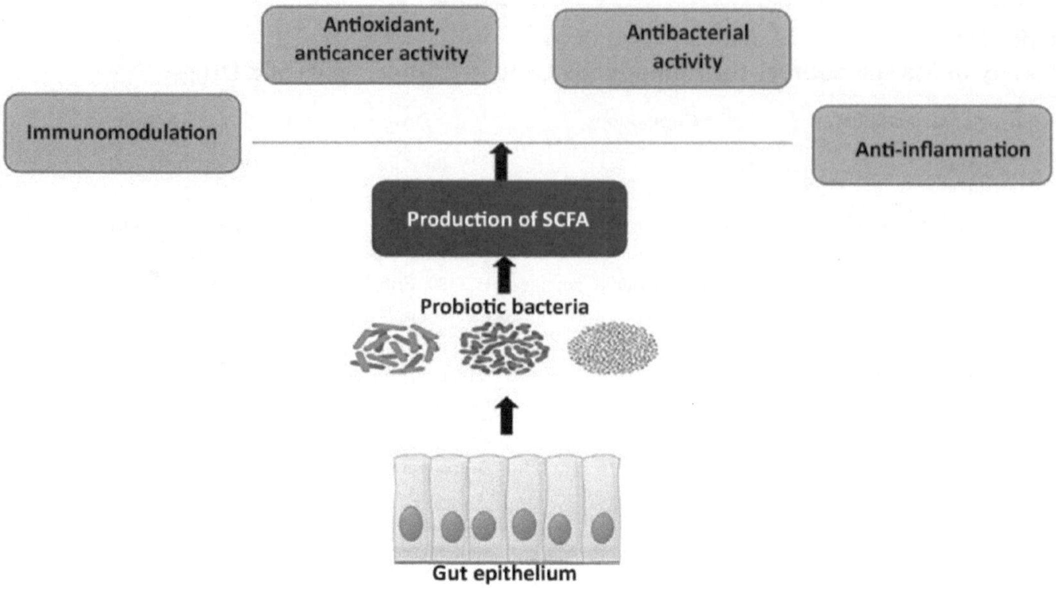

FIGURE 11.2 Applications of SCFA from probiotic bacteria.

TABLE 11.2
Drugs Derived from Microbial Metabolites

Diclofenac	Glycyrrhizin	Digoxin	Acetaminophen	Irinotecan
β - glucuronidase	18b-glycyrrhetic acid	Dihydrodigoxin Dihydrodigoxigenin	P-cresol	β-glucuronidase
Anti-inflammatory drug	Antiviral drug	Cardiotonics	Analgesics	Anticancer drug

11.5 TARGETS: ANTIMICROBIAL, ANTIDIABETIC, ANTIOXIDANT, AND ANTICANCER

The disease indication class defined by the latest International Classification of Diseases (Lancet 2018) released by World Health Organization (WHO) belong to varied super-classes that include infections (viral, bacterial, fungal etc.,), neoplasms (cancer, melanoma, leukemia) metabolic disorders (diabetes, etc.), circulatory system diseases (cardiovascular disease [CVD]), respiratory disorders (chronic obstructive pulmonary disease [COPD], pulmonary fibrosis, etc.), digestive diseases (ulcer, colitis etc.), musculoskeletal diseases, and genitourinary diseases (Figure 11.3).

11.6 DISEASE-SPECIFIC REGULATION OF MOLECULES AND PATHWAYS

It has been stated that regulated molecules belong to 71 biochemical classes namely GPCR (G-protein–coupled receptor), peptidase, transcription factor, microRNA, kinase, ABC transporter,

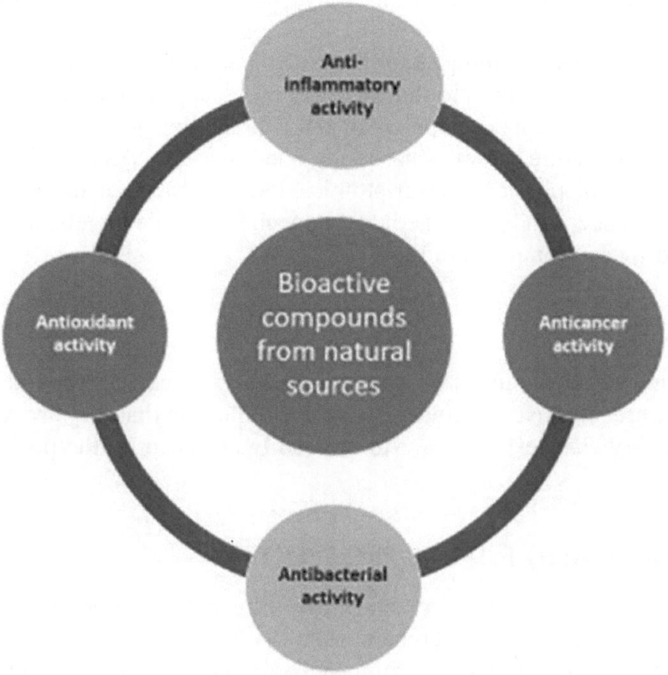

FIGURE 11.3 Significance of bioactive compounds from natural sources.

etc. The precise mechanism of regulation of the molecular pathway has been described as up-/down-regulation of gene expression–induced degradation of proteins, phosphorylation, and cleavage by some drug combination. In a similar manner, biological regulation of drug combinations has been demonstrated in ways such as induced cell cycle arrest, metabolite biosynthesis inhibition, over-accumulation of reactive oxygen species (ROS), delayed clotting time, induced DNA damage, etc. These regulation mechanisms have helped to widen our knowledge on the therapeutic benefits of natural products, thereby increasing their efficacy and employability.

The mechanism of molecular and biological regulation has been studied by means of various analytical and experimental techniques such as shRNA, siRNA, Western blot, quantitative polymerase chain reaction (qPCR), and so on. Apart from these, notable pathways that have undergone modification owing to drug action have also been enlisted that have been cross-linked to data available from well-known pathways that include KEGG, Reactome, Biocyc, SIGNOR, and Pathway Commons (Kanehisa et al. 2017; Jassal et al. 2020; Karp et al. 2019; Rodchenkov et al. 2020; Licata et al. 2020).

Analytical techniques such as nuclear magnetic resonance (NMR) spectroscopy, high resolution mass spectrometry (HRMS), and liquid chromatography (LC) have enabled the use of metabolite profiling of natural product extracts. A deeper understanding of the biosynthetic pathway for natural products has been obtained from analyzing and manipulating genomes. Genome mining is said to govern the biosynthesis of structural scaffolds that would be helpful in the identification of natural product biosynthetic gene clusters. Targeted genetic engineering of natural products' biosynthetic genes could increase the yield of natural products. Moreover, targeted gene manipulation could also help to bring about alterations or modifications in biosynthetic pathways in order to develop natural product analogues with improved pharmacokinetic properties. This chapter supports the notion that scientific and technological advances could help in the development of a strong framework for natural product–based drug discovery, which would further enhance the longevity and quality of human health in the near future (Viehrig et al. 2017; Surup et al. 2018; Yan et al. 2018; Kayrouz et al. 2020; Laureti et al. 2011).

Some of the commonly used techniques for the identification of targets against bioactive natural products have been broadly divided into the chemical probe approach and non-probe approach. It has been predicted that significant development of artificial intelligence could help arrive at decisive and conclusive results in future research based on scientific data. Additionally, the combination of ideas and inputs from biophysics, transcriptomics, bioinformatics, and other disciplines would enhance the improved understanding of methods to identify suitable targets, thereby enabling suitable success in the drug development process (Xia 2017).

11.7 SIGNIFICANCE

Drug discovery from natural resources is significant because natural compounds have a high level of biodiversity. They are bioactive in nature. This is due to the fact that they are produced as a result of constant evolutionary changes. They are more readily accepted by the public (Atanasov et al. 2015).

11.8 CONCLUSION AND FUTURE PROSPECTS

Finally, there is an ardent desire to maximize the utilization of natural products of medicinal value for the growth and development of human beings on this Earth. It is essential to shift our focus of research towards the visualization, identification, and development of inaccessible natural resources in order to explore their medicinal properties so that we could evade deadly diseases from this planet, prompting a safer and healthier environment. Moreover, natural products tend to occupy a superior position in the search for novel pharmaceutical compounds. Natural product–based drug discovery could be realized if we could perform molecular modification of natural compounds in order to deliver successful medicines in the pharmaceutical market. Structural modification or alteration of natural products helps to increase selectivity and potency, reformed biochemical and pharmacokinetic properties, a simplified structure, and a decrease in adverse side effects, eventually contributing to patentable compounds. It has to be noted that mutual understanding and cooperation combined with interaction among scientists might help in the delivery of modified natural products for commercial purpose with effectively garnered support from drug research and development studies.

BIBLIOGRAPHY

Akshada, A. K., R. C. Doijad, and S. M. Chandrakant Shripal Magdum. 2019. Pharmacognosy-medicinal plants. *Natural Products in Drug Discovery* 1–19. http://dx.doi.org/10.5772/intechopen.82860

Atanasov, A. G., S. B. Zotchev, and V. M. Dirsch, et al. 2021. Natural products in drug discovery: Advances and opportunities. *Nat Rev Drug Discov* 20: 200–216. https://doi.org/10.1038/s41573-020-00114-z

Bhanot, A., R. Sharma, and M. N. Noolvi. 2011. Natural sources as potential anti-cancer agents: A review. *Indian Journal of Phytomedice* 3: 9–26.

Burkle, F. M. (2020). Declining public health protections within autocratic regimes: Impact on global public health security, infectious disease outbreaks, epidemics, and pandemics. *Prehospital and Disaster Medicine* 35(3): 237–246.

Chopra, I. 2000. Drugs for the superbugs. *Microbiol Today* 27: 4–6.

Cowan, M. M. 1999. Plant products as antimicrobial agents. *Clinical Microbiology Reviews* 12(4): 564–582. https://doi.org/10.1128/CMR.12.4.564

Cragg, G. M., and D. J. Newman. 2013. Natural products: A continuing source of novel drug leads. *Biochimica et biophysica acta* 1830(6): 3670–3695. https://doi.org/10.1016/j.bbagen.2013.02.008

Destro, D. 2018. *Development of Phase Transfer Catalysed Amidation and Cyclisation* Reactions, Doctoral dissertation, The Royal College of Surgeons in Ireland, Ireland.

Demain, A. L., and S. Sanchez. 2009. Microbial drug discovery: 80 years of progress. *J Antibiot.* 62: 5–16.

Dias, D. A., S. Urban, and U. Roessner. 2012. A historical overview of natural products in drug discovery. *Metabolites* 2: 303–336.

Fellmann, C., B. G. Gowen, P. C. Lin, J. A. Doudna, and J. E. Corn. 2017. Cornerstones of CRISPR: Cas in drug discovery and therapy. *Nat. Rev. Drug Discov.* 16: 89–100.

Fokunang, C. N., V. Ndikum, O. Y. Tabi, R. B. Jiofack, B. Ngameni, N. M. Guedje, E. A. Tembe-Fokunang, P. Tomkins, S. Barkwan, F. Kechia, E. Asongalem, J. Ngoupayou, N. J. Torimiro, K. H. Gonsu, V. Sielinou, B. T. Ngadjui, F. Angwafor, A. Nkongmeneck, O. M. Abena, J. Ngogang, and Kamsu-Kom. 2011. Traditional medicine: Past, present and future research and development prospects and integration in the National Health System of Cameroon. *African Journal of Traditional, Complementary, and Alternative Medicines AJTCAM* 8(3): 284–295. https://doi.org/10.4314/ajtcam.v8i3.65276

Guo, Z. 2017. The modification of natural products for medical use. *Acta Pharmaceutica Sinica B* 7(2): 119–136. https://doi.org/10.1016/j.apsb.2016.06.003

Gurnani, N., D. Mehta, M. Gupta, and B. K Mehta. 2014. Natural products: Source of potential drugs. *Afr J Basic Appl Sci.* 6: 171–186.

Haefner, B. 2003. Drugs from the deep: Marine natural products as drug candidates. *Drug Discov Ther* 8: 536–544.

Harvey, A. L., R. Edrada-Ebel, and R. J Quinn. 2015. The re-emergence of natural products for drug discovery in the genomics era. *Nat. Rev. Drug Discov* 14: 111–129.

Jamison, D. T., J. G. Breman, A. R. Measham, et al., editors. 2006. "Disease Control Priorities in Developing Countries, 2nd Edition. Disease Control Priorities Project." The International Bank for Reconstruction and Development/The World Bank, Washington (DC); Oxford University Press, New York.

Jassal, B., L. Matthews, G. Viteri, C. Gong, P. Lorente, A. Fabregat, K. Sidiropoulos, J. Cook, M. Gillespie, and R. Haw. et al. 2020. The Reactome pathway knowledgebase. *Nucleic Acids Res* 48: D498–D503.

Ji, H. F., X. J. Li, and H. Y. Zhang. 2009. Natural products and drug discovery: Can thousands of years of ancient medical knowledge lead us to new and powerful drug combinations in the fight against cancer and dementia? *EMBO Reports* 10(3): 194–200. https://doi.org/10.1038/embor.2009.12

Kanehisa, M., M. Furumichi, M. Tanabe, Y. Sato, and K. Morishima. 2017. KEGG: New perspectives on genomes, pathways, diseases and drugs. *Nucleic Acids Res.* 45: D353–D361.

Karp, P. D., R. Billington, R. Caspi, C. A. Fulcher, M. Latendresse, A. Kothari, I. M. Keseler, M. Krummenacker, P. E. Midford, Q. Ong. et al. 2019. The BioCyc collection of microbial genomes and metabolic pathways. *Brief. Bioinform* 20: 1085–1093.

Kayrouz, C. M., Y. Zhang, T. M. Pham, and K. S. Ju. 2020. Genome mining reveals the phosphonoalamide natural products and a new route in phosphonic acid biosynthesis. *ACS Chem. Biol* 15: 1921–1929.

Khalifa, S., N. Elias, M. A. Farag, L. Chen, A. Saeed, M. F. Hegazy, M. S. Moustafa, A. Abd El-Wahed, S. M. Al-Mousawi, S. G. Musharraf, F. R. Chang, A. Iwasaki, K. Suenaga, M. Alajlani, U. Göransson, and H. R. El-Seedi. 2019. Marine natural products: A source of novel anticancer drugs. *Marine Drugs* 17(9): 491. https://doi.org/10.3390/md17090491

Krauze, M. 2021. Phytobiotics, a Natural Growth Promoter for Poultry. In L. Babinszky, J. Oliveira, & E. M. Santos (Eds.), *Advanced Studies in the 21st Century Animal Nutrition.* IntechOpen. doi:10.5772/intechopen.99030

Kulka, M. 2013. Using old solutions to new problems: Natural drug discovery in the 21st century. *Intech Open.* https://doi.org/10.5772/46069

Kumar, S. 2018. Reduction in antimicrobial resistance by the way of extensive vaccination. *Hum Vaccin Immunother* 14(12): 2955–2956. doi: 10.1080/21645515.2018.1506679. Epub Aug 17. PMID: 30060701; PMCID: PMC6343609.

Lancet, T. 2018. ICD-11: A brave attempt at classifying a new world. *Lancet* 391: 2476.

Laskar, Y. B., and P. B. Mazumder. 2020. Insight into the molecular evidence supporting the remarkable chemotherapeutic potential of Hibiscus sabdariffa L. *Biomed. Pharmacother* 127: 110153.

Laureti, L. et al. 2011. Identification of a bioactive 51-membered macrolide complex by activation of a silent polyketide synthase in Streptomyces ambofaciens. *Proc. Natl Acad. Sci.* USA 108: 6258–6263.

Lee, J. C., M. F. Hou, H. W. Huang, F. R. Chang, C. C. Yeh, J. Y. Tang et al. 2013. Marine algal natural products with anti-oxidative, anti-inflammatory, and anti-cancer properties. *Cancer Cell Int* 13: 55.

Licata, L., P. Lo Surdo, M. Iannuccelli, A. Palma, E. Micarelli, L. Perfetto, D. Peluso, A. Calderone, L. Castagnoli, and G. Cesareni. 2020. SIGNOR 2.0, the signaling network open resource 2.0: 2019 update. *Nucleic Acids Res* 48: D504–D510.

Mok, S. W., V. K. Wong, H. H. Lo, I. R. de Seabra Rodrigues Dias, E. L. Leung, B. Y. Law, and L. Liu. 2020. Natural products-based polypharmacological modulation of the peripheral immune system for the treatment of neuropsychiatric disorders. *Pharmacol. Ther* 208: 107480.

Mushtaq, S., B. H. Abbasi, B. Uzair, and R. Abbasi. 2018. Natural products as reservoirs of novel therapeutic agents. *EXCLI Journal* 17: 420–451. https://doi.org/10.17179/excli2018-1174

Newman, D. J., and G. M. Cragg. 2016. Natural products as sources of new drugs from 1981 to 2014. *J. Nat. Prod* 79(3): 629–661. doi: 10.1021/acs. jnatprod.5b01055

Newman, D. J., and G. M. Cragg. 2020. Natural products as sources of new drugs over the nearly four decades from 01/1981 to 09/2019. *J. Nat. Prod* 83: 770–803.

Noah, D., and G. Fidas. 2000. *The Global Infectious Disease Threat and Its Implications for the United States.* National Intelligence Council, Washington, DC.

Noble, R. L. 1990 The discovery of the vinca alkaloids—chemotherapeutic agents against cancer. *Biochem Cell Biol* 68:1344–51.

Pan, S. Y., G. Litscher, S. H. Gao, S. F. Zhou, Z. L. Yu, H. Q. Chen, S. F. Zhang, M. K. Tang, J. N. Sun, and K. M. Ko. 2014. Historical perspective of traditional indigenous medical practices: the current renaissance and conservation of herbal resources. *Evidence-based complementary and alternative medicine: eCAM, 2014*, 525340. https://doi.org/10.1155/2014/525340

Panda, H. 2004. *Handbook on Medicinal Herbs with Uses.* Asia Pacific Business Press, New Delhi. p. 564.

Parham, S., A. Z. Kharazi, H. R. Bakhsheshi-Rad, A. F. Nur, S. Ismail, S. Sharif, S. RamaKrishna, and F. Berto. 2020. Antioxidant, antimicrobial and antiviral properties of herbal materials. *Antioxidants (Basel, Switzerland)* 9(12): 1309. https//doi.org/10.3390/antiox9121309

Patridge, E., P. Gareiss, M. S. Kinch, and D. Hoyer. 2016. An analysis of FDA-approved drugs: Natural products and their derivatives. *Drug Discov Today* 21(2): 204–207. doi: 10.1016/j.drudis.2015.01.009. Epub 2015 Jan 21. PMID: 25617672.

Rodchenkov, I., O. Babur, A. Luna, B. A. Aksoy, J. V. Wong, D. Fong, M. Franz, M. C. Siper, M. Cheung, M. Wrana et al. 2020. Pathway commons 2019 update: Integration, analysis and exploration of pathway data. *Nucleic Acids Res* 48: D489–D497.

Sagarika Sarkar and Sauris Panda. 2021. Moringa oleifera Lam. – A Multipurpose and Great Therapeutic Tree: A Review. In: Indu Rani Sharma (Eds.), *Medicinal Plants: Herbal Wealth of India* (pp. 103–119). Agrobios, Jodhpur, India. Available from: https://www.researchgate.net/publication/356593294_Moringa_oleifera_Lam_-A_Multipurpose_and_Great_Therapeutic_Tree_A_Review [accessed Aug 27 2022].

Schirle, M., and J. L. Jenkins. 2016. Identifying compound efficacy targets in phenotypic drug discovery. *Drug Discov.* 21: 82–89.

Shi, J., J. Li, J. Li, R. Li, X. Wu, F. Gao, L. Zou, W. W. S. Mak, C. Fu, J. Zhang et al. 2021. Synergistic breast cancer suppression efficacy of doxorubicin by combination with glycyrrhetinic acid as an angiogenesis inhibitor. *Phytomedicine* 81: 153408.

Sircar, N. N. 1982. Medicinal plants. *The Eastern Pharmacist* 29(291):49–52[2].

Subramani, R., and W. Aalbersberg. 2012. Marine actinomycetes: An ongoing source of novel bioactive metabolites. *Microbiol Res* 167: 571–580.

Sultana, A., Hossain, M. J., Kuddus, M. R., Rashid, M. A., Zahan, M. S., Mitra, S., . . . & Naina Mohamed, I. 2022. Ethnobotanical uses, phytochemistry, toxicology, and pharmacological properties of *Euphorbia neriifolia* Linn. against infectious diseases: A comprehensive review. *Molecules* 27(14): 4374.

Sun, X., Y. Zhang, Y. Zhou, X. Lian, L. Yan, T. Pan, T. Jin, H. Xie, Z. Liang, W. Qiu, J. Wang, Z. Li, F. Zhu, and X. Sui. 2022. NPCDR: Natural product-based drug combination and its disease-specific molecular regulation. *Nucleic Acids Research* 50D1: D1324–D1333. https://doi.org/10.1093/nar/gkab913

Surup, F. et al. 2018. Crocadepsins-depsipeptides from the myxobacterium Chondromyces crocatus found by a genome mining approach. *ACS Chem. Biol.* 13: 267–272.

Veeresham, C. 2012. Natural products derived from plants as a source of drugs. *Journal of Advanced Pharmaceutical Technology & Research* 3(4): 200–201. https://doi.org/10.4103/2231-4040.104709

Ventola, C. L. 2015. The antibiotic resistance crisis: Part 1: Causes and threats. *P & T: A Peer-Reviewed Journal for Formulary Management* 40(4): 277–283.

Viehrig, K. et al. 2017. Structure and biosynthesis of crocagins: Polycyclic posttranslationally modified ribosomal peptides from Chondromyces crocatus. *Angew. Chem. Int. Ed. Engl* 56: 7407–7410.

Waltenberger, B., A. Mocan, K. Šmejkal, E. H. Heiss, and A. G. Atanasov. 2016. Natural products to counteract the epidemic of cardiovascular and metabolic disorders. *Molecules* 21: 807.

Wang, L., C. Hu, and L. Shao. 2017. The antimicrobial activity of nanoparticles: Present situation and prospects for the future. *International Journal of Nanomedicine* 12: 1227–1249. https://doi.org/10.2147/IJN.S121956

Xia, X. 2017. Bioinformatics and drug discovery. *Current Topics in Medicinal Chemistry* 17(15): 1709–1726. https://doi.org/10.2174/1568026617666161116143440

Yan, F. et al. 2018. Biosynthesis and heterologous production of vioprolides: Rational biosynthetic engineering and unprecedented 4-methylazetidinecarboxylic acid formation. *Angew. Chem. Int. Ed.* 57: 8754–8759.

Yuan, H., Q. Ma, L. Ye, and G. Piao. 2016. The traditional medicine and modern medicine from natural products. *Molecules* 21(5): 559. https://doi.org/10.3390/molecules21050559

Zhang, B., J. Jiang, P. Wu, J. Zou, J. Le, J. Lin, C. Li, B. Luo, Y. Zhang, R. Huang et al. 2021. A smart dual-drug nanosystem based on co-assembly of plant and food-derived natural products for synergistic HCC immunotherapy. *Acta Pharm. Sin. B.* 11: 246–257.

12 Recent Insights into Cannabinoids

Functions, Mechanisms, Clinical Trials, and Technological Advancements in Cancer and Neurological Diseases

Tarun Pant, Matea Juric, and Zeljko J. Bosnjak

CONTENTS

12.1 Introduction 225
12.2 The Functional Role of Cannabinoids in Cancer 226
12.2.1 Cannabinoids in Glioblastoma 226
12.2.2 Cannabinoids in Lung Cancer 228
12.2.3 Cannabidiol in Breast Cancer 229
12.3 The Functional Role of Cannabinoids in Neurological Diseases 230
12.4 Recent Clinical Trials of Cannabinoids in Cancer 232
12.5 Recent Clinical Trials of Cannibinoids in Neurological Diseases 233
12.6 Challenges for Cannabinoids Use in Cancer and Neurological Diseases 234
12.7 Conclusion and Future Prospects 234
Bibliography 234

12.1 INTRODUCTION

Cannabinoids are significant constituents of the *Cannabis sativa* plant, used for millennia as textile fibers (Skoglund, Nockert, and Holst 2013; Andre, Hausman, and Guerriero 2016). However, in the past couple of decades, technological/scientific advancements, with landmark changes in the law and policy, have substantially elevated the use of cannabinoids for chronic medical conditions (Lafaye et al. 2017; Bifulco and Pisanti 2015). The cannabinoid family comprises a repertoire of lipophilic compounds playing a critical role in regulating numerous pathophysiological processes (Richter et al. 2018). Among these, the most pharmacologically relevant cannabinoids are tetrahydrocannabinol (THC) and cannabidiol (CBD), which have been safe and broadly applicable against a spectrum of diseases in both animal and human patients (Kogan and Mechoulam 2007). Most of the recognized actions of THC and CBD have been attributed to members of the G protein–coupled receptor (GPCR) superfamily comprising endogenous cannabinoid receptors known as CB1R, predominantly expressed in the central nervous system (CNS) and to a certain extent on non-neuronal cells like adipocytes, and progenitor stem cells localized in the peripheral tissues (Pertwee 2012; Katona et al. 2001). On the contrary, CB2R depicts a relatively lower expression than CB1R in the brain and is exclusively expressed on the innate immune cells (specifically tissue-specific

DOI: 10.1201/b22917-12

macrophages like osteoclasts and osteoblasts) and cells of the peripheral nervous system (PNS) (Graham et al. 2010; Cabral and Griffin-Thomas 2009).

Accumulating lines of evidence from clinical and preclinical studies reported ameliorating effects of CBDs in numerous neurological diseases, including epilepsy, anxiety, and schizophrenia (Blessing et al. 2015; McGuire et al. 2018). Beyond neurological disorders, studies have also reported the palliative effect of cannabinoids in mitigating tumor growth in various animal cancer models (Velasco et al. 2016; Ladin et al. 2016). However, despite these successful clinical trials, the underlying molecular mechanisms for CBD's therapeutic benefit remain elusive.

Intriguingly, the recent surge in scientific publications has documented cannabinoid receptors to be expressed explicitly on the mitochondria membrane (mtCB1), modulating the mitochondrial processes (Hebert-Chatelain et al. 2016; Mendizabal-Zubiaga et al. 2016). The disturbance in the mitochondrial processes often contributes to the pathogenesis of various diseases, including neurodegenerative and multiple cancer types (Zong, Rabinowitz, and White 2016). Emerging studies have reported using cannabinoids to treat cancer and neurological diseases via modulating mitochondrial function. However, the underlying mechanisms via which CBD modulates mitochondrial functions remain enigmatic. A better understanding of the same would help enhance the therapeutic efficacy of CBD and its use in clinics. The current chapter highlights the understanding and mechanism underlying the therapeutic effects of cannabinoids. In addition, we also discuss the progress in understanding the mechanisms of CBD-mediated regulation of mitochondrial functions, clinical trials employing CBD in cancer therapy, and neurological disorders.

12.2 THE FUNCTIONAL ROLE OF CANNABINOIDS IN CANCER

The concept that cannabinoids can serve as therapeutic agents to mitigate cancer growth, metastasis, and induce cell death have been explored for decades thanks to Raphael Mechoulam, who was the first researcher to successfully identify the cannabinoid THC in 1964 (Pertwee 2006). His discovery opened the door for further progress and identification of the endocannabinoid system detailed in the introduction of this chapter. Over a decade later, in 1975, Munson et al. provided evidence that THC could inhibit tumor growth and increase survival of transgenic mice implanted with Lewis lung adenocarcinoma (Munson et al. 1975). Since then, researchers have explored the role of cannabinoids in mitigating tumor burden in various cancers, including glioblastoma multiforme (GBM) and breast and lung cancers. Here, we aim to review the underlying mechanism inhibiting tumor growth and metastasis and inducing cell death by the cannabinoids THC and CBD in the following cancers: GBM, breast, and lung cancer (Figure 12.1).

12.2.1 Cannabinoids in Glioblastoma

GBM is a deadly form of brain cancer derived from star-shaped neuroglial cells known as astrocytes responsible for maintaining the blood-brain barrier (BBB) (Hanif et al. 2017; Heithoff et al. 2021). GBM consists of a heterogeneous cell population employing multiple genetic mutations (EGFR, MDM2, PTEN) and spider-like appendages that extend deep into the brain matter and destroy healthy tissue, making it notoriously difficult to treat (Liu et al. 2016). Beyond the aggressive nature of this cancer, its location within the protective confines of the cerebrum surrounded by an almost impenetrable barrier makes the delivery of therapeutic agents nearly impossible (Noch, Ramakrishna, and Magge 2018). In the last two decades, treatment modalities have improved a bit, with surgical resection, ionizing radiation, and temozolomide (an alkylating agent that delivers a

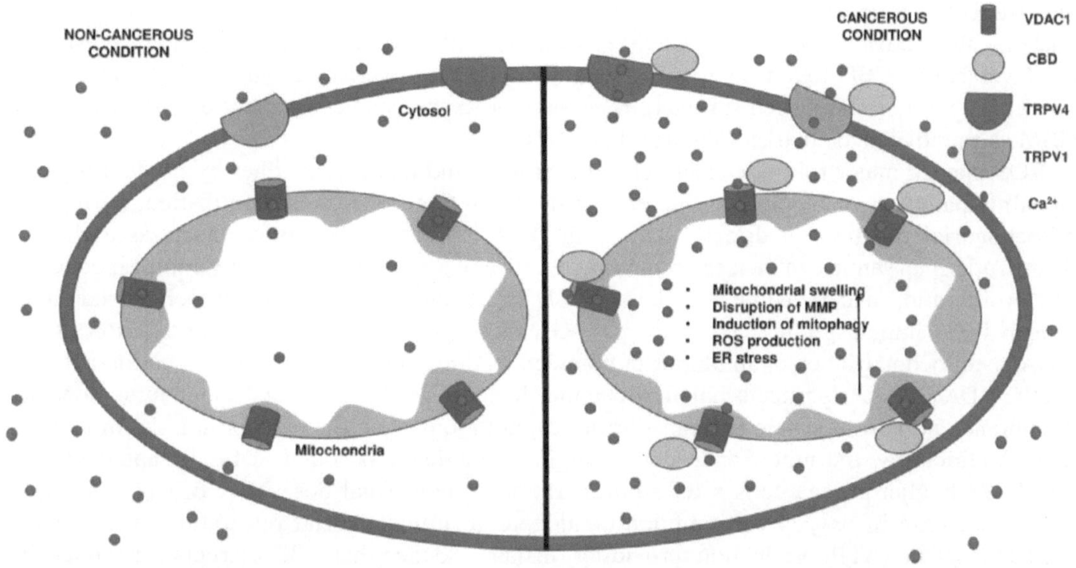

FIGURE 12.1 CBD mechanism of action in cancer: TRPV1 and TRPV4 calcium transporters homeostatically transport calcium into the cell. As an activator of the TRPV1/4 and VDAC1 channels, CBD can modulate the calcium ion influx into the cytosol and the mitochondria. As depicted in cancerous conditions, the influx of calcium ions leads to mitochondrial swelling, decreased mitochondrial membrane potential (MMP), increased ROS production, and downstream endoplasmic reticulum (ER) stress.

methyl group to guanine and adenine bases) being employed as the primary treatment strategies used in clinical settings (Shergalis et al. 2018). Unfortunately, these combined treatments provide only a 12-month survival advantage, leaving researchers desperately looking for better therapeutics. Cannabinoids have provided new hope for researchers and patients seeking long-term survival and improved quality of life following a glioblastoma diagnosis. Of note, one of the significant distinctions in GMB is aberrant nuclear factor-κB (NF-κB) activity (Bhat et al. 2013). NF-κB is a transcription factor composed of the heterodimeric subunits RELA (p65) and p50 responsible for cell proliferation, invasion, angiogenesis, apoptosis, and induction of a metabolic switch from oxidative phosphorylation to glycolysis in cancer cells. It is frequently found to be constitutively active in a wide range of cancers and has garnered interest as a molecular target for novel anticancer therapies (Oeckinghaus and Ghosh 2009). However, modulating the activity of NF-κB in cancer poses challenges, as designing drugs targeted to cancer cells while avoiding harm to healthy cells is still an up-and-coming science.

However, aiming to inhibit NF-κB activity without a known targeting strategy ultimately encourages chronic inflammation and inhibition of the many homeostatic functions of NF-κB, such as induction of apoptosis, response to cytokines involved in the immune response to pathogens, and DNA repair (Hoesel and Schmid 2013). Instead of completely blocking the activity of NF-κB, a more palpable route of modulating the pro-oncogenic effects of NF-κB is to induce a "switch" from its pro-oncogenic phenotype to that of its tumor suppressor phenotype within cancer cells. Interestingly, CBD was found to recapitulate the antitumor effects of NF-κB in transgenic mouse models harboring human GBM (hGBM) tumors by inducing a RELA isoform devoid of the phosphorylation on the serine-311 residue while allowing nuclear accumulation and DNA binding of inactive RELA. Additionally, it was uncovered that the effects of CBD in GBM were inversely correlated with reactive oxygen species (ROS) production, with lower ROS-producing hGBM responding favorably to CBD treatment both in vitro (evidenced by increased cytotoxicity) and in vivo (evidenced by more prolonged survival), necessitating further studies on the use of ROS levels in GBM as a predictive biomarker (Volmar et al. 2021).

ROS plays a major role in cancer cell proliferation and metastasis either by modulating cell signaling pathways, activating cell-defense systems involved in ROS dissemination, perturbing mitochondrial function, or directly damaging DNA. Targeting the mitochondria as a chemotherapeutic intervention in cancer therapy is an attractive route for selectively inhibiting cancer cell proliferation and inducing cell death. CBD has been shown to alter mitochondrial function in U373 human glioma cells and SDT3G canine glioma cell lines by modulation of the VDAC1-mitochondrial calcium uniporter complex leading to mitochondrial destruction and cell death. VDAC1 is a significant calcium channel located on the inner and outer mitochondrial membrane and is responsible for transporting calcium ions from the cytosol into the mitochondria. Additionally, the mitochondrial bioenergetic function was found to be disrupted in U87 and U373 human glioma cells after administration of non-lethal doses of CBD, as evidenced by the decrease in oxygen consumption (analogous to mitochondrial respiration) and adenosine triphosphate (ATP) production, providing further evidence that CBD disrupts mitochondrial homeostasis via perturbation of calcium influx (Seltzer et al. 2020; Gross et al. 2021). CBD's role in the modulation of glioma bioenergetics and mitochondrial structure via disruption of calcium homeostasis has been supported further following evidence that CBD induced mitophagy via direct binding to the mechanosensitive ion channel TRPV4 responsible for calcium and sodium ion transport. CBD serves as an agonist for TRPV4, binding to a hydrophobic pocket within the inner pore of the protein and modulating calcium influx. Following this binding modality, endoplasmic reticulum (ER) stress occurs as evidenced by upregulation of ER stress markers (DDIT3/CHOP, EIF2A, ATF4, EIF2AK3/PERK) leading to mitophagy and cell death in U87 and U251 cells (Huang et al. 2021).

12.2.2 Cannabinoids in Lung Cancer

According to the latest Centers for Disease Control and Prevention (CDC) data from 2021, lung cancer is the leading cause of cancer deaths in the United States (Siegel et al. 2021). Cancers of the lung are denoted by cell type and aggressiveness, with the distinction made between small cell lung cancer (SCLC) and non–small cell lung cancer (NSCLC). Of these two, approximately 80% of lung cancer diagnoses are attributed to NSCLC, which serves as the less-aggressive form. Still, the five-year survival rate for individuals diagnosed with metastatic NSCLC is <15.5%, a statistic complicated by the notion that 48% of NSCLC diagnoses are made after the disease has spread to distant organs (Toh et al. 2007; Oser et al. 2015). In addition, lung cancer is marred with genetic mutations, most notably *KRAS*, *ALK*, and *EGFR* overexpression (El-Telbany and Ma 2012). While targeted therapies have been developed for NSCLC tumors harboring *ALK* (crizotinib, ceritinib, and alectinib) and *EGFR* mutations (erlotinib and gefitinib), among others, resistance to these therapies eventually occurs (Burris 2009; Yuan et al. 2019). As such, new therapies are urgently needed to circumvent eventual resistance to these first-line treatments. Notably, THC and CBD were found to inhibit *EGFR* expression, leading to decreased proliferation and migration in A549, H460, and H1792 lung cancer cell lines, with additive effects observed when combining THC and CBD for the administration of a lower concentration of both compounds. Additionally, epithelial-to-mesenchymal transition (EMT) was inhibited by CBD and THC, as evidenced by increased expression of CDH1 and diminished expression of CDH2 and VIM in these cell lines (Milian et al. 2020). EMT is regarded as the first step in metastasis, as epithelial cells are phenotypically transformed into mesenchymal (stem-like) cells owing to downregulation of E-cadherin and upregulation of N-cadherin (Kim et al. 2015). Lung cancer recurrence is thought to occur consequently: of first-line therapies which fail to effectively kill cancer stem cells employing more aggressive and resistant phenotypes. Interestingly, CBD administration to the NSCLC cell lines A549 and H1299 as well as in the SCLC cell line H69, cultured as spheres after selecting for stemness (stem cell–like properties), significantly reduced viability, sphere formation, and the expression of stem-related genes, including *SOX2*, *POU5F1*, *CD44*, and *PROM1*. Furthermore, CBD-induced caspase 3/7–dependent apoptosis via upregulation of pro-apoptotic genes *BAX*, *BAD*, and *BAK1* and subsequent downregulation of the antiapoptotic gene *BCL2* in all three cancer stem-like cell lines. Further studies found that CBD administration significantly increased ROS levels in all three cell lines, and concomitant loss of mitochondrial membrane potential was observed (Hamad and Olsen 2021).

12.2.3 Cannabidiol in Breast Cancer

Breast cancer is the most prevalent worldwide among all the cancer types, with 2.3 million women diagnosed with this cancer globally (DeSantis et al. 2019). Breast cancer is graded by tumor location and size and further categorized based on the involvement of estrogen receptor (ER), progesterone receptor (PR), and epidermal growth factor receptor 2 (HER2) mutations. Eighty-five percent of metastatic breast tumors arise in the ducts, with the remaining 15% arising in the lobules (Al-Thoubaity 2020). Treatments for breast cancer have improved in the last 40 years with the advent of targeted hormone therapies such as tamoxifen (used for ER-positive breast cancers) and traztuzumab and pertuzumab (for HER2 positive breast cancers). These drugs have improved increased progression-free survival (PFS) and overall survival (OS) for patients. Still, they have unintended side effects such as an increased risk of endometrial cancer in postmenopausal women, advanced heart failure risk, and eventual resistance to therapy (Shien and Iwata 2020). There is an imminent need to develop novel therapies capable of inhibiting breast cancer progression. CBD has garnered attention as a potent anticancer therapeutic in breast cancer based on its distinct antiproliferative, antimetastatic, and apoptosis/mitophagy/autophagy-induction mechanisms of action (Sultan, Marie, and Sheweita 2018; Fraguas-Sanchez et al. 2020). Recently, CBD was shown to inhibit EMT in MCF-7 breast cancer cells selected for sensitivity to IL-1B via upregulation of B-catenin and E-cadherin. Furthermore, treatment with CBD downregulated proteins downstream of IL-1B signaling pathways involved in EMT such as BIRC3, TP63, and ID-1, in addition to inactivating AKT phosphorylation (a component of the PI3K/AKT signaling pathway). This resulted in the inhibition of cell migration and supports the notion that CBD can reverse the malignant phenotype of breast cancer, such as migration and drug resistance (Garcia-Morales et al. 2020). Similar to its mechanism of cancer inhibition described previously for glioblastoma, CBD congregates with the TRPV1 (vs. TRPV4 in glioblastoma) calcium ion transporter, inducing intracellular Ca^{2+} buildup, downstream oxidative stress, and subsequent activation of ER stress in MCF-7 breast cancer cells (de la Harpe, Beukes, and Frost 2021).

These studies provide mechanistic insight into underlying mechanisms in context to the antitumor effect of CBD in various cancer types. The results from these studies have paved the path for naïve clinical trials in specific cancer types. The various CBD-mediated cancer therapeutic mechanisms are shown in Table 12.1.

TABLE 12.1
The Activity of Cannabidiol (CBD) against Cancer

S. No.	Cancer Type	Cell Lines	Species Model	Physiological Changes	Mechanism of Action	References
1.	Brain cancer	U87 & U373 glioma cell line	Human	Mitochondrial function	CBD mediated Ca^{2+} influx via VDAC1	Gross et al. (2021)
2.	Brain cancer	U87 & U251 glioma cell line	Human	Mitophagy ER stress Cell viability	CBD mediated Ca^{2+} influx via TRPV4 aggravated ATF4-DDIT3-TRIB3-AKT-MTOR axis	Huang et al. (2021)
3.	Lung cancer	A549, H460 & H1792 cell line	Human	Cell proliferation	CBD attenuated cellular proliferation via inhibiting EGFR expression	Milian et al. (2020)
4.	Lung cancer	H1299, SCLC, and H69 cell line	Human	Cell viability Sphere formation	CBD mediated apoptosis via activation of pro-apoptotic genes *BAD*, *BAX*, and *BAK1*	Hamad and Olsen (2021)
5.	Breast cancer	MCF-7	Human	Cell migration Drug resistance EMT	CBD inhibits EMT via upregulating B-catenin and E-cadherin, further attenuated PI3K/AKT	Garcia-Morales et al. (2020)
6.	Breast cancer	MCF-7	Human	Oxidative stress ER stress Cell viability	CBD mediated Ca^{2+} influx via TRPVI	de la Harpe, Beukes, and Frost (2021)

12.3 THE FUNCTIONAL ROLE OF CANNABINOIDS IN NEUROLOGICAL DISEASES

Mitochondria, the key player in metabolic homeostasis, play a significant role in numerous neurobiological processes (Kann and Kovacs 2007). Subtle disruption in the neuronal metabolism can alter the redox and calcium homeostasis, activating apoptotic pathways and contributing to the pathogenesis of neurological diseases (Zhao et al. 2019; Norat et al. 2020). Emerging studies have provided substantial evidence regarding the distribution of CB1 receptors on the neuronal mitochondrial membranes (mtCB1) (Marsicano et al. 2003). However, the understanding of mtCB1 signaling mechanisms regarding neuroprotection have yet to be deciphered. Considering the importance of mitochondrial activity in neurons, further supported with studies demonstrating the role of CBD in modulating the same, makes it an exciting topic for discussion. In this section, we discuss recently published studies that depict the neuroprotective effect of CBD via modulating mitochondrial functions (Figure 12.2).

The brain-specific studies of CBD and its impact on mitochondrial dynamics were recently published. Xu and colleagues, using the traumatic brain injury (TBI) mice model, investigated the effect of the mtCB1 receptor on neuronal apoptosis. The results depict activation of mtCB1 imparting a dual role via activating the mitochondrial cAMP/PKA/complex I and AKT/complex V axis, with the former aggravating metabolic alteration, energy deficit, neuronal apoptosis, and later mitigating energy insufficiency and neuronal apoptosis (Xu et al. 2016).

In addition to TBI, the protective role of CBD has also been tested employing a mice model of cerebral ischemia/reperfusion (I/R) injury. First, the authors demonstrate that administration of CBD agonist arachidonyl-2-chloroethylamide (ACEA) enhanced mtCB1 expression, providing neuroprotection against cerebral I/R injury. In addition, ACEA-induced mtCB1 activation further

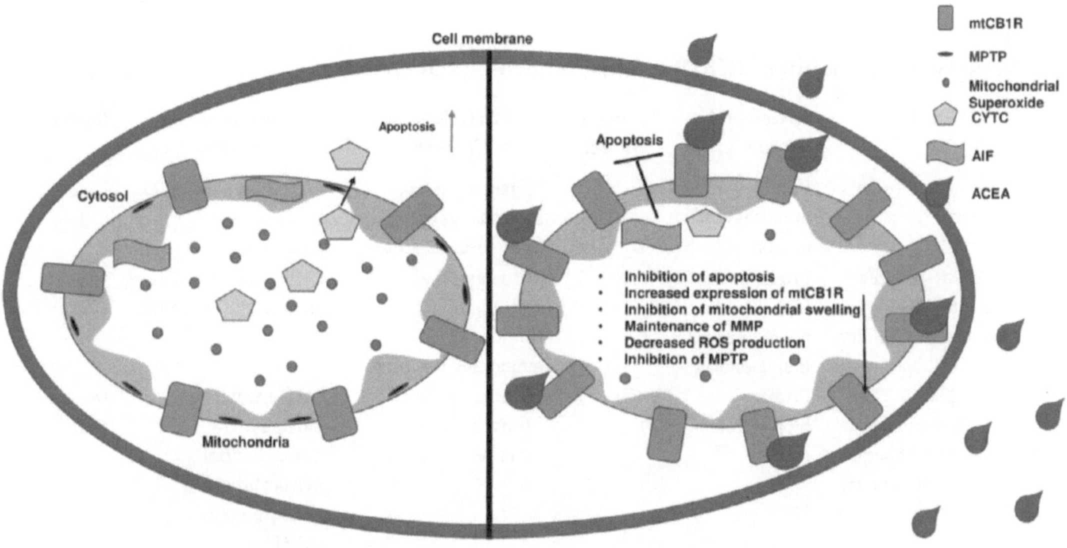

FIGURE 12.2 CBD mechanism of action in neurological diseases: CBDA agonist arachidonyl-2′-chloroethylamide (ACEA) acts as an activator of mtCB1R, attenuating mitochondrial apoptosis and restoring redox balance. The activation of mtCB1R by ACEA inhibits the release of cytochrome C by closing mitochondrial transition pores. In addition, the mitochondrial membrane potential and superoxide production are decreased, inhibiting apoptosis.

prevented Ca2+–induced injury in the purified mitochondria from the normal hippocampal neuron (Ma et al. 2015).

In a follow-up study, the same researchers performed bilateral common carotid artery occlusion (BCCAO) in mice and oxygen-glucose deprivation (OGD) in primary cultured neurons to mimic ischemic brain injury. Again, this demonstrated ACEA is able to protect ischemic neurons by inhibiting mitochondrial permeability transition pore (mPTP) opening (Ma et al. 2018).

The data from an additional study by Shuai and colleagues demonstrate the neuroprotective effect of ACEA in I/R injury and the OGD/R model. In the present study, HT22 hippocampal neurons were deprived of oxygen and glucose to induce neuronal injury. These conditions led to reduced cell viability and mitochondrial function as evidenced by increased lactate dehydrogenase (LDH) release and induction of cell apoptosis and mitochondrial fission. In addition, treatment of cells with 1 or 2 μM ACEAs inhibited OGD/R-induced mitochondrial fission as a consequence of decreasing the expression of Drpl, thereby restoring cell viability and mitochondrial function. Furthermore, authors observed similar outcomes in vivo, where ACEA treatment counteracted middle cerebral artery occlusion (MCAO)–induced I/R injury, decreasing brain infarct volume, neuronal apoptosis, and mitochondrial fission (Yang et al. 2020). Intriguingly, a research study by Sun and colleagues also supports the neuroprotective effect of CBD in the OGD/R model of HT22 hippocampal neurons via modulating mitochondrial respiration. The study demonstrates that treating cells with CBD could have a biphasic effect, restoring redox balance and energy via attenuating intracellular ROS generation, lipid peroxidation, and enhancing mitochondrial bioenergetics (Sun et al. 2017).

In a recently published preclinical study, ACEA treatment was found to mitigate acute restraint stress (ARS)–induced anxiety-like behavior and oxidative stress in the zebrafish brain. The animal subjected to ARS demonstrated anxiety with elevated lipid peroxidation and a low level of gamma-aminobutyric acid (GABA), specifically in the telencephalon. However, the mesencephalon region

TABLE 12.2
The Activity of Cannabidiol (CBD) against Neurological Diseases

S. No.	Disease	Cell and Tissue Types	Species Model	Physiological Changes	Mechanism of Action	References
1.	Traumatic brain injury	Primary hippocampal neuron	Mice	Metabolic defect Neuronal apoptosis Energy Production	cAMP/PKA/complex I AKT/complex V	Xu et al. (2016)
2.	OGDR-induced cerebral ischemia reperfusion injury	Primary hippocampal neuron derived mitochondria	Mice	Mitochondrial swelling Mitochondrial membrane potential	CBD agonist ACEA attenuated Ca^{2+} influx	Ma et al. (2015)
3.	BCCAO/ OGDR-induced cerebral ischemia reperfusion injury	Primary hippocampal neuron	Mice	Neuronal apoptosis Mitochondrial function	CBD mediated closing of mitochondrial permeability transition pore (MPTP)	Ma et al. (2018)
4.	OGDR-induced ischemia reperfusion injury	HT22-hippocampal neuron cell line	Mice	Mitochondrial fission Mitochondrial function Neuronal viability	CBD agonist ACEA attenuated Drp1 expression	Yang et al. (2020)
5.	OGDR-induced ischemia reperfusion injury	HT22-hippocampal neuronal cell line	Mice	ROS generation Lipid peroxidation Mitochondrial bioenergetics	CBD mediated restored mitochondrial respiration	Sun et al. (2017)
6.	Acute restrain stress	Telencephalic tissues	Zebrafish	ROS generation Lipid peroxidation	CBD agonist ACEA restored redox balance	Drori et al. (2019)

of the brain displayed no alteration regarding GABA. The CB1 stimulation was found to restore the homeostasis of the GABAergic synapse via maintaining the redox equilibrium (Lucas Luz et al. 2020).

Studies investigating the effect of CBD on the modulation of mitochondria dynamics in context of other metabolic diseases have also begun. Adi and colleagues determined that activation of CB1R prevents mitochondrial fission via modulating the ERK1/2/ DRP1 axis, preventing renal proximal tubular cell (RPTC) dysfunction, further mitigating kidney damage (Drori et al. 2019).

Taken together, these studies provide substantial evidence that the therapeutic effect of CBD is primarily mediated via activation of mtCB1 receptors modulating mitochondrial dynamics in numerous neurodegenerative diseases. However, further studies will be needed to understand the therapeutic mechanism. The various CBD mediated neuroprotective mechanisms are shown in Table 12.2.

12.4 RECENT CLINICAL TRIALS OF CANNABINOIDS IN CANCER

Despite advancements in medical care, patients undergoing cancer treatment experience substantial distress. The preliminary data from studies in cells and animals demonstrating the effectiveness of CBD in numerous malignancies has elevated the use of medical cannabinoids. However, the knowledge in terms of efficacy, safety, and acceptability in cancer patients undergoing standard palliative care is limited, underscoring the urgency to conduct more controlled clinical trials.

Recently, a randomized, double-blinded, placebo-controlled phase II/III trial included 81 patients experiencing chemotherapy-induced nausea and vomiting (CINV). Grimison et al. investigated the impact of one cycle of 1–4 self-titrated capsules of oral THC 2.5 mg/CBD 2.5 mg (TN-TC11M) three times daily, from days 1 to 5, and 1 cycle of matching placebo in a crossover design, having blinded patient preference for a third cycle. The addition of THC:CBD to standard antiemetics was well tolerated. Interestingly, most participants experienced less nausea and vomiting with no adverse side effects. Based on the present study's encouraging results, the same group plans to recruit 170 patients for a randomized placebo-controlled parallel phase III trial (Grimison et al. 2020).

A phase II/randomized clinical trial was recently conducted to determine the tolerability of differing ratios of medicinal cannabis involving adult patients diagnosed with high-grade glioma. Patients had moderate side effects, including dry mouth, tiredness at night, dizziness, and drowsiness. Overall, no adverse events were reported, and each participant tolerated well a single nightly dose of THC-containing medicinal cannabis (Schloss et al. 2021).

In 2021, a phase Ib randomized, placebo-controlled trial was published investigating the safety and efficacy of nabiximols oromucosal cannabinoid spray plus dose-intense temozolomide (DIT) to treat patients with glioblastoma. The researcher found that nabiximols was well accepted with a limited level of side effects, including vomiting, dizziness, fatigue, nausea, and headache. The trial is being extended into phase II to recruit 230 patients (Twelves et al. 2021).

The results of these recent clinical trials are encouraging. However, future studies are warranted to address questions like which cancer type would be treated the best using cannabinoids and a clinically relevant dose to mitigate cancer development and metastasis with minimum side effects on other cell types. In addition, cannabinoid supplementation with a traditional anticancer drug could enhance or mitigate the chemotherapeutic efficacy.

12.5 RECENT CLINICAL TRIALS OF CANNIBINOIDS IN NEUROLOGICAL DISEASES

In recent decades, numerous publications have depicted the therapeutic potential of cannabinoids in neurological disorders (Solimini et al. 2017). Recently the data gathered from in vitro and in vivo models have further proven that cannabis-derived compounds like CBD are beneficial in treating numerous neurodegenerative diseases (Cristino, Bisogno, and Di Marzo 2020).

Recently, Devinsky et al conducted a randomized, double blinded, placebo-controlled trial to establish the safety and efficacy of CBD for drop seizures in patients suffering from Lennox-Gastaut syndrome and Dravet syndrome. These two diseases are the result of severe developmental epileptic encephalopathy. The data from both studies demonstrate that addition of CBD resulted in a greater reduction of seizure frequency compared to the placebo group. The data gathered from all these studies have garnered acceptance of CBDs across the country, as evidenced by its Food and Drug Administration (FDA) approval as Epidiolex, a CBD oral solution available to patients suffering from seizures linked to Lennox-Gastaut syndrome or Dravet syndrome (Devinsky et al. 2017; Devinsky et al. 2018). In addition, GW Pharmaceuticals conducted the first phase of a clinical trial using Epidiolex/Epidyolex (>99% CBD), a purified form of CBD. Consequently, this was followed by the FDA and the European Medicines Agency (EMA) approval of CBD.

These preliminary studies have provided substantial evidence that cannabinoids can effectively treat different neurological diseases. However, to robustly test the potential benefits of cannabinoids, it is necessary to conduct clinical trials with a large sample size that would help generate reliable data and pave the path to personalized medicine.

12.6 CHALLENGES FOR CANNABINOIDS USE IN CANCER AND NEUROLOGICAL DISEASES

Interestingly, numerous scientific publications have reported the beneficial effect of cannabinoids in cancer and neurological diseases. However, there are only a few which got translated to the clinic. Unfortunately, the inaccessibility to human tissues and patients makes it nearly impossible to robustly assess the pharmacological relevance of cannabinoids in human diseases, leaving us with an option to rely on the in vitro and in vivo models. Although the data gathered using these models are meaningful, providing us with an early indication regarding the altered pathway and benefits, it is not necessary that it always have a similar effect on the patients. Most importantly, how a patient's genetic background affects cannabinoids-based therapeutics is worth discussing. Finally, since many of these protective mechanisms are mitochondrial driven, an efficient in vivo delivery might be beneficial. We strongly feel that emphasizing the following questions would lay a firm foundation for future clinical trials.

12.7 CONCLUSION AND FUTURE PROSPECTS

Alteration in mitochondrial bioenergetics plays a pivotal role in the pathogenesis of cancer and neurological diseases. Modulating specific pathways with CBDs to restore the mitochondrial dynamics can impact the development of these diseases. Despite the limited data on the effect of CBD on the mitochondria, the preliminary results seem to be encouraging and indicate its relevance. Mitochondria and ER mediate communication via mitochondria-associated endoplasmic reticulum membranes (MAMs), and a few of the previously discussed studies demonstrate the therapeutic effect of CBD via attenuating ER stress; hence, future studies should be designed to decipher the cross-talk of these different signaling pathways in the context of cancer and neurological diseases. Additionally, researchers need to further explore the effect of CBD in the context of mitochondria-related diseases using model systems that are closer to human patients. The human-induced pluripotent stem cells with unlimited potential to differentiate into different cell types and recapitulate the genetic background of different patients can be an option worth exploring. This could pave a path for personalized medicine testing the patient-specific response to CBD. In addition, the data from these studies can be used to design clinical trials enrolling more patients to determine the efficacy of cannabinoids in clinics.

BIBLIOGRAPHY

Al-Thoubaity, F. K. 2020. 'Molecular classification of breast cancer: A retrospective cohort study', *Ann Med Surg (Lond)*, 49: 44–48.

Andre, C. M., J. F. Hausman, and G. Guerriero. 2016. 'Cannabis sativa: The plant of the thousand and one molecules', *Front Plant Sci*, 7: 19.

Bhat, K. P. L., V. Balasubramaniyan, B. Vaillant, R. Ezhilarasan, K. Hummelink, F. Hollingsworth, K. Wani, L. Heathcock, J. D. James, L. D. Goodman, S. Conroy, L. Long, N. Lelic, S. Wang, J. Gumin, D. Raj, Y. Kodama, A. Raghunathan, A. Olar, K. Joshi, C. E. Pelloski, A. Heimberger, S. H. Kim, D. P. Cahill, G. Rao, W. F. A. Den Dunnen, Hwgm Boddeke, H. S. Phillips, I. Nakano, F. F. Lang, H. Colman, E. P. Sulman, and K. Aldape. 2013. 'Mesenchymal differentiation mediated by NF-kappaB promotes radiation resistance in glioblastoma', *Cancer Cell*, 24: 331–346.

Bifulco, M., and S. Pisanti. 2015. 'Medicinal use of cannabis in Europe: The fact that more countries legalize the medicinal use of cannabis should not become an argument for unfettered and uncontrolled use', *EMBO Rep*, 16: 130–132.

Blessing, E. M., M. M. Steenkamp, J. Manzanares, and C. R. Marmar. 2015. 'Cannabidiol as a potential treatment for anxiety disorders', *Neurotherapeutics*, 12: 825–836.

Burris, H. A., 3rd. 2009. 'Shortcomings of current therapies for non-small-cell lung cancer: Unmet medical needs', *Oncogene*, 28 Suppl 1: S4–13.

Cabral, G. A., and L. Griffin-Thomas. 2009. 'Emerging role of the cannabinoid receptor CB2 in immune regulation: Therapeutic prospects for neuroinflammation', *Expert Rev Mol Med*, 11: e3.

Cristino, L., T. Bisogno, and V. Di Marzo. 2020. 'Cannabinoids and the expanded endocannabinoid system in neurological disorders', *Nat Rev Neurol*, 16: 9–29.

de la Harpe, A., N. Beukes, and C. L. Frost. 2021. 'CBD activation of TRPV1 induces oxidative signaling and subsequent ER stress in breast cancer cell lines', *Biotechnol Appl Biochem*, 69: 420–430.

DeSantis, C. E., J. Ma, M. M. Gaudet, L. A. Newman, K. D. Miller, A. Goding Sauer, A. Jemal, and R. L. Siegel. 2019. 'Breast cancer statistics, 2019', *CA Cancer J Clin*, 69: 438–451.

Devinsky, O., J. H. Cross, L. Laux, E. Marsh, I. Miller, R. Nabbout, I. E. Scheffer, E. A. Thiele, S. Wright, and Group Cannabidiol in Dravet Syndrome Study. 2017. 'Trial of cannabidiol for drug-resistant seizures in the dravet syndrome', *N Engl J Med*, 376: 2011–2020.

Devinsky, O., A. D. Patel, J. H. Cross, V. Villanueva, E. C. Wirrell, M. Privitera, S. M. Greenwood, C. Roberts, D. Checketts, K. E. VanLandingham, S. M. Zuberi, and Gwpcare Study Group. 2018. 'Effect of cannabidiol on drop seizures in the lennox-gastaut syndrome', *N Engl J Med*, 378: 1888–1897.

Drori, A., A. Permyakova, R. Hadar, S. Udi, A. Nemirovski, and J. Tam. 2019. 'Cannabinoid-1 receptor regulates mitochondrial dynamics and function in renal proximal tubular cells', *Diabetes Obes Metab*, 21: 146–159.

El-Telbany, A., and P. C. Ma. 2012. 'Cancer genes in lung cancer: Racial disparities: Are there any?', *Genes Cancer*, 3: 467–480.

Fraguas-Sanchez, A. I., A. Fernandez-Carballido, R. Simancas-Herbada, C. Martin-Sabroso, and A. I. Torres-Suarez. 2020. 'CBD loaded microparticles as a potential formulation to improve paclitaxel and doxorubicin-based chemotherapy in breast cancer', *Int J Pharm*, 574: 118916.

Garcia-Morales, L., A. M. Castillo, J. Tapia Ramirez, H. Zamudio-Meza, M. D. C. Dominguez-Robles, and I. Meza. 2020. 'CBD reverts the mesenchymal invasive phenotype of breast cancer cells induced by the inflammatory cytokine IL-1beta', *Int J Mol Sci*, 21.

Graham, E. S., C. E. Angel, L. E. Schwarcz, P. R. Dunbar, and M. Glass. 2010. 'Detailed characterisation of CB2 receptor protein expression in peripheral blood immune cells from healthy human volunteers using flow cytometry', *Int J Immunopathol Pharmacol*, 23: 25–34.

Grimison, P., A. Mersiades, A. Kirby, N. Lintzeris, R. Morton, P. Haber, I. Olver, A. Walsh, I. McGregor, Y. Cheung, A. Tognela, C. Hahn, K. Briscoe, M. Aghmesheh, P. Fox, E. Abdi, S. Clarke, S. Della-Fiorentina, J. Shannon, C. Gedye, S. Begbie, J. Simes, and M. Stockler. 2020. 'Oral THC: CBD cannabis extract for refractory chemotherapy-induced nausea and vomiting: A randomised, placebo-controlled, phase II crossover trial', *Ann Oncol*, 31: 1553–1560.

Gross, C., D. A. Ramirez, S. McGrath, and D. L. Gustafson. 2021. 'Cannabidiol induces apoptosis and perturbs mitochondrial function in human and canine glioma cells', *Front Pharmacol*, 12: 725136.

Hamad, H., and B. B. Olsen. 2021. 'Cannabidiol induces cell death in human lung cancer cells and cancer stem cells', *Pharmaceuticals (Basel)*, 14.

Hanif, F., K. Muzaffar, K. Perveen, S. M. Malhi, and U. Simjee Sh. 2017. 'Glioblastoma multiforme: A review of its epidemiology and pathogenesis through clinical presentation and treatment', *Asian Pac J Cancer Prev*, 18: 3–9.

Hebert-Chatelain, E., T. Desprez, R. Serrat, L. Bellocchio, E. Soria-Gomez, A. Busquets-Garcia, A. C. Pagano Zottola, A. Delamarre, A. Cannich, P. Vincent, M. Varilh, L. M. Robin, G. Terral, M. D. Garcia-Fernandez, M. Colavita, W. Mazier, F. Drago, N. Puente, L. Reguero, I. Elezgarai, J. W. Dupuy, D. Cota, M. L. Lopez-Rodriguez, G. Barreda-Gomez, F. Massa, P. Grandes, G. Benard, and G. Marsicano. 2016. 'A cannabinoid link between mitochondria and memory', *Nature*, 539: 555–559.

Heithoff, B. P., K. K. George, A. N. Phares, I. A. Zuidhoek, C. Munoz-Ballester, and S. Robel. 2021. 'Astrocytes are necessary for blood-brain barrier maintenance in the adult mouse brain', *Glia*, 69: 436–472.

Hoesel, B., and J. A. Schmid. 2013. 'The complexity of NF-kappaB signaling in inflammation and cancer', *Mol Cancer*, 12: 86.

Huang, T., T. Xu, Y. Wang, Y. Zhou, D. Yu, Z. Wang, L. He, Z. Chen, Y. Zhang, D. Davidson, Y. Dai, C. Hang, X. Liu, and C. Yan. 2021. 'Cannabidiol inhibits human glioma by induction of lethal mitophagy through activating TRPV4', *Autophagy*, 17: 3592–3606.

Kann, O., and R. Kovacs. 2007. 'Mitochondria and neuronal activity', *Am J Physiol Cell Physiol*, 292: C641–C657.

Katona, I., E. A. Rancz, L. Acsady, C. Ledent, K. Mackie, N. Hajos, and T. F. Freund. 2001. 'Distribution of CB1 cannabinoid receptors in the amygdala and their role in the control of GABAergic transmission', *J Neurosci*, 21: 9506–9518.

Kim, E. J., H. J. Kim, M. K. Park, G. J. Kang, H. J. Byun, H. Lee, and C. H. Lee. 2015. 'Cardamonin suppresses TGF-beta1-induced epithelial mesenchymal transition via restoring protein phosphatase 2A expression', *Biomol Ther (Seoul)*, 23: 141–148.

Kogan, N. M., and R. Mechoulam. 2007. 'Cannabinoids in health and disease', *Dialogues Clin Neurosci*, 9: 413–430.

Ladin, D. A., E. Soliman, L. Griffin, and R. Van Dross. 2016. 'Preclinical and clinical assessment of cannabinoids as anti-cancer agents', *Front Pharmacol*, 7: 361.

Lafaye, G., L. Karila, L. Blecha, and A. Benyamina. 2017. 'Cannabis, cannabinoids, and health', *Dialogues Clin Neurosci*, 19: 309–316.

Liu, A., C. Hou, H. Chen, X. Zong, and P. Zong. 2016. 'Genetics and epigenetics of glioblastoma: Applications and overall incidence of IDH1 mutation', *Front Oncol*, 6: 16.

Lucas Luz, W., M. Santos-Silva, P. B. Cardoso, N. Assad, Erds Moraes, A. B. A. Grisolia, D. V. Braga, L. K. R. Leao, S. A. S. de Moraes, A. D. C. Passos, E. J. O. Batista, A. Gouveia, Jr., Krhm Oliveira, and A. M. Herculano. 2020. 'Putative activation of the CB1 cannabinoid receptors prevents anxiety-like behavior, oxidative stress, and GABA decrease in the brain of zebrafish submitted to acute restraint stress', *Front Behav Neurosci*, 14: 598812.

Ma, L., J. Jia, W. Niu, T. Jiang, Q. Zhai, L. Yang, F. Bai, Q. Wang, and L. Xiong. 2015. 'Mitochondrial CB1 receptor is involved in ACEA-induced protective effects on neurons and mitochondrial functions', *Sci Rep*, 5: 12440.

Ma, L., W. Niu, S. Yang, J. Tian, H. Luan, M. Cao, W. Xi, W. Tu, J. Jia, and J. Lv. 2018. 'Inhibition of mitochondrial permeability transition pore opening contributes to cannabinoid type 1 receptor agonist ACEA-induced neuroprotection', *Neuropharmacology*, 135: 211–222.

Marsicano, G., S. Goodenough, K. Monory, H. Hermann, M. Eder, A. Cannich, S. C. Azad, M. G. Cascio, S. O. Gutierrez, M. van der Stelt, M. L. Lopez-Rodriguez, E. Casanova, G. Schutz, W. Zieglgansberger, V. Di Marzo, C. Behl, and B. Lutz. 2003. 'CB1 cannabinoid receptors and on-demand defense against excitotoxicity', *Science*, 302: 84–88.

McGuire, P., P. Robson, W. J. Cubala, D. Vasile, P. D. Morrison, R. Barron, A. Taylor, and S. Wright. 2018. 'Cannabidiol (CBD) as an adjunctive therapy in schizophrenia: A multicenter randomized controlled trial', *Am J Psychiatry*, 175: 225–231.

Mendizabal-Zubiaga, J., S. Melser, G. Benard, A. Ramos, L. Reguero, S. Arrabal, I. Elezgarai, I. Gerrikagoitia, J. Suarez, F. Rodriguez De Fonseca, N. Puente, G. Marsicano, and P. Grandes. 2016. 'Cannabinoid CB1 receptors are localized in striated muscle mitochondria and regulate mitochondrial respiration', *Front Physiol*, 7: 476.

Milian, L., M. Mata, J. Alcacer, M. Oliver, M. Sancho-Tello, J. J. Martin de Llano, C. Camps, J. Galbis, J. Carretero, and C. Carda. 2020. 'Cannabinoid receptor expression in non-small cell lung cancer: Effectiveness of tetrahydrocannabinol and cannabidiol inhibiting cell proliferation and epithelial-mesenchymal transition in vitro', *PLoS One*, 15: e0228909.

Munson, A. E., L. S. Harris, M. A. Friedman, W. L. Dewey, and R. A. Carchman. 1975. 'Antineoplastic activity of cannabinoids', *J Natl Cancer Inst*, 55: 597–602.

Noch, E. K., R. Ramakrishna, and R. Magge. 2018. 'Challenges in the treatment of glioblastoma: Multisystem mechanisms of therapeutic resistance', *World Neurosurg*, 116: 505–517.

Norat, P., S. Soldozy, J. D. Sokolowski, C. M. Gorick, J. S. Kumar, Y. Chae, K. Yagmurlu, F. Prada, M. Walker, M. R. Levitt, R. J. Price, P. Tvrdik, and M. Y. S. Kalani. 2020. 'Mitochondrial dysfunction in neurological disorders: Exploring mitochondrial transplantation', *NPJ Regen Med*, 5: 22.

Oeckinghaus, A., and S. Ghosh. 2009. 'The NF-kappaB family of transcription factors and its regulation', *Cold Spring Harb Perspect Biol*, 1: a000034.

Oser, M. G., M. J. Niederst, L. V. Sequist, and J. A. Engelman. 2015. 'Transformation from non-small-cell lung cancer to small-cell lung cancer: Molecular drivers and cells of origin', *Lancet Oncol*, 16: e165–e172.

Pertwee, R. G. 2006. 'Cannabinoid pharmacology: The first 66 years', *Br J Pharmacol*, 147 Suppl 1: S163–S171.

Pertwee, R. G. 2012. 'Targeting the endocannabinoid system with cannabinoid receptor agonists: Pharmacological strategies and therapeutic possibilities', *Philos Trans R Soc Lond B Biol Sci*, 367: 3353–3363.

Richter, J. S., V. Quenardelle, O. Rouyer, J. S. Raul, R. Beaujeux, B. Geny, and V. Wolff. 2018. 'A systematic review of the complex effects of cannabinoids on cerebral and peripheral circulation in animal models', *Front Physiol*, 9: 622.

Schloss, J., J. Lacey, J. Sinclair, A. Steel, M. Sughrue, D. Sibbritt, and C. Teo. 2021. 'A phase 2 randomised clinical trial assessing the tolerability of two different ratios of medicinal cannabis in patients with high hrade gliomas', *Front Oncol*, 11: 649555.

Seltzer, E. S., A. K. Watters, D. MacKenzie, Jr., L. M. Granat, and D. Zhang. 2020. 'Cannabidiol (CBD) as a promising anti-cancer drug', *Cancers (Basel)*, 12.

Shergalis, A., A. Bankhead, 3rd, U. Luesakul, N. Muangsin, and N. Neamati. 2018. 'Current challenges and opportunities in treating glioblastoma', *Pharmacol Rev*, 70: 412–445.

Shien, T., and H. Iwata. 2020. 'Adjuvant and neoadjuvant therapy for breast cancer', *Jpn J Clin Oncol*, 50: 225–229.

Siegel, R. L., K. D. Miller, H. E. Fuchs, and A. Jemal. 2021. 'Cancer statistics, 2021', *CA Cancer J Clin*, 71: 7–33.

Skoglund, G., M. Nockert, and B. Holst. 2013. 'Viking and early middle ages northern scandinavian textiles proven to be made with hemp', *Sci Rep*, 3: 2686.

Solimini, R., M. C. Rotolo, S. Pichini, and R. Pacifici. 2017. 'Neurological disorders in medical use of cannabis: An update', *CNS Neurol Disord Drug Targets*, 16: 527–533.

Sultan, A. S., M. A. Marie, and S. A. Sheweita. 2018. 'Novel mechanism of cannabidiol-induced apoptosis in breast cancer cell lines', *Breast*, 41: 34–41.

Sun, S., F. Hu, J. Wu, and S. Zhang. 2017. 'Cannabidiol attenuates OGD/R-induced damage by enhancing mitochondrial bioenergetics and modulating glucose metabolism via pentose-phosphate pathway in hippocampal neurons', *Redox Biol*, 11: 577–585.

Toh, C. K., F. Gao, W. T. Lim, S. S. Leong, K. W. Fong, S. P. Yap, A. A. Hsu, P. Eng, H. N. Koong, A. Thirugnanam, and E. H. Tan. 2007. 'Differences between small-cell lung cancer and non-small-cell lung cancer among tobacco smokers', *Lung Cancer*, 56: 161–166.

Twelves, C., M. Sabel, D. Checketts, S. Miller, B. Tayo, M. Jove, L. Brazil, S. C. Short, and Gwca study group. 2021. 'A phase 1b randomised, placebo-controlled trial of nabiximols cannabinoid oromucosal spray with temozolomide in patients with recurrent glioblastoma', *Br J Cancer*, 124: 1379–1387.

Velasco, G., S. Hernandez-Tiedra, D. Davila, and M. Lorente. 2016. 'The use of cannabinoids as anticancer agents', *Prog Neuropsychopharmacol Biol Psychiatry*, 64: 259–266.

Volmar, M. N. M., J. Cheng, H. Alenezi, S. Richter, A. Haug, Z. Hassan, M. Goldberg, Y. Li, M. Hou, C. Herold-Mende, C. L. Maire, K. Lamszus, C. Fluh, J. Held-Feindt, G. Gargiulo, G. J. Topping, F. Schilling, D. Saur, G. Schneider, M. Synowitz, J. A. Schick, R. E. Kalin, and R. Glass. 2021. 'Cannabidiol converts NF-kappaB into a tumor suppressor in glioblastoma with defined antioxidative properties', *Neuro Oncol*, 23: 1898–1910.

Xu, Z., X. A. Lv, Q. Dai, Y. Q. Ge, and J. Xu. 2016. 'Acute upregulation of neuronal mitochondrial type-1 cannabinoid receptor and it's role in metabolic defects and neuronal apoptosis after TBI', *Mol Brain*, 9: 75.

Yang, S., B. Hu, Z. Wang, C. Zhang, H. Jiao, Z. Mao, L. Wei, J. Jia, and J. Zhao. 2020. 'Cannabinoid CB1 receptor agonist ACEA alleviates brain ischemia/reperfusion injury via CB1-Drp1 pathway', *Cell Death Discov*, 6: 102.

Yuan, M., L. L. Huang, J. H. Chen, J. Wu, and Q. Xu. 2019. 'The emerging treatment landscape of targeted therapy in non-small-cell lung cancer', *Signal Transduct Target Ther*, 4: 61.

Zhao, X. Y., M. H. Lu, D. J. Yuan, D. E. Xu, P. P. Yao, W. L. Ji, H. Chen, W. L. Liu, C. X. Yan, Y. Y. Xia, S. Li, J. Tao, and Q. H. Ma. 2019. 'Mitochondrial dysfunction in neural injury', *Front Neurosci*, 13: 30.

Zong, W. X., J. D. Rabinowitz, and E. White. 2016. 'Mitochondria and cancer', *Mol Cell*, 61: 667–676.

13 Anti-Cancer Phytopharmaceuticals
A Green Approach Towards Treating Cancer

Priyanka Mudaliar, Apoorva Jagannath Nalawade, Priyamvada Arte, Akanksha Dhuliya Deepak Iyer, Kanchanlata Tungare, Mustansir Bhori, and Jyotirmoi Aich

CONTENTS

13.1 Introduction: Anti-Cancer Agents 239
13.2 Drug Resistance and Mechanism of Drug Resistance 240
13.3 Natural Herbal Compounds: Future Anti-Cancer Drugs 244
13.3.1 Phenolic Compounds and Their Derivatives 244
13.3.2 Vinca Alkaloid and Derivatives 245
13.3.3 Epipodophyllotoxin and Its Derivatives 245
13.3.4 Camptothecin and Its Derivatives 245
13.3.5 Flavonoids and Its Derivatives 246
13.3.6 Isothiocyanate and Its Derivatives 246
13.3.7 Triterpenoids and Its Derivatives 247
13.3.8 Carotenoid and Its Derivatives 247
13.3.9 Diindolylmethane/Indole-3-Carbinol 247
13.3.10 Epigallocatechin Gallate Catechin 248
13.3.11 Resveratrol and Stilbenoid 248
13.3.12 Vitamin D and Vitamin E 248
13.3.13 Taxanes 248
13.3.14 Indirubin 248
13.3.15 Cannabinoids 249
13.4 Development, Challenges, and Opportunities 249
13.5 Conclusion 250
Bibliography 250

13.1 INTRODUCTION: ANTI-CANCER AGENTS

Presently cancer is one of the world's most serious health problems, and its prevalence is expected to escalate in the future. Some substantial progress has been achieved in the treatment and management of cancer growth, but there are still significant flaws and space for improvements. Several of the chemotherapeutic medications used to treat this condition are toxic and have substantial adverse effects (Phillips & Currow, 2010). Furthermore, the development of resistance to currently available chemical therapeutic medications has necessitated the implementation of enhanced substitute treatment options involving novel agents, such as using plant derivatives. Plant-derived natural

DOI: 10.1201/b22917-13

anti-cancer medications have been shown to be more efficacious towards several malignancies with minimal adverse impacts. Nature has provided the basis for more than half of the anti-cancer medication compounds currently in clinical use (Fridlender et al., 2015). A lot of major marketed cancer treatments have been derived from natural sources, either through structural alteration of native compounds or by the synthesis of new compounds based on a natural component as a model. Exploration and harnessing nature, particularly for the discovery of new anti-cancer medicines, has seen a boom in the search for new and better pharmaceuticals (Habtemariam & Lentini, 2018). Plants continue to be a significant source of numerous clinically effective anti-cancer drugs currently in use across the world. A lot of plant-derived drugs have been identified and are in clinical studies (Iqbal et al., 2017). The current chapter encapsulates the most recent findings in cancer prevention and treatment using phytochemical constituents from plant sources, with a focus on plant-derived organic chemicals that have shown great potential as anti-cancer agents in recent years, as well as their potential mechanisms of actions and development challenges and opportunities (Table 13.1).

13.2 DRUG RESISTANCE AND MECHANISM OF DRUG RESISTANCE

Cancer therapy has been studied and developed for several decades where drug-based therapy such as chemotherapy has played a major role. Over the years, chemotherapy has presented certain side effects and challenges, which have encouraged the exploration of new compounds that could be more effective and beneficial for therapeutic purposes. One such "green" approach is the use of plant-derived phytochemicals. Because of their safety, low side effects, and the capacity to display chemo-preventive and anti-carcinogenic effects in a wide range of tumors via multistep targeting of signal transduction pathways, they may play an integral role in the future of cancer therapy (Ranjan et al., 2019; Iqbal et al., 2017; Fridlender et al., 2015). One of the major predicaments that arise with chemotherapy is the innate or acquired multidrug resistance (MDR) exhibited by cancer cells when they become tolerant to the administered chemotherapeutic drugs. Drug resistance of cancer cells can be attributed to a plethora of mechanisms. These include drug inactivation by upregulation of specific enzymes and detoxification of drugs, rendering them less potent. In chemoresistant cells, overexpression of ATP-binding cassette (ABC) drug transporters (p-gp) causes efflux of chemotherapeutics, limiting drug uptake. Genetic mutations or epigenetic changes in oncogenes and their expression may lead to alterations of drug targets, which affects drug metabolism, thus contributing to chemoresistance of cancer cells to former treatments. Downregulation or mutation of p53 (a tumor suppressor gene) contributes to inhibition of apoptosis. Consequently, mutations and epigenetic silencing in DNA repair systems enable these malignant cells to tolerate severe DNA damage while also showing evidence of improved DNA repair capabilities. Cancer cells adapt to changes in their microenvironment. Studies have indicated that high quantities of reactive oxygen species (ROS) are found in resistant cancer cells because of chemotherapy or radiotherapy, creating the optimal environment for tumor cell proliferation and DNA alteration. Currently, 90% of cases of chemotherapy failures arise during the invasion and metastasis of cancers caused by drug resistance (Mansoori et al., 2017; Cree & Charlton, 2017). These revelations suggest that by emphasizing these mechanisms, tumor cells can be sensitized to chemotherapy. Recent studies have indicated that use of plant-derived chemicals alone or in combination with chemotherapeutic drugs can potentially reverse MDR in cancer cells by targeting several such contributing pathways. Phenolic compounds like curcumin (CUR) and their derivatives, in combination with anti-neoplastic drugs, reverse MDR by regulating related signaling pathways, transcription factors, proteins, or genes in a wide range of cancer cells, e.g., reversal of doxorubicin and cisplatin resistance in colon cancer and breast cancer cells (Xu et al., 2020; Zhang et al., 2021; Xiao-ai et al., 2017). Flavonoids are cost-effective and multifunctional molecules which exert their sensitizing effects by inhibiting certain ABC transporters; suppressing Nrf2-dependent antioxidant enzymes; and regulating cancer stem

TABLE 13.1
Clinical Status and Mechanism of Action of Various Phytopharmaceuticals in Different Cancer Types

Compound Name	Kind of Compound	Cancer Type	Mechanism of Action	Clinical Status	Reference
Curcumin	Phenolic compound derivative, and a component of curcuminoids	Pancreatic, brain, head and neck, breast, colon, prostate and gastric cancer	It induces apoptosis by downregulating cellular signaling pathways consisting of transcription factors, activator proteins, cyclooxygenase II, NO synthase, etc.	Study completed (as drug: curcumin + gemcitabine for pancreatic cancer, as dietary supplement: curcumin for prostate cancer) and recruiting (as drug: placebo for breast cancer)	(Tomeh et al., 2019; Najafi et al., 2020) (ClinicalTrials.gov, NCT03769766) (ClinicalTrials.gov, NCT03980509) (ClinicalTrials.gov, NCT03211104) (ClinicalTrials.gov, NCT03211104)
Gingerol	Phenolic compound	Colorectal cancer, solid tumor, neuroblastoma, human myeloid leukemia, and breast cancer	Modulates cell signaling pathways NF-KB, COX-2, Nrf2, reduced expression of factors inducing apoptosis, etc.	Phase II, randomized trials, chemotherapeutic adjuvant	(Garcia-Oliveira et al., 2021)
Rosmarinic acid	Aqueous phenolic compound, rosemary extract derivatives	Colon, colorectal, breast, prostate, and gastric carcinoma	Prevention of tumor development and proliferation of the cell, arresting of the cell cycle, induces apoptosis, stimulates translocation of Nrf2, inhibits transcriptional (HIF-α) factor, increases multidrug resistance–associated protein 2 efflux, decreases levels of important processes involving COX-2 and TNF, modulates HDAC expression and expression of p65 protein	Under study	(Nadeem et al., 2019) (Wei & Chan, 2022)
Vinca alkaloids and derivatives	Terpenoid indole alkaloids, asymmetric dimeric compounds	Lymphoma, breast, glioma and colorectal cancer, tumors, and sarcoma	Inhibit the cell signaling pathways and (tubulin) polymerization of tumor cells along with destruction of mitotic spindle	Under combination clinical trials	(Garcia-Oliveira et al., 2021) (Lee et al., 2015)
Etoposide	Semi-synthetic derivative of epipodophyllotoxin	Stomach, ovary, Hodgkin's lymphoma, acute leukemia, neuroblastoma and sarcomas (soft tissues)	Inhibits DNA topoisomerase II by binding to it	Clinically approved (trade name: Toposar /VM-16) and in use commercially as a chemotherapeutic drug	(Zhang et al., 2018; D'yakonov et al., 2017)

(Continued)

TABLE 13.1 (Continued)

Compound Name	Kind of Compound	Cancer Type	Mechanism of Action	Clinical Status	Reference
Teniposide	Semi-synthetic derivative of epipodophyllotoxin	Breast, pancreatic, lung, stomach, and ovarian cancers	Inhibits DNA topoisomerase by binding to it	Clinically approved (trade name: Vumon/VM-26) and in use commercially as a chemotherapeutic drug	(Chakraborthy et al., 2021; de Jong et al., 1993; Ardalani et al., 2017)
Irinotecan	Camptothecin derivative	Metastatic colorectal cancer	Topoisomerase I inhibition leading to induction of cellular apoptosis; effects elevated in combination with chemotherapeutics	Phase III clinical trials, clinically and commercially approved	(Heinemann, 2014; Fujita et al., 2015)
Belotecan and topotecan	Camptothecin derivative	Ovarian cancer	Topoisomerase I inhibition leading to induction of cellular apoptosis	Phase II, randomized trials. clinically and commercially approve	(Kim et al., 2021)
Genistein	Soy-derived isoflavone	Bladder cancer, metastatic colorectal cancer	Inhibition of EGFR phosphorylation and induction of PI3K/Akt pathway	Phase I and II clinical trials	(Tuli et al., 2019; Javed et al., 2021)
2-Phenethyl isothiocyanate	Isothiocyanate with phenethyl group	Prevention of lung cancer	Inhibition of carcinogenesis by NNK via metabolic activation of cytochrome p450	Phase II clinical trials, chemopreventive	(Yuan et al., 2016) (University of Minnesota, 2017)
Ginsenoside G-Rg3	Triterpenoid saponin	Advanced non–small cell lung cancer, advanced esophageal cancer, gastric cancer, hepatocellular carcinoma	Induction of apoptosis, cell cycle arrest, autophagy, effects elevated in combination with chemotherapeutics	In clinical use, combination trials	(Yu et al., 2019)
Crocetin	Diterpenoid and natural carotenoid	Colon, hepatocellular carcinoma	Cell death via p53-dependent and -independent mechanisms, reducing the human telomerase reverse transcriptase	In clinical use; combination trials	(Li et al., 2012; Noureini & Wink, 2012)
Lycopene	Organic compound belonging to the isoprenoid series	Prostate cancer	It induces apoptosis in different cell lines by inhibiting or downregulating intracellular transcription factors	Phase II	(Dobrzyńska et al., 2019)

Diindolylmethane (DIM)	Diterpenoid	Breast cancer	Increased production of the BRCA1 protein in *BRCA1* mutation carriers	Phase I	(clinicaltrials.gov, NCT01022333)
Epigallocatechin-3-gallate	Ester of epigallocatechin and gallic acid	Lung cancer	Reducing grade II esophagitis	Phase II	(clinicaltrials.gov, NCT02577393)
Vitamin D	Fat-soluble secosteroids	Lymphoma, colorectal, breast cancer	Prevention of tumor development and proliferation of cells	Phase II	(Lappe et al., 2017)
Vitamin E	Plant-derived, lipid-soluble substance	Prostate	Inhibition of DNA topoisomerase II by binding to it	Phase III	(Klein et al., 2011) (Lippman et al., 2009)
Taxanes	A class of diterpenes	Breast, ovary, and lung cancers	Affect the normal activity of microtubules by acting as microtubulin-stabilizing agents	Clinically approved (paclitaxel and docetaxel) and commercially used as chemotherapeutic drugs	(Mosca et al., 2021)
Indirubin	A chemical component of Indigo naturalis	Acute promyelocytic leukemia	Inhibits cyclin-dependent kinases in tumor cells	Recruiting	(ClinicalTrials.gov, NCT02200978)
Cannabinoids	Chemicals compounds extracted from *Cannabis sativa*	Chemotherapy-induced peripheral neuropathy	Stimulates cannabinoid receptors	Clinically approved (dronabinol and nabilone) for managing chemotherapy-associated adverse effects	(H. Meng et al., 2020; Ward et al., 2021).

cells, hypoxia-inducible factors (HIFs), apoptosis, autophagy, etc. (Ye et al., 2019; Liskova et al., 2021). Ginkgo biloba extracts, triterpenoids, and vitamin E downregulate p-gp (efflux membrane transporter) expressions by decreasing efflux and increasing absorption of chemotherapeutic drugs. Similarly, in melanoma cells, cannabinoids encourage autophagy-dependent apoptosis, while ginseng and its derivatives in combination with docetaxel, doxorubicin, cisplatin, or paclitaxel have been demonstrated to enhance the chemosensitivity of prostate cancer and human colon cancer cells (Yan et al., 2014; Talib et al., 2021). Thus, abundant evidence is available supporting the fact that plant-derived chemicals and subsequent drugs can help cancer cells overcome chemotherapy-induced drug resistance.

13.3 NATURAL HERBAL COMPOUNDS: FUTURE ANTI-CANCER DRUGS

13.3.1 Phenolic Compounds and Their Derivatives

Compounds derived from both medicinal and dietary plants, consisting of aromatic rings bonded with at least one or several hydroxyl groups, are called phenolics. They have been researched for decades for their cancer-protective properties. One such compound is the curcuminoids, which are basically a ferulic acid derivative (a conjugated system of 4-hydroxy-3-methoxycinnamic acid bonded to a dihydridocarbon with a diketon-like structure) belonging to the Zingiberaceae family. It contains three major compounds, namely curcumin, demethoxycurcumin, and bis-demethoxycurcumin, imparting a distinctive yellow color to the turmeric plant (Huang et al., 2010). Curcumin is a compound obtained from the basal part of *Curcuma longa* L. It is recognized to be a safe, non-poisonous, and notable biologically active component, exhibiting a variety of biological actions including anti-neoplastic activity. It constrains cell proliferation and metastasis, along with improving the cessation of the cell cycle or apoptosis within numerous cancer-causing cells. It has shown potential therapeutic activity against miscellaneous types of malignancy like breast, prostate, colon, pancreatic, brain, head and neck, and gastric cancer (Mbese et al., 2019; Tomeh et al., 2019; Najafi et al., 2020).

Another phenolic compound with essential cancer-treating properties is gingerol. It can be extracted from roots as well as the rhizome of the ginger plant (*Zingiber officinale*). A clinical experiment involving the administration of gingerol as a supplement to cancer patients indicated that it may have benefited and aided in reducing recurrence of the condition. Gingerol's mechanism of action includes controlling certain important pathways of cells, including nuclear determinants (NF-KB), inflammatory mediators (COX-2), cellular resistance regulator (Nrf2), reduced expression of factors inducing apoptosis, etc. These actions have provided evidence of its usefulness against several cancers, namely colorectal cancer (decreases risk of cancer growth), solid tumor (improvement of cancer in patient receiving chemotherapy), neuroblastoma, human myeloid leukemia, and breast cancer (Garcia-Oliveira et al., 2021).

Rosmarinus officinalis L., commonly called rosemary, is an extensively used aromatic herb in the culinary world belonging to the Lamiaceae family. It has some beneficial medicinal properties which have been rigorously researched for many years. The phytoconstituents present in its extracts have shown to possess anti-cancer properties. Rosemary extracts are thought to be cytostatic against miscellaneous types of malignancy, including breast, bladder, liver, pancreatic, lung, colon, bladder, ovarian, and prostate cancer. Its phytochemicals have abundant antioxidants, making them apt for further research as cytotoxic agents (Wei & Chan, 2022). Such polyphenols, like rosmarinic acid and hydroxycinnamic acid, are found to exhibit superior antioxidant activity. It acts as a cytostatic agent against several tumor malignancies. The mechanism of action includes preventing cell proliferation and tumor development, reducing peroxidation of lipids, arresting of the cell cycle, inducing apoptosis, stimulating translocation of Nrf2 to the nucleus from cytoplasm, inhibiting transcriptional HIF-α factor, increasing MDR-associated protein 2 efflux, decreasing levels of important processes involving COX-2 and tumor necrosis factor (TNF), and modulating histone deacetylases (HDACs) and p65 expression. A growing body of evidence has shown the anti-cancer effect of rosmarinic acid against several cancer lines, including colon, colorectal, breast, prostate, and gastric carcinoma (Nadeem et al., 2019).

13.3.2 Vinca Alkaloid and Derivatives

A major focus of plant-based drug research is the development of diverse anti-carcinogenic medicines from alkaloids (five-membered aromatic rings attached by a nitrogen atom) and their derivatives. One such terpenoid, i.e., indole alkaloids naturally obtained from the aerial (leaves) part of the plant belonging to the family of Apnaocyceae, is called vinca alkaloids. They are the primary alkaloids applied for making anti-neoplastic therapeutics. At present, numerous of them are in use for cancer management, namely vinblastine and vincristine, and semi-synthetic derivatives like vinorelbine, vindesine, or vinflunine are under study. These compounds are known to be small molecules that interact with the protein and evade its polymer formation into small tubules, ultimately inhibiting the cell division process. Therefore, the cells are in a ceased phase leading to expiration (Garcia-Oliveira et al., 2021). Vincristine is a natural alkaloid that has been exploited for years as an anti-carcinogenic agent against numerous malignancies (breast, colorectal, glioma, different tumors). Vinblastine, a mitotic inhibitor, has been utilized as a clinical treatment remedy for leukemia, breast, testicular, and lung cancers. It confines the lump growth along with malignant development. Vinorelbine, a semi-synthetic vinca alkaloid, has a broad antineoplastic spectrum activity, as reported in breast and lung cancer. It has less neurotoxicity as compared to the other two mentioned alkaloids. Vindesine has shown similar implications as that of vinblastine. Vinflunine is currently under clinical evaluation. It is likely to be an effective remedy for curing cancer due to its high anti-carcinogenic activity and low neurotoxicity. Vinca alkaloids, although effective against cancer, show other types of toxic reactions or side effects that have made researchers construct various drug delivery systems for the same. Combination therapy is also under evaluation for decreasing such side effects shown by vinca alkaloids (Lee et al., 2015).

13.3.3 Epipodophyllotoxin and Its Derivatives

Epipodophyllotoxin is a natural substance extracted from *Podophyllum peltatum* (American mayapple) plant root. It has shown antineoplastic potential along with its semi-synthetic derivatives, namely teniposide (VP~26) and etoposide (VP~16). Etoposide specifically inhibits topoisomerase II in humans by interacting and forming an enzyme-drug complex. Due to this interaction, there is a strong inhibition of numerous essential cell cycle processes such as transcription, replication, and DNA repair resulting in injury or death of the cell. For several years, etoposide has been used for therapy against malignancies like stomach, ovary, Hodgkin's lymphoma, acute leukemia, neuroblastoma, and sarcomas (soft tissues) (D'yakonov et al., 2017). Teniposide is a D-glucose derivative of podophyllotoxin glycoside. (Shah et al., 2021). It has significant bioactivity, primarily as anti-viral and anti-cancer medications. Its anti-cancer activity has been investigated in several clinical trials with beneficial effect in treating a multitude of cancers, such as testicular, breast, lung, stomach, pancreatic, and ovarian cancers (Chakraborthy et al., 2021; Meng et al., 2021). Teniposide has been reported to suppress the growth of cancer cells by inhibiting DNA topoisomerase II (de Jong et al., 1993; Ardalani et al., 2017).

13.3.4 Camptothecin and Its Derivatives

Camptothecin is a quinolone alkaloid isolated from the stem wood of the Chinese tree, *Camptotheca acuminata.* It is a wide-spectrum anti-tumor agent that exerts its effects by inhibiting nuclear enzyme DNA topoisomerase I, which prevents resealing of DNA after synthesis, thereby halting cell development and inducing apoptosis in cancer cells during the S phase of cell cycle (Khaiwa et al., 2021). Camptothecin analogs are used in the treatment of an array of solid tumors. Irinotecan, belotecan, and topotecan are camptothecin derivatives that are clinically licensed for anti-cancer treatment. Irinotecan is widely used for advanced colon and metastatic colorectal cancer. Enhanced anti-cancer effects are observed when combined with oxaliplatin and 5-fluorouracil (Heinemann, ClinicalTrials.gov, NCT00433927) (Fujita et al., 2015). SN38, an irinotecan active metabolite with 1000 times the activity of the parent molecule, is effective in treating extremely metastatic cancer cells. SN38-loaded liposomes have proven to extend survival of patients with metastatic pancreatic cancer without compromising their quality of life (Bailly, 2019). Topotecan and belotecan are most effective against pre-treated and untreated ovarian cancer (Kim et al., 2021). To improve their bioavailability, bioaccumulation, and efficacy, development of novel camptothecin-based liposomal nano-delivery systems is currently being pursued (Wen et al., 2017).

13.3.5 Flavonoids and Its Derivatives

Flavonoids are versatile polyphenolic compounds, which are abundantly found in various plants. They exert anti-carcinogenic activities by influencing the activity of ROS-scavenging enzymes, participating in cell cycle arrest, inducing autophagy and apoptosis, and reducing cancer cell proliferation and invasiveness. Apigenin is a trihydroxy flavone, largely found in parsley. It promotes apoptosis in ovarian and cervical cancer and produces necroptosis and mitochondrial malfunction, as well as adenosine triphosphate (ATP) exhaustion in malignant mesothelioma cells via ROS signaling (Lee et al., 2020; Kopustinskiene et al., 2020). Cyanidin is an anthocyanidin found in red-purple pigmented fruits and vegetables like berries, grapes, apples, red onion, cabbage, etc. Cyanidin-3-glucoside (C3G) has very low toxicity and has been demonstrated to decrease tumor growth in in vitro and in vivo models (Baster et al., 2020; Cho et al., 2017). Genistein is a soy-derived isoflavone and phytoestrogen. Its ability to target caspases, Bcl-2 regulator proteins, transcription factors, and protein kinases and inhibit Wnt/β-catenin, NF-κB, and PI3K/Akt signaling pathways makes it a potent anti-cancer compound (Spagnuolo et al., 2015). It is the most widely studied and clinically tried flavonoid with successful clinical trials for colorectal, prostate, and bladder cancer (Tuli et al., 2019; Javed et al., 2021). Like genistein, fisetin is a bioactive plant flavanol with pleiotropic pharmacological activities whose toxicity and interaction with chemotherapeutic agents are being explored in clinical settings (Imran et al., 2021). The extract from ginkgo biloba leaves consists of several medicinal bioactive compounds, including flavonoids like kaempferol that have been studied in a variety of cancer cell lines and animal models (Imran et al., 2019).

13.3.6 Isothiocyanate and Its Derivatives

Isothiocyanates are hydrolytic products of secondary metabolites called glucosinolates. Phenethyl isothiocyanate (PEITC) and sulforaphane (SFN) are two such isothiocyanates that are found at high concentrations in cruciferous vegetables like broccoli, kale, radish, and watercress. They are nontoxic with minimal side effects and have been extensively studied for skin, breast, lung, colon, and prostate cancers (Ranjan et al., 2019). In vitro studies have suggested that PEITC and SFN exhibit chemo-preventive and anti-carcinogenic effects via stable epigenetic modifications. SNF upregulates cytoprotective genes by activating a transcription factor called nuclear factor erythroid-2-(NF-E2-) related factor 2 (Nrf2), whereas both PEITC and SNF restrict tumor progression by targeting DNA methyltransferases (DNMTs), polycomb-group (PcG) proteins, HDACs, and non-coding RNAs, thus ultimately suppressing tumor progression (Su et al., 2018; Park et al., 2017; Hudlikar et al., 2021). SNF can potentiate the effects of chemotherapeutics like doxorubicin and cisplatin while mitigating their cytotoxic effects (Calcabrini et al., 2020). Clinical trials using 2-PEITC on rat and mice models provided the basis for carrying out phase II clinical trials on humans with a history of smoking, which indicated that 2-PEITC successfully inhibited lung tumor formation by tobacco carcinogenic agent 4-(methylnitrosamino)-1-(3-pyridyl)-1-butanone (NNK) through metabolic activation of cytochrome p450 enzymes (Yuan et al., 2016; Clinicaltrials.gov, NCT00691132).

13.3.7 Triterpenoids and Its Derivatives

Triterpenoids represent the largest group of phytochemicals, with more than 20,000 terpenoids existing in nature. Triterpenes are oxygen derivatives of triterpenoids, and triterpenes that belong to saponin compounds are known as triterpenoid saponins. Ginsenosides are triterpenoid saponins found in the roots of *Panax ginseng* (Garg et al., 2020). Ginsenosides have exhibited anti-cancer and anti-metastatic properties in a wide array of cancers, including breast, brain, liver, gastric, and lung cancers, by inhibiting cancer cell invasion, migration, proliferation, and survival, as well as promoting apoptosis, cell cycle arrest, and autophagy. Furthermore, numerous ginsenoside derivatives have been discovered, including G-Rg1, which exhibits hepatoprotective properties in liver cancer cells; G-Rg2, which induces autophagy in breast cancer cells; and G-Rg3, which improves sensitivity and suppresses resistance to several chemotherapeutic drugs along with anti-carcinogenic properties. Ginsenosides, when combined with other chemotherapeutics, enhanced efficacy in patients with non–small cell lung cancer, colon cancer, esophageal cancer, and gastric cancer, according to clinical trials. However, despite positive in vitro results, the majority of current research has been done on cell lines and animal models, with only a few human trials being conducted (Hong et al., 2021; Yu et al., 2019).

13.3.8 Carotenoid and Its Derivatives

Carotenoids, also known as tetraterpenoids, are biological pigments (lemon, magenta, and crimson) produced by plants, plankton, microbes, and fungus. Carotenoids are polyunsaturated organic compounds chains with 9–11 double bonds and perhaps terminated in rings (Koklesova et al., 2020). They have been reported to limit tumor cell proliferation by intervening with various stages of the cell cycle. β-carotene showed anti-cancer effects against acute myeloid leukemia among the carotenoids (Niranjana et al., 2015). Crocetin slows tumor cell proliferation by decreasing nucleic acid production, boosting the immune system, triggering cell death, and blocking transcription factor signaling pathways (Gutheil et al., 2012). Crocetin is currently under many clinical trials for colon and hepatocellular carcinoma (Li et al., 2012; Noureini & Wink, 2012). Lycopene not only lowers the likelihood of cancer in several organs but also slows tumor growth and possesses chemoprotective benefits against diseases like arthritis and sexual dysfunction. Lycopene is currently under clinical trials for prostate cancer (Dobrzyńska et al., 2019).

13.3.9 Diindolylmethane/Indole-3-Carbinol

Diindolylmethane (DIM) is a repercussion of the breakdown of indole-3-carbinol (I3C), which is present in leafy greens like spinach, lettuce, and collards (Gong et al., 2006). DIM has been demonstrated to have some disease-preventative benefits and certain anti-tumor characteristics. DIM also reduces IB-phosphorylation and inhibits beclin-2, X-linked apoptosis, and nuclear factor kappa B in CNE-2 cells in vitro in a dose-dependent manner (Xu et al., 2013). In recent years DIM and I3C have caught the attention of researchers because of their inhibitory impact on breast cancer (Ashok et al., 2001; ClinicalTrials.gov, NCT01022333).

13.3.10 Epigallocatechin Gallate Catechin

Epigallocatechin gallate (EGCG) is a kind of epicatechin mainly found in tea such as herbal, green and black tea (Chen et al., 2020). The most prevalent polyphenol in tea is EGCG, which has the potential to alter human health and illness. In vivo and in vitro studies have shown that EGCG can induce cell death and decrease tumor progression in a variety of malignancies, such as colorectal, renal, breast, and brain tumors, as well as myeloma and lung cancer (Yang et al., 2009; Chen et al., 2020). Epigallocatechin is in a phase I clinical trial on lung cancer (ClinicalTrials.gov, NCT02577393).

13.3.11 Resveratrol and Stilbenoid

Resveratrol and stilbenoid belong to the phenolic family of chemicals and possess a C6–C2–C6 structure. It is found in the skin of grapes, as well as nuts, red wine, fruits, and other foods and berries. Recently, it has been found that stilbenoid supports tumor cell death and growth arrest, implying that this phytochemical could be a superior therapy option for breast cancer (Salehi et al., 2018). Resveratrol has been proven to kill numerous types of cancer cells, including stomach, colorectal, epidermal, mammary, and rectal cancer cells, in in vitro and in vivo studies (Subedi et al., 2019).

13.3.12 Vitamin D and Vitamin E

Vitamin D is among the essential components required by our body and is responsible for absorbing calcium, phosphorous, magnesium, and other biological components. On other hand, vitamin E is fat-soluble and it is important for eliminating free electrons that can cause cellular injury (Stampfer et al., 1993). Vitamin D and vitamin E both act as a promising anti-cancer agent, as they possess an anti-tumor effect by inhibiting cancer cell proliferation and causing apoptosis of cancerous cells. A plethora of reports have demonstrated the anti-cancer activity of vitamins D and E. They have been used to treat a variety of malignancies, with the most significant results in colorectal, breast, and prostate cancers (Mizoue et al., 2008; Chakraborti, 2011; ClinicalTrials.gov, NCT01052051).

13.3.13 Taxanes

Taxanes are diterpenoids (a kind of diterpene) extracted from the *Taxus brevifolia* species (Mosca et al., 2021). Taxanes such as paclitaxel and docetaxel, also known by their brand names Taxol and Taxotere, respectively, are widely regarded as important chemotherapy drugs, particularly in gynecological cancers. Both medications work as anti-microtubule agents by encouraging tubulin polymerization into very stable intracellular microtubules, causing mitosis and normal cell division to be disrupted, and eventually cell death (Sibaud et al., 2016). Despite their side effects and low solubilities, taxanes are currently the first line of therapy for metastatic malignancies of the breast, ovary, lung, and other organs, according to several ongoing clinical and research investigations (Fauzee et al., 2011; Sisodiya, 2013).

13.3.14 Indirubin

Indirubin is a chemical component of qing dai (*Indigo naturalis*) extracted from the *Isatis* root and leaf, which is utilized in combination formulations and herbal medicinal remedies to treat chronic ailments (Li et al., 2018). It is a constitutional isomer of the indigo dye. It is prepared through accumulating waste products from the bacterial decomposition of particular types of plants. Indirubin and its derivatives have been reported to be strong cyclin-dependent kinase (CDK) inhibitors, eliciting G2/M arrest and apoptosis in cancer cells by impeding ATP binding at the active center (Chen et al., 2018). Indirubin has not been reported to prevent or manage cancer in humans, although it is being tested in vitro and in vivo as a treatment for small cell lung cancer, chronic myeloid leukemia, and other cancers, alone or in combination with more traditional therapies (Williams et al., 2011; Kim et al., 2011). In addition, a clinical trial in children with acute promyelocytic leukemia (APL) has begun to assess the efficacy and quality of indirubin in association with other anti-cancer medications (ClinicalTrials.gov, NCT02200978).

13.3.15 Cannabinoids

Cannabinoids are chemical compounds (meroterpenes) contained in the *Cannabis sativa* plant, which has been utilized for ages as an herbal treatment (Sledziński et al., 2018). Cannabinoids are beneficial for a variety of cancer-related symptoms, including nausea, pain, and sleeplessness, induced due to chemotherapy, to name a few, and large, randomized, controlled clinical studies are being conducted to establish their benefits and risks in cancer patients (ClinicalTrials.gov NCT01489826, NCT03984214). Cannabis' analgesic effects may be mediated via inhibition of presynaptic neurotransmitter and neuropeptide release, modulation of postsynaptic neurotransmission, stimulation of the downstream inhibitory pain pathway, and decreased inflammatory response of neurons (Starowicz & Finn, 2017). Cannabinoids such as dronabinol and nabilone, which are commercially accessible, are licensed medications utilized to manage cancer-related adverse effects (Meng et al., 2020; Ward et al., 2021).

13.4 DEVELOPMENT, CHALLENGES, AND OPPORTUNITIES

Phytopharmaceuticals have played an essential part in cancer therapy, and plant sources have produced some of the most prospective and superior medications. Before their anti-cancer treatments may be approved for clinical use, they must have considerable proof of efficacy acquired from sufficient clinical trials (Chakraborthy et al., 2021). Even though the compounds discussed here have considerable therapeutic potential, they also have practical limits that must be conquered before further medical evaluation, in the instance of phytochemicals in the clinical stage, or to improve the qualities of chemotherapy, in the case of drugs already on the market. Compounds such as podophyllotoxin derivatives, for example, had restricted applications due to adverse effects (Choudhari et al., 2020a). Vinka alkaloids have a minor impact in some circumstances and are used in conjunction with other drugs (Sisodiya, 2013). Poor solubility in water, poor uptake into recipient cells, uptake into healthy cells, and restricted remedial value, as well as severe negative consequences, have all been important considerations to keep in mind while using phytochemical substances to treat cancer. Apart from these difficulties, the identification and development of potential phytopharmaceuticals for clinical use must also overcome challenges associated with anti-cancer compound extraction, production, refinement, and characterization (Garcia-Oliveira et al., 2021). Given the growing worldwide popularity of phytopharmaceuticals, their production is fraught with challenges. In particular, the herbal medicine industry's foundation has not been adequately considered. The toxicity of the drug has poor quality assurance and medical observation. There seems to be no global standard specifying the process and dosage of these medications, nor is there any assessment of their side effects. Furthermore, people have no understanding of the dangers of self-medication, and the probable combinations of phytopharmaceuticals with conventional pharmaceuticals is lacking, and the appropriate and regular use of herbal medicines has not been prioritized. These are only a few of the issues that could influence herbal medicine's health-care impacts (Mirzaeian et al., 2021).

The accessibility of traditional medicinal records, which allows researchers to specify the wide range of different findings towards the most desirable ones, is among the essential benefits of phytopharmaceutical research (Choudhari et al., 2020). To fully exploit the potential of phytochemicals, a novel approach to integrated drug discovery is required in which traditional medicinal knowledge is endorsed by complex and multifaceted forces that include ethnomedical chemistry, etc. In addition, developments in analytical technology and computational approaches will make it easier to find new bioactive key agents for pharmaceutical testing.

13.5 CONCLUSION

The development of novel medications that are effective against resistant cancers is a critical method for enhancing chemotherapy. Medicinal herbs are a great place to look for new pharmacological active ingredients. Plant-based medications are making a resurgence, as evidenced by the growing number of compounds in clinical trials, mostly for cancer therapy. Many innovative plant-based therapeutic candidates are currently undergoing preclinical testing. Since many medicinal plants have yet to be investigated and can provide a variety of chemicals of various structural kinds and pharmacological potency in the design of novel medications, more research is needed in the discovery for effective anti-cancer agents. The current chapter encapsulates the most recent findings in cancer prevention and treatment using phytochemical constituents from plant sources, with a focus on plant-derived organic chemicals that have shown great potential as anti-cancer agents in recent years, as well as their potential mechanisms of actions, development challenges, and opportunities.

BIBLIOGRAPHY

Ardalani, H., Avan, A., & Ghayour-Mobarhan, M. (2017). Podophyllotoxin: A novel potential natural anti-cancer agent. *Avicenna Journal of Phytomedicine* 7(4), 285–294.

Ashok, B. T., Chen, Y., Liu, X., Bradlow, H. L., Mittelman, A., & Tiwari, R. K. (2001). Abrogation of estrogen-mediated cellular and biochemical effects by indole-3-carbinol. *Nutrition and Cancer* 41(1–2), 180–187.

Bailly, C. (2019). Irinotecan: 25 years of cancer treatment. *Pharmacological Research* 148, 104398.

Baster, Z., Li, L., Kukkurainen, S., Chen, J., Pentikäinen, O., Győrffy, B., Hytönen, V. P., Zhu, H., Rajfur, Z., & Huang, C. (2020). Cyanidin-3-glucoside binds to talin and modulates colon cancer cell adhesions and 3D growth. *The FASEB Journal* 34(2), 2227–2237.

Calcabrini, C., Maffei, F., Turrini, E., & Fimognari, C. (2020). Sulforaphane potentiates anti-cancer effects of doxorubicin and cisplatin and mitigates their toxic effects. *Frontiers in Pharmacology* 11, 567.

Chakraborthy, G. S., Ukirde, R., Nerkar, A. G., Chakraborthy, G. S., & Ukirde, R. D. (2021). Current trends in pharmacy and pharmaceutical chemistry anti-cancer agents from natural sources: A review. *Current Trends in Pharmacy and Pharmaceutical Chemistry* 3(4), 38–46.

Chakraborti, C. (2011). Vitamin D as a promising anti-cancer agent. *Indian Journal of Pharmacology* 43(2), 113.

Chen, B. H., Hsieh, C. H., Tsai, S. Y., Wang, C. Y., & Wang, C. C. (2020). Anti-cancer effects of epigallocatechin-3-gallate nanoemulsion on lung cancer cells through the activation of AMP-activated protein kinase signaling pathway. *Scientific Reports* 10(1), 1–11.

Chen, L., Wang, J., Wu, J., Zheng, Q., & Hu, J. (2018). Indirubin suppresses ovarian cancer cell viabilities through the STAT3 signaling pathway. *Drug Design, Development and Therapy* 12, 3335.

Cho, E., Chung, E. Y., Jang, H. Y., Hong, O. Y., Chae, H. S., Jeong, Y. J., Kim, S. Y., Kim, B. S., Yoo, D. J., Kim, J. S., & Park, K. H. (2017). Anti-cancer effect of cyanidin-3-glucoside from mulberry via caspase-3 cleavage and DNA fragmentation *in vitro* and *in vivo*. *Anti-Cancer Agents in Medicinal Chemistry* 17(11), 1519–1525.

Choudhari, A. S., Mandave, P. C., Deshpande, M., Ranjekar, P., & Prakash, O. (2020). Corrigendum: Phytochemicals in cancer treatment: From preclinical studies to clinical practice. *Frontiers in Pharmacology* 11, 175.

Clinicaltrials.gov. National Library of Medicine (US). University of Minnesota. (5 June 2008–12 May, 2017). Randomized Trial of PEITC as a Modifier of NNK Metabolism in Smokers. Identifier: NCT00691132. https://clinicaltrials.gov/ct2/show/NCT00691132.

ClinicalTrials.gov. National Library of Medicine (US). (June 2009–August 2015). Clinical Trial of Vitamin D3 to Reduce Cancer Risk in Postmenopausal Women (CAPS). Identifier: NCT01052051. https://clinicaltrials.gov/ct2/show/NCT01052051.

ClinicalTrials.gov. National Library of Medicine (US). (December 1, 2009-Present). The Potential for Oral Diindolylmethane (DIM) Supplementation to Increase the Production of the BRCA1 Protein in BRCA1 Mutation Carriers. Identifier: NCT01022333. https://clinicaltrials.gov/ct2/show/NCT01022333.

ClinicalTrials.gov. National Library of Medicine (U.S). (September 2011-). A Study for Improving the Outcome of Childhood Acute Promyeloid Leukemia. Identifier: NCT02200978. https://clinicaltrials.gov/ct2/show/NCT02200978.

ClinicalTrials.gov. National Library of Medicine (U.S). (January 2012–July 2015). A Phase 1 Study of Dexanabinol in Patients with Advanced Solid Tumours. Identifier: NCT01489826. https://clinicaltrials.gov/ct2/show/NCT01489826.

ClinicalTrials.gov. National Library of Medicine (US). (November 20, 2019–January 17, 2020). Study of Epigallocatechin-3-Gallate (EGCG) for Esophagus Protection in Patients with Lung Cancer Receiving Radial Radiotherapy. Identifier: NCT02577393. https://clinicaltrials.gov/ct2/show/NCT02577393.

ClinicalTrials.gov. National Library of Medicine (U.S). (December 16, 2019-Present). Efficacy and Safety of Dronabinol in the Improvement of Chemotherapy-Induced and Tumor-Related Symptoms in Advanced Pancreatic Cancer. Identifier: NCT03984214. https://clinicaltrials.gov/ct2/show/NCT03984214.

Cree, I. A., & Charlton, P. (2017). Molecular chess? Hallmarks of anti-cancer drug resistance. *BMC Cancer* 17(1), 1–8.

De Jong, S., Kooistra, A. J., de Vries, E. G. E., Mulder, N. H., & Zijlstra, J. G. (1993). Topoisomerase II as a target of VM-26 and 4f-(9-acridinylamino) methanesulfon-m-aniside in atypical multidrug resistant human small cell lung carcinoma cells. *Cancer Research* 53(5), 1064–1071.

Dobrzyńska, M. M., Gajowik, A., & Radzikowska, J. (2019). The effect of lycopene supplementation on radiation-induced micronuclei in mice reticulocytes in vivo. *Radiation and Environmental Biophysics* 58(3), 425–432.

D'yakonov, V. A., Dzhemileva, L. U., & Dzhemilev, U. M. (2017). Chapter 2—Advances in the Chemistry of Natural and Semisynthetic Topoisomerase I/II Inhibitors. In Atta-ur-Rahman (Ed.), *Studies in Natural Products Chemistry* (Vol. 54, pp. 21–86).

Fauzee, N. J., Dong, Z., & Wang, Y. L. (2011). Taxanes: Promising anti-cancer drugs. *Asian Pac J Cancer Prev* 12(4), 837–851.

Fridlender, M., Kapulnik, Y., & Koltai, H. (2015). Plant derived substances with anti-cancer activity: From folklore to practice. *Frontiers in Plant Science* 6, 799.

Fujita, K., Kubota, Y., Ishida, H., & Sasaki, Y. (2015). Irinotecan, a key chemotherapeutic drug for metastatic colorectal cancer. *World Journal of Gastroenterology* 21(43), 12234–12248.

Garcia-Oliveira, P., Otero, P., Pereira, A. G., Chamorro, F., Carpena, M., Echave, J., Fraga-Corral, M., Simal-Gandara, J., & Prieto, M. A. (2021). Status and challenges of plant-anti-cancer compounds in cancer treatment. *Pharmaceuticals* 14(2), 1–28.

Garg, A., Sharma, R., Dey, P., Kundu, A., Kim, H. S., Bhakta, T., & Kumar, A. (2020). Analysis of triterpenes and triterpenoids. *Elsevier-Recent Advances in Natural Products Analysis* 393–426.

Gong, Y., Sohn, H., Xue, L., Firestone, G. L., & Bjeldanes, L. F. (2006). 3,3'-diindolylmethane is a novel mitochondrial H+-ATP synthase inhibitor that can induce p21Cip1/Waf1 expression by induction of oxidative stress in human breast cancer cells. *Cancer Research* 66(9), 4880–4887.

Gutheil, G. W., Reed, G., Ray, A., Anant, S., & Dhar, A. (2012). Crocetin: An agent derived from saffron for prevention and therapy for cancer. *Current Pharmaceutical Biotechnology* 13(1), 173.

Habtemariam, S., & Lentini, G. (2018). Plant-derived anti-cancer agents: Lessons from the pharmacology of geniposide and its aglycone, genipin. *Biomedicines* 6(2), 39.

Heinemann, V., von Weikersthal, L. F., Decker, T., Kiani, A., Vehling-Kaiser, U., Al-Batran, S.-E., Heintges, T., Lerchenmüller, C., Kahl, C., Seipelt, G., Kullmann, F., Stauch, M., Scheithauer, W., Hielscher, J., Scholz, M., Müller, S., Link, H., Niederle, N., Rost, A., … Stintzing, S. (2014). FOLFIRI plus cetuximab versus FOLFIRI plus bevacizumab as first-line treatment for patients with metastatic colorectal cancer (FIRE-3): A randomised, open-label, phase 3 trial. *The Lancet. Oncology*, 15(10), 1065–1075.

Hong, H., Baatar, D., & Hwang, S. G. (2021). Anti-cancer activities of ginsenosides, the main active components of ginseng. *Evidence-Based Complementary and Alternative Medicine: ECAM*, 8858006.

Huang, W. Y., Cai, Y. Z., & Zhang, Y. (2010). Natural phenolic compounds from medicinal herbs and dietary plants: Potential use for cancer prevention. *Nutrition and Cancer* 62(1), 1–20.

Hudlikar, R., Wang, L., Wu, R., Li, S., Peter, R., Shannar, A., Chou, P. J., Liu, X., Liu, Z., Kuo, H.-C. D., & Kong, A.-N. (2021). Epigenetics/epigenomics and prevention of early stages of cancer by isothiocyanates. *Cancer Prevention Research* 14(2), 151–164.

Imran, M., Saeed, F., Gilani, S. A., Shariati, M. A., Imran, A., Afzaal, M., Atif, M., Tufail, T., & Anjum, F. M. (2021). Fisetin: An anti-cancer perspective. *Food Science & Nutrition* 9(1), 3–16.

Imran, M., Salehi, B., Sharifi-Rad, J., Aslam Gondal, T., Saeed, F., Imran, A., Shahbaz, M., Tsouh Fokou, P. V., Umair Arshad, M., & Khan, H. (2019). Kaempferol: A key emphasis to its anti-cancer potential. *Molecules* 24(12), 2277.

Iqbal, J., Abbasi, B. A., Mahmood, T., Kanwal, S., Ali, B., Shah, S. A., & Khalil, A. T. (2017). Plant-derived anti-cancer agents: A green anti-cancer approach. *Asian Pacific Journal of Tropical Biomedicine* 7(12), 1129–1150.

Javed, Z., Khan, K., Herrera-Bravo, J., Naeem, S., Iqbal, M. J., Sadia, H., Qadri, Q. R., Raza, S., Irshad, A., Akbar, A., Reiner, Ž., Al-Harrasi, A., Al-Rawahi, A., Satmbekova, D., Butnariu, M., Bagiu, I. C., Bagiu, R. V., & Sharifi-Rad, J. (2021). Genistein as a regulator of signaling pathways and microRNAs in different types of cancers. *Cancer Cell International* 21(1), 388.

Khaiwa, N., Maarouf, N. R., Darwish, M. H., Alhamad, D. W. M., Sebastian, A., Hamad, M., Omar, H. A., Orive, G., & Al-Tel, T. H. (2021). Camptothecin's journey from discovery to WHO Essential Medicine: Fifty years of promise. *European Journal of Medicinal Chemistry* 223, 113639.

Kim, H. S., Park, S.-Y., Park, C.-Y., Kim, Y. T., Kim, B.-J., Song, Y. J., Kim, B.-G., Kim, Y. B., Cho, C.-H., Kim, J.-H., & Song, Y. S. (2021). A multicentre, randomised, open-label, parallel-group Phase 2b study of belotecan versus topotecan for recurrent ovarian cancer. *British Journal of Cancer* 124(2), 375–382.

Kim, S. A., Kwon, S. M., Kim, J. A., Kang, K. W., Yoon, J. H., & Ahn, S. G. (2011). 5'-nitro-indirubinoxime, an indirubin derivative, suppresses metastatic ability of human head and neck cancer cells through the inhibition of Integrin β1/FAK/Akt signaling. *Cancer Letters* 306(2), 197–204.

Klein, E. A., Thompson, I. M., Tangen, C. M., Crowley, J. J., Lucia, M. S., Goodman, P. J., Minasian, L. M., Ford, L. G., Parnes, H. L., Gaziano, J. M., Karp, D. D., Lieber, M. M., Walther, P. J., Klotz, L., Parsons, J. K., Chin, J. L., Darke, A. K., Lippman, S. M., Goodman, G. E., ... Baker, L. H. (2011). Vitamin E and the risk of prostate cancer: The Selenium and Vitamin E Cancer Prevention Trial (SELECT). *JAMA*, 306(14), 1549–1556.

Koklesova, L., Liskova, A., Samec, M., Buhrmann, C., Samuel, S. M., Varghese, E., Ashrafizadeh, M., Najafi, M., Shakibaei, M., Büsselberg, D., Giordano, F. A., Golubnitschaja, O., & Kubatka, P. (2020). Carotenoids in cancer apoptosis: The road from bench to bedside and back. *Cancers* 12(9), 1–41.

Kopustinskiene, D. M., Jakstas, V., Savickas, A., & Bernatoniene, J. (2020). Flavonoids as anti-cancer agents. *Nutrients* 12(2), 457.

Lappe, J., Watson, P., Travers-Gustafson, D., Recker, R., Garland, C., Gorham, E., Baggerly, K., & McDonnell, S. L. (2017). Effect of Vitamin D and Calcium Supplementation on Cancer Incidence in Older Women: A Randomized Clinical Trial. *JAMA*, 317(12), 1234–1243.

Lee, C. T., Huang, Y. W., Yang, C. H., & Huang, K. S. (2015). Drug delivery systems and combination therapy by using vinca alkaloids. *Current Topics in Medicinal Chemistry* 15(15), 1491–1500.

Lee, Y.-J., Park, K.-S., Nam, H.-S., Cho, M.-K., & Lee, S.-H. (2020). Apigenin causes necroptosis by inducing ROS accumulation, mitochondrial dysfunction, and ATP depletion in malignant mesothelioma cells. *The Korean Journal of Physiology & Pharmacology: Official Journal of the Korean Physiological Society and the Korean Society of Pharmacology* 24(6), 493–502.

Li, C. Y., Huang, W. F., Wang, Q. L., Wang, F., Cai, E., Hu, B., Du, J. C., Wang, J., Chen, R., Cai, X. J., Feng, J., & Li, H. H. (2012). Crocetin induces cytotoxicity in colon cancer cells via p53-independent mechanisms. *Asian Pacific Journal of Cancer Prevention* 13(8), 3757–3761.

Li, Z., Zhu, C., An, B., Chen, Y., He, X., Qian, L., Lan, L., & Li, S. (2018). Indirubin inhibits cell proliferation, migration, invasion and angiogenesis in tumor-derived endothelial cells. *OncoTargets and Therapy* 11, 2937.

Lippman, S. M., Klein, E. A., Goodman, P. J., Lucia, M. S., Thompson, I. M., Ford, L. G., Parnes, H. L., Minasian, L. M., Gaziano, J. M., Hartline, J. A., Parsons, J. K., Bearden, J. D., Crawford, E. D., Goodman,

G. E., Claudio, J., Winquist, E., Cook, E. D., Karp, D. D., Walther, P., … Coltman, C. A. (2009). Effect of selenium and vitamin E on risk of prostate cancer and other cancers: The selenium and vitamin E cancer prevention trial (SELECT). *JAMA – Journal of the American Medical Association*, 301(1), 39–51.

Liskova, A., Samec, M., Koklesova, L., Brockmueller, A., Zhai, K., Abdellatif, B., Siddiqui, M., Biringer, K., Kudela, E., Pec, M., Gadanec, L. K., Šudomová, M., Hassan, S. T. S., Zulli, A., Shakibaei, M., Giordano, F. A., Büsselberg, D., Golubnitschaja, O., & Kubatka, P. (2021). Flavonoids as an effective sensitizer for anti-cancer therapy: Insights into multi-faceted mechanisms and applicability towards individualized patient profiles. *EPMA Journal* 12(2), 155–176.

Mansoori, B., Mohammadi, A., Davudian, S., Shirjang, S., & Baradaran, B. (2017). The different mechanisms of cancer drug resistance: A brief review. *Advanced Pharmaceutical Bulletin* 7(3), 339.

Mbese, Z., Khwaza, V., & Aderibigbe, B. A. (2019). Curcumin and its derivatives as potential therapeutic agents in prostate, colon and breast cancers. *Molecules* 24(23), 4386.

Meng, H., Dai, T., Hanlon, J. G., Downar, J., Alibhai, S. M. H., & Clarke, H. (2020). Cannabis and cannabinoids in cancer pain management. *Current Opinion in Supportive and Palliative Care* 14(2), 87–93.

Mirzaeian, R., Sadoughi, F., Tahmasebian, S., & Mojahedi, M. (2021). The role of herbal medicines in health care quality and the related challenges. *Journal of HerbMed Pharmacology* 10(2), 156–165.

Mizoue, T., Kimura, Y., Toyomura, K., Nagano, J., Kono, S., Mibu, R., & Imaizumi, N. (2008). Calcium, dairy foods, vitamin D, and colorectal cancer risk: The fukuoka colorectal cancer study. *Cancer Epidemiology and Prevention Biomarkers* 17(10), 2800–2807.

Mosca, L., Ilari, A., Fazi, F., Assaraf, Y. G., & Colotti, G. (2021). Taxanes in cancer treatment: Activity, chemoresistance and its overcoming. *Drug Resistance Updates* 54, 100742.

Nadeem, M., Imran, M., Aslam Gondal, T., Imran, A., Shahbaz, M., Muhammad Amir, R., Wasim Sajid, M., Batool Qaisrani, T., Atif, M., Hussain, G., Salehi, B., Adrian Ostrander, E., Martorell, M., Sharifi-Rad, J. C., Cho, W., & Martins, N. (2019). Therapeutic potential of rosmarinic acid: A comprehensive review. *Applied Sciences* 9(15), 3139.

Najafi, M., Mortezaee, K., Rahimifard, M., Farhood, B., & Haghi-Aminjan, H. (2020). The role of curcumin/curcuminoids during gastric cancer chemotherapy: A systematic review of non-clinical study. *Life Sciences* 257, 118051.

Niranjana, R., Gayathri, R., Nimish Mol, S., Sugawara, T., Hirata, T., Miyashita, K., & Ganesan, P. (2015). Carotenoids modulate the hallmarks of cancer cells. *Journal of Functional Foods* 18, 968–985.

Noureini, S. K., & Wink, M. (2012). Antiproliferative effects of crocin in HepG2 cells by telomerase inhibition and hTERT down-regulation. *Asian Pacific Journal of Cancer Prevention* 13(5), 2305–2309.

Park, J. E., Sun, Y., Lim, S. K., Tam, J. P., Dekker, M., Chen, H., & Sze, S. K. (2017). Dietary phytochemical PEITC restricts tumor development via modulation of epigenetic writers and erasers. *Scientific Reports* 7(1), 40569.

Phillips, J. L., & Currow, D. C. (2010). Cancer as a chronic disease. *Collegian* 17(2), 47–50.

Ranjan, A., Ramachandran, S., Gupta, N., Kaushik, I., Wright, S., Srivastava, S., Das, H., Srivastava, S., Prasad, S., & Srivastava, S. K. (2019). Role of phytochemicals in cancer prevention. *International Journal of Molecular Sciences* 20(20), 4981.

Salehi, B., Mishra, A. P., Nigam, M., Sener, B., Kilic, M., Sharifi-Rad, M., Fokou, P., Martins, N., & Sharifi-Rad, J. (2018). Resveratrol: A double-edged sword in health benefits. *Biomedicines* 6(3), 91.

Shah, Z., Gohar, U. F., Jamshed, I., Mushtaq, A., Mukhtar, H., Zia-Ui-haq, M., Toma, S. I., Manea, R., Moga, M., & Popovici, B. (2021). Podophyllotoxin: History, recent advances and future prospects. *Biomolecules* 11(4), 603.

Sibaud, V., Lebœuf, N. R., Roche, H., Belum, V. R., Gladieff, L., Deslandres, M., . . . Lacouture, M. E. (2016). Dermatological adverse events with taxane chemotherapy. *European Journal of Dermatology* 26(5), 427–443.

Sisodiya, P. S. (2013). Plant derived anti-cancer agents : A review. *International Journal of Research and Development in Pharmacy and Life Sciences* 2(2), 293–308.

Śledziński, P., Zeyland, J., Słomski, R., & Nowak, A. (2018). The current state and future perspectives of cannabinoids in cancer biology. *Cancer Medicine* 7(3), 765–775.

Spagnuolo, C., Russo, G. L., Orhan, I. E., Habtemariam, S., Daglia, M., Sureda, A., Nabavi, S. F., Devi, K. P., Loizzo, M. R., Tundis, R., & Nabavi, S. M. (2015). Genistein and cancer: Current status, challenges, and future directions 12. *Advances in Nutrition* 6(4), 408–419.

Stampfer, M. J., Hennekens, C. H., Manson, J. E., Colditz, G. A., Rosner, B., & Willett, W. C. (1993). Vitamin E consumption and the risk of coronary disease in women. *New England Journal of Medicine* 328(20), 1444–1449.

Starowicz, K., & Finn, D. P. (2017). Cannabinoids and Pain: Sites and Mechanisms of Action. *Advances in Pharmacology (San Diego, Calif.)*, 80, 437–475.

Su, X., Jiang, X., Meng, L., Dong, X., Shen, Y., & Xin, Y. (2018). Anti-cancer activity of sulforaphane: The epigenetic mechanisms and the Nrf2 signaling pathway. *Oxidative Medicine and Cellular Longevity* 5438179.

Subedi, L., Teli, M. K., Lee, J. H., Gaire, B. P., Kim, M. H., & Kim, S. Y. (2019). A stilbenoid isorhapontigenin as a potential anti-cancer agent against breast cancer through inhibiting sphingosine kinases/tubulin stabilization. *Cancers* 11(12), 1947.

Talib, W. H., Alsayed, A. R., Barakat, M., Abu-Taha, M. I., & Mahmod, A. I. (2021). Targeting drug chemoresistance in cancer using natural products. *Biomedicines* 9(10), 1353.

Tomeh, M. A., Hadianamrei, R., & Zhao, X. (2019). A review of curcumin and its derivatives as anti-cancer agents. *International Journal of Molecular Sciences* 20(5), 1033.

Tuli, H. S., Tuorkey, M. J., Thakral, F., Sak, K., Kumar, M., Sharma, A. K., Sharma, U., Jain, A., Aggarwal, V., & Bishayee, A. (2019). Molecular mechanisms of action of genistein in cancer: Recent advances. *Frontiers in Pharmacology* 10, 1336.

Ward, S. J., Lichtman, A. H., Piomelli, D., & Parker, L. A. (2021). Cannabinoids and Cancer Chemotherapy-Associated Adverse Effects. *Journal of the National Cancer Institute. Monographs*, 2021(58), 78–85.

Wei, E., & Chan, C. (2022). An overview of the chemistry and anti-cancer properties of rosemary extract and its diterpenes. *Journal of Herbmed Pharmacology* 11(1), 10–19.

Wen, Y., Wang, Y., Liu, X., Zhang, W., Xiong, X., Han, Z., & Liang, X. (2017). Camptothecin-based nanodrug delivery systems. *Cancer Biology & Medicine* 14(4), 363–370.

Williams, S. P., Nowicki, M. O., Liu, F., Press, R., Godlewski, J., Abdel-Rasoul, M., Kaur, B., Fernandez, S. A., Chiocca, E. A., & Lawler, S. E. (2011). Indirubins decrease glioma invasion by blocking migratory phenotypes in both the tumor and stromal endothelial cell compartments. *Cancer Research* 71(16), 5374–5380.

Xiao-ai, L., Bei, W., Xiao-hong, X., Lei, P., Bin, W., Xiao-xue, D., Chen-hui, Z., & Qi-wei, D. (2017). Curcumin re-sensitizes multidrug resistant (MDR) breast cancer to cisplatin through inducing autophagy by decreasing CCAT1 expression. *RSC Advances* 7(53), 33572–33579.

Xu, T., Guo, P., He, Y., Pi, C., Wang, Y., Feng, X., Hou, Y., Jiang, Q., Zhao, L., & Wei, Y. (2020). Application of curcumin and its derivatives in tumor multidrug resistance. *Phytotherapy Research* 34(10), 2438–2458.

Xu, Y., Zhang, J., Shi, W., & Liu, Y. (2013). Anti-cancer effects of 3,3′-diindolylmethane are associated with G1 arrest and mitochondria-dependent apoptosis in human nasopharyngeal carcinoma cells. *Oncology Letters* 5(2), 655.

Yan, X. J., Gong, L. H., Zheng, F. Y., Cheng, K. J., Chen, Z. S., & Shi, Z. (2014). Triterpenoids as reversal agents for anti-cancer drug resistance treatment. *Drug Discovery Today* 19(4), 482–488.

Yang, C. S., Wang, X., Lu, G., & Picinich, S. C. (2009). Cancer prevention by tea: Animal studies, molecular mechanisms and human relevance. *Nature Reviews. Cancer* 9(6), 429.

Ye, Q., Liu, K., Shen, Q., Li, Q., Hao, J., Han, F., & Jiang, R. W. (2019). Reversal of multidrug resistance in cancer by multi-functional flavonoids. *Frontiers in Oncology* 9, 487.

Yu, S. E., Mwesige, B., Yi, Y. S., & Yoo, B. C. (2019). Ginsenosides: The need to move forward from bench to clinical trials. *Journal of Ginseng Research* 43(3), 361–367.

Yuan, J. M., Stepanov, I., Murphy, S. E., Wang, R., Allen, S., Jensen, J., Strayer, L., Adams-Haduch, J., Upadhyaya, P., Le, C., Kurzer, M. S., Nelson, H. H., Yu, M. C., Hatsukami, D., & Hecht, S. S. (2016). Clinical trial of 2-phenethyl isothiocyanate as an inhibitor of metabolic activation of a tobacco-specific lung carcinogen in cigarette smokers. *Cancer Prevention Research* 9(5), 396–405.

Zhang, N., Gao, M., Wang, Z., Zhang, J., Cui, W., Li, J., Zhu, X., Zhang, H., Yang, D. H., & Xu, X. (2021). Curcumin reverses doxorubicin resistance in colon cancer cells at the metabolic level. *Journal of Pharmaceutical and Biomedical Analysis* 201, 114129.

14 Synergism between Anticancer Drugs and Seaweed Metabolites for Cancer Treatment

Priscilla Arockiasamy, Sriram Srinivasan, Daniel A Gideon, Abhinav Parashar, Shivadharshini Ramachandran, and Janet Ann Netto

CONTENTS

14.1 Introduction 255
14.2 Seaweeds Extracts' Anticancer Activity 256
 14.2.1 Seaweeds' Cytotoxicity Against Skin Cancer 256
 14.2.2 Seaweeds' Cytotoxicity Against Breast Cancer 257
 14.2.3 Seaweeds' Cytotoxicity Against Colon and Bladder Cancer 258
14.3 Synergistic Effects of Metabolites From Seaweeds and Anticancer Drugs Against Cancer Cells 259
 14.3.1 Fucosterol 259
 14.3.2 Fucoxanthin and Phloroglucinol 259
 14.3.3 Porphyran 261
 14.3.4 Fucoidan 262
14.4 Anticancer Activity of Nanoparticles Synthesized From Seaweed Compounds 263
14.5 Conclusion 263
Bibliography 264

14.1 INTRODUCTION

Cancer is a disease which is characterized by loss of control in cell division, metastasis (movement of cancer cells to the neighboring tissue), and tumor formation. Different factors such as environmental toxins, inherited genetic mutation, unhealthy diet, drug use, and infectious organisms can act in a sequence or together to cause cancer. Surgery, radiation therapy, and chemotherapy were used to treat the cancer. In the case of chemotherapy, side effects caused by the anticancer drugs led to find an alternative bioactive compound from phytochemicals, which are highly toxic to cancer cells but showed no or a less toxic effect on the normal cells. Moreover, some cancer cells gain resistance to anticancer drugs, which hinders the cancer treatment using chemotherapy. Resistance to chemotherapy denotes the importance of identifying bioactive metabolites from phytochemicals which can act against the resistant cancer cells and also to sensitize the resistant cancer cells to chemotherapy.

Seaweeds or macroalgae are rich in phytochemicals that exhibit various biological activities such as anticancer, antioxidant, anti-inflammatory, and immunomodulatory function. Seaweeds lack true leaves, roots, or stems but have a higher growth rate and photosynthetic efficiency

DOI: 10.1201/b22917-14

when compared with land plants. Bioactive compounds from seaweeds such as phlorotannins, carotenoids, and polysaccharides exhibited cytotoxic activity against various cancer cells with no/less cytotoxic activity against normal cells. Bioactive compounds extracted from seaweeds showed anticancer activity against drug-resistant cancer cell lines. Moreover, seaweed bioactive compounds were used in combination with anticancer drugs such as cisplatin, gefitinib, lapatinib, 5-fluorouracil, and doxorubicin to enhance the anticancer property of drugs against sensitive and drug-resistant cancer cell lines. In addition, the use of bioactive compounds along with anticancer drugs reduced the side effects caused by chemotherapy. Nanoparticles synthesized using seaweed extracts also exhibited anticancer activity against different cancer cell lines. Moreover, nanoparticles synthesized using seaweed metabolites were used as vehicles to deliver the anticancer drug into the target site. Bioactive metabolites from seaweeds exhibited anticancer activity by inducing apoptosis and by decreasing and increasing the expression of mesenchymal and epithelial cell markers, respectively. In this chapter, seaweed crude extracts' anticancer activity and synergistic effects of seaweed metabolites and anticancer drugs against various cancer cell lines are discussed in detail. Moreover the anticancer activity of nanoparticles synthesized using the seaweed metabolites are also discussed in this chapter.

14.2 SEAWEEDS EXTRACTS' ANTICANCER ACTIVITY

14.2.1 Seaweeds' Cytotoxicity Against Skin Cancer

Skin cancer is a prevalent cancer requiring an alternative to chemotherapy treatment that is less damaging to normal cells. Seaweed polysaccharides were popular for their anticancer, anti-inflammatory, anticoagulant, and antidislipidmeic properties. Formulated agar from seaweeds was found to be effective in the treatment of skin cancer in in vivo conditions. Formulated agar from *Laminaria digitata* exhibited anti–skin cancer activity in a mice model. 7,12-Dimethyl Benz [a] anthracene (DMBA) is used to induce skin cancer in a mice model. Increased water intake, decreased food intake, and reduction in weight and physical movement were observed in the DMBA-injected mice. Pretreatment of mice with formulated agar prepared from *L. digitata* abrogated the reduction in weight, decrease in food intake, increase in water intake, and slow movement observed in the DMBA-injected mice model. In in vitro conditions, an increase in the concentration of formulated agar above 300 μg/mL exhibited significant cytotoxicity against mouse fibroblast cells (George et al. 2021). Polysaccharide from seaweeds showed no effect on the viability of B16-F10 cells but affected the metastasis. Sulfated polysaccharide (SP) from *Caulerpa cupressoides* var. flabellata showed no effect on B16-F10 cell viability but had inhibitory effects on cell migration, colony formation, and melanin production. The contact time between sulfated polysaccharide from *C. cupressoides* and the B16-F10 cell line affects cell migration. Incubation of B16-F10 cells with SP from *C. cupressoides* for 24 h did not inhibit the cell migration, but incubation of B16-F10 cells with SP for 48 h inhibited cell migration. On the other hand, an increase in the concentration of SP from *C. cupressoides* led to a decrease in the B16-F10 cell colony formation in a dose-dependent manner. Melanin production in melanoma cells plays an important role in conferring resistance to chemotherapy. The melanin production in B16-F10 cells decreased with SP treatment, which indicates that SP from seaweeds could be used along with anticancer drugs to enhance its cytotoxicity against resistant cancer cells. The melanoma development is related to melanin production which inturn is related to the Reactive Oxygen Species (ROS) status of the cell. The SP from seaweeds could alleviate melanin production by regulating the ROS status of the cell (daSilva Barbosa et al. 2021). Sulfated galactan from *Gracilaria fisheri* showed no effect on viability of B16-F10 cells but decreased the melanin production by decreasing the expression of genes involved in the melanogenesis pathway (tyrosinase, TRP-1 and TRP-2). Moreover, expression of the transcription regulator (MITF) of the melanogenesis pathway also decreased with sulfated galactan treatment (Pratoomthai et al. 2018).

14.2.2 Seaweeds' Cytotoxicity Against Breast Cancer

Intake of seaweeds through diet is correlated with a decrease in cancer disease progression. SP present in the seaweed plays a key role in decreasing cancer progression. SP extracted from *Laurencia papillosa* decreased the viability of MDA-MB-231 cells through the induction of apoptosis. Different concentrations of SP from *L. papillosa* affects the growth and viability of MDA-MB-231 cells. SP causes cell cycle arrest in MDA-MB-231 cells at lower concentrations through downregulation of cyclins and CDK_S genes. On the other hand, SP induces apoptosis in MDA-MB-231 cells at higher concentrations through the downregulation of *Bcl-2* and upregulation of *Bax* genes (Murad et al. 2016). Carrageenan, a type of SP fractionated from *L. papillosa*, exhibited cytotoxicity against MCF-7 cells. The three types of carrageenan isolated from *L. papillosa* such as kappa, iota and lambda were referred to as LP-W1, LP-W2, and LP-W3, respectively. Cytotoxicity of LP-W2 and LP-W3 against MCF-7 cells was higher when compared with LP-W1. Carrageenan from *L. papillosa* decreased the viability of MCF-7 cells in a time- and dose-dependent manner. The SPs from *L. papillosa* affect the cell viability of MCF-7 cells through the induction of apoptosis. Increase in the gene expression of *PARP*, p53, caspase 3, caspase 8, and caspase 9 observed in SP-treated MCF-7 cells corroborated the apoptosis induction property of LP-W2 and -W3 (Ghannam et al. 2018). *Sargassum linearifolium* and *Cystoseira crinita* extraction temperatures affect the secondary metabolites profile, which inturn affects the bioactivity. A higher total phenolic content was observed in a hot aqueous extract of *Sargassum linearifolium* (SHAE) and in a cold methanolic extract of *Cystoseira crinita* (CCME). The bioactive secondary metabolites were rich in CCME when compared with SHAE, although CCME showed a differential effect on different cancer cell lines. The MCF-7 cell line was highly sensitive to CCME and SHAE, whereas MIA PaCa-2 cancer cells were highly resistant to CCME and SHAE. On the other hand, CCME and SHAE slightly decreased the HepG2 cell viability. The higher sensitivity of MCF-7 cells to CCME and SHAE is correlated with the elevated expression of proapoptotic (Bax) and decreased expression of anti-apoptotic (Bcl-2) markers. Among the extracts, CCME was found to be very effective in enhancing the expression of proapoptotic and anti-apoptotic markers, which demonstrates that the extraction temperature plays a vital role in isolating bioactive metabolites (Abu-Khudir et al. 2021). Breast cancer cells were found to be resistant not only to anticancer drugs but also to some class of secondary metabolites from seaweeds. Fucoidan and alginate from *Colpomenia sinuosa* showed a variable effect on different cancer cell lines. The HCT-116 cells were highly sensitive to fucoidan and alginate when compared with HeLa and MCF-7 cell lines (Al Monla et al. 2022). On the contrary, fucoidan from *Laminaria japonica* inhibited the growth of MDA-MB-231 and HCC1806 cells but induced cell death in MDA-MB-231 cells but not in HCC1806 cells. In addition, fucoidan exhibited the antiangiogenic property by inhibiting the expression of MMPs, VEGF-A, and IGF-I. Moreover, fucoidan effectively inhibited the micrometastasis of MDA-MB-231 cells in the zebrafish model (Hsu et al. 2020). Ethanol extract of *Chaetomorpha* sp. inhibited the growth of MDA-MB-231 cell lines but did not show any effect on MCF-7 cell lines. On the other hand, aqueous extract of *Chaetomorpha* sp. showed no inhibitory effect on the growth MDA-MB-231 and MCF-7 cell lines. The gas chromatography–mass spectrometry (GC-MS) analysis of the bioactive extract of *Chaetomorpha* sp. showed the presence of antitumor compounds such as oxime and dichloroacetic acid. Moreover, total flavonoids, tannins, and phenolic content was higher in the ethanolic extract of *Chaetomorpha* sp. when compared with the aqueous extract (Haq et al. 2019). Phlorotannins from brown algae were found to possess various medicinal properties. Three phlorotannins such as 2,7′-phloroglucinol-6, 6′-bieckol, and dieckol were derived from *Ecklonia cava*. Among the three phlorotannins, dieckol showed effective anticancer activity against MCF-7 cells. In addition, dieckol effectively inhibited the MCF-7 cell migration and also decreased the messenger RNA (mRNA) and protein expression of vascular endothelial growth factor (VEGF) in MCF-7 cells. In contrast, dieckol effectively decreased the protein expression of MMP-9 but did not affect the mRNA expression of MMP-9 in MCF-7 cells. Along with a decrease in MMP-9, dieckol increased the protein expression of TIMP-1 and TIMP-2, which are the natural matrix metalloproteinase (MMP) inhibitors. Decrease in MMP-9 and VEGF expression and increase in TIMPs expression in MCF-7 cells corroborated the antimetastatic and antiangiogenic activity of dieckol isolated from *E. cava* (Kim et al. 2015a).

14.2.3 Seaweeds' Cytotoxicity Against Colon and Bladder Cancer

The colorectal cancer incidence rate is rapidly increasing because of smoking, obesity, and alcohol consumption. A search for less toxic natural metabolites against colorectal cancer is in high demand. SP from *Capsosiphon fulvescens* decreased the viability of HT-29 cells in a dose-dependent manner upto 500 μg/mL. The SP from *C. fulvescens* decreased the viability of HT-29 cells through induction of apoptosis. Treatment of HT-29 cells with SP from *C. fulvescens* induced DNA fragmentation, cleavage of caspase 9 and caspase 3, and an increase in the level of cleaved poly (ADP-ribose) polymerase (PARP). In the in vivo condition, SP decreased tumor growth without affecting the mice's body weight. Activation of PARP, caspase 3, caspase 8, and caspase 9 was observed in SP-treated tumor tissue corroborated the apoptosis induction by SP in the in vivo condition. Moreover, the presence of a sulfate group in SP has proved to be essential for its anticancer activity against HT-29 cells, because the absence of sulfate group in SP did not affect the growth of HT-29 cells (Choi et al. 2019). Polysaccharides extracted from *Porphyra haitanensis* were fractionated into three fractions, referred to as PHP-F1, PHP-F2, and PHP-F3. Colon cancer cell lines showed differential sensitivity to polysaccharides from *P. haitanensis*. Fractionated polysaccharides from *P. haitanensis* exhibited cytotoxicity against colon cancer cell lines such as HT-29, LoVo, and SW-480 but showed no toxic effect on normal cell lines. Among the colon cancer cells, HT-29 is highly sensitive to polysaccharide fractions from *P. haitanensis*. Moreover, PHP-F2 and PHP-F3 showed higher inhibitory activity against HT-29 when compared with PHP-F1. PHP-F2 and PHP-F3 decreased the viability of HT-29 cells by inducing ROS generation and apoptosis (Yao et al. 2020). Polysaccharides from seaweeds showed differential effects on colon cancer cells when used in combination with vitamin C. Fucoidan and alginate extracted from *Colpomenia sinuosa* effectively decreased the viability of HCT-116 cell lines in a dose-dependent manner, although fucoidan showed higher cytotoxic activity against HCT-116 cells when compared with alginate. However, HeLa and MCF-7 cell lines were least sensitive to fucoidan and alginate when compared with HCT-116. ROS plays an important role in the induction of apoptosis. Fucoidan and alginate exhibited cytotoxicity against HCT-116 cells through the induction of ROS generation. Moreover, the anticancer activity of fucoidan from *C. sinuosa* against the HCT-116 cell line was enhanced when used in combination with vitamin C. In contrast, alginate from *C. sinuosa* did not show any enhancement in anticancer activity against the HCT-116 cell line when used in combination with vitamin C (Al Monla et al. 2022). Bladder cancer is the most fatal in males and elderly people. SP from *Hizikia fusiforme* exhibited cytotoxicity against human bladder cancer cells. The anticancer activity exhibited by SP from *H. fusiforme* is comparable with the cisplatin anticancer activity against bladder cells (positive control). In addition, SP from *H. fusiforme* showed no toxic effect on the normal bladder cells (BdFC), whereas the positive control (cisplatin) had toxic effects on the normal bladder cells. The SP from *H. fusiforme* exhibited anticancer activity by modulating the cell cycle regulators. The expression of cell cycle regulators such as CDK2 and CDK4 and their corresponding cyclins were downregulated, whereas p21WAF1 expression was upregulated in EJ cells treated with SP. Moreover, migratory and invasive properties of EJ cells are reduced with SP treatment. The MMPs play an important role in bladder cancer cell metastasis. The SP from *H. fusiforme* exhibited an antimetastatic property by reducing the enzyme activity of MMP-2 and MMP-9. In in vivo conditions, SP treatment significantly reduced the tumor weight without causing weight loss in the mice model, whereas the weight loss was accompanied by tumor weight reduction in cisplatin-treated mice (Song et al. 2020).

14.3 SYNERGISTIC EFFECTS OF METABOLITES FROM SEAWEEDS AND ANTICANCER DRUGS AGAINST CANCER CELLS

14.3.1 Fucosterol

Fucosterol, a phytosterol present in microalgae and seaweeds, showed various medicinal properties. Fucosterol showed differential cytotoxicity on different cancer cell lines. Its cytotoxicity against cervical cancer cells (HeLa) was significantly higher when compared with gastric, lung, breast, prostate, and pancreatic cancer cell lines. Moreover, fucosterol also reduced the ovarian cancer cell viability in a dose-dependent manner. Fucosterol decreased the viability of HeLa and ovarian cancer cells through induction of apoptosis by increasing the ROS production and decreasing the mitochondria membrane potential. Fucosterol also affects the growth of the cancer cells through induction of cell cycle arrest. Treatment of ovarian cancer cells with fucosterol causes the decrease in phosphorylated cyclin D level (cell cycle regulator), which demonstrates the role of fucosterol in cell cycle regulation. In addition, fucosterol affects the growth of cancer cells by deactivating the proteins involved in various signaling pathways (Jiang et al. 2018; Bae et al. 2020). Fucosterol from brown algae was used in combination with anticancer drugs to reduce the side effects and to enhance the cytotoxicity of anticancer drugs. Treatment of ovarian cancer cells with fucosterol along with cisplatin increased the level of cleaved caspases when compared with fucosterol- or cisplatin-alone treatment (Bae et al. 2020). Moreover, the interactive effect of fucosterol from seaweed and 5-fluorouracil (5-Fu) was analyzed in human colorectal cancer cell lines. Compared with 5-Fu, fucosterol reduced the cell viability of HT-29 cells but not HCT-116 cell. Compared with fucosterol or 5-Fu treatment alone, fucosterol cytotoxicity against HT-29 cells is further enhanced when treated in combination with 5-Fu. However, fucosterol + 5-Fu showed no effect on the viability of HCT-116. Fucosterol alone did not show a significant effect on the spheroid structures and colony-forming ability of HT-29 cells. On the other hand, 5-Fu alone did not affect the HT-29 cell migration. Synergistic effects of fucosterol and 5-Fu decreased the spheroid growth, colony-forming ability, and cell migration of HT-29 cells (Ramos et al. 2019). Doxorubicin was found to enhance the anticancer activity of secondary metabolites from seaweeds. Doxorubicin alone decreased the viability of MDA-MB-231 cells. Fucosterol alone did not show any cytotoxic activity against MDA-MB-231 cell line in monolayer as well as in 3D culture, but fucosterol along with doxorubicin exhibited cytotoxic activity against MDA-MB-231 cells. The synergistic effect of doxorubicin and fucosterol enhanced the anticancer activity of fucosterol against the MDA-MB-231 cell line (Malhão et al. 2021).

14.3.2 Fucoxanthin and Phloroglucinol

Fucoxanthin decreased the proliferation of different types of cancer cells by arresting cell cycle either through increasing the protein expression of cyclin-dependent kinase inhibitory proteins or by decreasing the level of cyclins such as cyclin D1, cyclin D2, and cyclin E. Fucoxanthin induces DNA damage in the cancer cells either through induction of ROS or by modulating the expression of protein involved in the DNA repair mechanism. Fucoxanthin affects the viability of different types of cancer cells by inducing apoptosis through modulating the level of cleaved caspases, cleaved PARP, mortalin, and survivin. In addition, fucoxanthin exhibits antimetastatic activity. Fucoxanthin decreased the metastasis of lung cancer cells by downregulating the expression of mesenchymal markers and epithelial to mesenchymal transition transcription factors and by increasing the expression of TIMP-2. Treatment of lung cancer cells with fucoxanthin and interference RNA for TIMP-2 did not inhibit the metastasis of lung cancer cells, which demonstrates that fucoxanthin mediates antimetastatic activity through upregulating the TIMP-2 expression (Ming et al. 2021). Fucoxanthin decreased the viability of KB cells and CMT-U27 cells by inducing apoptosis (Iyappan et al. 2021; Jang et al. 2021).

Phloroglucinol, a phenolic compound present in brown algae, has been found to possess various biological activities and is commercially used in cosmetics, medicine, cements, and dyeing. Breast cancer patients are highly resistant to therapies because of the presence of cancer stem-like cells in tumor tissue. Phloroglucinol is found to decrease the proportion of $CD44^{+}$ cells and also the regulators of stemness (Kim et al. 2015b). Cisplatin, ionizing radiation, and etoposide did not significantly decrease the viability of breast cancer stem cells (BCSCs). Phloroglucinol also did not significantly decrease the BCSCs viability, whereas phloroglucinol, when used in combination with cisplatin, ionizing radiation, and etoposide, decreased the viability of BCSCs, which demonstrates that phloroglucinol can increase the sensitivity of BCSCs to anticancer drugs and radiation therapy (Kim et al. 2015b). Phloroglucinol did not affect the viability of breast cancer cells but decreased the invasive and migratory properties by increasing the epithelial markers and by decreasing the mesenchymal markers. In addition, phloroglucinol regulates the EMT markers by downregulating the expression of SLUG, a transcription factor, in breast cancer cells. Similarly, phloroglucinol inhibits the metastasis of breast cancer cells in in vivo conditions (Kim et al. 2015c).

Fucoxanthin and phloroglucinol present in brown algae effectively enhanced the 5-Fu cytotoxicity against HCT-116 and HT-29 cells (Table 14.1). The cytotoxicity of fucoxanthin against HCT-116 and HT-29 cells increased with an increase in the concentration of fucoxanthin. Lower concentrations of fucoxanthin enhanced the 5-Fu cytotoxicity against HCT-116 and HT-29 cells. Moreover, viability of the normal cell line CCD-18Co decreased only when the fucoxanthin concentration increased above 50 µM. Combination treatment of fucoxanthin along with 5-Fu did not affect the cell viability of CCD-18Co cells. On the other hand, phloroglucinol cytotoxicity against HCT-116 and HT-29 cells was observed at higher concentration (300 µM). Treatment of HCT-116 and HT-29 cells with phloroglucinol + 5-Fu enhanced the cytotoxic activity of 5-Fu but not the phloroglucinol. Higher concentrations of phloroglucinol alone or in combination treatment with 5-Fu did not show any cytotoxicity against the normal cell line CCD-18Co. The DNA damage was observed in fucoxanthin-treated HCT-116 and HT-29 cells, whereas DNA damage was observed in phloroglucinol-treated HT-29 cells but not in phloroglucinol-treated HCT-116 cells. Secondary metabolites from seaweed showed a differential effect on cancer cells when used in combination with anticancer drugs. The synergistic effect of fucoxanthin and 5-Fu induced nuclear chromatin condensation in HT-29 cells, but fucoxanthin and 5-Fu alone did not induce the nuclear condensation in HT-29 cells. On the other hand, nuclear chromatin condensation was not observed in HCT-116 cells treated with fucoxanthin + 5-Fu. Phloroglucinol alone or in combination with 5-Fu induced nuclear chromatin condensation in HCT-116 and HT-29 cells (Lopes-Costa et al. 2017).

Cancer cells gain resistance to cisplatin treatment because of the drug efflux caused by the transporter protein. Fucoxanthin is found to enhance the sensitivity of cancer cells to cisplatin by inhibiting the transporter protein expression. Cisplatin exhibits anticancer activity by causing DNA damage and inducing apoptosis in different cancer cells. Cisplatin alone did not inhibit the growth of A549 cells but inhibited the growth of A549 cells when used in combination with different concentrations of fucoxanthin, which indicates that the fucoxanthin decreased the resistance of cancer cells to cisplatin. On the other hand, fucoxanthin did not reduce the resistance of HeLa cells to cisplatin at lower concentrations. Fucoxanthin can act as both an antioxidant and pro-oxidant, depending upon the concentration and cell type. Cisplatin causes cell death in cancer cells by inducing ROS generation. Lower fucoxanthin concentrations used along with cisplatin could have exhibited antioxidant properties in HeLa cells, which in turn further decreases the sensitivity of HeLa cells to cisplatin (Nurcahyanti et al. 2021). Fucoxanthin enhanced the cytotoxicity of doxorubicin to doxorubicin-resistant cancer cells by inhibiting the transporter protein expression, which is responsible for drug efflux in doxorubicin-resistant cancer cell lines. Doxorubicin IC50 against resistant breast, hepatic, and ovarian cancer cell lines were reduced when used in combination with fucoxanthin. Among the tested resistant cancer cells, sensitivity of breast cancer cells to doxorubicin was significantly enhanced by fucoxanthin treatment, followed by hepatic and ovarian cancer cells. In

TABLE 14.1
Synergistic Cytotoxic Effect of Seaweed Metabolites and Anticancer Drugs against Cancer Cell Lines

Seaweed Metabolite	Anticancer Drug	Synergistic Cytotoxic Effect	Cancer Cell Lines	Mode of Action
Fucosterol	5-Fu	Yes	HT-29 cells	Decreased the spheroid growth, colony-forming ability, and cell migration
Phloroglucinol	Cisplatin	Yes	BCSCs	-
Phloroglucinol	5-Fu	Yes	HCT-116 and HT-29 cells	Induced nuclear chromatin condensation
Fucoxanthin	Cisplatin	Yes	A549 cells	-
Fucoxanthin	Doxorubicin	Yes	Resistant breast, hepatic, and ovarian cancer cell lines	Downregulation of transporter protein and upregulation of apoptotic genes
Fucoxanthin	Gefitinib	Yes	PC9/G cells	Induction of apoptosis
Fucoidan	Cisplatin	Yes	LLC1 cells	Induction of apoptosis
Fucoidan	Lapatinib	Yes	Melanoma cells	Decreasing the activity of AKT and NFκB
Fucoidan	Gefitinib	Yes	Resistant lung cancer cells	Degradation of SLUG and inducing apoptosis

addition to downregulation of transporter protein, upregulation of apoptotic genes observed in fucoxanthin + doxorubicin–treated resistant breast cancer cells indicates that fucoxanthin enhanced the apoptosis-inducing ability of doxorubicin in resistant cancer cell lines (Eid et al. 2020). Resistance against gefitinib by lung cancer cells was reduced with fucoxanthin treatment. Gefitinib treatment decreased the viability of PC9 cells (gefitinib sensitive). On the other hand, PC9/G cells' (gefitinib-resistant cells) viability was not decreased with gefitinib treatment. The viability of PC9 and PC9/G cells was decreased with fucoxanthin treatment. Moreover, cytotoxicity of gefitinib against PC9/G was enhanced by combination treatment with fucoxanthin. In in vivo conditions also fucoxanthin along with gefitinib treatment reduced the tumor weight in PC9 cell–bearing mice when compared with mice treated with gefitinib alone. Decreased viability of PC9 cells could be because of induction of apoptosis in combination treatment (fucoxanthin + Gefitinib) (Ming et al. 2021).

14.3.3 Porphyran

Metabolites from seaweeds were conjugated with anticancer drugs to decrease its side effects. Porphyran, a sulfated polysaccharide from red algae, was used for slow release of the drug to reduce its side effects. Reduction in the side effects caused by 5-Fu and enhancement in anticancer activity of 5-Fu were obtained by conjugating the 5-Fu with polysaccharide from red algae. Low-molecular-weight porphyran (LP) from red algae was conjugated with 5-Fu to obtain the slow release of 5-Fu for reducing its side effects. The tumor weight in mice was significantly reduced with LP+5-Fu treatment when compared with LP or 5-Fu alone. The inhibition rate of 5-Fu conjugated with LP on the tumor was higher than the combination treatment (LP+5-Fu). Moreover, the reduction in the spleen weight caused by 5-Fu was recovered with conjugation of 5-Fu with LP (Wang and Zhang 2014). Lack of drug specificity to target sites leads to the cause of side effects. Gold nanoparticles synthesized using seaweed metabolites were used as drug delivery vehicles for doxorubicin to enhance its anticancer activity. Porphyran-capped gold nanoparticles were used as drug delivery vehicles to deliver doxorubicin. Porphyran exhibits anticancer activity against cancer cells through activating proapoptotic proteins and by inhibiting anti-apoptotic proteins. Porphyran-capped gold nanoparticles' cytotoxic activity against cancer cells was higher than the native porphyran cytotoxic activity. Moreover, doxorubicin-loaded gold nanoparticles' cytotoxic activity against cancer cells was higher than cytotoxicity of native doxorubicin. Enhanced uptake of doxorubicin-capped gold nanoparticles through an endocytosis mechanism might result in higher cytotoxicity of doxorubicin-loaded gold nanoparticles when compared with native doxorubicin (Venkatpurwar et al. 2011).

14.3.4 Fucoidan

Fucoidan, a sulfated polysaccharide extracted from brown algae, has been found to possess antitumor, antioxidant, immunomodulatory, and antiviral properties. Fucoidan combined with different anticancer drugs such as gefitinib, cisplatin, and lapatinib was used to enhance the cytotoxicity of anticancer drugs against cancer cell lines as well as to drug-resistant cancer cell lines (Table 14.1). In addition, fucoidan is found to decrease the side effects of anticancer drugs. Fucoidan decreased the cancer cell viability by inducing apoptosis. Non-coding RNAs with more than 200 nucleotides are referred as lncRNAs. Tumor progression and resistance of cancer cells to anticancer drugs were significantly affected by lncRNAs. Fucoidan significantly decreased the viability and proliferation of MHCC-97H and Hep3B cell lines in dose-dependent manner. In addition, invasion ability of the MHCC-97H cells was significantly inhibited by fucoidan treatment. Fucoidan decreased the viability of MHCC-97H cells by inducing apoptosis. Fucoidan inhibited the MHCC-97H cell proliferation by upregulating the expression of lncRNAs named LINC00261, which can act as tumor suppressor genes. Moreover, LINC00261 inhibits cancer cell proliferation by downregulating the miR-522–3p. The MHCC-97H cell proliferation increased and decreased in the presence of a miR-522–3p mimic and miR-522–3p inhibitor, respectively, which demonstrates the role of miR-522–3p in cell proliferation. Thus, fucoidan exhibited anticancer activity through increasing the expression of LINC00261, which in turn downregulated the miR-522–3p (Ma et al. 2021). Moreover, fucoidan affected the viability of cancer cells through regulating the JAK-STAT3 pathway. The viability, proliferation, and migration of DU-145 were inhibited by fucoidan treatment in in vitro conditions. In in vivo conditions (DU-145–injected mice), fucoidan decreased the tumor growth and downregulated the expression of CD31 and CD105 in tumor tissue. In addition, fucoidan decreased the expression of VEGF, cyclin D1, and Bcl-xL in tumor tissue by downregulating the phosphorylated levels of STAT3 (Rui et al. 2017).

Fucoidan is used in combination with cisplatin to reduce its side effects and to enhance its cytotoxicity against lung cancer cells. In in vitro conditions, cisplatin cytotoxicity against LLC1 cells was enhanced significantly when used in combination with fucoidan (sequential treatment of cisplatin followed by fucoidan). In in vivo conditions, tumor volume reduction in LLC1-bearing mice by cisplatin was enhanced when used in combination with fucoidan, whereas fucoidan treatment alone significantly reduced tumor volume in mice when compared with sequential treatment (cisplatin followed by fucoidan). Cisplatin treatment causes a change in mice behavior and weight reduction. Sequential supplementation of cisplatin followed by fucoidan to the LLC1 cell–bearing mice did not cause any change in mice behavior and weight reduction, which demonstrates that the fucoidan alleviates the side effects of cisplatin. Fucoidan enhanced the cytotoxic activity of cisplatin against human lung cancer cells through induction of apoptosis, whereas fucoidan did not affect the cisplatin-induced DNA damage in human lung cancer cells. Knockdown of C/EBP homologous protein (CHOP) using shRNA did not enhance the tunicamycin (TM) and combination treatment (fucoidan + cisplatin)–induced apoptosis. Elimination of apoptosis enhancement in TM and fucoidan + cisplatin–treated cells by CHOP knockdown indicates that CHOP expression plays a vital role in mediating the apoptosis (Hsu et al. 2018).

Signaling through the ERRB3 pathway is responsible for cancer cell survival. Lapatinib, a tyrosine kinase inhibitor, is used for cancer treatment. However, lapatinib use causes side effects such as gastrointestinal disorders, skin rashes, and weight loss. Supplementation of fucoidan along with lapatinib enhanced the tumor weight reduction ability of lapatinib and ameliorated the side effects of lapatinib in in vivo conditions. Fucoidan alone showed no significant effect on the viability of melanoma cells in in vitro conditions and tumor growth in in vivo conditions. Treatment of melanoma cells with fucoidan + lapatinib decreased the viability of melanoma cells (in vitro) and tumor growth (in vivo). Fucoidan enhanced lapatinib cytotoxicity against the melanoma cells by suppressing the activity of AKT and NFκB (Thakur et al. 2017). Fucoidan enhanced gefitinib cytotoxicity against drug-resistant lung cancer cell lines by enhancing apoptosis. The SLUG level modulates the resistance of lung cancer cells to gefitinib. Fucoidan reduced the resistance of lung cancer cells to gefitinib through activating the proteasome-mediated degradation of SLUG. In in vivo conditions, tumor volume reduction observed in LLC1 cell–bearing mice with combination treatment (fucoidan + gefitinib) was not significantly different from gefitinib alone. Moreover, combination treatment of fucoidan along with gefitinib decreased the tumor growth without any side effects (Qiu et al. 2020).

14.4 ANTICANCER ACTIVITY OF NANOPARTICLES SYNTHESIZED FROM SEAWEED COMPOUNDS

Silver, gold, and zinc oxide nanoparticles synthesized using seaweed extracts exhibited cytotoxicity against cancer cell lines when compared with seaweed extracts. Nanoparticles synthesized using seaweed extracts showed differential effects on different cancer cell lines and decreased the viability of cancer cells through induction of apoptosis. Silver nanoparticles (AgNPs) synthesized using *Chaetomorpha linum* extract showed higher cytotoxicity against HCT-116 cells when compared with the cytotoxicity of silver nitrate and *C. linum*. The AgNPs from *C. linum* cause cell death in HCT-116 cells through the accumulation of ROS. Moreover, AgNPs from *C. linum* enhanced the expression of Bax, Bid, caspase 3, and caspase 9 and decreased the expression of Bcl-2 and Bcl-xl (Acharya et al. 2021). Similarly, AgNPs produced using *Sargassum myriocystum* aqueous extract showed dose-dependent cytotoxic activity against HeLa cells through the induction of apoptosis (Balaraman et al. 2020). *Sargassum glaucescens* aqueous extract is not toxic to HeLa, HepG2, MDA-MB-231, and CEM-ss cell lines. However, gold nanoparticles (AuNPs) synthesized using *S. glaucescens* extract showed cytotoxicity against HeLa, HepG2, MDA-MB-231, and CEM-ss cell lines but not on normal cell lines such as MCF-10A. Among the cell lines, HeLa was found to be most sensitive to gold nanoparticles synthesized using *S. glaucescens*. Similar to AgNPs, AuNPs synthesized using *S. glaucescens* extract cause cell death in HeLa cells by inducing apoptosis through enhancing the caspase-3 and caspase-9 activities (Ajdari et al. 2016). Zinc oxide nanoparticles (ZnONPs) synthesized using the aqueous extract of *Gracilaria edulis* decreased the cell viability of SiHa cells through the induction of apoptosis but showed no significant cytotoxicity on peripheral blood mononuclear cells (PBMCs). Apoptosis induction in SiHa cells by ZnONPs is characterized by chromatin fragmentation, nucleus shrinkage, and condensed cytoplasm (MA et al. 2019). Gold nanoparticles synthesized using *Acanthophora spicifera* aqueous extract effectively decreased the viability of HT-29 cells through the induction of apoptosis (Babu et al. 2020). Gold and silver nanoparticles synthesized using the *Ulva lactuca* aqueous extract exhibited differential effects on different cancer cell lines. The AuNPs and AgNPs from *U. lactuca* showed significant toxicity against HT-29 cells but not on Caco-2 cells and normal cell lines such as PCS-201–010 (González-Ballesteros et al. 2019).

14.5 CONCLUSION

Seaweed metabolites effectively decreased the viability of breast and colon cancer cells. The toxicity caused by the anticancer drugs against the normal cells was effectively decreased when used in combination with seaweed metabolites. Moreover, drug-resistant cancer cell lines' sensitivity to anticancer drugs was enhanced when used in combination with seaweed metabolites. The synergism between anticancer drugs and seaweed metabolites could be used for treating cancer with fewer side effects and enhanced cytotoxicity.

BIBLIOGRAPHY

Abu-Khudir, R., Ismail, G.A. and Diab, T. 2021. Antimicrobial, antioxidant, and anti-tumor activities of Sargassum linearifolium and Cystoseira crinita from Egyptian Mediterranean Coast. *Nutrition and Cancer* 73: 829–844.

Acharya, D., Satapathy, S., Somu, P., Parida, U.K. and Mishra, G. 2021. Apoptotic effect and anticancer activity of biosynthesized silver nanoparticles from marine algae Chaetomorpha linum extract against human colon cancer cell HCT-116. *Biological Trace Element Research* 199: 1812–1822.

Ajdari, Z., Rahman, H., Shameli, K., Abdullah, R., Abd Ghani, M., Yeap, S., Abbasiliasi, S., Ajdari, D. and Ariff, A. 2016. Novel gold nanoparticles reduced by Sargassum glaucescens: Preparation, characterization and anticancer activity. *Molecules* 21: 123.

Al Monla, R., Dassouki, Z., Sari-Chmayssem, N., Mawlawi, H. and Gali-Muhtasib, H. 2022. Fucoidan and alginate from the Brown Algae colpomenia sinuosa and their combination with vitamin C trigger apoptosis in colon cancer. *Molecules* 27: 358.

Babu, B., Palanisamy, S., Vinosha, M., Anjali, R., Kumar, P., Pandi, B., Tabarsa, M., You, S. and Prabhu, N.M. 2020. Bioengineered gold nanoparticles from marine seaweed Acanthophora spicifera for pharmaceutical uses: Antioxidant, antibacterial, and anticancer activities. *Bioprocess and Biosystems Engineering* 43: 2231–2242.

Bae, H., Lee, J.Y., Song, G. and Lim, W. 2020. Fucosterol suppresses the progression of human ovarian cancer by inducing mitochondrial dysfunction and endoplasmic reticulum stress. *Marine Drugs* 18: 261.

Balaraman, P., Balasubramanian, B., Kaliannan, D., Durai, M., Kamyab, H., Park, S., Chelliapan, S., Lee, C.T., Maluventhen, V. and Maruthupandian, A. 2020. Phyco-synthesis of silver nanoparticles mediated from marine algae Sargassum myriocystum and its potential biological and environmental applications. *Waste and Biomass Valorization* 11: 5255–5271.

Choi, J.W., Lee, J., Kim, S.C., You, S., Lee, C.W., Shin, J. and Park, Y.I. 2019. Glucuronorhamnoxylan from capsosiphon fulvescens inhibits the growth of HT-29 human colon cancer cells in vitro and in vivo via induction of apoptotic cell death. *International Journal of Biological Macromolecules* 124: 1060–1068.

daSilva Barbosa, J., Palhares, L.C.G.F., Silva, C.H.F., Sabry, D.A., Chavante, S.F. and Rocha, H.A.O. 2021. In vitro antitumor potential of sulfated polysaccharides from seaweed caulerpa cupressoides var. flabellata. *Marine Biotechnology* 23: 77–89.

Eid, S.Y., Althubiti, M.A., Abdallah, M.E., Wink, M. and El-Readi, M.Z. (2020). The carotenoid fucoxanthin can sensitize multidrug resistant cancer cells to doxorubicin via induction of apoptosis, inhibition of multidrug resistance proteins and metabolic enzymes. *Phytomedicine* 77: 153280.

George, J., Thabitha, A., Vignesh, N., Manigandan, V., Saravanan, R., Daradkeh, G. and Qoronfleh, M.W. 2021. Antiskin cancer and antioxidant activities of formulated agar from Brown Seaweed Laminaria digitata (Hudson) in dimethyl benzanthracene-induced swiss albino mice. *International Journal of Polymer Science* 2021: 9930777.

Ghannam, A., Murad, H., Jazzara, M., Odeh, A. and Allaf, A.W. 2018. Isolation, structural characterization, and antiproliferative activity of phycocolloids from the red seaweed Laurencia papillosa on MCF-7 human breast cancer cells. *International Journal of Biological Macromolecules* 108: 916–926.

González-Ballesteros, N., Rodríguez-Argüelles, M.C., Prado-López, S., Lastra, M., Grimaldi, M., Cavazza, A., Nasi, L., Salviati, G. and Bigi, F. 2019. Macroalgae to nanoparticles: Study of Ulva lactuca L. role in biosynthesis of gold and silver nanoparticles and of their cytotoxicity on colon cancer cell lines. *Materials Science and Engineering: C* 97: 498–509.

Haq, S.H., Al-Ruwaished, G., Al-Mutlaq, M.A., Naji, S.A., Al-Mogren, M., Al-Rashed, S., Ain, Q.T., Al-Amro, A.A. and Al-Mussallam, A. 2019. Antioxidant, anticancer activity and phytochemical analysis of green algae, Chaetomorpha collected from the Arabian Gulf. *Scientific Reports* 9: 1–7.

Hsu, H.Y., Lin, T.Y., Hu, C.H., Shu, D.T.F. and Lu, M.K. 2018. Fucoidan upregulates TLR4/CHOP-mediated caspase-3 and PARP activation to enhance cisplatin-induced cytotoxicity in human lung cancer cells. *Cancer Letters* 432: 112–120.

Hsu, W.J., Lin, M.H., Kuo, T.C., Chou, C.M., Mi, F.L., Cheng, C.H. and Lin, C.W. 2020. Fucoidan from Laminaria japonica exerts antitumor effects on angiogenesis and micrometastasis in triple-negative breast cancer cells. *International Journal of Biological Macromolecules* 149: 600–608.

Iyappan, P., Bala, M.D., Sureshkumar, M., Veeraraghavan, V.P. and Palanisamy, A. 2021. Fucoxanthin induced apoptotic cell death in oral squamous carcinoma (KB) cells. *Bioinformation*17: 181.

Jang, H., Choi, J., Park, J.K., Won, G. and Seol, J.W. 2021. Fucoxanthin exerts anti-tumor activity on canine mammary tumor cells via tumor cell apoptosis induction and angiogenesis inhibition. *Animals* 11: 1512.

Jiang, H., Li, J., Chen, A., Li, Y., Xia, M., Guo, P., Yao, S. and Chen, S. 2018. Fucosterol exhibits selective antitumor anticancer activity against HeLa human cervical cell line by inducing mitochondrial mediated apoptosis, cell cycle migration inhibition and downregulation of m-TOR/PI3K/Akt signalling pathway. *Oncology Letters* 15: 3458–3463.

Kim, E.K., Tang, Y., Kim, Y.S., Hwang, J.W., Choi, E.J., Lee, J.H., Lee, S.H., Jeon, Y.J. and Park, P.J. 2015a. First evidence that Ecklonia cava-derived dieckol attenuates MCF-7 human breast carcinoma cell migration. *Marine Drugs*13: 1785–1797.

Kim, R.K., Suh, Y., Yoo, K.C., Cui, Y.H., Hwang, E., Kim, H.J., Kang, J.S., Kim, M.J., Lee, Y.Y. and Lee, S.J. 2015c. Phloroglucinol suppresses metastatic ability of breast cancer cells by inhibition of epithelial-mesenchymal cell transition. *Cancer Science*106: 94–101.

Kim, R.K., Uddin, N., Hyun, J.W., Kim, C., Suh, Y. and Lee, S.J. 2015b. Novel anticancer activity of phloroglucinol against breast cancer stem-like cells. *Toxicology and Applied Pharmacology* 286: 143–150.

Lopes-Costa, E., Abreu, M., Gargiulo, D., Rocha, E. and Ramos, A.A. 2017. Anticancer effects of seaweed compounds fucoxanthin and phloroglucinol, alone and in combination with 5-fluorouracil in colon cells. *Journal of Toxicology and Environmental Health, Part A*80: 776–787.

Ma, D., Wei, J., Chen, S., Wang, H., Ning, L., Luo, S.H., Liu, C.L., Song, G. and Yao, Q. 2021. Fucoidan inhibits the progression of hepatocellular carcinoma via causing lncRNA LINC00261 overexpression. *Frontiers in Oncology* 11.

MA, R. and MJ, M. 2019. Anticancer potential of zinc oxide nanoparticles against cervical carcinoma cells synthesized via biogenic route using aqueous extract of Gracilaria edulis. *Materials Science & Engineering. C, Materials for Biological Applications*103: 109840–109840.

Malhão, F., Ramos, A. and Rocha, E. 2021. Cytotoxic and anti-proliferative effects of fucosterol, alone and in combination with doxorubicin. 2D and 3D Cultures of Triple-Negative Breast Cancer Cells. doi.org/10.3390/CAHD2020-08600

Ming, J.X., Wang, Z.C., Huang, Y., Ohishi, H., Wu, R.J., Shao, Y., Wang, H., Qin, M.Y., Wu, Z.L., Li, Y.Y. and Zhou, S.C. 2021. Fucoxanthin extracted from Laminaria Japonica inhibits metastasis and enhances the sensitivity of lung cancer to Gefitinib. *Journal of Ethnopharmacology* 265: 113302.

Murad, H., Hawat, M., Ekhtiar, A., AlJapawe, A., Abbas, A., Darwish, H., Sbenati, O. and Ghannam, A. 2016. Induction of G1-phase cell cycle arrest and apoptosis pathway in MDA-MB-231 human breast cancer cells by sulfated polysaccharide extracted from Laurencia papillosa. *Cancer Cell International*16: 1–11.

Nurcahyanti, A.D.R., Kusmita, L. and Wink, M. 2021. Bixin and fucoxanthin sensitize human lung cancer and cervical cancer cell to cisplatin in vitro. *BMC Research Notes* 14: 1–8.

Pratoomthai, B., Songtavisin, T., Gangnonngiw, W. and Wongprasert, K. 2018. In vitro inhibitory effect of sulfated galactans isolated from red alga Gracilaria fisheri on melanogenesis in B16F10 melanoma cells. *Journal of Applied Phycology* 30: 2611–2618.

Qiu, W.L., Tseng, A.J., Hsu, H.Y., Hsu, W.H., Lin, Z.H., Hua, W.J. and Lin, T.Y. 2020. Fucoidan increased the sensitivity to gefitinib in lung cancer cells correlates with reduction of TGFβ-mediated Slug expression. *International Journal of Biological Macromolecules* 153: 796–805.

Ramos, A.A., Almeida, T., Lima, B. and Rocha, E. 2019. Cytotoxic activity of the seaweed compound fucosterol, alone and in combination with 5-fluorouracil, in colon cells using 2D and 3D culturing. *Journal of Toxicology and Environmental Health, Part A* 82: 537–549.

Rui, X., Pan, H.F., Shao, S.L. and Xu, X.M. 2017. Anti-tumor and anti-angiogenic effects of Fucoidan on prostate cancer: Possible JAK-STAT3 pathway. *BMC Complementary and Alternative Medicine*17: 1–8.

Song, J.H., Won, S.Y., Hwang, B., Jung, S., Choi, C., Park, S.S., Choi, Y.H., Kim, W.J. and Moon, S.K. 2020. In vitro and in vivo antitumor efficacy of Hizikia fusiforme celluclast extract against bladder cancer. *Nutrients*12: 2159.

Thakur, V., Lu, J., Roscilli, G., Aurisicchio, L., Cappelletti, M., Pavoni, E., White, W.L. and Bedogni, B. 2017. The natural compound fucoidan from New Zealand Undaria pinnatifida synergizes with the ERBB inhibitor lapatinib enhancing melanoma growth inhibition. *Oncotarget* 8: 17887.

Venkatpurwar, V., Shiras, A. and Pokharkar, V. 2011. Porphyran capped gold nanoparticles as a novel carrier for delivery of anticancer drug: In vitro cytotoxicity study. *International Journal of Pharmaceutics* 409: 314–320.

Wang, X. and Zhang, Z. 2014. The antitumor activity of a red alga polysaccharide complexes carrying 5-fluorouracil. *International Journal of Biological Macromolecules* 69: 542–545.

Yao, W.Z., Veeraperumal, S., Qiu, H.M., Chen, X.Q. and Cheong, K.L. 2020. Anti-cancer effects of Porphyra haitanensis polysaccharides on human colon cancer cells via cell cycle arrest and apoptosis without causing adverse effects in vitro.*3 Biotech*10: 1–11.

15 Recent Perspectives on the Medicinal Potential of Plant-Based Adaptogens

Kajal Pandey, Manvi Arora, and Sonam Chawla

CONTENTS

15.1 Introduction 267
15.2 Overview of Protective Actions of Adaptogens 269
15.3 Adaptogen Plants From Indian Traditional Medicine 270
  15.3.1 *Withania somnifera* 270
  15.3.2 *Panax ginseng* 270
  15.3.3 *Bacopa monnieri* 270
  15.3.4 *Emblica officinalis* 271
  15.3.5 *Ginkgo biloba* 271
  15.3.6 *Rhodiola rosea* 272
  15.3.7 *Cordyceps sinensis* 272
  15.3.8 *Glycyrrhiza glabra* 272
  15.3.9 *Tinospora cordifolia* 273
15.4 Global Commercial Value of Adaptogens 273
15.5 Challenges and Opportunities in Plant-Based Adaptogens Research 276
15.6 Conclusion 276
Bibliography 277

15.1 INTRODUCTION

The term "adaptogen" was coined by a Soviet scientist, Dr. Nicolai V. Lazarev, when he was studying the body's response to stress (Liao et al. 2018). Adaptogens are herbs that have been used in the Ayurvedic medicine system for over hundreds of years but now they are highly accessible in western world. Adaptogens stimulate the inherent adaptive mechanisms of an organism facilitating indirect resistance, allowing for optimal physiological, physical and cognitive performance even under stress. These herbal agents act through complex mechanisms and demonstrate a multitude of activities such as anti-fatigue, neuroprotective, anti-depressant, physical performance enhancer, life span–increasing and central nervous system (CNS) stimulating effects as shown in the figure given below (Figure 15.1) (Murray et al. 2020). Most adaptogens target and affect the neuroendocrine-immune system by regulating the onset of various adaptive signaling pathways that assist cell survival and bolster stamina for stress regulation, metabolism and homeostasis by the release of specific hormones/signaling molecules and their cognate receptors (Maimes 2007). Adaptogens can be classified into two major categories: plant adaptogens and actoprotectors. Plant adaptogens are composed of a rich complex of biologically active compounds produced by plants. The prime advantage of plant adaptogen consumption is that the chance of any serious side effects is low (Todorova et al. 2021). Actoprotectors, on the other hand, are regarded as synthetic adaptogens having the capability to enhance physical performance. Additionally, they promote vasodilation and reduce blood glucose and lactate levels. For example, the

DOI: 10.1201/b22917-15

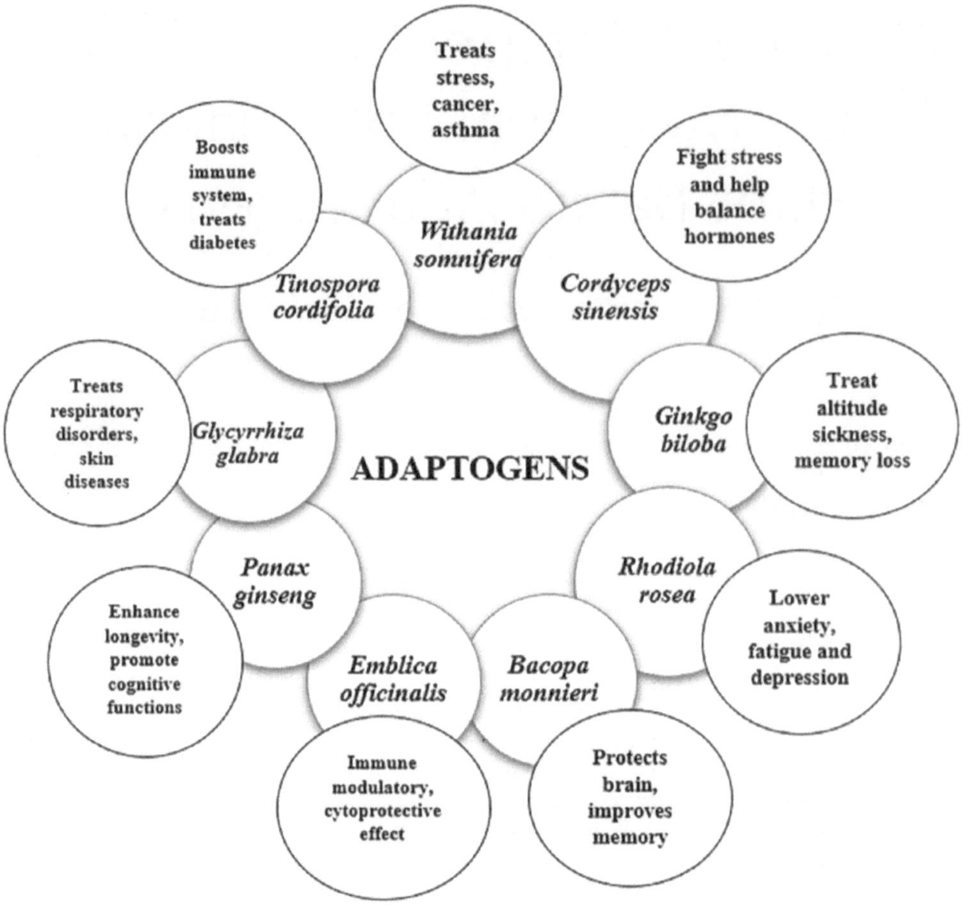

FIGURE 15.1 Health benefits of adaptogens.

benzimidazole derivative bemitil and adamantane derivatives bromantane and chlodantane are considered actoprotectors (Oliynyk et al. 2012).

Focusing on plant adaptogens, *Withania somnifera* is one of the most studied and widely used adaptogen followed by *Panax ginseng*. The diverse adaptogenic potential of *W. somnifera* includes control of blood sugar, triglycerides and cholesterol levels; reduces anxiety and stress; and aids in insomnia (Firdous and Marwah 2020a, 2020b). Similarly, *P. ginseng* is found to enhance immunity, promote muscle growth and treat anxiety, hypothyroidism and diabetes. Mixtures of their extracts can be used as geriatric tonics, and they are found to enhance hematopoiesis (Aphale et al. 1998; Firdous and Marwah 2020a, 2020b). It is notable here that the conventional pharmacological paradigm—"one disease, one drug"—is not applicable to adaptogens as they exhibit multiple biological actions and are more inclined to adhere to the emerging field of polypharmacology.

Up to recent times, a limited number of herbs were pursued as adaptogens; however, the pharmaceutical industry, as well as practitioners of traditional medicine, are increasingly classifying and proposing more herbs as adaptogens. An accumulating literature advocates *Bacopa monnieri, Emblica officinalis, Ginkgo biloba, Rhodiola rosea, Cordyceps sinensis, Glycyrrhiza glabra* and *Tinospora cordifolia* as adaptogens. These herbs have been reported to exhibit multivalent favorable effects against aging-related diseases, neurodegenerative and cognitive impairment, adrenal fatigue, arthritis, insomnia, chronic inflammation, as shown in Figure 15.1

(Panossian et al. 2008; Panossian et al. 2010; Khanum et al. 2005; Leung et al. 2010). However, these newly endorsed adaptogenic plants need extensive experimental validation and exploration in terms of their curative potential and the underlying molecular mediators responsible for the benefits.

15.2 OVERVIEW OF PROTECTIVE ACTIONS OF ADAPTOGENS

In order to attain homeostasis, adaptogens target different organs and tissues in the human body. The herbal extract regulates adrenal gland secretions and hormonal balance to reduce stress and targets damaged tissues and joints to show the anti-inflammatory effect, reducing the pain and swelling caused by arthritis; in order to increase adaptability and survival, they target the neuroendocrine system and enhance the sugar, oxygen and protein utilization rates (Liao et al. 2018).

The major pharmacological actions of adaptogens include reducing stress-related behavioral disorders, mental illness, regulation of the hypothalamic–pituitary–adrenal (HPA) axis, glucocorticoid receptors and production of cortisol (Liao et al. 2018). Phytoadaptogens are herbal pharmaceuticals acting on numerous targets and disease pathways and fit perfectly into the example of polypharmacology. Therefore, different molecular network interactions of multifold adaptogens are responsible for lessending the negative impact that stress has on human health. The traditional methods being dispiriting, network pharmacology of adaptogens is the contemporary way, with the potential to provide treatment for complex diseases and to increase survival and adaptability to stress.

Hans Selye's theory of stress uses a model that divides stress into three phases that are compared to a normal level of resistance the body offers: The alarm phase is the initial reaction of an individual subjected to a new environment and is characterized by the fight-or-flight response. In this phase the body copes up by releasing adrenaline and noradrenaline. The phase of resistance is characterized by how the body learns and adapts to the stressor, and the phase of exhaustion is characterized by loss of the body's adaptogenic ability, also sometimes referred as adrenal maladaptation. When an individual is exposed to stressful conditions such as electric shock or an accident, they must have the ability to respond and also predict these events to adapt and survive. Adaptogens suppress the stress response by delaying the exhaustion stage and by reducing the signs in the alarm phase (Panossian et al. 2010).

There is abundant scientific evidence obtained from studies on animal models and isolated neuronal cells that proves adaptogens offer neuroprotection, anti-fatigue, anti-stress, anti-inflammatory, anti-viral, antioxidant, anti-depressive, anxiolytic and CNS-stimulating activity (Panossian et al. 2010). Some other practical utilities of adaptogens are as therapeutic agents for remedying neurological diseases, metabolic disorders, acute respiratory diseases and as an adjuvant in the treatment of tuberculosis and cancer. All of these diseases affect the body's ability to cope with stress and disrupt homeostasis (Panossian 2017). Identification of molecular targets and signaling pathways common to adaptogens showed association with genes responsible for stress response and regulation of homeostasis (molecular chaperone Hsp70, neuropeptide Y, corticotropin-releasing hormone [CRH], urocortin [UCN], G protein–coupled receptors, dopamine, NF-kB, protein kinases MAPK10, FLT1, stress-activated kinase JNK, forkhead box [FOX, FOXO3], cortisol, estrogens, nitric oxide synthase) and adaptogens also showed increased expression of the prolactin receptor (PRLR) and toll-like receptor (TLR9) genes (Panossian et al. 2018). Scientific evidence was obtained from a study by Asea et al. (2013) by administering extract obtained from adaptogens (*E. senticosus*, *S. chinensis*, *R. rosea*) in neuroglia cells, which showed increased expression and release of neuropeptide Y (NPY) and molecular chaperone Hsp72. Both of the genes responsible for the stress response are known to play immunomodulatory and neuroprotective roles, and it was thus concluded that NPY and Hsp72 are molecular biomarkers for adaptogenic activity.

15.3 ADAPTOGEN PLANTS FROM INDIAN TRADITIONAL MEDICINE

15.3.1 *Withania somnifera*

A member of the family Solanaceae, *Withania somnifera* (ashwagandha) is commonly known as Indian winter cherry, Sattvic Kapha Rasayana or Indian ginseng (Singh et al. 2011; Sharma 1938). Ashwagandha is a potent adaptogen available in a powdered form and can be taken with water or honey and is known to have a wide variety of therapeutic potentials such as immunomodulatory, anti-proliferative, neuroprotective, hepatoprotective and hypoglycemic effects (Kaul and Wadhwa 2017). The phytoactive chemical components of *W. somnifera* include alkaloids, steroidal lactones (withanolides and withaferins) and saponins. Sitoindosides and withaferin A, have been shown to have significant anti-stress activity against acute models of experimental stress (Bhattacharya et al. 1987). The main active component of *W. somnifera* is glycowithanolides, with the capacity to upgrade the human defense system against many diseases, increase longevity and enhance physical and mental health (Srivastava et al. 2019). Many scientific studies have demonstrated the use of *W. somnifera* in reducing stress and hypertension, increasing swimming performance and averting stress-related ulcers (Rathi et al. 2021; Baliga et al. 2015). Ashwagandha reportedly prevents the decrease in ascorbic acid content and adrenal cortisol content, and the root extracts of *W. somnifera* show various health benefits in cognitive and neurological disorders, anxiety-related behavioral disorders, depression and primarily in the management of stress by decreasing the serum cortisol level in the human body (Salve et al. 2019). In a 60-day randomized, double-blind, placebo-controlled trial a dosage of 240 mg *W. somnifera* extract in stressed healthy adults compared with the placebo was investigated for stress-relieving activity of an ashwagandha extract. Ashwagandha intake was found to be effective, with a significant reduction in the DASS-21 (Depression Anxiety Stress Scale-21). Ashwagandha intake was also associated with a decrease in levels of cortisol and dehydroepiandrosterone sulfate (DHEA-S) compared with the placebo (Lopresti et al. 2019)

15.3.2 *Panax ginseng*

Panax ginseng, also known as the king of all herbs, belongs to the family Araliaceae and is known to possess many pharmacological properties like enhancing work performance, boosting the immune system and stimulating psychological functioning (Kaifer and Pantuso 2003). It reportedly exhibits antioxidant, anti-microbial, anti-inflammatory, anti-obesity, anti-cancer, anti-diabetic and anti-cardiovascular disease activities (Ratan et al. 2021). The major bioactive compounds found in *P. ginseng* are ginsenosides which show various activities in pathological and physiological conditions; for instance, it enhances nitric oxide (NO) production, stimulates CNS functions and reduces reactive oxygen species (ROS) production (Lu et al. 2009). *P. ginseng* is believed to increase the function of adrenocorticotropic hormone in the brain without entailing any adrenal glands, thereby developing a generalized memory and sensory awareness (Rodriguez et al. 2019). In a study conducted by Rai et al. *Gingko biloba* and *P. ginseng* were found most effective in the 30 mg/kg p.o. and 100 mg/kg p.o. dose, respectively, for anti-stress activities of these extract.

15.3.3 *Bacopa monnieri*

Bacopa monnieri, also known as brahmi (Hindi) and water hyssop (English), is a member of the Plantaginaceae family, and the part of the perennial plant that has medicinal value is its leaves. Reports suggest that it contains alkaloids such as brahmin, nicotine and herpestine; potassium salts, sterols like beta sterols; and saponins like herasapononin, bacosides and bacosaponins (Pandey 2022). Bacosides A is mainly responsible for the antioxidant and neuropharmacological effects by helping it to cross the blood-brain barrier (BBB) (Anbarasi et al. 2006). Scientific evidence suggests the use of methanolic and ethanolic extracts of *B. monnieri* is effective in treatment of a wide range of diseases like Alzheimer's disease (Chaudhari et al. 2017), asthma, cancer, inflammation, etc. Rai et al. (2003) investigated a standardized preparation of *B. monnieri* for its adaptogenic response to acute and chronic stress in rats by exposing them to immobilization stress. It was found that *B. monnieri* extract alleviates the stress-induced increase in the ulcer index, weight of the adrenal gland, blood sugar, alanine aminotransferase, aspartate aminotransferase and creatine kinase. Chowdhuri et al. (2002) conducted a study on rats that showed the adaptogenic potential of the bacosides of *B. monnieri* by altering the activities of Hsp70, P450 and superoxide dismutase (SOD), which prepares the brain to mount an appropriate response at the time of an emergency.

15.3.4 *EMBLICA OFFICINALIS*

Ebblica officinalis, also known as Indian gooseberry or amla, is a deciduous tree that belongs to the Euphorbiaceae family and has been used as a medicinal plant in the Indian traditional system of medicine as rasayana (Ayurveda) to treat constipation and jaundice, alleviate fever, strengthen the heart, reduce inflammation and as a digestive (Baliga and Dsouza 2011). The pharmacological effects of *E. officinalis* are attributed to the presence of a number of biologically active compounds, tannins, alkaloids and phenols. Emblicanin A and B are two important hydrolysable tannins present in its fruit which show antioxidant effects similar to ascorbic acid (cardioprotective), which on hydrolysis yields gallic acid, ellagic acid (phenols) and glucose. Ellagic acid is reported to prevent mutations at the gene level and chromosomal abnormalities, whereas gallic acid offers anti-diabetic and cardioprotective potential (Dasaroju et al. 2014). Along with these, amla is a rich source of vitamin C (478.56 mg/100 mL) when compared to other fruits such as oranges, tangerines or lemons (Variya et al. 2016). Neelima et al. (2014) evaluated the antistress activity of *E. officinalis* on rats subjected to acute physical stress (forced swimming endurance stress), and it reduced stress by decreasing the high glucose levels, which shows its inhibitory effect on adrenal cortex activity. It also leads to a reduction in liver, brain, kidney and adrenal gland weight, which occurs due to the induced adrenomedullary response.

15.3.5 *GINKGO BILOBA*

G. biloba, a member of the Ginkgoceae family, is commonly known as kew tree, fossil tree and yinhsing. Major phytochemicals that can be extracted from the leaves of this tree are flavonoids and terpene lactones. The flavonoids play a major role in the antioxidant properties by scavenging the free radicals directly (Smith and Luo 2003). The terpene lactones present in gingko are of two types: the ginkgolides and bilobalides. The ginkgolides enhance coronary blood flow through platelet-activating factor (PAF) antagonists' activity by reducing blood clotting in cardiovascular diseases (Mahady 2002). Bilobalide aids in the prevention of neurodegenerative diseases by increasing cerebral blood flow by activating norepinephrine secretion (Yang et al. 2005). Ginkgolides A and B were reported to decrease the activity of the peripheral benzodiazepine receptor (PBR) by reducing the ligand-binding ability of the receptor and decreasing its messenger RNA (mRNA) expression, which further reduced corticosteroid levels, leading to a decreased level of glucocorticoids in the systemic circulation (Amri et al. 1996). The ginkgo leaf extract (EGb 761) is a standardized preparation composed of flavonoid glycosides (24%) and terpene lactones (6%), ginkgolic acid (<5%) tested EGb 761 for its possible role in laboratory rats on stress and anxiety when subjected to the Morris Water Maze. The results showed a decrease in stress levels, as EGb 761 is a serotonin antagonist and stimulates 5-HT1A receptors. It has also been shown to reduce the basal levels of monoamine oxidase.

15.3.6 *Rhodiola rosea*

Rhodiola rosea, commonly known as golden root, rose root or Arctic root, belongs to the plant family Crassulaceae and has long been used as a medicinal plant in Europe. The rose root is used as a therapeutic agent, while the green aerial part of the plant is used as a raw ingredient in the food industry. Results from recent studies have revealed its wide variety of functions such as stimulating the nervous system and memory, decreasing depression, reducing stress, enhancing work productivity, ameliorating fatigue and preventing high-altitude sickness. The roots contain a range of phytochemicals which include flavonoids, rosavins, tannins and phenolic glycosides. The adaptogenic properties (neuroprotective, cardioprotective, anti-fatigue, anti-stres, anxiolytic) of *R. rosea* is ascribed primarily to salidroside and rosavin (Chiang et al. 2015). Evidence suggests that salidroside offers protection against beta-amyloid peptide–induced oxidative stress by dephosphorylating JNK and p38 MAP kinase, but not ERK1/2, which suggests the usefulness of salidroside for treating or preventing neurodegenerative diseases (Panossian et al. 2007). The adaptogenic capability can be attributed to its ability to modulate monoamines and opioid peptides such as beta-endorphin levels and activity (Kelly 2001). Many scientific studies have supported that administration of *R. rosea* extract SHR-5 exerts an anti-fatigue and anti-stress effect that enhances mental performance and minimizes the chances of experiencing burnout symptoms (Olsson et al. 2009). Students treated with *R. rosea* extract (RRE) for self-diagnosed anxiety and stress were randomized to receive either RRE or a control (no treatment) and reported significant reduction in anxiety and stress as compared to the control group (Cropley et al. 2015).

15.3.7 *Cordyceps sinensis*

Cordyceps sinensis, a fungus, is a member of the family Clavicipitaceae and commonly known as winter worm summer grass, caterpillar fungus and traditionally called herbal Viagra (Shashidhar et al. 2013). Nucleosides, polysaccharides and sterols are considered the main bioactive components of this ascomycete (Liu et al. 2015). Cordycepin is a nucleoside with significant therapeutic potential which includes anti-tumor and pro-apoptotic activity (on cancer cells) by activating adenosine receptors, which further trigger the protein kinase C pathway by the action of either of these compounds: diacylglycerol, phospholipase C and calcium ions. This signaling cascade positively regulates steroid production, which leads to increased progesterone production that exerts an anti-tumor effect (Qin et al. 2019). *C. sinensis* has been shown to be effective in renal and liver diseases and also serves as an anti-inflammatory, antioxidant, anti-tumor and anti-apoptotic agent (Shashidhar et al. 2013). Koh et al. (2003) conducted a study to investigate the anti-fatigue and anti-stress effects of *C. sinensis* in rats and mice by administering a hot water fraction of its mycelia with carbohydrate (78.9%) and moisture (5%) content. The swimming endurance capacity of mice was found to increase significantly from 75 to 90 min and fatigue decreased. Rats exposed to 48 h stress when administered with the HW fraction (150 mg/kg/d) for 8 d saw an increase in cholesterol levels and decrease in alkaline phosphatase and showed changes in weight of the adrenal gland, spleen, thymus and thyroid, which indicates decreased stress levels.

15.3.8 *Glycyrrhiza glabra*

Glycyrrhiza glabra, popularly known as licorice, mulahatti (Hindi) and sweet wood, is a hardy perennial shrub that belongs to the Leguminosae family (Saxena 2004). The roots of the shrub are a reservoir for a saponin that offers 60 times more sweetness than sucrose, i.e., glycyrrhizin, including flavonoids, isoprenoid-substituted phenolic constituents, isoflavones and coumarins (Kaur et al. 2013). Many clinical studies have provided scientific proof of the pharmacological properties such as hepatoprotective, anti-inflammatory, anti-viral, anti-ulcer and immunoregulatory effects. Glycyrrhizic acid and glycyrrhetic acid form the two major metabolites of licorice extract. Glycyrrhetic acid exists in two isomeric forms, i.e.,18a and 18b glycyrrhetic acid, which have different roles. 18a-glycyrrhetinic acid inhibits only the 11b-HSD1 enzyme but not 11b-HSD2, whereas 18b-glycyrrhetinic acid has an inhibitory effect on both enzymes 11bHSD1 and 11b-HSD2. These two 11b-HSD isozymes catalyze the interconversion between cortisol and cortisone. The enzyme 11b-HSD1 converts inactive cortisone into active cortisol in glucocorticoid target tissues such as adipose tissue, muscle and liver. Thus, 11b-HSD1 inhibition by glycyrrhetic acid could ameliorate insulin resistance and type 2 diabetes, whereas the enzyme 11b-HSD2 converts active cortisol into inactive cortisone and plays a significant role in increased activity of mineralocorticoid tissue by cortisol, which is responsible for the anti-inflammatory, immunomodulatory and antioxidant properties of licorice (Kwon et al. 2019).

15.3.9 *Tinospora cordifolia*

Tinospora cordifolia, also called guduchi (Sanskrit) and giloy (Hindi), is a deciduous climbing shrub that belongs to the Menispermaceae family. The plant has a succulent stem and contains various classes of phytoactive molecules such as alkaloids, terpenoids, steroids, lactones, glycoside, polyphenols and so on. It is widely used in Indian Ayurvedic medicine because of its anti-inflammatory, neuroprotective, immunomodulatory, antioxidant, anti-diabetic, anti-spasmodic, anti-stress, anti-malarial, hepatoprotective, anti-allergic and anti-arthritic properties (Joshi and Kaur 2016). A study involving the pretreatment of animals exposed to stress with a blend of extract of *Asparagus racemosus* and *T. cordifolia* (administered orally 400 mg /kg b.wt.) was conducted to evaluate the anti-stress activity by studying the biochemical changes induced by the Cold-Water Swim Stress. Diazepam (5 mg/ kg b.wt.) was used as a gold standard. Serum lipid, serum glucose, plasma cortisol, plasma glutathione, total plasma NO and lipid peroxidation in terms of malondialdehyde (MDA) were used as the stress indices. The results showed increased levels of total plasma NO, MDA, a lipid peroxidation marker, and cortisol in rats with chronic stress compared with a healthy control group (Biswas and Saha 2015).

15.4 GLOBAL COMMERCIAL VALUE OF ADAPTOGENS

The era of increasing stress, anxiety, diseases and the desire for healthy food has led to an influential growth in the global adaptogens market. The extracts of different adaptogens are used to make herbal tonics, oils, tablets, drinks, cereals, snacks, cosmetics and many other products are listed in Table 15.1. The food and beverage industry holds the biggest market share followed by the pharmaceutical and nutraceutical industries (Maximize Market Research Pvt. Ltd 2021). According to the Global Organization of Stress, in a report of April 2021 almost 75% of the population has experienced high to moderate levels of stress, which further leads to health risks like headache, heart disease, asthma, gastrointestinal problems, accelerated aging and premature death (Knowledge Sourcing Intelligence 2021). Adaptogens and their products reportedly show favorable results and can be used as a novel pharmaceutical remedy for increasing stress and related diseases. Based on regional analysis, North America is the leader of the adaptogenic business market because of the majority of residents switching to natural herbal products for health treatments, followed by Asia Pacific, Europe and the rest of the world (New Hope Network 2017). On the basis of application, the extracts are used in small amounts in meals and as supplements brewed with tea or milk. Ashwagandha and ginseng are the prime adaptogens being used in major herbal products in the medical industry and are consumed to boost immunity, lower stress and anxiety and to regulate blood pressure (Verified Market Research 2020). There is an increasing demand for adaptogenic products in the sports nutrition sector to reduce the mental stress and enhance the immunity of athletes. Companies are investing in adaptogens in the Ayurvedic sectors for product development, but

TABLE 15.1
List of Commercial Adaptogen Products and Their Pharmacological Activities

S. No.	Adaptogen	Commercial Products	Pharmacological Activity	Link to Products
1.	*Withania somnifera*	Ashwagandharista Stresscom Vital Plus	Memory and cognition improvement Relieves anxiety and mental stress Recovery from general weakness and fatigue	www.dabur.com/in/en-us/ayurvedic-herbal-products/dabur-ashwagandharishta www.dabur.com/in/en-us/ayurvedic-herbal-products/stresscom https://mukthipharma.com/?product=vital-plus-granule
2.	*Panax ginseng*	Healthy Hey Nutrition Panax Ginseng Extract GNC Herbal Plus Triple Ginseng Root Veda Oils Ginseng Liquid Extract	Supports physical and intellectual work capacity Improves strength and stamina, promotes vitality and supports mental alertness Antioxidant and anti-inflammatory properties, strengthens immune system	www.healthyhey.com/products/healthyhey-nutrition-panax-ginseng-400mg-90-vegetable-capsules www.guardian.in/gnc-herbal-plus-triple-ginseng-root-90-vegetarian-capsules.html?gclid=Cj0KCQiAoY-PBhCNARIsABcz771VWs69T-qoPboXaGKbTS9FfbJB0XRZauAxeoPbnEZPgWm_onWV-QUaApXBEALw_wcB www.vedaoils.com/products/ginseng-liquid-extract
3.	*Bacopa monnieri*	Baidyanath Junior Chyawanprash Amrita Drugs *Brahmi* Powder Dabur Chyawanprash	Provides brain and body nourishment Improves memory and concentration level, relieves stress Ensures immunity, mental and physical fitness, provides antioxidants and strengthens immune system	www.baidyanath.com/junior-prash-1kg.html http://amritadrugs.com/Ingredient/b-brahmi.php www.dabur.com/in/en-us/ayurvedic-herbal-products/dabur-chyawanprash
4.	*Emblica officinalis*	Way 4 Organic Emblica Officinalis Raw Powder Herbal and Natural Products Amla Powder H&C Amla Capsules	Boosts protein metabolism, improves eyesight, strengthens heart muscles, prevents early aging and graying of hair Improves memory and brain health, improves skin health and hair growth Antioxidant property, supports healthy metabolism and digestion	www.indiamart.com/proddetail/organic-emblica-officinalis-11549992233.html www.flipkart.com/natural-herbal-products-amla-powder-indian-gooseberry-emblica-officinalis-hair-growth-skin-care-face-mask-lose-weight-200gram/p/itm34eef1a3dd94e https://arabianorganics.com/h-c-amla-capsules.html
5.	*Ginkgo biloba*	Healthvit Ginkgo biloba (60 mg) capsule Forever ginkgo plus 60 tablets Vista Nutritions Ginkgo biloba 60mg	Enhances healthy brain function and alertness Boosts energy level and supports circulation Improves memory and mental alertness	https://healthvit.com/products/healthvit-ginkgo-biloba-supports-memory-focus-clarity-60mg-60-capsules https://shopforeverproducts.com/shop/nutritional/forever-ginkgo-plus www.healthkart.com/sv/vista-nutrition-ginkgo-biloba-60-mg/SP-8795?navKey=VRNT-13666

S. No.	Adaptogen	Commercial Products	Pharmacological Activity	Link to Products
6.	*Rhodiola rosea*	NOW Rhodiola 500mg Gaia Herbs, Rhodiola Rosea Liquid Phyto-caps Green Naturals Rhodiola rosea	Helps body adapt to everyday stress Supports healthy stamina and relieves stress Stimulates nervous system, improves performance and reduces stress	www.nowfoods.com/products/supplements/rhodiola-500-mg-veg-capsules www.gaiaherbs.com/products/rhodiola-rosea https://kiwla.com/products/rhodiola-rosea-120-capsules-yy-500mg-german-production-100-vegan-and-without-additives-4-month-stock-stress-management-gift-guide-ebook
7.	*Cordyceps sinensis*	CORDYCEPS-MRL Oxford Vitality Cordyceps Sinensis Extract 240mg Four Sigmatic Mushroom Cacao Mix with Cordyceps	Support body's immune system Stimulates body's natural mechanisms and relieves symptoms of mental and physiological stress Increases energy and physical performance	www.mycologyresearch.com/products/cordyceps https://oxfordvitality.co.uk/products/cordyceps-sinensis-tablets?variant=34728284684454 https://us.foursigmatic.com/products/fsf-xoco-red
8.	*Glycyrrhiza glabra*	Mulethi Licorice Powder by PureNaturals Deve Herbes Pure Liquorice Oil Glycyrrhiza glabra Glycyrrhiza glabra MT Tablets	Anti-inflammatory, antioxidant effects and cures skin ailments Enhances immunity, anti-allergic, anti-depressant, anti-pyretic and promotes longevity Anti-inflammatory, inhibits the growth of prostate cancer and useful in sore throat, colds and bronchitis	www.bypurenaturals.com/collections/organic-herbs/products/organic-mulethi-licorice-powder-bypurenaturals www.deveherbes.com/Pure-Liquorice-Oil-Glycyrrhiza%20-glabra www.schwabeindia.com/speciality-formulations/84-glycyrrhiza-glabra-1x.html
9.	*Tinospora cordifolia*	Daivik Moringa's Giloy Capsules JIVA Ayurveda Giloy Immunity Enhancer Tridosha Balancing Capsule Giloy by Arogyam Ayurveda	Boosts immunity, provides relief from muscle pain, reduces body pain, fatigue and inflammation Improves immunity, purifies blood, reduces colds and cough and provides anti-inflammatory benefits Immunity booster, antioxidant activity	https://daivikmoringa.com/product/giloy-veg-capsules/ https://store.jiva.com/products/jiva-giloy-capsule-2/ https://store.arogyamayurveda.com/product/giloy-by-arogyam-ayurveda-60-tablets/

certain regulatory considerations and high costs of research and development (R&D) activities are still a drawback to the research and trials of adaptogenic extracts.

15.5 CHALLENGES AND OPPORTUNITIES IN PLANT-BASED ADAPTOGENS RESEARCH

- **Acceptability of the term "adaptogen":** In 2008 the European Medicines Agency published a report by the Committee on Herbal Medicinal Products titled "Reflection Paper on the Adaptogenic Concept".[1] The article reviewed the adaptogenic action of the herb *Eleutherococcus* published over five decades and suggested that the mechanism of action associated with adaptogenic activity is equivocal and requires further validation in various preclinical and clinical studies. Specifically, the clinical studies need to increase stringency in terms of study design, exclusion criteria and sample size, in light of the unreliability of the mentioned criteria in the published clinical investigations for *Eleutherococcus*. However, the term "adaptogen" has gained scientific acceptability, as evident in the published literature in PubMed (818 results retrieved on 19 March 2022 in PubMed on the keyword "adaptogen").
- **Reproducibility of the clinical studies:** Ascertaining reproducible pharmacological activity is a challenge in herbal medicine, and there are numerous instances of contradictory/no observable benefit of adaptogen supplementation in stress-related disorders. For instance, *R. rosea* was established to have a significant preclinical benefit in alleviating symptoms associated with hypobaric hypoxia (high-altitude hypoxia) exposure; however, extending its usage for clinical use was observed to confer no significant benefit against the incidence of common high-altitude disease—acute mountain sickness (Chiu et al. 2013).
- **Network pharmacology-based strategy for unraveling the mechanistic mediators of adaptogens:** Hopkins, in his seminal paper in 2007, introduced the term "network pharmacology". Network pharmacology shifts the focus from the oversimplified approach of "one drug, one target" to the more realistic systems approach, which profiles the interaction of one molecule with multiple protein targets and interactomes thereof (Hopkins 2007; Zhou et al. 2020). The adaptogens are excellent candidates for analytical approaches based on network pharmacology, as their molecular targets include hormones, key upstream signaling proteins and transcription factors, which are bound to perturb a larger spectrum of protein-protein interactions as compared to conventional small-molecule drugs. Thus, the authors endorse the utility of network pharmacology for unraveling the molecular mechanisms of adaptogens.

15.6 CONCLUSION

Plant-based adaptogens have been demonstrated to show an extensively large variety of medicinal properties like enhancing the immune system; stress regulation; treatment and prevention of anxiety disorders, hypertension and memory impairment; and in curing many neurodegenerative and cardiovascular diseases. However, the "adaptogen" research has obstacles like acceptability of the adaptogenic activity concept, reliable and reproducible clinical studies and the equivocal knowledge of molecular mechanisms for the basis of adaptogenic activity. Therefore, further investigations are warranted to establish the molecular mechanisms of action of various adaptogenic herbs/extracts/active principal ingredients to unravel and exploit the therapeutic potential and curative effects on human health.

1 Committee on herbal medicinal products (2008).

BIBLIOGRAPHY

Adaptogens Market: Insights, and Forecast Analysis (2021–2027) by Nature, Source, Application, Distribution Channel, and Region (2021). *Maximize Market Research Pvt. Ltd.* https://www.maximizemarketresearch.com/market-report/adaptogens-market/123228/

Adaptogens Market Size and Forecast (2020). *Verified Market Research.* https://www.verifiedmarketresearch.com/product/adaptogens-market/

Adaptogens Power Up Stress and Sleep Solutions (2017). *New Hope Network.* https://www.newhope.com/ingredients/adaptogens-power-stress-and-sleep-solutions-next-forecast-2017

Amri, H., S. O. Ogwuegbu, N. Boujrad, K. Drieu, V. Papadopoulos. 1996. In vivo regulation of peripheral-type benzodiazepine receptor and glucocorticoid synthesis by Ginkgo biloba extract EGb 761 and isolated ginkgolides. *Endocrinology.* 137: 5707–5718.

Anbarasi, K., G. Vani, K. Balakrishna, C. S. Devi. 2006. Effect of bacoside-A on brain antioxidant status in cigarette smoke exposed rats. *Life Sci.* 78: 1378–1384.

Aphale, A. A., A. D. Chhibba, N. R. Kumbhakarna, M. Mateenuddin, S. H. Dahat. 1998. Subacute toxicity study of the combination of ginseng (Panax ginseng) and ashwagandha (Withania somnifera) in rats: A safety assessment. *Indian J Physiol Pharmacol.* 42(2): 299–302.

Asea, A., P. Kaur, A. Panossian, K. G. Wikman. 2013. Evaluation of molecular chaperons Hsp72 and neuropeptide Y as characteristic markers of adaptogenic activity of plant extracts. *Phytomedicine.* 20(14): 1323–1329.

Baliga, M. S., J. J. Dsouza. 2011. Amla Emblica officinalis Gaertn, a wonderberry in the treatment and prevention of cancer. *European Journal of Cancer Prevention.* 20(3): 225–239.

Baliga, M. S., S. Meera, A. R. Shivashankara, P. L. Palatty, R. Haniadka. 2015. The health benefits of Indian traditional ayurvedic rasayana (anti-aging) drugs: A review. *Foods and Dietary Supplements in the Prevention and Treatment of Disease in Older Adults: 151–161.*

Belwal, T., L. Giri, A. Bahukhandi, R. Rawal, *et. al.* 2018. *Ginkgo Biloba: Nonvitamin and Nonmineral Nutritional Supplements.*

Bhattacharya, S. K., R. K. Goel, R. Kaur, S. Ghosal. 1987. Anti-stress activity of Sitoindosides VII and VIII: New acylsterylglucosides from withania somnifera. *Phytother Res.* 1: 32–37.

Biswas, P., A. Saha. 2015. Evaluation of antistress activity of extract of *Tinospora Cordifolia & Asparagus Racemosus* in rats. *International Journal of Ayurveda and Pharma Research.* 3(8).

Chaudhari, K. S., N. R. Tiwari, R. R. Tiwari. 2017. Neurocognitive effect of nootropic drug brahmi (bacopa monnieri) in Alzheimer's disease. *Annals of Neuroscience.* 24(1): 111–122.

Chiang, H. S., H. C. Chen, C. S. Wu, P. Y. Wu, K. C. Wen. 2015. Rhodiola plants: Chemistry and biological activity. *Journal of Food and Drug Analysis.* 23(3): 359–369.

Chiu, T. F., L. L. Chen, D. H. Su, H. Y. Lo, C. H. Chen, S. H. Wang, W. L. Chen. 2013, Oct 31. Rhodiola crenulata extract for prevention of acute mountain sickness: A randomized, double-blind, placebo-controlled, crossover trial. *BMC Complement Altern Med.* 13: 298.

Chowdhuri, D. K., D. Parmar, P. Kakkar, R. Shukla, P. K. Seth, R.C. Srimal. 2002. Antistress effects of bacosides of Bacopa monnieri: Modulation of Hsp70 expression, superoxide dismutase and cytochrome P450 activity in rat brain. *Phytotherapy Research.* 16(7): 639–645.

Cropley, M., A. P. Banks, J. Boyle. 2015. The effects of Rhodiola rosea L. extract on anxiety, stress, cognition and other mood symptoms. *Phytotherapy Research.* 29: 1934–1939.

Committee on herbal medicinal products (2008). *European Medicines Agency Evaluation Of Medicines For Human Use.* https://www.ema.europa.eu/en/documents/scientific-guideline/reflection-paper-adaptogenic-concept_en.pdf

Dasaroju, S., K. Gottumukkala. 2014. Current trends in the research of Emblica officinalis (Amla): A pharmacological perspective. *International Journal of Pharmaceutical Sciences Review and Research.* 24: 150–159.

Firdous, H., S. Marwah. (2020a). *Ashwag Andha Benefits and Its Side Effects.* https://www.lybrate.com/topic/ashwagandha-benefits-and-side-effects

Firdous, H., S. Marwah. (2020b). *Benefits of Indian Ginseng and Its Side Effects.* https://www.lybrate.com/topic/benefits-of-indian-ginseng-and-its-side-effects

Global Adaptogens Market Size, Share, Opportunities, COVID-19 Impact, and Trends by Nature, by Source, Applications, by Distribution Channel and by Geography- Forecast (2020 To 2025), (2021). *Knowledge Sourcing Intelligence.* https://www.knowledge-sourcing.com/report/global-adaptogens-market

Hopkins, A. L. 2007, Oct. Network pharmacology. *Nat Biotechnol.* 25(10): 1110–1111.

Joshi, G., R. Kaur. 2016. Tinospora cordifolia: A phytopharmacological review. *International Journal of Pharmaceutical Sciences and Research.* 7(3): 890–897.

Kaul, S. C., R. Wadhwa. 2017. *Science of Ashwagandha: Preventive and Therapeutic Potentials.* Springer International Publishing, New Delhi.

Kaur, R., H. Kaur, A. S. Dhindsa. 2013. Glycyrrhiza glabra: A phytopharmacological review. *Int J Pharm Sci Res.* 4(7): 2470–2477.

Kelly, G. S. 2001. Rhodiola rosea: A possible plant adaptogen. *Altern Med.* 6: 293–302.

Khanum, F., A. S. Bawa, B. Singh. 2005. Rhodiola rosea: A versatile adaptogen. *Comprehensive Reviews in Food Science and Food Safety.* 4: 55–62.

Kiefer D., T. Pantuso. 2003. Panax ginseng. *Am Fam Physician.* 15;68(8):1539–42.

Koh, J. H., K. M. Kim, J. M. Kim, J. C. Song, H. J. Suh. 2003. Anti Fatigue and antistress effect of the hot-water fraction from mycelia of Cordyceps sinensis. *Biological and Pharmaceutical Bulletin.* 26: 691–694.

Kwon, Y. J., D. H. Son, T. H. Chung, Y.-J. Lee. 2019. A review of the pharmacological efficacy and safety of licorice root from corroborative clinical trial findings. *Journal of Medicinal Food.* 23(1): 1–9.

Leung, K. W., K. S. Wong. 2010. Pharmacology of ginsenosides: A literature review. *Chinese Med.* 5(20): 1–7.

Liao, L. Y., Y. He, L. Li, H. Meong, Y. Dong, *et al.* 2018. A preliminary review of studies on adaptogens: Comparison of their bioactivity in TCM with that of ginseng-like herbs used worldwide. *Chinese Medicine* 13: 57.

Liu, Y., J. Wang, W. Wang, H. Zhang, X. Zhang, C. Han. 2015. The chemical constituents and pharmacological actions of cordyceps sinensis. *Evidence-Based Complementary and Alternative Medicine.* 2015: 1–12.

Lopresti, A. L., S. J. Smith, H. Malvi, R. Kodgule. 2019. An investigation into the stress-relieving and pharmacological actions of an ashwagandha (*Withania somnifera*) extract. *Medicine.* 98(37).

Lu, J. M., Q. Yao, C. Chen. 2009. Ginseng compounds: An update on their molecular mechanisms and medical applications. *Current Vascular Pharmacology.* 7(3): 293–302.

Mahady, G. B. 2002. Ginkgo biloba for the prevention and treatment of cardiovascular disease: A review of the literature. *J Cardiovasc Nurs.* 16: 21–32.

Maimes, S. 2007. *Adaptogens: Herbs for Strength, Stamina, and Stress Relief.*

Murray ND, M. T., J. Nowicki ND. 2020. 99: *Panax ginseng* (Korean Ginseng). *Textbook of Natural Medicine.* 5: 756–767.

Neelima, N., M. Sudhakar, B. V. S. Lakshmi. 2014. Adaptogenic activity of ethanolic extract of emblica officinalis fruits. *Pharmnest.* 5(1): 1859–1865.

Oliynyk, S., S. Oh. 2012. The pharmacology of actoprotectors: Practical application for improvement of mental and physical performance. *Biomolecules & Therapeutics.* 20(5): 446–456.

Olsson, E. M., B. Scheele, A. G. Panossian. 2009. A randomised, double-blind, placebo-controlled, parallel-group study of the standardised extract shr-5 of the roots of Rhodiola rosea in the treatment of subjects with stress-related fatigue. *Planta Med.* 75(2): 105–112.

Pandey, A. K. 2022. Bacopa monnieri (Linn.) pennell: A possible plant for impossible diseases (A review). *Advances in Pharmacology and Pharmacy.* 10(1): 35–53.

Panossian, A. 2017. Understanding adaptogenic activity: Specificity of the pharmacological action of adaptogens and other phytochemicals. *Annals of the New York Academy of Sciences* 1401: 49–64.

Panossian, A., M. Hambardzumyan, A. G. Hovhanissyan. 2007. The adaptogens rhodiola and schizandra modify the response to immobilization stress in rabbits by suppressing the increase of phosphorylated stress-activated protein kinase, nitric oxide and cortisol. *Drug Target Insights.* 2: 39–54.

Panossian, A., E. J. Seo, T. Efferth. 2018. Novel molecular mechanisms for the adaptogenic effects of herbal extracts on isolated brain cells using systems biology. *Phytomedicine.* 50: 257–284.

Panossian, A., H. Wagner. 2011. Adaptogens: A review of their history, biological activity, and clinical benefits. *Herbalgram.* 90: 52–63.

Panossian, A., G. Wikman. 2008. Pharmacology of schisandra chinensis bail: An overview of Russian research and uses in medicine. *J Ethnopharmacol.* 118(2): 183–212.

Panossian, A., G. Wikman. 2010. Effects of adaptogens on the central nervous system and the molecular mechanisms associated with their stress-protective activity. *Pharmaceuticals*. 3(1): 188–224.

Panossian, A., G. Wikman, J. Sarris. 2010. Rosenroot (Rhodiola rosea): Traditional use, chemical composition, pharmacology and clinical efficacy. *Phytomedicine* 7:481–493.

Patel, S., P. Birwal, S. Basu, G. Deshmukh, *et al.* 2018. Brahmi (*Bacopa monnieri*) as functional food ingredient in food processing industry. *Journal of Pharmacognosy and Phytochemistry*. 7: 189–194.

Qin, P., X. Li, H. Yang, Z.-Y. Wang, D. Lu. 2019. Therapeutic potential and biological applications of cordycepin and metabolic mechanisms in cordycepin-producing fungi. *Molecules*. 24: 22–31.

Rai, D., G. Bhatia, G. Palit, R. Pal, S. Singh, H. K. Singh. 2003. Adaptogenic effect of Bacopa monniera (Brahmi). *Pharmacol Biochem Behav*. 75(4): 823–830.

Ratan, Z. A., M. F. Haidere, Y. H. Hong, S. H. Park, J. Lee, *et al.* 2021. Pharmacological potential of ginseng and its major component ginsenosides. *Journal of Ginseng Research*. 45(2): 199–210.

Rathi, V., A. K. Dhingra, B. Chopra. 2021. Withania somnifera. *Naturally Occurring Chemicals Against Alzheimer's Disease*. 401–407.

Rodriguez, E., M. Petruzzello, G. Lotha, Y. Chauhan, S. Singh, *et al.* 2019. Ginseng. *Encyclopedia Britannica*.

Salve, J., S. Pate, K. Debnath, D. Langade. 2019. Adaptogenic and anxiolytic effects of ashwagandha root extract in healthy adults: A double-blind, randomized, placebo-controlled clinical study. *Cureus*. 11(12).

Sangwan, R. S., N. D. Chaurasia, L. Misra, P. Lal, *et al.* 2004. Phytochemical variability in commercial herbal products and preparations of *Withania somnifera* (Ashwagandha). *Current Science*. 86(3): 461–465.

Saxena, S. 2004. Glycyrrhiza glabra: Medicine over the millennium. *Natural Product Radiance*. 4(5): 358–367.

Sharma, G. C. 1938. Ashwagandharishta—Rastantra Sar Evam Sidhyaprayog Sangrah—*Krishna-Gopal Ayurveda Bhawan (Dharmarth Trust)* Nagpur. 743–744.

Shashidhar, M. G., P. Giridhar, K. Shankar, P. Manohar. 2013. Bioactive principles from cordyceps sinensis: A potent food supplement: A review. *Journal of Functional Foods*. 5(3): 1013–1030.

Singh, N., M. Bhalla, P. D. Jager, M. Gilca. 2011. An overview on ashwagandha: A rasayana (rejuvenator) of ayurveda. *African Journal of Traditional, Complementary, and Alternative Medicines: AJTCAM*. 8(5): 208–213.

Smith, J., Y. Luo. 2003. Elevation of oxidative free radicals in Alzheimer's disease models can be attenuated by Ginkgo biloba extract EGb 761. *J Alzheimer's Dis*. 5: 287–300.

Srivastava, A., P. Srivastava, A. Pandey, V. K. Khanna, A. B. Pant. 2019. Phytomedicine: A potential alternative medicine in controlling neurological disorders. *New Look to Phytomedicine*. 625–655.

Todorova, V., K. Ivanov, C. Delattre, V. Nalbantova, D. Karcheva-Bahchevanska, S. Ivanova. 2021. Plant adaptogens-history and future perspectives. *Nutrients*. 13(8): 2861.

Variya, B. C., A. K. Bakrania, S. S. Patel. 2016. Emblica officinalis (Amla): A review for its phytochemistry, ethnomedicinal uses and medicinal potentials with respect to molecular mechanisms. *Pharmacological Research*. 111: 180–200.

Yang, Y. L., Y. W. Su, M. C. Ng, C. L. Chang, K. T. Lu. 2005. Extract of Ginkgo biloba EGb 761 facilitates fear conditioning measured by fear-potentiated startle. *Neurosci Lett*. 383: 145–150.

Zhou, Z., B. Chen, S. Chen, M. Lin, Y. Chen, S. Jin, W. Chen, Y. Zhang. 2020, Feb 14. Applications of network pharmacology in traditional Chinese medicine research. *Evid Based Complement Alternat Med*. 2020: 1646905.

16 Workflow and Methods of In Silico Analysis of Compounds Isolated From Medicinal Plants

An Overview

Rosita A. Sherlin, Pushparaj Annadurai, Thomas Jebastin, Sriram Srinivasan, Priscilla Arockiasamy, S. Danushri, Vijay Nirusimhan, Abhinav Parashar, and Daniel A. Gideon

CONTENTS

16.1 Introduction 282
16.1.1 Phytochemicals and Their Usefulness 282
16.2 Methodological Workflow of Bioinformatics-Based Phytochemical Investigations 283
16.2.1 Overview of Workflow in Phytochemistry 283
16.2.2 Spectroscopic Investigations 283
16.2.3 Identification of Components Using NIST (Gas Chromatography-Mass Spectrometry) 284
16.2.4 Analysis of MS/MS and LC-MS Using Global Natural Product Social Molecular Networking 284
16.2.5 Other Databases/Phytochemical Repositories 286
16.2.6 Bioinformatics Tools for the Process of Drug Discovery 289
16.2.7 QSAR in Phytochemistry Research 289
16.3 Bioinformatics Tools Used in Phytomedicinal Research 294
16.3.1 Bioactivity Molecular Networking 294
16.3.2 Ligand-Based Bioactivity Prediction Tools 294
16.3.3 Effective Target Prediction Tools for the Analyzed Phytochemicals 295
16.3.4 Plant Genomics and EST Databases 295
16.3.5 Reverse Screening 296
16.3.6 Network Building and Pathway Advancement 297
16.3.7 Evaluation Through Docking and Molecular Dynamics Simulations 297
16.4 Conclusion 298
Bibliography 298

DOI: 10.1201/b22917-16

16.1 INTRODUCTION

16.1.1 Phytochemicals and Their Usefulness

Plant-based natural products have been used in ethnic medicine for several millennia (Atanasov et al., 2021; Bernardini et al., 2018). There has been a "revival" of traditional medicine forms such as Chinese medicine, Indian traditional medicine (Ayurveda, Siddha, Unani), traditional Korean medicine, Kampo and other indigenous varieties from all over the globe (Yuan et al., 2016). It is estimated that about one-third of adults in developed countries and as much as 80% of the adult population in developing countries use traditional/ethnic medicine in some shape or form to treat ailments (Ekor, 2014). Nearly 70% of modern medicines in India are derived from natural sources (Verma and Singh, 2008), and almost 25% of them come from higher plants. In a comprehensive review of the importance of natural products in medicine, Newman and Cragg (2016) showed that most of the drug compounds that have been approved by drug regulatory agencies around the world (such as the Food and Drug Administration [FDA] in the United States and categorized as biological natural products, unmodified natural compounds, natural product derivatives and natural product mimics) are mostly from natural sources, derived from natural products or are synthetic mimetics; only about 30% of the compounds are purely synthetic (Newman and Cragg, 2016). Plant compounds/plant medicinal formulations have both pleiotropic (largely beneficial actions at several levels simultaneously) and synergistic effects (Singh et al., 2016), and therefore, it had been close to impossible to completely uncover all possible physiological/cellular targets of plant metabolites. Such long-standing problems are becoming easier to address through computational approaches. As more new investigations are in progress in the field of plant secondary metabolite research and phytochemistry, new databases of plants and their metabolites have emerged (Xie et al., 2015; Nguyen-Vo et al., 2019). A plethora of computational methods are increasingly becoming efficient at addressing such complex gaps/lacunae; indeed, in silico methods help us to rapidly decipher, predict, explore and pinpoint the exact molecular targets (and mechanisms of action) and absorption, distribution, metabolism and excretion (ADME) profiles of plant compounds. In vitro methods are being supported with bioinformatics analyses/tools, and these investigations lead to deep insights on the molecular mechanisms of phytochemical action. While the mechanisms of alleviation of many ailments by these phytochemical compounds are still being uncovered, many phytochemicals and their formulations are poorly tested, and their beneficial uses in the human body have not been holistically/comprehensively studied (Bode and Dong, 2015; Thakur et al., 2020); for example, focused studies on a particular signal transduction pathway are usually commonplace, and it is difficult to get the entire (and larger) picture of what plant metabolites with medicinal properties may do to the whole body in terms of positive as well as negative effects. Hence, a compound that is reckoned to be beneficial for a particular ailment may possess undesirable side effects when administered systemically. With the advent of bioinformatics in drug design and development, natural product research has become more feasible and presents a very attractive avenue for the discovery of novel bioactive compounds.

Computers have revolutionized human life and positively impacted almost every realm of human existence. Bioinformatics tools have become a boon for phytomedicine and phytochemistry research. In the field of phytochemistry and phytotherapy, bioinformatics applications have made a tremendous impact on our knowledge and application of ethnic and traditional knowledge by accelerating and enhancing diverse research areas like -omics research (metabolomics, genomics, epigenomics, microbomics, proteomics, lipidomics, transcriptomics, etc.) in both plants and humans, plant metabolite discovery, bioactivity prediction, drug discovery, safety profiling, toxicology studies and disease therapy (Lagunin et al., 2014; Sumner et al., 2003). Several plant databases and phytochemical repositories are emerging, and this is because a considerable number of plant species ranging from 35,000 to 70,000 have been explored for their medicinal properties (Fukushima and Kusano, 2013; Lagunin et al., 2014; Nakamura et al., 2013; Nguyen-Vo et al., 2019), allowing for virtual screening of phytochemicals for manifold cellular targets in the attempt to prevent as well as halt both progression and pathophysiological manifestation of many communicable and non-communicable human ailments. Chemo- and bio-informatics advances and systems biology approaches have been guideposts in mankind's attempt to study the molecular mechanisms of disease pathogenesis, progression and drug action/treatment (Kriete and Eils, 2013). Artificial intelligence (AI) and mathematical modeling (Sarker and Nahar, 2018) have been incorporated to the galaxy of bioinformatics tools/methods, and newer AI-based approaches have heralded the use of big data analytics in the -omics fields (Chen et al., 2020).

16.2 METHODOLOGICAL WORKFLOW OF BIOINFORMATICS-BASED PHYTOCHEMICAL INVESTIGATIONS

16.2.1 Overview of Workflow in Phytochemistry

Plant metabolites are separated with the aid of gas chromatography-mass spectrometry (GC-MS) and liquid chromatography-mass spectrometry (LC-MS) techniques (Wagner et al., 2003); the metabolites are fractionated/purified in bulk with the help of preparative high-performance liquid chromatography (HPLC) and high-performance thin layer chromatography (HPTLC) (Das and Gezici, 2018). Then, mass spectrometric analysis of the compound fractions leads to structural elucidation of the separated analytes (Levsen et al., 2005; Matsuda et al., 2011; Fiehn et al., 2000). These metabolites of plants are stored in databases and their structure-function properties, ADME profiles and quantitative structure–activity relationships (QSAR) studies are carried out (Dey et al., 2021; Lasker et al., 2021; Mahanta et al., 2021). With the help of high-throughput screening (HTS) as well as virtual high-throughput screening (VHTS) processes, the purified metabolites can be screened for bioactivity either through in vitro or in silico approaches. Phytomedicine research can be performed by either taking a novel plant whose metabolite composition is unknown, or it can be executed in silico with the structures obtained from either metabolite/plant databases. Structure-based screening and ligand-based screening approaches (Vázquez et al., 2020) lead to identification of suitable and effective lead compounds, which can be further investigated in vitro and in vivo for diverse beneficial bioactivities like antioxidant, anticancer, antimicrobial, anti-inflammatory, anti-aging, etc. This chapter makes an attempt to collate the recent databases and tools used for bioinformatics-based plant metabolite research and present a methodological workflow on approaches which are currently employed in phytochemical research. Plant metabolites which are obtained naturally are tested for their in vitro/in vivo efficacy, and bioinformatics tools greatly aid in discovering the bioactivity and enhancing the efficacy of these compounds.

16.2.2 Spectroscopic Investigations

The identity and composition of phytochemicals are solved using combinations of spectroscopic studies. GC-MS is used for separating and characterizing volatile compounds and is particularly useful in the analysis of compounds which dissolve in organic compounds like methanol, ethanol, ethylacetate, chloroform and acetone (and a range of other organic solvents). GC-MS is useful for separation and characterization of the mass (and structural elucidation) of compounds which can dissolve in aqueous as well as organic solvents. High-resolution mass spectrometry (HR-MS) is particularly quite helpful in phytochemical structure elucidations because it amalgamates high resolution as well as high mass accuracy and, hence, is an advanced tool which is advantageous for determining elemental compositions as well as identification of novel/unknown compounds (Stock, 2017). Solution nuclear magnetic resonance (NMR) spectroscopy has long been a method of choice for elucidation of phytochemical structures, and the high sensitivity and resolution of NMR-based metabolomics profiling are noteworthy (Verpoorte et al., 2007). Ultraviolet (UV)-visible spectroscopy is used to identify pi-conjugated systems such as aromatic rings and is particularly useful in identification of electronic transitions of molecules such as carotenoids. Fourier-transform infrared spectroscopy (FT-IR) has been used to identify the functional groups based on the vibrational energies of specific groups present in the compounds. Combinations of these methods lead to deciphering the structure and properties of the molecules. Earlier, methodologies such as mass spectrometry were used for separation and mass profiling of plant secondary metabolites. However, starting nearly a decade ago, automated elucidation of metabolites from LC-MS metabolomes has been accomplished in tandem with integration of phytochemical databases. The HR-MS spectra data are integrated with KNApSAcK, ReSpect and PRIMe databases, and the search outputs from these databases are used for putative identification of metabolites in an attempt to explore phytochemical diversity (Matsuda et al., 2011). A novel in situ label-free imaging method termed as mass spectrometry imaging was employed to visualize spatially resolved biotransformations by mapping the major bioactive polyphenols of green tea. This method enhances the knowledge of multicomponent medicinal herbs (Fujimura et al., 2017). However, it is difficult to directly identify/apprehend the biochemical targets of plant bioactive compounds (Polya, 2003) because in the cellular milieu and in vivo (whole organisms), there are thousands of proteins, and the probability that a molecule simultaneously interacts with hundreds or thousands of other molecules (such as enzymes, membrane proteins, ion channels, nuclear receptors, etc.) is quite high. Fortunately, computational chemistry and bioinformatics tools have made it possible to delve deeper into the molecular mechanisms by which a given phytochemical can proffer beneficial biochemical activity against a range of human ailments. Figure 16.1 represents a process flow in conventional phytomedicine research, where the discovery process moves from phytochemical/secondary metabolite profiling to in silico studies, in vitro (cell line studies and biochemical assays) and later, to in vivo studies (using animal models). In this sequence, bioinformatics studies enable researchers to quickly zero in on the most preferred targets and the modes of bonding/interactions between a given receptor and the phytochemical ligand of interest, and this forms the basis for molecular exploration of how phytochemicals work.

16.2.3 Identification of Components Using NIST (Gas Chromatography-Mass Spectrometry)

The NIST-2008 (National Institute of Standards and Technology) dataset is a useful tool in the quest to identify the structure and molecular mass of compounds; indeed, the NIST/EPA (Environmental Protection Agency)/NIH (National Institute of Health) information databases are software used to examine obtained mass spectrographic peaks. The NIST (Diallo et al.) library can be utilized to identify the parts of the peak (Darmasiwi et al., 2015). The location can be precisely analyzed by contrasting it with mass spectral standard reference information, and NIST owns over 62,000 patterns.

16.2.4 Analysis of MS/MS and LC-MS Using Global Natural Product Social Molecular Networking

The recently launched Global Natural Product Social Molecular Networking (GNPS) platform (http://gnps.ucsd.edu) is specially advantageous, as it can automatically mine thousands of samples and run several experiments in a short time (about two hours) (Wang et al., 2016). Phytochemicals are annotated using MS/MS spectra using GNPS, an online mass spectrometry ecosystem which provides open-access knowledge to users through community sharing of mass spectrometric data that is either in raw or processed form. Any newly identified molecules are annotated using "analogues search mode," and the MS/MS spectral data are expanded through accelerated dereplication

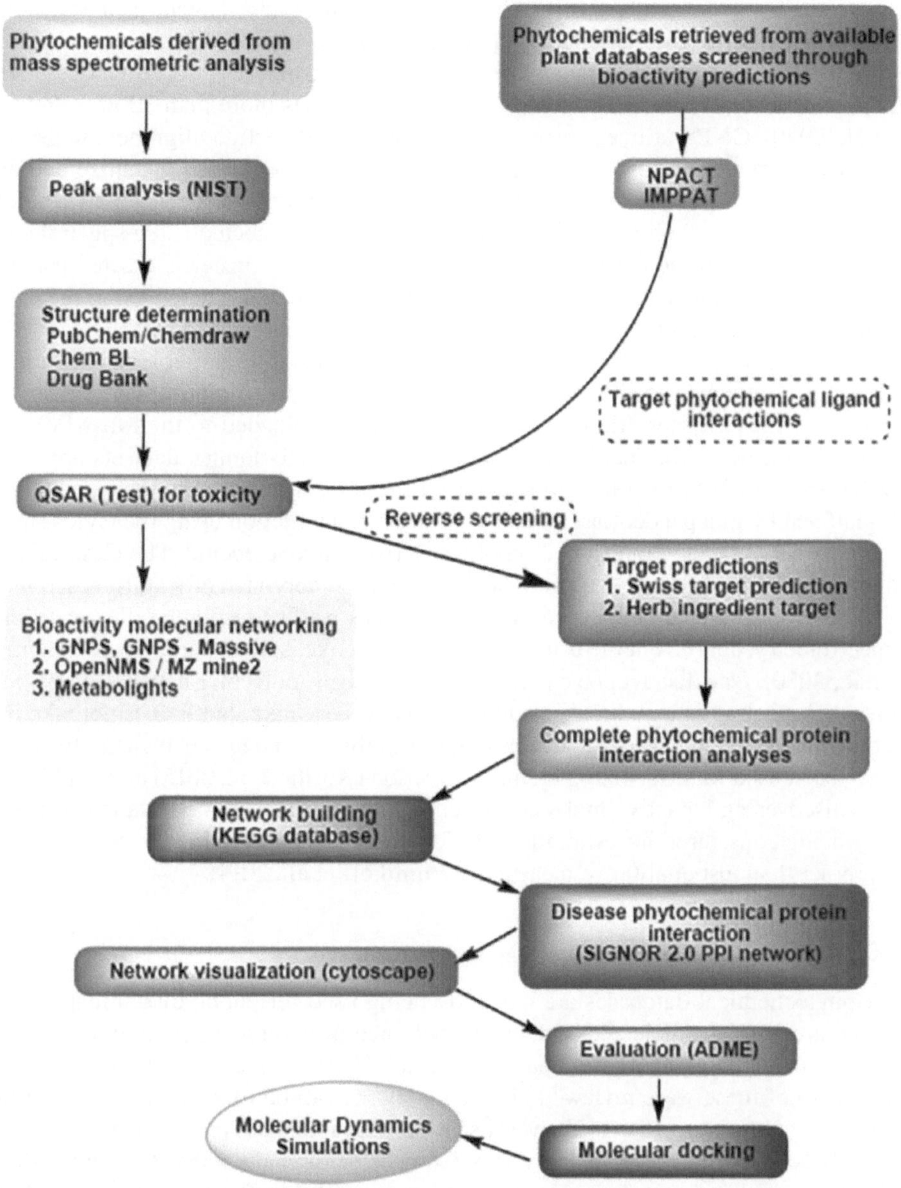

FIGURE 16.1 Workflow of bioinformatics tools in phytochemical research: Phytochemical structures are elucidated either directly from plant sources (through mass spectrometric studies) and by screening phytochemical databases. A series of bioinformatics methods like NPACT/IMPPACT analysis, QSAR studies, reverse screening (and target phytochemical ligand identification), phytochemical interaction analyses and identification of PPI networks (and network visualization), determination of ADME properties, molecular docking (MD) and MD simulations are commonly carried out to zero in on the molecular targets of phytochemical compounds.

capabilities to analogous structures. Investigators can interpret any defects in mass estimations because the fragmentation patterns and the spectra are also available. Moreover, unique clusters of MS/MS spectra are called "molecular families" (Quinn et al., 2017). The GNPS platform is housed at the University of San Diego, California, and comprises more than 3000 computer cores (Nothias et al., 2020). GNPS utilizes chromatographic feature detection alignment tools; the overall method involves qualitative assessment and resolution of monomers to visualize and annotate non-targeted mass spectrometry data (MS/MS or MS^2). Tandem mass spectrometric data/spectra are aligned against one another, and the similar molecules are connected using spectral similarity matching a hierarchical cosine clustering algorithm. A consensus cluster is created using an MS cluster algorithm, and a massive interactive network can then be constructed using the Cytoscape tool. To a novice, the procedure of GNPS takes 90–120 minutes, and the methodology is as follows (Aron et al., 2020): i) collection/acquisition of MS^2 data using high-performance liquid chromatography (HPLC) coupled to the mass spectrometer; ii) the instrument-specific mass spectral data are converted into an open file format; iii) the data are uploaded to the Mass-IVE portal, a repository which integrates the mass data of >1000 public metabolomics datasets and comprises a high number (>221,083) of reference MS^2 data; iv) the mass spectral data jobs are submitted to the GNPS platform by independent users; and v) network construction using the Cytoscape tool to analyze complex networks and interpretation of the network image occurs. The classical approach of molecular networking (MN) can be contrasted with a novel approach of feature-based molecular networking (FBMN) to afford better qualitative analysis as well as resolution of isomers from ion mobility spectrometry, apart from the usually obtained data like isotope patterns and retention time (Nothias et al., 2020). One disadvantage is that GNPS involves molecular networking based on the results from MS^2 spectral data and makes comparisons of these data, but it does not take into consideration the relative abundance of molecules that are analyzed, and hence, bioactivity and LC-MS methods have been used for prioritizing biomass selection (Kurita et al., 2015) and to afford a faster approach for discovering bioactive molecules (Kellogg et al., 2016). Hence, statistical approaches are currently being considered for estimation of bioactivity scores by using the relative abundance of samples, rather than just qualitative measures (Winnikoff et al., 2014).

16.2.5 Other Databases/Phytochemical Repositories

Many other phytochemical databases are currently being used for public dissemination of phytochemical structures, and the molecules that can be obtained from these repositories are available for retrieval of plant metabolites. In a recent paper (Vo et al., 2020), 30 plant metabolite databases were mentioned and 18 of these were reviewed. These databases can be used to find the drug likeness, potency, selectivity/target specificity of several scaffolds/fragments of phytochemicals and to identify biologically active structures. An analysis of Chinese medicinal databases revealed that nearly 80% of all natural products possessed drug likeness and that the scaffolds/fragments could be used for de novo as well as fragment-based drug design (Tian et al., 2013). Terrestrial natural products were phenomenally different in their structural properties when compared to marine natural products; moreover, terrestrial natural products possessed better potency against human ailments than did the marine natural products (Tian et al., 2012).

A succinct description has been provided in Table 16.1 for each of the phytochemical databases, and more detailed explanations for the underlying algorithms and working methodology can be obtained from the respective webpages and the references provided in Table 16.1. More details about the databases can be obtained from other comprehensive reviews (Nguyen-Vo et al., 2018; Sorokina and Steinbeck, 2020; Xie et al., 2015). Using the representative databases listed here (and a host of other databases which could not be included herein for the sake of brevity), medicinal plants with potent bioactivity against a range of human ailments can be identified and their bioactive secondary metabolites can be recognized. Moreover, the structure of the phytochemicals and their

TABLE 16.1
Phytochemical/Plant Metabolite Databases

Database	Descriptions	References
BIOFACQUIM	Mexican plants and fungi database 90 species and ~ 6.7k compounds bring out the unique features of Mexican plants to the world. http://zinc15.docking.org/catalogs/biofacquimnp/	(Pilón-Jiménez et al., 2019)
CancerHSP	System pharmacology of anticancer compounds and well-defined ADME metrics; these anti-carcinogens have been tested against 500 different cell lines, and 800 cancer-related types have been reported using computational methods. http://lsp.nwsuaf.edu.cn/load_intro.php?site=CancerHSP&id=60	(Tao et al., 2015)
CVDHD	Virtual screening and drug discovery database specifically for natural products against cardiovascular disease. 2000 cardiovascular-related targets and optimized 3D structures with the aid of MMFF94 force field. http://pkuxxj.pku.edu.cn/CVDHD	(Gu et al., 2013)
TCMSP	TCMSP is used for a drug-target network and drug-disease network and can help to decipher the mechanism of action of Chinese herbal compounds. http://sm.nwsuaf.edu.cn/lsp/tcmsp.php	(Ru et al., 2014)
DNP	DNP contains over 200,000 natural products and their related knowledge. DNP is a sub-database of CHEMnetBASE, which now allows users to concurrently search across all databases integrated into CHEMnetBASE. www.routledge.com/Dictionary-of-Natural-Products-Supplement 1/ Buckingham/p/book/9780412577802	(Buckingham, 1997)
HerDing	HerDing provides information on herbs for treating illness through integrating existing several databases and a text mining approach. HerDing mainly helps researchers to get new insights for identifying new drugs. http://combio.gist.ac.kr/herding	(Choi et al., 2016)
HIT	HIT is a comprehensive anhd fully curated database of plant compounds and their protein targets. Details pertaining to upregulation and downregulation of target genes are provided. http://lifecenter.sgst.cn/hit/	(Ye et al., 2010)
Interbioscreen	IBS is one of the top companies providing chemicals. IBS possess over 485,000 synthetic structures, 68,000 natural products and 13,000 building blocks from both natural and synthetic sources. www.chemblink.com/webs/www.ibscreen.com.htm	(Roy et al., 2015)
IMPPAT	This is a manually curated database for Indian medicinal plant compounds. This database promotes traditional Indian medicinal literature. Three different records for each plant, *viz*., phytochemical constituents, therapeutics and formulations are available. https://cb.imsc.res.in/imppat	(Mohanraj et al., 2018)
KNApSAcK Core System	This database comprises 24 sub-databases. The core system database contains thousands of plant compound structures. http://kanaya.naist.jp/KNApSAcK_Family	(Afendi et al., 2012)
NuBBE DB	NuBBE DB is designed by Sao Paulo State University and the Laboratory of Computational and Medicinal Chemistry (LQMC) at the University of Sao Paulo. It is an attractive knowledge source in the field of natural products, medicinal chemistry and natural drug discovery. (https://nubbe.iq.unesp.br/portal/nubbedb.html)	(Pilon et al., 2017)

(Continued)

TABLE 16.1 (Continued)

Database	Descriptions	References
NPACT	NPACT was created from more than 700 full-text research articles on PubMed, which allows nearly 1600 plant-derived anticancer natural compounds, their structures, properties and inhibitory potentials. Structural similarity studies in association with human metabolites and the FDA is also supported. http:// crdd.osdd.net/raghava/npact/	(Mangal et al., 2013)
NPASS	NPASS is a source of natural products, bioactivities and biological targets. With the help of publications and online databases, the species data have been collected. http://bidd2.nus.edu.sg/NPASS/	(Zeng et al., 2018)
NPCARE	NPCARE is a library of natural product–cancer gene regulation. It allows researchers to identify the potential of anticancer medicines and their mechanisms of action. http://silver.sejong.ac.kr/npcare	(Choi et al., 2017)
Reaxys	This is a commercial drug library providing scientists with chemical data. This database has over 200 years of chemistry and over 7 million structures of organic and inorganic compounds. https://guides.lib.utexas.edu/chemistry/reaxys	(Goodman, 2009)
SANCDB	SANCDB was specially designed for South African natural product sources. It is manually collected from >200 scientific documents, including books and publications. The website interface of this database is accessible with a searching tool that contains a wide range of medicinal knowledge. https://sancdb.rubi.ru.ac.za/	(Diallo et al., 2021)
Super Natural II	Super Natural II covers around 46,000 natural products and vendor information. After eight years of maintenance and development, it was updated into Super Natural II72, with more than 300,000 natural products. Structural similarity comparison drug molecules in conjunction with their corresponding target proteins is also available. http://bioinformatics.charite.de/supernatural	(Banerjee et al., 2015)
TCMID	TCMID records TCM-related updates obtained from various resources of drugs and diseases, as well as DrugBank, OMIM and PubChem. The database also gives online-based software with a network for combining interaction with herbs and treated diseases. www.megabionet.org/tcmid/	(Xue et al., 2012)
TCM Database@ Taiwan	TCM Database@Taiwan was developed to connect TCM and western advances. It is a repository of every known Chinese herb and contains built 2D/3D structures for all the TCM compounds. http://tcm.cmu.edu.tw/	(Chen, 2011)
TradiMed	TradiMed is a paid database combining traditional information of Chinese and Korean medicinal plants. This database records over 3200 herbal plants and covers >20,000 compounds. TradiMed is available in three different languages—English, Japanese and Korean. www.tradimed.com/	(Chang, 2001)
UNPD	UNPD holds enormous resources of natural products to support virtual and high-throughput screening. http://pkuxxj.pku.edu.cn/UNPD/	(Kind and Fiehn, 2017)
VIETHERB	VIETHERB contains Vietnamese plant information along with herbal details of metabolites, diseases, morphologies and geographical locations. http://vietherb.com.vn/	(Nguyen-Vo et al., 2018)

cellular targets can also be identified. With thousands of compounds being made available and a host of cellular macromolecules that can serve as receptors/targets, it is imperative that bioinformatics tools help identify the best fit and most robust statistical/probabilistic combinations of receptors and ligands. This can further enhance the process of potential drug candidates against specific ailments in the process of drug discovery.

16.2.6 Bioinformatics Tools for the Process of Drug Discovery

Other tools which can be of great use in phytomedicinal/phytochemical research (mentioned in Table 16.2) are i) mass spectral analysis tools (e.g. NIST and GNPS), ii) network construction tools (e.g. Cytoscape), iii) protein databases (e.g. PDB), iv) metabolomics databases (e.g. MPGR and Metabolights, v) small molecule/phytochemical databases (examples given in Table 16.1 and ChEBI, KNIME, PubChem, CMKb and ChEMBL), vi) comprehensive suites for molecular modeling and combinations of applications (e.g. Schrödinger suite and Discovery Studio), vii) molecular docking tools (e.g. AUTODOCK and Schrödinger), viii) bioactivity prediction and ADME assessment tools (e.g. PASSONLINE, SwissADME, Molinspiration, etc.), ix) molecular viewers/explorers (e.g. PyMoL, ChimeraX/Chimera, Rasmol, Discovery Studio Viewer, etc.) and x) QSAR and QCAR (e.g. Schrödinger, Discovery Studio, QSAR tool box 4.5, etc.). Other tools like LigandScout and StarDrop aid in clarification and optimization of drug-receptor interactions. Tools such as POCASA and CAVER aid in identification of topological perturbations and identification of druggable sites/pockets on the protein surface. A compendium of these tools is available for use all over the world, and these tools effectively serve as the avenues by which one may arrive at lead compounds/drug candidates for a disease. Using QSAR and ADME tools, assessment of the physicochemical properties of molecules and the effects of group alterations/substitutions on a given natural compound (and the overall effect on the activity of a compound) can be predicted. There are thousands of these tools which have made life easier for researchers in almost all fields of research. Phytochemistry, in particular, can be challenging because of the complex structures of natural products and the wide structural diversities that they possess.

16.2.7 QSAR in Phytochemistry Research

Phytochemicals are deemed to be the most prolific sources of lead compounds for discovery and formulation of new drug compounds (Jain et al., 2021). The structure-function correlations of natural compounds and their diversities in chemical groups are closely related to functional variations, and hence, structural features of a phytochemical dictate the functionalities of the compounds. QSAR is one of the most important facets in modern-day drug discovery. Natural compounds are subjected to manipulations using computational drug design tools, and these molecules serve as scaffolds or starting points for synthesis and in vitro/in vivo evaluation of novel compounds with improved safety/low toxicity, better drug likeness and greater potency. For example, a natural compound may be isolated and characterized using biophysical approaches (NMR/LC-MS/GC-MS) and if either the in silico or in vitro K_i/IC_{50} value of a compound is in the micromolar ranges, QSAR-aided drug modifications can help bring those values to nanomolar concentrations. QSAR-based phytochemical screening is another important avenue by which the toxicity of phytochemicals (and ADME properties) can be assessed and improved. For example, an ensemble docking-based hybrid QSAR was found to be useful in the identification of novel phytochemical ligands of beta-secretase (BACE1) to tackle Alzheimer's disease (Chakraborty et al., 2014). Due to its flexibility and conformational transitioning from an open to closed form upon ligand binding, it was difficult to identify ligands for this enzyme. Chakraborty et al. designed a hybrid structure-activity model of BACE1 by incorporating ligand-based descriptors as well as energetic descriptors based on molecular docking and then screened a 3D library of 731 phytochemical query compounds; of these, around 24 very

TABLE 16.2
Bioinformatics Tools Discussed in this Chapter

Name of the Software/ Tool	Use	Example Reference for Paper Citing the Software and Phytochemical Compounds Studied
NIST (National Institute of Standards and Technology)	Tandem mass spectral library comprising 31,000 compounds, 186,000 precursor ions and 1.3 million spectra and used for analyzing and interpreting mass spectrometric data.	Cyclomethicone, 4-vinylphenol, 4-methylcatechol, indole, tyrosol, 2,6-dimethoxyphenol, dihydroactinidiolide, capric acid, phytol, methyl palmitate, stearic acid, palmitic acid, glyceryl 2-palmitate, coprostanol, etc. (Annadurai et al., 2022)
GNPS (Global Natural Product Social Molecular Networking)	An online mass spectrometry ecosystem for community-wide and open-access sharing of raw/processed/annotated fragmentation MS^2 data. Used for construction of phytochemical networks using Cytoscape.	Glycoalkaloids (Soares et al., 2019)
Jupyter	A Python-based environment which uses Pandas and Pandas profiling to analyze data	2-nonenal, 2-pentyl, 9-octadecenoic acid (Z)-, dodecane, heneicosane, icosanoic acid, nonacosane, octadecanoic acid, phytol stigmasterol and tridecane etc. (Sangeet and Khan, 2021)
Cytoscape	Open-source platform for visualization of complex networks and for network construction and analysis of mainly protein-protein interaction networks.	Protocatechuic aldehyde, osmundacetone, eriodictyol, naringenin, sakuranetin, inoscavin A, phelligrin A and phelligrin B (Dong et al., 2019)
Metabolights	Metabolomics database that contains metabolite structures, their spectra, biological roles, locations and concentrations. Also, experimental data from users in the community are available.	Longipilin acetate, polymatin B, quercitrin, fluctuadin, polymatin B, putative triterpene, longipilin acetate, pinobanksin, kaempferol, hesperetin, 3-methoxy quercetin etc. (Padilla-González et al., 2017)
SMILES Generator (Simplified Molecular Input Line Entry System)	An online tool used for interconversion of chemical smiles into molecular structures.	Agastenol, agastinol, curcumin, tectoridin, vibsanin C and inophyllum E (Chopdar et al., 2020)
QSAR toolbox 4.5 (Quantitative Structure Activity Relationship)	A free, comprehensive tool for QSAR studies of compounds; comprises 59 databases containing 100,000 chemicals and 3 million measured data points.	Anthraquinones, salicylic acid, sanguinarine, terpenoids, flavonoids, and alkaloids, etc. (Liu, 2018)
PubChem	World's largest collection of chemicals (111 million molecules, 280 million substances, 295 million bioactivities and 34 million references). Both natural and synthetic compound structures are available in 2D and 3D format, and chemical structures can be downloaded in different file formats.	Anthraquinone, 2-phenylchromenylium (Anthocyanins), hemlock tannin, sitogluside (glycoside) and A-phenolic steroid (Zainab et al., 2020)
ChEBI (Chemical Entities of Biological Interest)	Databases for small molecules with bioactivity, hosted by Open Biomedical Ontologies at the European Bioinformatics Institute.	Protochlorophyllide, flavonoid, myristate phytyl ester, phytolaccinic acid, phytyl phosphate (Scalbert et al., 2011)
LigandScout	Software for creating 3D pharmacophore structures or models of macromolecule-ligand complexes.	Beta-sitosterol, xanthoangelol E, curcumin, dihydrotanshinone-I, ellagic acid, etc. (Swargiary et al., 2020)

Name of the Software/ Tool	Use	Example Reference for Paper Citing the Software and Phytochemical Compounds Studied
Schrödinger Suite	A leading suite which can be used for molecular modeling, drug design and materials science. Also used for docking, dynamics, QSAR, ADME prediction and molecular modeling of macromolecular structures.	Akummidine, antirhine, condylocarpine, isositsirikine, leuconalm, rhazinliam, etc. (Obaid et al., 2017)
TTD (Therapeutic Target Database)	A comprehensive database for known therapeutic protein and nucleic acid targets and correlation with disease, pathway and efficacy of drugs against those specific targets.	N-hexacosane, cosanyl octadecanoate ungeremine, zefbetaine hydroxyoctacosan-9-one, hydroxytriacontan-9-one, methyl palmitate, p-cymene, alpha-terpinene, beta-elemene, (penta-1,3-diyn-1-yl)benzene, etc. (Raafat, 2021)
PDB (Protein Data Bank)	PDB is the most comprehensive repository of protein structures obtained using XRD, NMR, Cryo-EM and other methods. The PDB files can be downloaded and used for docking and dynamics studies.	5-Methylundecane, 7-Hydroxyfavone, alpha pinene, arachidonic acid, benzenemethanol, beta-cubebene, campesterol, camphene, etc. (Nag et al., 2021)
CMKb	Web-based prototype designed to integrate Australian Aboriginal ethnomedical knowledge.	Berberin (Gaikwad et al., 2008)
EGassembler	A web-based tool for EST and genomic fragment analysis developed by Kyoto University.	Alkaloids (de Bernonville et al., 2015)
MPGR	A database of metabolome and transcriptome information obtained from taxonomically diverse medicinal plants.	Alkaloids (de Bernonville et al., 2015)
SwissProt	High-quality annotated and non-redundant protein sequence database.	N-hexacosane, cosanyl octadecanoate ungeremine, zefbetaine hydroxyoctacosan-9-one, hydroxytriacontan-9-one, methyl palmitate, etc. (Raafat, 2021)
PathPred	Web-based server for predicting plausible enzyme-catalyzed reaction pathways from a query compound with the aid of RDM patterns and chemical structure alignments of substrate-product pairs.	Monolignols, lignans, coumarins, flavonoids, stilbenoids, quinones, alkaloids, etc. (Tokimatsu et al., 2011)
QCAR (Quantitative Composition Activity Relationship)	Aids in the discovery of active components in traditional Chinese medicine (TCM) and helps to predict integral bioactivity from the chemical composition.	Benzoic acid, caffeic acid, ferulic acid, hydrocinnamic acid, cinnamic acid, ellagic acid, quercetin, catechin, etc. (Virak et al., 2019)
admetLab 2.0	An online portal for systematically evaluating absorption, distribution, metabolism and toxicity (ADMET), medicinal chemistry and safety profiling of compounds, both natural and synthetic.	Silymarin, flavonoids, isosilybin, isopomiferin, mundulinole, derrisin, etc. (Qaddir et al., 2017)
KNIME	Open-source and free data analytics, reporting and integration platform for machine learning and data mining.	Flavonoids, hesperidin, naringin, nobiletin, tangeretin, etc. (Wardani et al., 2022)

(Continued)

TABLE 16.2
(Continued)

Name of the Software/ Tool	Use	Example Reference for Paper Citing the Software and Phytochemical Compounds Studied
StarDrop	Software suite to deliver optimally balanced and effective drugs.	Myristic acid, syringaldehyde, eugenol, vanillin, 2,4-di-tert-butylphenol, lauric acid, palmitic acid methyl ester and γ-sitosterol (Singla and Dubey, 2019)
Discovery Studio	This is a leading molecular viewer/ visualization tool and is used for viewing, sharing and analyzing protein structures.	Curcumin, resveratrol, capsaicin, piperine, carnosol, quercetin, carnosic acid, flavanone, etc. (Rampogu et al., 2019)
POCASA (POcket-CAvity Search Application)	Predict binding sites by detecting pockets and cavities of proteins with a rolling sphere.	Neoandrographolide, andrographidine C, andrographoside, myristic acid, paniculide A, anagrelide, bisandrographolide, etc. (Shivashankar et al., 2019)
PASSOnline	For predicting a wide range (4000 types) of bioactivity, targets and adverse effects.	Alkaloids, flavonoid, glycosides, phenol, reducing sugar, saponin, steroids, tannin, terpenoids, volatile oil, etc. (NaAllah et al., 2021)
ChEMBL	Conglomerate of chemical, bioactivity and genomic data; manually cured database of bioactive molecules.	Cedeodarin, deodarin, hematoxylin, matairesinol, quercetin and taxifolin (Qazi and Raza, 2021)
SwissADME	Program for predicting ADME parameters, pharmacokinetic properties and drug likeness and to compute medicinal chemistry friendliness.	Ursolic acid, berberastine, cleomiscosin-A, berberine, enterolactone, anastrozole, cyclophosphamide, everolimus, exemestane, fluorouracil, lapatinib, letrozole, methotrexate, tamoxifen, etc. (jha et al., 2022)
AutoDock	A free Python-based program to prepare and dock macromolecules with ligands to explore molecular interactions between them.	Sarsasapogenin, ursonic acid, novobiocin, aranotin, ajmalicine, beta-sitosterol, alpha amyrin, silymarin, rutin, etc. (Kumar et al., 2021)
ChimeraX	A molecular viewer used for visualizing molecular structures, molecular modeling and preparation of movies and images of macromolecules.	2,3-dichloro-1-propene, 4-methylcatechol, indole, coprostanol, pyrocatechol, etc. (Annadurai et al., 2022)

potent ones (Ki ≤50 nM) were identified. This is why bioinformatics holds significant sway in biomedical science and in pharmacology.

Combinatorial chemistry approaches involve the mingling of QSAR with molecular docking, e-pharmacophore generation, virtual screening/HTVS and other methods to search for and perform virtual structural alterations of compounds to obtain in silico metrics of bioactivity, efficacy and safety (Kaur et al., 2012). These molecules can then be investigated in vitro and in suitable animal models of disease to corroborate their efficacy in more realistic scenarios. One of the most critical aspects in pharmacology is the investigation of toxicological characteristics of a potential drug molecule. This is because binding-based scenarios can often involve the binding of drug compounds to proteins other than the desired/actual target, and such off-target binding can elicit serious side effects or cause pleiotropic effects. When well-studied and approved drugs can exhibit side effects, drug candidates may be less safe and have potential risks and may also cause death. Hence, safety profiling is a key facet that is given significant attention in the drug industry. The U.S. FDA and the U.S. Department of Health and Human Services enforce several regulatory mechanisms to ensure that drug molecules are safe for human consumption when they are released in the market. Some phases of preclinical toxicology testing that are associated with safety testing of chemical medications are i) intense studies, ii) rehashed dose review, iii) reproductive toxicity review, iv) review of carcinogenic effects and iv) toxicokinetics review (Health et al., 2006). Despite the fact that plant materials are used widely for food and in ethnic/traditional medicine, most of the plant compounds have not been comprehensively evaluated for toxicity and harmful side effects. Such tests are expensive and time-consuming, and therefore, it is practically impossible to perform in vitro screening and evaluation of all potential test compounds. For example, an estimated 3 million U.S. dollars would be required for each standardized study, and therefore, without funding, drug development will be relegated to multibillion-dollar big pharma enterprises and will remain inaccessible to much less-funded researchers. This is where combinatorial approaches like HTVS, reverse screening, molecular docking and e-pharmacophore generation greatly narrow down the best potential hits and allow for more rigorous in vitro and in vivo evaluation of those best hits. These methods provide a cost-effective and fast-track alternative to the expensive and time-consuming approaches of the past. QSAR is a convenient replacement for the preliminary steps in conventional/traditional drug discovery. QSAR-based tools such as LMA and MC4PC have been widely used for predicting carcinogenicity of phytochemicals in rodents (Valerio Jr et al., 2010). Despite the advantages that QSAR-based tools present, they are still not close to the real in vivo scenarios, and hence, these methods cannot totally replace animal toxicity testing and human clinical trials. This is where statistical methods bring tremendous advantages to the table; the application of statistics to bioinformatics drug safety testing tools need not put restrictions on the structural features or molecule types that are studied (Mosier et al., 2003). Coupled with toxicogenomics studies, wherein, genomics and toxicology are combined together to provide a cheaper alternative to the erstwhile toxicology screening approaches, bioinformatics-based drug discovery has become much more viable, and the precision of toxicological assessments has been improved (Snijesh et al., 2014).

Herbal dietary supplements are known to cause liver and kidney failure as well as cardiotoxicity. One QSAR study (Liu, 2018) assessed the safety of herbal dietary supplements and their potential to exhibit unsolicited pharmacokinetic and toxicological properties. The tiered approach of i) absorption/permeability evaluation, ii) cytochrome P450 phase I metabolism prediction (including P450 type involved and assessment of whether a phytochemical can proffer inhibition/activation/drug-drug contraindications) and iii) liver tissue toxicity assessment (e.g. transaminase elevation) was used to screen 255 phytochemical compounds that were obtained from 20 different plant species that were presumed to be associated with clinical reports of liver injury. Of the 255 compounds, 48 were identified to have liver-damaging attributes and 19 of these were already known to have such actions (e.g. pyrrolizidine alkaloids), while some sesquiterpenes were newly attributed to possess liver toxicity. QSAR has significantly impacted phytochemical and phytomedicinal research, as observed by the barrage of publications which mention the utilization of QSAR approaches for improving phytochemical ADME predictions, structure–activity relationships and activity improvisation. In conjunction with phytochemical databases, QSAR prediction has provided greater clarity and simplified the overall process of target identification and greatly aided in HTVS. Chopdar et al. demonstrated this by investigating the usefulness of QSAR analysis in conjunction with NPACT as a test case and found that prediction of pIC50 and docking score estimations of the phytochemicals in the NPACT database were reckoned to be applicable to in silico drug discovery using the database of choice (Chopdar et al., 2020).

The key usefulness of QSAR is to mathematically link different molecular descriptors and the respective biological activities. Further, connection of biomolecular function with certain physicochemical parameters is useful for exploring ADMET characteristics during the drug formulation processes. QSAR modeling is used in conjunction with KNIME, GUSAR, ADMET Predictor[M], StarDrop, SYBYL-X, KNIME, ChemBench, Accelrys and Discovery Studio (Rudik et al., 2014).

16.3 BIOINFORMATICS TOOLS USED IN PHYTOMEDICINAL RESEARCH

16.3.1 Bioactivity Molecular Networking

There are three key stages in bioactivity molecular networking, which range from stage I to II. In stage I, the spectral information is coupled with the OpenNMS algorithm-based Optimus workflow (https://github.com/MolecularCartography/Optimus) (Röst et al., 2016), or alternatively, utilization of the MZmine2 toolbox (Pluskal et al., 2010). Through these methods, there is a relative quantification of many spectral features of LC-MS/MS data, such as ions across the different chromatographic fractions. In the subsequent step (step II), bioactivity scores and feature intensities are estimated using a Jupyter Notebook, which is based on R programming (Ihaka and Gentleman, 1996). The Jupyter Notebook (http://jupyter.org/) (Perez and Granger, 2015) is a freely accessible and reliable open-source web application which can be used by data scientists and scientific researchers to analyze, create and share documents. Notebook comprises documentation, live codes, graphs, plots and visualizations and can support more than 40+ programming languages (e.g. Python and Julia). Stage III involves the detection of MS/MS data with the aid of the GNPS platform (Wang et al., 2016) and the construction of a molecular network model with the aid of the Cytoscape tool (Shannon et al., 2003). The GNPS platform has aided in greatly improving phytochemical research. GNPS is a web platform that serves both as a data analysis tool and information archive which contains data analysis infrastructure, spectral library search and molecular networking resources. Because of these superior advantages presented by this platform, it is one of the most sought-after tools in metabolomics studies and natural products research. Since it is a crowd-sourced platform, it allows for dereplication of known molecules among the spectral libraries, and since it enables construction of molecular networks, it allows annotation propagation to identify both unknown and unrelated molecules and thus, it greatly aids in the discovery of new compounds. The final step of the workflow allows for annotation of detected molecules based on the spectral library annotations as well as network topology. Mainly, molecules with significantly greater predicted bioactivity scores are highlighted, and this can serve as the starting point for new in-depth in vitro or in vivo research. Hence, the curation of data using GNPS relies on two different complementary approaches: first, the targeted approach involves use of prior knowledge (e.g. chemotaxonomic information) for previously described/annotated molecules. Second, an untargeted approach is used, wherein network-based scrutiny of bioactive candidates is performed based on the high frequency of bioactive candidates is available and can be used for generation of a bioactive pharmacophore. The GNPS tool is freely accessible and can be used even without the need for advanced computer skills. The work process can enable advancement of new extensions of bioinformatics tools which aim at improving LC-MS identification (Pluskal et al., 2010) or spectral (Scheubert et al., 2017) and in silico annotation (Gerlich and Neumann, 2013). Bioassay-guided fractionation of plant extracts is facilitated, and GNPS can be used to store MS/MS data for the benefit of other researchers.

16.3.2 Ligand-Based Bioactivity Prediction Tools

Chemical SMILES (Simplified Molecular Input Line Entry System) of the entire set of atoms can be retrieved from the popular website PubChem (Kim et al., 2019). Classification and arrangement of phytochemicals can be done based on the Medical Subject Headings Tree (McDonald and Humphreys, 2021) or Chemical Entities of Biological Interest (ChEBI) requirements/philosophy (Hastings et al., 2016). Programs like ACD/ChemSketch are used to re-enact the metabolic pathway; for example, glycosides are hydrolyzed and the aglycone non-sugar units are also catalogued.

ACD/ChemSketch is a molecular modeling program that is used for the creation and modification of chemical structures; moreover, molecules and molecular models are displayed in 2D or 3D, and the chemical bonds as well as the nature of functional groups can be analyzed. GC-MS compound diagrams can be copied atom by atom with the aid of drawing tools such as ChemDraw from CambridgeSoft (www.cambridgesoft.com), which is widely used for modeling, drawing and designing chemical structures. There are many other tools like MedChem Designer, a hybrid tool which can combine innovative molecule drawing features and ADMET property predictions. There are some more tools that greatly aid ligand-based drug design to recognize ligands or targets, such as Accelrys (https://accelrys.com), LigandScout (Wolber and Langer, 2005) and Schrodinger's small-molecule drug design suite (www.schrodinger.com/suites/small-molecule-drug-discovery-suite). ADME assessment and QSAR/SAR prediction tools are often used to identify the structure–activity relationships and bioactivity of phytochemical compounds.

16.3.3 Effective Target Prediction Tools for the Analyzed Phytochemicals

One of the key sources which yields reliable information pertaining to bioactive herbal medicinal compounds is the Herbal Ingredients' Targets (HIT) tool (http://lifecenter.sgst.cn/hit/) (Ye et al., 2010). The dataset in this resource is derived from over 3250 literature sources; there are about 5208 chemical entries and 1301 known protein targets, which were derived from information such as directly/indirectly activated or inhibited protein binders and enzymes. The genes which are up/downregulated by particular phytochemical agents are also included in the analyses; 221 are direct protein targets, which were affected by 586 herbal compounds; these compounds are obtained from many other databases like 1300 Chinese herbs, 280 therapeutic targets (from Therapeutic Targets Database [TTD]), and there were 445 protein targets (which were affected by 1488 molecules) which were derived from Protein Data Bank (PDB). Hence, HIT database is an amalgamation, combining the independent advantages/features of different sources like the TTD (Liu et al., 2011), KEGG (Kanehisa and Goto, 2000a), PDB (Bernstein, 1978), Uniprot (Magrane, 2011), Pfam (Punta et al., 2012), NCBI (Sayer et al., 2010) and TCM-ID (Wang et al., 2005). Very few datasets particularly focus on topography. For example, CMKb is a source of Aboriginal medicinal knowledge from Australia (Gaikwad et al., 2008), and it contains data on scientific classification, biogeography, phytochemistry, organic exercises, etc., as well as photographs or pictures of the plants from which these phytochemicals are derived from. The bioactive metabolites are linked to PubChem, and there are additional features which allow for altering the representations of atoms. Databases like Raintree include unambiguous data of ethnic and medicinal uses of plants belonging to the Amazon rainforest (http://downpour tree.com/ethnic.htm). This dataset possesses structured information, phytochemical information sets, ethnobotanical data and links to clinical modified works (Verma et al., 2021).

16.3.4 Plant Genomics and EST Databases

The NCBI website contains Plant Genome Central, a massive suite of general-purpose and integrated tools which allow for access, analysis and visualization of biological data. The tools include BLAST (Basic Local Alignment Search Tool), a suite of tools for the sequence alignment MapViewer, which is a system to search as well as display genomic data by UniGene (even to the detail of chromosomal position), and is used for creation of GenBank sequences that are related to individual genes. Entrez, a text-based search and retrieval system, is used for all NCBI biological data. This database also contains expressed sequence tags (EST) data. The EGENES data provides an avenue for examination of plant ESTs (Masoudi-Nejad et al., 2007). The general methodology/workflow involves arrangement cleaning, rehash veiling, vector covering and group gathering, as well as KEGG explanation. EGENES is used for collecting all responses, as well as metabolic pathways in plant species that are connected to EST data. However, the widespread utility of this database is limited owing to the low number of datasets that it contains. Another fascinating database is the Medicinal Plants Genomics Resource (MPGR, http://medicinalplantgenomics.msu.edu/). The main objective of this database is to make the transcriptome as well as metabolome accessible to researchers from several organized and assorted herbal medicinal plants.

Very limited whole-genome sequences of plants are available, and as of now, the majority of plant genome resources are only ESTs. Transcriptomics data are valuable in molecular biology research of medicinal plants because these data allow for identification of putative genes as well as gene networks that are involved in secondary metabolite information (Li et al., 2010). After generation and assembly of the EST data, the next step in the process is annotation. Several other resources like KEGG genes, TAIR and SwissProt provide a convenient platform for annotation of sequence data (Legrand et al., 2010). Apart from EST data, simple sequence repeat (SSR) markers are arguably very favorable for their multiallelic nature, as well as co-dominance and reproducibility of plant genomic data, apart from broad genetic information (Varshney et al., 2007). SSRLocator is used as a computational approach which aids in detection as well as characterization of SSRs and mini-satellite motifs. The bioinformatics techniques are used for making expression networks from transcriptomic data and provide valuable information pertaining to specific genes in plant species which control phytochemical and secondary metabolite production. The use of near genomics data serves as a platform for exploration of variations of genomic sequences among various plant species. Plant informational indexes can be obtained from PLEXdb (Wise et al., 2007), GEO (Barrett and Edgar, 2006) and EBI ArrayExpress (Rocca-Serra et al., 2003). By combining co-expression of genes, it is feasible to discover genes of potential interest in phytochemical/secondary metabolite production and clone, manipulate and overexpress these genes in plant cells/microbial cells to improve phytochemical production. The KEGG pathway provides a valuable avenue for investigating auxiliary metabolites from some living organisms (Kanehisa et al., 2010). The KEGG database contains a large dataset which contains two different molecular network types: i) association of various metabolites and structural similarities and ii) biological synthetic routes in different living organisms. The KEGG database is particularly useful for studying the effect of compounds (and their curative properties) that are retrieved from other databases which encompass traditional Chinese medicine (TCM) (Kanehisa et al., 2019). These data are potentially useful for drug discovery based on plant genomics data. Another key tool in KEGG is PathPred. The KEGG RPAIR database contains information from several biochemical structure transformation patterns, which are known as RDM patterns, and these structure alignments of structure-product pairs are quite well-known in enzymatic reactions. This web-based server is useful in predicting the plausible signaling cascades that can begin with a query compound, based on the local RDM pattern match and global chemical structure alignment, which is matched to the information provided in the complete reactant pair library (Moriya et al., 2019).

16.3.5 Reverse Screening

This method involves searching of protein targets of chemopreventive compounds or drugs. The usual types of compounds that are studied in phytomedicine are phytochemicals identified in TCM, natural compounds and FDA-approved compounds. While these compounds are known to be selective for their targets, these compounds can bind to multiple protein targets which are widely dispersed in the human proteome. While conventional virtual screening/VHTS involves identification of ligands of targeted proteins from compound databases, reverse screening is useful for identification of potential and even unintended targets of a given compound from a massive set of receptors by examining their crystal structures or known ligands. This method is otherwise referred to as "in silico/computational target fishing," and it is quite useful in target receptor discovery, exploration of molecular mechanisms of action and identification of other alternative/similar structures via drug repositioning, detection of adverse drug reaction and drug toxicity. The three major types of reverse screening are i) shape screening, ii) pharmacophore screening and iii) reverse docking. Conventional docking involves taking a ligand and finding the molecular interactions between ligand(s) with a molecular target. Reverse docking involves the docking of small molecules/ligands to a set of clinically relevant macromolecular targets like proteins/enzymes and DNA (Kharkar et al., 2014; Lee et al., 2016). Many software packages like Schrödinger and Discovery Studio and other software or tools like ChemMapper, idTarget, PharmMapper and INVDOCK are useful in reverse screening. ChEMBL, Binding DB and PDB are also useful tools which provide the resources such as protein structures which can be used for reverse docking (Huang and Li, 2018). Reverse screening can be done using SwissTargetPrediction (www.swisstargetprediction.ch), a program used for shape screening. It utilizes ligand-based protein association information from the ChEMBL database (Mendez et al., 2019). A target molecule query search leads to ascertainment of 2D and 3D comparability scores with ligands from the dataset. These two scores are linked to obtain the likelihood that a given query particle matches the ligands (Daina et al., 2019).

16.3.6 Network Building and Pathway Advancement

The process of network building comprises four stages: i) pathway advancement, ii) expansion of protein-protein interactions (PPIs) and glycoside-aglycone connections, iii) allocation of molecular areas, and iv) pharmacokinetic examination of phytochemicals. The program g: Profiler (https://biit.cs.ut.ee/gprofiler/gost) (Reimand et al., 2007) comprises g:GOSt, g:convert, g:Orth and g:SNPense. The program g:GOSt is utilized to perform functional enrichment analysis by recognizing the excessively represented sets of metabolic reaction steps in the arrangement of anticipated protein focuses from the "Complete PCPIs" list. In a sense, the program is used for overrepresentation analysis, or gene set enrichment analysis, and it can help in mapping of genes to known functional information sources. It detects the most significantly enhanced/enriched terms. The protein names are placed as a question, and the search is completed with humans (*Homo sapiens*) as the chosen organic entity, and the Kyoto Encyclopaedia of Genes and Genomes (KEGG) (Kanehisa and Goto, 2000b) can be the reference dataset. Results are downloaded as a comma-separated value (CSV) record, which records generally advanced pathways and crossed proteins per pathway derived from the record, and the item of interest is found. The converged proteins under the ailment are utilized to channel the "Complete PCPIs" rundown to just the "Disease-specific PCPIs." The SIGNOR 2.0 database is used as a source of PPIs. SIGNOR 2.0 is a biological network of literature-based causal interactions between proteins. The entire network is directed from source to target node (Licata et al., 2020). The "Disease-specific PCPIs" with PPIs from a SIGNOR 2.0 using the disease-specific protein targets as a join clause. Glycoside-aglycone relationships are added to the network as interactions directed from the parent glycoside to child aglycone. The subsequent record is the "PCPI-SIGNOR disease network." In g:Profiler, data are retrieved from Ensembl database and from several databases which are specific for fungi, plants and metazoans or parasites. The KEGG reactome, WikiPathways, miRNA targets from miRTarBase and TRANSFAC regulatory motif matches are all involved in this program, and it comprises 500 organisms and, hence, is an exhaustive resource for data retrieval (Singh et al., 2019).

16.3.7 Evaluation Through Docking and Molecular Dynamics Simulations

Molecular docking is one of the most vital components in phytochemical/phytomedicine research. By taking ligands from plant sources which are obtained either through HTVS or from biophysical methods like NMR, GC-MS and LC-MS/MS, docking is performed to identify the molecular interactions between the potential molecules of interest and a receptor macromolecule. The evaluation of associations with "SIGNOR disease network visualization" might be done by various computational methods, in laboratories or within the cell. PCPIs in silico technology can be accessed through molecular docking with Autodock Vina, which outperforms the previous AutoDock 4 in terms of speed and precision (Trott and Olson, 2010). Protein structures can be downloaded from the PDB (Berman et al., 2000). The Auto in silico Consensus Inverse Docking (ACID) server can be used to direct PDB structure choice (Wang and Wang, 2019). ACID has an arrangement of protein focuses, as indicated by the accompanying limitations: i) no structures with resolution greater than 3.0 Å; ii) no structures solved by NMR (structures are all solved by x-ray diffraction for uniformity); iii) no structures with ligands containing nonstandard atoms (e.g. Si, Be); and iv) structures must have only one drug-like ligand bound in the active site. Structures of the bound inhibitors were obtained from PDB, while phytochemical structures were obtained from PubChem. Protein and ligand structures were prepared for docking with Autodock Tools. Grid boxes are prepared for three-dimensional coordinates using AutoDock Tools, centering the box on the bound ligand in the active site. In the absence of a bound inhibitor, the protein structure was analyzed with Aquaria (http://aquaria.ws/), which aligns the UniProt sequence with a chosen PDB structure and highlights features such as the binding site (O'donoghue and Candes, 2015). For each protein, the inhibitors/phytochemical ligands are docked to the prepared PDB structures as a positive control before docking the phytochemicals. The results are usually visualized in 3D with ChimeraX (Pettersen et al., 2021) and in 2D with LigPlot+ (Laskowski and Swindells, 2011). Using molecular dynamics tools, the stability of specific receptor-ligand complexes can scrutinized through assessment of root mean square deviation (RSMD) and root mean square fluctuation (RMSF) using GROMACS (Van Der Spoel et al., 2005), Desmond (Schrödinger) (AlAjmi et al., 2018) and LAMMPS (Sharma, 2019), among a host of several other programs.

16.4 CONCLUSION

This chapter encapsulates the important bioinformatics tools that are used for phytochemical or phytomedicine research. It is certain that bioinformatics and computer-aided programs have enabled effective research in phytomedicine and have made it easier to identify molecular explanations for phytochemical action and specificity. Using a combination of these methods, and by utilizing the given methodological workflow, it is feasible to take specific phytochemical structures and to evaluate their bioactivity. Many phytochemical databases exist, and these contain plant metabolites and their plant species sources. Traditional and ethnomedical knowledge is ratified by molecular docking, dynamics and network analysis. ADME tools enable toxicity assessment of phytochemicals, and QSAR tools help to investigate the structure-activity correlations of phytochemicals. Bioinformatics is rapidly improving, and hence, it plays a central role in the field of phytochemistry and phytomedicine.

BIBLIOGRAPHY

Afendi, F.M., Okada, T., Yamazaki, M., Hirai-Morita, A., Nakamura, Y., Nakamura, K., Ikeda, S., Takahashi, H., Altaf-Ul-Amin, M., and Darusman, L.K. (2012). KNApSAcK family databases: Integrated metabolite: Plant species databases for multifaceted plant research. Plant and Cell Physiology *53*, e1–e1.

AlAjmi, M.F., Rehman, M.T., Hussain, A., and Rather, G.M. (2018). Pharmacoinformatics approach for the identification of Polo-like kinase-1 inhibitors from natural sources as anti-cancer agents. International Journal of Biological Macromolecules *116*, 173–181.

Annadurai, P., Gideon, D.A., Nirusimhan, V., Sivaramakrishnan, R., Dhandayuthapani, K., and Pugazhendhi, A. (2022). Deciphering the pharmacological potentials of Aganosma cymosa (Roxb.) G. Don using in vitro and computational methods. Process Biochemistry *114*, 119–133.

Aron, A.T., Gentry, E.C., McPhail, K.L., Nothias, L.-F., Nothias-Esposito, M., Bouslimani, A., Petras, D., Gauglitz, J.M., Sikora, N., and Vargas, F. (2020). Reproducible molecular networking of untargeted mass spectrometry data using GNPS. Nature Protocols *15*, 1954–1991.

Atanasov, A.G., Zotchev, S.B., Dirsch, V.M., and Supuran, C.T. (2021). Natural products in drug discovery: Advances and opportunities. Nature Reviews Drug Discovery *20*, 200–216.

Banerjee, P., Erehman, J., Gohlke, B.-O., Wilhelm, T., Preissner, R., and Dunkel, M. (2015). Super natural II: A database of natural products. Nucleic Acids Research *43*, D935–D939.

Barrett, T., and Edgar, R. (2006). Mining microarray data at NCBI's Gene Expression Omnibus (GEO). In *Gene mapping, discovery, and expression* pp. 175–190. Humana Press, Danvers, NJ, USA.

Berman, H.M., Westbrook, J., Feng, Z., Gilliland, G., Bhat, T.N., Weissig, H., Shindyalov, I.N., and Bourne, P.E. (2000). The protein data bank. Nucleic Acids Research *28*, 235–242.

Bernardini, S., Tiezzi, A., Laghezza Masci, V., and Ovidi, E. (2018). Natural products for human health: An historical overview of the drug discovery approaches. Natural Product Research *32*, 1926–1950.

Bernstein, I.L. (1978). Learned taste aversions in children receiving chemotherapy. Science *200*, 1302–1303.

Bode, A.M., and Dong, Z. (2015). Toxic phytochemicals and their potential risks for human cancer. Cancer Prevention Research *8*, 1–8.

Buckingham, J. (1997). Dictionary of natural products, supplement 4, Vol. 11. Chapman & Hall, London, UK.

Chakraborty, S., Ramachandran, B., and Basu, S. (2014). Encompassing receptor flexibility in virtual screening using ensemble docking-based hybrid QSAR: Discovery of novel phytochemicals for BACE1 inhibition. Molecular BioSystems *10*, 2684–2692.

Chang, I.M. (2001). Anti-aging and health-promoting constituents derived from traditional oriental herbal remedies: Information retrieval using the TradiMed 2000 DB. Annals of the New York Academy of Sciences *928*, 281–286.

Chen, B., Garmire, L., Calvisi, D.F., Chua, M.-S., Kelley, R.K., and Chen, X. (2020). Harnessing big 'omics' data and AI for drug discovery in hepatocellular carcinoma. Nature Reviews Gastroenterology & Hepatology *17*, 238–251.

Chen, C.Y.-C. (2011). TCM database@ Taiwan: The world's largest traditional Chinese medicine database for drug screening in silico. PloS One *6*, e15939.

Choi, H., Cho, S.Y., Pak, H.J., Kim, Y., Choi, J.-Y., Lee, Y.J., Gong, B.H., Kang, Y.S., Han, T., and Choi, G. (2017). NPCARE: Database of natural products and fractional extracts for cancer regulation. Journal of Cheminformatics *9*, 1–9.

Choi, W., Choi, C.-H., Kim, Y.R., Kim, S.-J., Na, C.-S., and Lee, H. (2016). HerDing: Herb recommendation system to treat diseases using genes and chemicals. Database, https://doi.org/10.1093/database/baw011

Chopdar, K.S., Dash, G.C., Mohapatra, P.K., Nayak, B., and Raval, M.K. (2020). Monte-Carlo method-based QSAR model to discover phytochemical urease inhibitors using SMILES and GRAPH descriptors. Journal of Biomolecular Structure and Dynamics, 1–10.

Daina, A., Michielin, O., and Zoete, V. (2019). SwissTargetPrediction: Updated data and new features for efficient prediction of protein targets of small molecules. Nucleic Acids Research *47*, W357–W364.

Darmasiwi, S., Indriani, V., Innata, D., and Semiarti, E. (2015). The potential production of aromatic compounds in flowers of vanda tricolor. Paper presented at: AIP Conference Proceedings (AIP Publishing LLC), New York, USA.

Das, K., and Gezici, S. (2018). Secondary plant metabolites, their separation and identification, and role in human disease prevention. Ann Phytomedicine *7*, 13–24.

de Bernonville, T.D., Clastre, M., Besseau, S., Oudin, A., Burlat, V., Glévarec, G., Lanoue, A., Papon, N., Giglioli-Guivarc'h, N., and St-Pierre, B. (2015). Phytochemical genomics of the Madagascar periwinkle: Unravelling the last twists of the alkaloid engine. Phytochemistry *113*, 9–23.

Dey, D., Paul, P.K., Al Azad, S., Al Mazid, M.F., Khan, A.M., Sharif, M.A., and Rahman, M.H. (2021). Molecular optimization, docking, and dynamic simulation profiling of selective aromatic phytochemical ligands in blocking the SARS-CoV-2 S protein attachment to ACE2 receptor: An in silico approach of targeted drug designing. Journal of Advanced Veterinary and Animal Research *8*, 24.

Diallo, B.N.T., Glenister, M., Musyoka, T.M., Lobb, K., and Bishop, Ö.T. (2021). SANCDB: An update on South African natural compounds and their readily available analogs. Journal of Cheminformatics *13*, 1–14.

Dong, Y., Qiu, P., Zhu, R., Zhao, L., Zhang, P., Wang, Y., Li, C., Chai, K., Shou, D., and Zhao, H. (2019). A combined phytochemistry and network pharmacology approach to reveal the potential antitumor effective substances and mechanism of Phellinus igniarius. Frontiers in Pharmacology *10*, 266.

Ekor, M. (2014). The growing use of herbal medicines: issues relating to adverse reactions and challenges in monitoring safety. Frontiers in Pharmacology *4*, 177.

Fiehn, O., Kopka, J., Dörmann, P., Altmann, T., Trethewey, R. N., and Willmitzer, L. (2000). Metabolite profiling for plant functional genomics. Nature Biotechnology *18*, 1157–1161.

Fujimura, Y., Miura, D., and Tachibana, H. (2017). A phytochemical-sensing strategy based on mass spectrometry imaging and metabolic profiling for understanding the functionality of the medicinal herb green tea. Molecules *22*, 1621.

Fukushima, A., and Kusano, M. (2013). Recent progress in the development of metabolome databases for plant systems biology. Frontiers in Plant Science *4*, 73.

Gaikwad, J., Khanna, V., Vemulpad, S., Jamie, J., Kohen, J., and Ranganathan, S. (2008). CMKb: A web-based prototype for integrating Australian Aboriginal customary medicinal plant knowledge. BMC Bioinformatics *9*, 1–8.

Gerlich, M., and Neumann, S. (2013). MetFusion: Integration of compound identification strategies. Journal of Mass Spectrometry *48*, 291–298.

Goodman, J. (2009). Computer software review: Reaxys. Journal of Chemical Information and Modeling *49*, 2897–2898.

Gu, J., Gui, Y., Chen, L., Yuan, G., and Xu, X. (2013). CVDHD: A cardiovascular disease herbal database for drug discovery and network pharmacology. Journal of Cheminformatics *5*, 1–6.

Hastings, J., Owen, G., Dekker, A., Ennis, M., Kale, N., Muthukrishnan, V., Turner, S., Swainston, N., Mendes, P., and Steinbeck, C. (2016). ChEBI in 2016: Improved services and an expanding collection of metabolites. Nucleic Acids Research *44*, D1214–D1219.

Health, U.D.O., Evaluation, H.S.F.C.f.D., Research, Health, U.D.o., Evaluation, H.S.F.C.f.B., and Research (2006). US department of health and human services FDA center for devices and radiological health. Guidance for industry: Patient-reported outcome measures: Use in medical product development to support labeling claims: Draft guidance. Health Qual Life Outcomes *4*, 79.

Huang, J., and Li, Y. (2018). BN embedded polycyclic π-conjugated systems: Synthesis, optoelectronic properties, and photovoltaic applications. Frontiers in Chemistry, 341.

Ihaka, R., and Gentleman, R. (1996). R: A language for data analysis and graphics. Journal of Computational and Graphical Statistics *5*, 299–314.

Jain, S., Kaur, J., Prasad, S., and Roy, I. (2021). Nucleic acid therapeutics: A focus on the development of aptamers. Expert Opinion on Drug Discovery *16*, 255–274.

Jha, V., Devkar, S., Gharat, K., Kasbe, S., Matharoo, D.K., Pendse, S., Bhosale, A., and Bhargava, A. (2022). Screening of phytochemicals as potential inhibitors of breast cancer using structure based multitargeted molecular docking analysis. Phytomedicine Plus, 100227.

Kanehisa, M., and Goto, S. (2000a). Comprehensive gene and pathway analysis of cervical cancer progression. Nucleic Acids Res *28*, 27–30.

Kanehisa, M., and Goto, S. (2000b). KEGG: Kyoto encyclopedia of genes and genomes. Nucleic Acids Research *28*, 27–30.

Kanehisa, M., Goto, S., Furumichi, M., Tanabe, M., and Hirakawa, M. (2010). KEGG for representation and analysis of molecular networks involving diseases and drugs. Nucleic Acids Research *38*, D355–D360.

Kanehisa, M., Sato, Y., Furumichi, M., Morishima, K., and Tanabe, M. (2019). New approach for understanding genome variations in KEGG. Nucleic Acids Research *47*, D590–D595.

Kaur, G., Panini, P., Chopra, D., and Roy Choudhury, A. (2012). Structural investigation of weak intermolecular interactions in fluorine substituted isomeric N-benzylideneanilines. Crystal Growth & Design *12*, 5096–5110.

Kellogg, J.J., Todd, D.A., Egan, J.M., Raja, H.A., Oberlies, N.H., Kvalheim, O.M., and Cech, N.B. (2016). Biochemometrics for natural products research: Comparison of data analysis approaches and application to identification of bioactive compounds. Journal of Natural Products *79*, 376–386.

Kharkar, P.S., Warrier, S., and Gaud, R.S. (2014). Reverse docking: A powerful tool for drug repositioning and drug rescue. Future Medicinal Chemistry *6*, 333–342.

Kim, S., Chen, J., Cheng, T., Gindulyte, A., He, J., He, S., Li, Q., Shoemaker, B.A., Thiessen, P.A., and Yu, B. (2019). PubChem 2019 update: Improved access to chemical data. Nucleic Acids Research *47*, D1102–D1109.

Kind, T., and Fiehn, O. (2017). Strategies for dereplication of natural compounds using high-resolution tandem mass spectrometry. Phytochemistry Letters *21*, 313–319.

Kriete, A., and Eils, R. (2013). *Computational systems biology: From molecular mechanisms to disease.* Academic Press, Oxford, UK.

Kumar, S., Kashyap, P., Chowdhury, S., Kumar, S., Panwar, A., and Kumar, A. (2021). Identification of phytochemicals as potential therapeutic agents that binds to Nsp15 protein target of coronavirus (SARS-CoV-2) that are capable of inhibiting virus replication. Phytomedicine *85*, 153317.

Kurita, K.L., Glassey, E., and Linington, R.G. (2015). Integration of high-content screening and untargeted metabolomics for comprehensive functional annotation of natural product libraries. Proceedings of the National Academy of Sciences *112*, 11999–12004.

Lagunin, A.A., Goel, R.K., Gawande, D.Y., Pahwa, P., Gloriozova, T.A., Dmitriev, A.V., Ivanov, S.M., Rudik, A.V., Konova, V.I., and Pogodin, P.V. (2014). Chemo-and bioinformatics resources for in silico drug discovery from medicinal plants beyond their traditional use: A critical review. Natural Product Reports *31*, 1585–1611.

Lasker, K., Boeynaems, S., Lam, V., Stainton, E., Jacquemyn, M., Daelemans, D., Villa, E., Holehouse, A.S., Gitler, A.D., and Shapiro, L. (2021). A modular platform for engineering function of natural and synthetic biomolecular condensates. bioRxiv. https://doi.org/10.1101/2021.02.03.429226

Laskowski, R.A., and Swindells, M.B. (2011). LigPlot+: multiple ligand–protein interaction diagrams for drug discovery. Journal of Chemical Information and Modeling *51*, 2778–2786.

Lee, D.E., Ayoub, N., and Agrawal, D.K. (2016). Mesenchymal stem cells and cutaneous wound healing: Novel methods to increase cell delivery and therapeutic efficacy. Stem Cell Research & Therapy *7*, 1–8.

Legrand, S., Valot, N., Nicolé, F., Moja, S., Baudino, S., Jullien, F., Magnard, J.-L., Caissard, J.-C., and Legendre, L. (2010). One-step identification of conserved miRNAs, their targets, potential transcription factors and effector genes of complete secondary metabolism pathways after 454 pyrosequencing of calyx cDNAs from the Labiate Salvia sclarea L. Gene *450*, 55–62.

Levsen, K., Schiebel, H.-M., Behnke, B., Dötzer, R., Dreher, W., Elend, M., and Thiele, H. (2005). Structure elucidation of phase II metabolites by tandem mass spectrometry: An overview. Journal of Chromatography A *1067*, 55–72.

Li, Y., Luo, H.-M., Sun, C., Song, J.-Y., Sun, Y.-Z., Wu, Q., Wang, N., Yao, H., Steinmetz, A., and Chen, S.-L. (2010). EST analysis reveals putative genes involved in glycyrrhizin biosynthesis. BMC Genomics *11*, 1–11.

Licata, L., Lo Surdo, P., Iannuccelli, M., Palma, A., Micarelli, E., Perfetto, L., Peluso, D., Calderone, A., Castagnoli, L., and Cesareni, G. (2020). SIGNOR 2.0, the SIGnaling network open resource 2.0: 2019 update. Nucleic Acids Research *48*, D504–D510.

Liu, Y. (2018). Incorporation of absorption and metabolism into liver toxicity prediction for phytochemicals: A tiered in silico QSAR approach. Food and Chemical Toxicology *118*, 409–415.

Liu, X., Zhu, F., Ma, X., Tao, L., Zhang, J., Yang, S., Wei, Y., and Chen, Y.Z. (2011). The therapeutic target database: An internet resource for the primary targets of approved, clinical trial and experimental drugs. Expert Opinion on Therapeutic Targets *15*, 903–912.

Magrane, M. (2011). Consortium U. UniProt Knowledgebase: A hub of integrated protein data. Database (Oxford) *2011*, 009.

Mahanta, S., Gogoi, B., Chetia, P., Tanti, B., and Handique, P.J. (2021). Identification of active phytochemical from traditional herbal knowledge-base targeting pantothenate synthetase for anti-tuberculosis therapy. Letters in Drug Design & Discovery *18*, 859–871.

Mangal, M., Sagar, P., Singh, H., Raghava, G.P., and Agarwal, S.M. (2013). NPACT: Naturally occurring plant-based anti-cancer compound-activity-target database. Nucleic Acids Research *41*, D1124–D1129.

Masoudi-Nejad, A., Goto, S., Jauregui, R., Ito, M., Kawashima, S., Moriya, Y., Endo, T.R., and Kanehisa, M. (2007). EGENES: Transcriptome-based plant database of genes with metabolic pathway information and expressed sequence tag indices in KEGG. Plant Physiology *144*, 857–866.

Matsuda, F., Nakabayashi, R., Sawada, Y., Suzuki, M., Hirai, M.Y., Kanaya, S., and Saito, K. (2011). Mass spectra-based framework for automated structural elucidation of metabolome data to explore phytochemical diversity. Frontiers in Plant Science 2, 40.

McDonald, C.J., and Humphreys, B.L. (2021). The US National Library of Medicine and standards for electronic health records: One thing led to another. Information Services & Use, 1–14.

Mendez, D., Gaulton, A., Bento, A.P., Chambers, J., De Veij, M., Félix, E., Magariños, M.P., Mosquera, J.F., Mutowo, P., and Nowotka, M. (2019). ChEMBL: Towards direct deposition of bioassay data. Nucleic Acids Research *47*, D930–D940.

Mohanraj, K., Karthikeyan, B.S., Vivek-Ananth, R., Chand, R., Aparna, S., Mangalapandi, P., and Samal, A. (2018). IMPPAT: A curated database of Indian medicinal plants, phytochemistry and therapeutics. Scientific Reports *8*, 1–17.

Moriya, Y., Kawano, S., Okuda, S., Watanabe, Y., Matsumoto, M., Takami, T., Kobayashi, D., Yamanouchi, Y., Araki, N., and Yoshizawa, A.C. (2019). The jPOST environment: An integrated proteomics data repository and database. Nucleic Acids Research *47*, D1218–D1224.

Mosier, P.D., Jurs, P.C., Custer, L.L., Durham, S.K., and Pearl, G.M. (2003). Predicting the genotoxicity of thiophene derivatives from molecular structure. Chemical Research in Toxicology *16*, 721–732.

NaAllah, A., Ayipo, Y.O., Komolafe, D.I., Solihu, S., Bamidele, B., Alabi, M.A., Balogun, A.-A., Abdulazeez, A.T., and Mordi, M.N. (2021). Phytochemical screening and in silico pharmacological profiling of ethanolic extract of Aframomum melegueta for prostate carcinoma. Journal of Applied Pharmaceutical Science *11*, 132–145.

Nag, A., Banerjee, R., Chowdhury, R.R., and Krishnapura Venkatesh, C. (2021). Phytochemicals as potential drug candidates for targeting SARS CoV 2 proteins, an in silico study. VirusDisease *32*, 98–107.

Nakamura, K., Shimura, N., Otabe, Y., Hirai-Morita, A., Nakamura, Y., Ono, N., Ul-Amin, M.A., and Kanaya, S. (2013). KNApSAcK-3D: A three-dimensional structure database of plant metabolites. Plant and Cell Physiology *54*, e4–e4.

Newman, D.J., and Cragg, G.M. (2016). Natural products as sources of new drugs from 1981 to 2014. Journal of Natural Products *79*, 629–661.

Nguyen-Vo, T.-H., Le, T., Pham, D., Nguyen, T., Le, P., Nguyen, A., Nguyen, T., Nguyen, T.-N., Nguyen, V., and Do, H. (2018). VIETHERB: A database for vietnamese herbal species. Journal of Chemical Information and Modeling *59*, 1–9.

Nguyen-Vo, T.-H., Nguyen, L., Do, N., Nguyen, T.-N., Trinh, K., Cao, H., and Le, L. (2019). Plant metabolite databases: From herbal medicines to modern drug discovery. Journal of Chemical Information and Modeling *60*, 1101–1110.

Nothias, L.-F., Petras, D., Schmid, R., Dührkop, K., Rainer, J., Sarvepalli, A., Protsyuk, I., Ernst, M., Tsugawa, H., and Fleischauer, M. (2020). Feature-based molecular networking in the GNPS analysis environment. Nature methods *17*, 905–908.

Obaid, A.Y., Voleti, S., Bora, R.S., Hajrah, N.H., Omer, A.M.S., Sabir, J.S., and Saini, K.S. (2017). Cheminformatics studies to analyze the therapeutic potential of phytochemicals from Rhazya stricta. Chemistry Central Journal *11*, 1–21.

O'donoghue, B., and Candes, E. (2015). Adaptive restart for accelerated gradient schemes. Foundations of Computational Mathematics *15*, 715–732.

Padilla-González, G.F., Diazgranados, M., and Da Costa, F.B. (2017). Biogeography shaped the metabolome of the genus Espeletia: A phytochemical perspective on an Andean adaptive radiation. Scientific Reports *7*, 1–11.

Perez, F., and Granger, B.E. (2015). Project Jupyter: Computational narratives as the engine of collaborative data science. Retrieved September *11*, 108.

Pettersen, E.F., Goddard, T.D., Huang, C.C., Meng, E.C., Couch, G.S., Croll, T.I., Morris, J.H., and Ferrin, T.E. (2021). UCSF ChimeraX: Structure visualization for researchers, educators, and developers. Protein Science *30*, 70–82.

Pilon, A.C., Valli, M., Dametto, A.C., Pinto, M.E.F., Freire, R.T., Castro-Gamboa, I., Andricopulo, A.D., and Bolzani, V.S. (2017). NuBBEDB: An updated database to uncover chemical and biological information from Brazilian biodiversity. Scientific Reports *7*, 1–12.

Pilón-Jiménez, B.A., Saldívar-González, F.I., Díaz-Eufracio, B.I., and Medina-Franco, J.L. (2019). BIOFACQUIM: A Mexican compound database of natural products. Biomolecules *9*, 31.

Pluskal, T., Castillo, S., and Villar-Briones, A. (2010). Oreši c M. (2010) (MZmine).

Polya, G. (2003). *Biochemical targets of plant bioactive compounds: A pharmacological reference guide to sites of action and biological effects.* CRC Press, London, UK.

Punta, M., Coggill, P.C., Eberhardt, R.Y., Mistry, J., Tate, J., Boursnell, C., Pang, N., Forslund, K., Ceric, G., and Clements, J. (2012). The Pfam protein families database. Nucleic Acids Research *40*, D290–D301.

Qaddir, I., Rasool, N., Hussain, W., and Mahmood, S. (2017). Computer-aided analysis of phytochemicals as potential dengue virus inhibitors based on molecular docking, ADMET and DFT studies. Journal of Vector Borne Diseases *54*, 255.

Qazi, S., and Raza, K. (2021). Phytochemicals from Ayurvedic plants as potential medicaments for ovarian cancer: An in silico analysis. Journal of Molecular Modeling *27*, 1–14.

Quinn, R.A., Nothias, L.-F., Vining, O., Meehan, M., Esquenazi, E., and Dorrestein, P.C. (2017). Molecular networking as a drug discovery, drug metabolism, and precision medicine strategy. Trends in Pharmacological Sciences *38*, 143–154.

Raafat, K. (2021). Identification of phytochemicals from North African plants for treating Alzheimer's diseases and of their molecular targets by in silico network pharmacology approach. Journal of Traditional and Complementary Medicine *11*, 268–278.

Rampogu, S., Park, C., Ravinder, D., Son, M., Baek, A., Zeb, A., Bavi, R., Kumar, R., Lee, G., and Parate, S. (2019). Pharmacotherapeutics and molecular mechanism of phytochemicals in alleviating hormone-responsive breast cancer. Oxidative Medicine and Cellular Longevity *2019*. https://doi.org/10.1155/2019/5189490

Reimand, J., Kull, M., Peterson, H., Hansen, J., and Vilo, J. (2007). g: Profiler: A web-based toolset for functional profiling of gene lists from large-scale experiments. Nucleic Acids Research *35*, W193–W200.

Rocca-Serra, P., Brazma, A., Parkinson, H., Sarkans, U., Shojatalab, M., Contrino, S., Vilo, J., Abeygunawardena, N., Mukherjee, G., and Holloway, E. (2003). ArrayExpress: A public database of gene expression data at EBI. Comptes Rendus Biologies *326*, 1075–1078.

Röst, H.L., Sachsenberg, T., Aiche, S., Bielow, C., Weisser, H., Aicheler, F., Andreotti, S., Ehrlich, H.-C., Gutenbrunner, P., and Kenar, E. (2016). OpenMS: A flexible open-source software platform for mass spectrometry data analysis. Nature Methods *13*, 741–748.

Roy, S., Kumar, A., Baig, M.H., Masařík, M., and Provazník, I. (2015). Virtual screening, ADMET profiling, molecular docking and dynamics approaches to search for potent selective natural molecules based inhibitors against metallothionein-III to study Alzheimer's disease. Methods *83*, 105–110.

Ru, J., Li, P., Wang, J., Zhou, W., Li, B., Huang, C., Li, P., Guo, Z., Tao, W., and Yang, Y. (2014). TCMSP: A database of systems pharmacology for drug discovery from herbal medicines. Journal of Cheminformatics *6*, 1–6.

Rudik, A.V., Dmitriev, A.V., Lagunin, A.A., Filimonov, D.A., and Poroikov, V.V. (2014). Metabolism site prediction based on xenobiotic structural formulas and PASS prediction algorithm. Journal of Chemical Information and Modeling *54*, 498–507.

Sangeet, S., and Khan, A. (2021). In-silico studies of Neurocognitive and Neuropharmacological effect of Bacopa monnieri (L.). bioRxiv. https://doi.org/10.1101/2021.01.20.427542

Sarker, S.D., and Nahar, L. (2018). An introduction to computational photochemistry. In *Computational Phytochemistry*, pp. 1–41. Elsevier, San Diego, California, USA.

Sayer, C.D., Burgess, A., Kari, K., Davidson, T.A., Peglar, S., Yang, H., and Rose, N. (2010). Long-term dynamics of submerged macrophytes and algae in a small and shallow, eutrophic lake: Implications for the stability of macrophyte-dominance. Freshwater Biology *55*, 565–583.

Scalbert, A., Andres-Lacueva, C., Arita, M., Kroon, P., Manach, C., Urpi-Sarda, M., and Wishart, D. (2011). Databases on food phytochemicals and their health-promoting effects. Journal of Agricultural and Food Chemistry *59*, 4331–4348.

Scheubert, K., Hufsky, F., Petras, D., Wang, M., Nothias, L.-F., Dührkop, K., Bandeira, N., Dorrestein, P.C., and Böcker, S. (2017). Significance estimation for large scale metabolomics annotations by spectral matching. Nature Communications *8*, 1–10.

Shannon, P., Markiel, A., Ozier, O., Baliga, N.S., Wang, J.T., Ramage, D., Amin, N., Schwikowski, B., and Ideker, T. (2003). Cytoscape: A software environment for integrated models of biomolecular interaction networks. Genome Research *13*, 2498–2504.

Sharma, S. (2019). Molecular dynamics simulation of nanocomposites using BIOVIA materials studio, lammps and gromacs. Elsevier, Amsterdam, Netherlands.

Shivashankar, S., Murali, A., and Sangeetha, M. (2019). Molecular interaction of phytochemicals with snake venom: Phytochemicals of Andrographis paniculata inhibits phospholipase A2 of Russell's viper (Daboia russelli). Biocatalysis and Agricultural Biotechnology *18*, 101058.

Singh, A., Sengar, R. S., Singh, A. K., Singh, S., Kumar, A., & Ahmad, K. (2019). Impact of genetic engineering on agriculture food crops. *Biotech Today: An International Journal of Biological Sciences* 9(1), 17–24.

Singh, G., Passsari, A.K., Leo, V.V., Mishra, V.K., Subbarayan, S., Singh, B.P., Kumar, B., Kumar, S., Gupta, V.K., and Lalhlenmawia, H. (2016). Evaluation of phenolic content variability along with antioxidant, antimicrobial, and cytotoxic potential of selected traditional medicinal plants from India. Frontiers in Plant Science *7*, 407.

Singla, R.K., and Dubey, A.K. (2019). Phytochemical profiling, GC-MS analysis and α-amylase inhibitory potential of ethanolic extract of Cocos nucifera Linn. Endocarp. Endocrine, Metabolic & Immune Disorders-Drug Targets (Formerly Current Drug Targets-Immune, Endocrine & Metabolic Disorders) *19*, 419–442.

Snijesh, V. P., & Singh, S. (2014). Molecular modeling and network based approach in explaining the medicinal properties of nyctanthes arbortristis, lippia nodiflora for rheumatoid arthritis. *Journal of Bioinformatics and Intelligent Control 3*(1), 31–38.

Soares, V., Taujale, R., Garrett, R., da Silva, A.J.R., and Borges, R.M. (2019). Extending compound identification for molecular network using the LipidXplorer database independent method: A proof of concept using glycoalkaloids from Solanum pseudoquina A. St.-Hil. Phytochemical Analysis *30*, 132–138.

Sorokina, M., and Steinbeck, C. (2020). Review on natural products databases: Where to find data in 2020. Journal of Cheminformatics *12*, 1–51.

Stock, N.L. (2017). Introducing graduate students to high-resolution mass spectrometry (HRMS) using a hands-on approach. Journal of Chemical Education *94*, 1978–1982.

Sumner, L.W., Mendes, P., and Dixon, R.A. (2003). Plant metabolomics: Large-scale phytochemistry in the functional genomics era. Phytochemistry *62*, 817–836.

Swargiary, A., Mahmud, S., and Saleh, M.A. (2020). Screening of phytochemicals as potent inhibitor of 3-chymotrypsin and papain-like proteases of SARS-CoV2: An in silico approach to combat COVID-19. Journal of Biomolecular Structure and Dynamics, 1–15.

Tao, W., Li, B., Gao, S., Bai, Y., Shar, P.A., Zhang, W., Guo, Z., Sun, K., Fu, Y., and Huang, C. (2015). CancerHSP: Anticancer herbs database of systems pharmacology. Scientific Reports *5*, 1–6.

Thakur, M., Singh, K., and Khedkar, R. (2020). Phytochemicals: Extraction process, safety assessment, toxicological evaluations, and regulatory issues. In *Functional and preservative properties of phytochemicals*. pp. 341–361. Elsevier, Oxford, UK.

Tian, S., Sun, H., Li, Y., Pan, P., Li, D., and Hou, T. (2013). Development and evaluation of an integrated virtual screening strategy by combining molecular docking and pharmacophore searching based on multiple protein structures. Journal of Chemical Information and Modeling *53*, 2743–2756.

Tian, S., Wang, J., Li, Y., Xu, X., and Hou, T. (2012). Drug-likeness analysis of traditional Chinese medicines: Prediction of drug-likeness using machine learning approaches. Molecular Pharmaceutics *9*, 2875–2886.

Tokimatsu, T., Kotera, M., Goto, S., and Kanehisa, M. (2011). KEGG and GenomeNet resources for predicting protein function from omics data including KEGG PLANT resource. In *Protein function prediction for omics era*, pp. 271–288. Springer, Indiana, USA.

Trott, O., and Olson, A.J. (2010). AutoDock Vina: Improving the speed and accuracy of docking with a new scoring function, efficient optimization, and multithreading. Journal of Computational Chemistry *31*, 455–461.

Valerio Jr, L.G., Arvidson, K.B., Busta, E., Minnier, B.L., Kruhlak, N.L., and Benz, R.D. (2010). Testing computational toxicology models with phytochemicals. Molecular Nutrition & Food Research *54*, 186–194.

Van Der Spoel, D., Lindahl, E., Hess, B., Groenhof, G., Mark, A.E., and Berendsen, H.J. (2005). GROMACS: Fast, flexible, and free. Journal of Computational Chemistry *26*, 1701–1718.

Varshney, R.K., Mahendar, T., Aggarwal, R.K., and Börner, A. (2007). Genic molecular markers in plants: Development and applications. Genomics-Assisted Crop Improvement, 13–29.

Vázquez, J., López, M., Gibert, E., Herrero, E., and Luque, F.J. (2020). Merging ligand-based and structure-based methods in drug discovery: An overview of combined virtual screening approaches. Molecules *25*, 4723.

Verma, N., Sao, P., Srivastava, A., & Singh, S. (2021). Physiological and molecular responses to drought, submergence and excessive watering in plants. In *Harsh environment and plant resilience*, pp. 305–321. Springer, Cham, Switzerland.

Verma, S., and Singh, S. (2008). Current and future status of herbal medicines. Veterinary World *1*, 347.

Verpoorte, R., Choi, Y., and Kim, H. (2007). NMR-based metabolomics at work in phytochemistry. Phytochemistry Reviews *6*, 3–14.

Virak, V., Maarof, H., Jamil, S., and Basar, N. (2019). Phytochemicals study and biological activities of anacardium occidentale. eProceedings Chemistry *4*.

Vo, N.D., Oh, D.H., Kang, J.-H., Oh, M., and Lee, C.-H. (2020). Dynamic-model-based artificial neural network for H2 recovery and CO2 capture from hydrogen tail gas. Applied Energy *273*, 115263.

Wagner, C., Sefkow, M., and Kopka, J. (2003). Construction and application of a mass spectral and retention time index database generated from plant GC/EI-TOF-MS metabolite profiles. Phytochemistry *62*, 887–900.

Wang, J., Li, Z., Cai, C., and Chen, Y.Z. (2005). Assessment of approximate string matching in a biomedical text retrieval problem. Computers in Biology and Medicine *35*, 717–724.

Wang, J., and Wang, X. (2019). Structural equation modeling: Applications using Mplus, Second edition. John Wiley & Sons, NJ, USA.

Wang, Q., Wang, Z., Awasthi, M.K., Jiang, Y., Li, R., Ren, X., Zhao, J., Shen, F., Wang, M., and Zhang, Z. (2016). Evaluation of medical stone amendment for the reduction of nitrogen loss and bioavailability of heavy metals during pig manure composting. Bioresource Technology *220*, 297–304.

Wardani, R.K., Rhamandana, I.M., Gono, C.M.P., and Ikawati, M. (2022). Phytochemical and bioinformatic studies of citrus flavonoids as chemopreventive agents targeting GGPS1 for liver cancer. Indonesian Journal of Cancer Chemoprevention *12*, 137–147.

Winnikoff, J.R., Glukhov, E., Watrous, J., Dorrestein, P.C., and Gerwick, W.H. (2014). Quantitative molecular networking to profile marine cyanobacterial metabolomes. The Journal of Antibiotics *67*, 105–112.

Wise, R.P., Caldo, R.A., Hong, L., Shen, L., Cannon, E., and Dickerson, J.A. (2007). Barleybase/plexdb. In *Plant bioinformatics*, pp. 347–363. Springer, NJ, USA.

Wolber, G., and Langer, T. (2005). LigandScout: 3-D pharmacophores derived from protein-bound ligands and their use as virtual screening filters. Journal of Chemical Information and Modeling *45*, 160–169.

Xie, B., Zhang, C., Kang, K., and Jiang, S. (2015). miR-599 inhibits vascular smooth muscle cells proliferation and migration by targeting TGFB2. PLoS One *10*, e0141512.

Xue, R., Fang, Z., Zhang, M., Yi, Z., Wen, C., and Shi, T. (2012). TCMID: Traditional Chinese medicine integrative database for herb molecular mechanism analysis. Nucleic Acids Research *41*, D1089–D1095.

Ye, H., Ye, L., Kang, H., Zhang, D., Tao, L., Tang, K., Liu, X., Zhu, R., Liu, Q., and Chen, Y.Z. (2010). HIT: Linking herbal active ingredients to targets. Nucleic Acids Research *39*, D1055–D1059.

Yuan, H., Ma, Q., Ye, L., and Piao, G. (2016). The traditional medicine and modern medicine from natural products. Molecules *21*, 559.

Zainab, B., Ayaz, Z., Alwahibi, M.S., Khan, S., Rizwana, H., Soliman, D.W., Alawaad, A., and Abbasi, A.M. (2020). In-silico elucidation of Moringa oleifera phytochemicals against diabetes mellitus. Saudi Journal of Biological Sciences *27*, 2299–2307.

Zeng, X., Zhang, P., He, W., Qin, C., Chen, S., Tao, L., Wang, Y., Tan, Y., Gao, D., and Wang, B. (2018). NPASS: Natural product activity and species source database for natural product research, discovery and tool development. Nucleic Acids Research *46*, D1217–D1222.

17 Basics of Photobiology and Clinical Indications for Phototherapy in Anti-Aging

Namrata Verma, Poornima Yadav, Pankhudi Gupta, Mahima Bhardwaj, Subhan Danish, Rahul Datta, and Sachidanand Singh

CONTENTS

17.1 Introduction 307
17.2 Low-Level Laser (Light) Therapy 308
17.2.1 LLLT's Evolution 308
17.2.2 LLLT's Mode of Action 309
17.2.3 LLLT System Based on Laser Diodes 311
17.2.4 LLLT System Based on Laser Emitting Diode 312
17.3 LLLT for Photorejuvenation 313
17.4 Conclusion 314
Bibliography 314

17.1 INTRODUCTION

On this planet Earth, almost all kinds of life require sunlight in each form, i.e., heat and light. Sunlight is important for photosynthesis in plants and in turn, these plants are utilized by man and animals for their biological and essential processes. Photobiology deals with the study of detrimental and useful interactions of non-ionizing radiation in living organisms. In the past centuries, the sun was considered God and was utilized for its therapeutic properties, and this ancient technology is termed phototherapy in today's modern society (Grosswe iner et al., 2005). In the early 1900s, Niels Finsen a physician, a scientist, and a Nobel Prize winner, invented a therapeutic arc lamp system to be used as light therapy in place of the sun and became the modern founder of phototherapy. Phototherapy has been used for a long time since its discovery for many clinical purposes. It is composed of two words—the first is photo, meaning light, and second is therapy, meaning treatment or cure, where at a certain wavelength, low levels of photons are used against the target tissue to attain clinical results without damaging the tissue (Calderhead and Tanaka, 2017). For the past several years, it has been supposed that there is a relation between human skin cancer with its exposure to sunlight based on the mathematical relationship between the average intensity of sunlight and skin cancer incidence. In the last few years, the advancement in research has to lead to the important proof against the DNA lesions generated due to the recurring exposure to sunlight which causes skin cancer. Hence further studies show that light can function as an essential drug, both solely and in union with different chemical agents. So, the aim of phototherapy is to make the best use of light for the betterment of human and animal health by minimizing the detrimental effects when treating disease through light (Grossweiner et al., 2005). The term "anti-aging" and its idea have gained a lot of popularity over the past few years because

DOI: 10.1201/b22917-17

of the rising demand to decelerate the effects of progressive aging and aging due to light within the population of this planet Earth, especially in the developing countries. A combination of medical and surgical techniques can lead to the concept of anti-aging where drastic change can be brought about by slowing down the process of aging in which these techniques are used to elevate the body's complete natural potential of maintaining homeostasis even in the damaging environmental exposure under the light. Phototherapy can assist in the deceleration or stoppage of lines and wrinkles even before they start to appear on the skin of the face. Just like in vaccination where we immunize the body with a less harmful form of the virus, which eventually prevents the following infection by the same virus by generating antibodies for the same, in photo-anti-aging, we can use certain doses of the specifically chosen light to prevent the damaging effect of sunlight exposure on the skin in the early to mid-20s to increase the skin resistance against photoaging. Photo-anti-aging does wonders to the skin when integrated with lotion, creams, and ointments along with nutritive dietary intake, drinking a good amount of water, and daily exercises. Phototherapy aids clinicians to maintain their patients' looks and make them feel younger for a long time, preventing them to undergoing ablative resurfacing and non-ablative skin rejuvenation (Trelles, 2006).

17.2 LOW-LEVEL LASER (LIGHT) THERAPY

In 1988, the term low-level laser (light) therapy (LLLT) was coined by Ohshiro and Calderhead during its clinical applications (Ohshiro et al., 1988) comprising photoactivation or photobiomodulation—the activity of light at the molecular and cellular level (Kim and Calderhead, 2011). It is called 'low-level' due to the use of low-intensity laser in comparison to other therapies (Barolet, 2008). This is a technology that is rapidly proliferating, as it can be used in a large number of different cases requiring soothing in pain and swelling, fast healing and regeneration, function revival, and tissue damage prevention (Chung et al., 2012). Skin cells contain mitochondrial chromophores that absorb photons produced by sunlight, so in turn, it activates multiple signaling pathways along with the increase in the reactive oxygen species, blood flow, electron transport, and adenosine triphosphate (ATP) nitric oxide release. Despite the skin being the largest organ of the body, which is naturally the most exposed part to the sunlight than the other body organs, its reaction to red and near-infrared wavelengths is pretty good (Gupta et al., 2013). The increase in tissue restoration and healing is the result of the activated stem cells. LLLT has been very helpful in skin issues like discoloration and defects from acne scars, wrinkles, and hypertrophic scars and also in the curing of rashes and burns as a result of infection and irritation. It stimulates the fast growth of melanocytes, increasing the pigmentation in vitiligo, which is a pigmentary disorder, and inhibits hypersensitivity reactions, decreasing depigmentation and showing that whether it's a treatment or prevention action against any disease, LLLT is beneficial in the reduction of damage from ultraviolet (UV) light (Avci et al., 2013).

17.2.1 LLLT's Evolution

During the emergence of the use of laser as a clinical practice at the end of the 1960s, there was an assumption regarding the side effect of the light energy to be dangerous, causing cancer when used in medicine and surgery. In the 1968 Professor Endrè Mester of the Semmelweis University, Budapest, did an experiment using a defocused ruby laser to deliver low incident levels of energy to the upper surface of shaved rats, finding no carcinogenic activity, but he found out that these rats which were showered with LLLT showed hair regrowth, making him the father of phototherapy of that time. This experiment using low incident levels of laser energy also led to the discovery of using it for a lot of important and helpful results such as curing a wound, decrease in inflammation, and reduction in postoperative pain (Kim and Calderhead, 2011). In the target tissues, different ranges of photothermal reactions take place simultaneously, but at the margins of where these reactions are occurring crucial ones are the non-thermal reactions, shown in Figure 17.1 (Trelles, 2006; Kim and Calderhead, 2011).

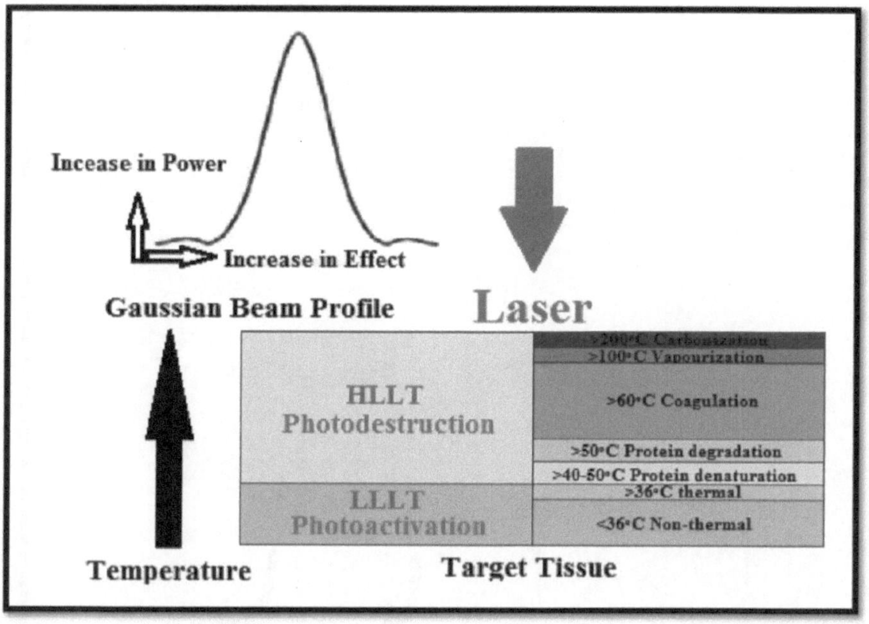

FIGURE 17.1 Range of temperatures and photobioreactions attached with a surgical laser showing damage caused by photoalterations to a thermal photobiomodulation where light/tissue interactions take place. Non-ablative skin rejuvenation and photo-anti-aging involve protein denaturations.

17.2.2 LLLT's Mode of Action

LLLT enhances biological activity using photons at a non-thermal irradiance by either using lasers or LEDs or a combination of both to promote regeneration and enhances tissue repair, preventing tissue damage, as it has varying degrees of effect at the cellular, molecular, and tissue level along with treating photoaged skin, scars, and fine wrinkles. In this, the cytochrome C oxidase, which is located within the mitochondria of the respiratory chain, and photoreceptors in cells' plasma membrane absorbs the red and near infrared (NIR) light and in turn stimulates different processes through a series of incidents that take place within the mitochondria (Oron, 2011). The action spectra for the response of the biological activity towards light were observed to be equivalent to that of the noted absorption spectra of the cytochrome C oxidase. It was assumed that light energy absorption causes breakdown of inhibitory nitric oxide from cytochrome C oxidase with the help of photons, also termed photolysis, which causes an increase in the transport of electrons, ATP production, mitochondrial respiration, and amplification in the enzyme activity, and so ultimately LLLT leads to cell proliferation, its survival, regeneration, and repair, which are the result of alterations caused by the affinity of the transcriptional factors because of the alterations in the cellular redox state activating various pathways (Avci et al., 2013), as shown in Figure 17.2 (Karu et al., 1995; Harris, 1991; Avci et al., 2013). There are pros and cons to everything, and the same goes for LLLT. Though LLLT has a wide range of applications in medicine, there are two main issues with it. First and the foremost is that when cells or tissues are exposed to the light, some biological effects are seen in these cells or tissues, but what molecular or cellular mechanism is involved behind its transduction of the signal is still doubtful, and the other issue is the different range of radiation measurement parameters like wavelength, coherence, irradiance, and many more. The low range of variables makes the treatment process less effective, whereas the high range can cause tissue damage (Chung

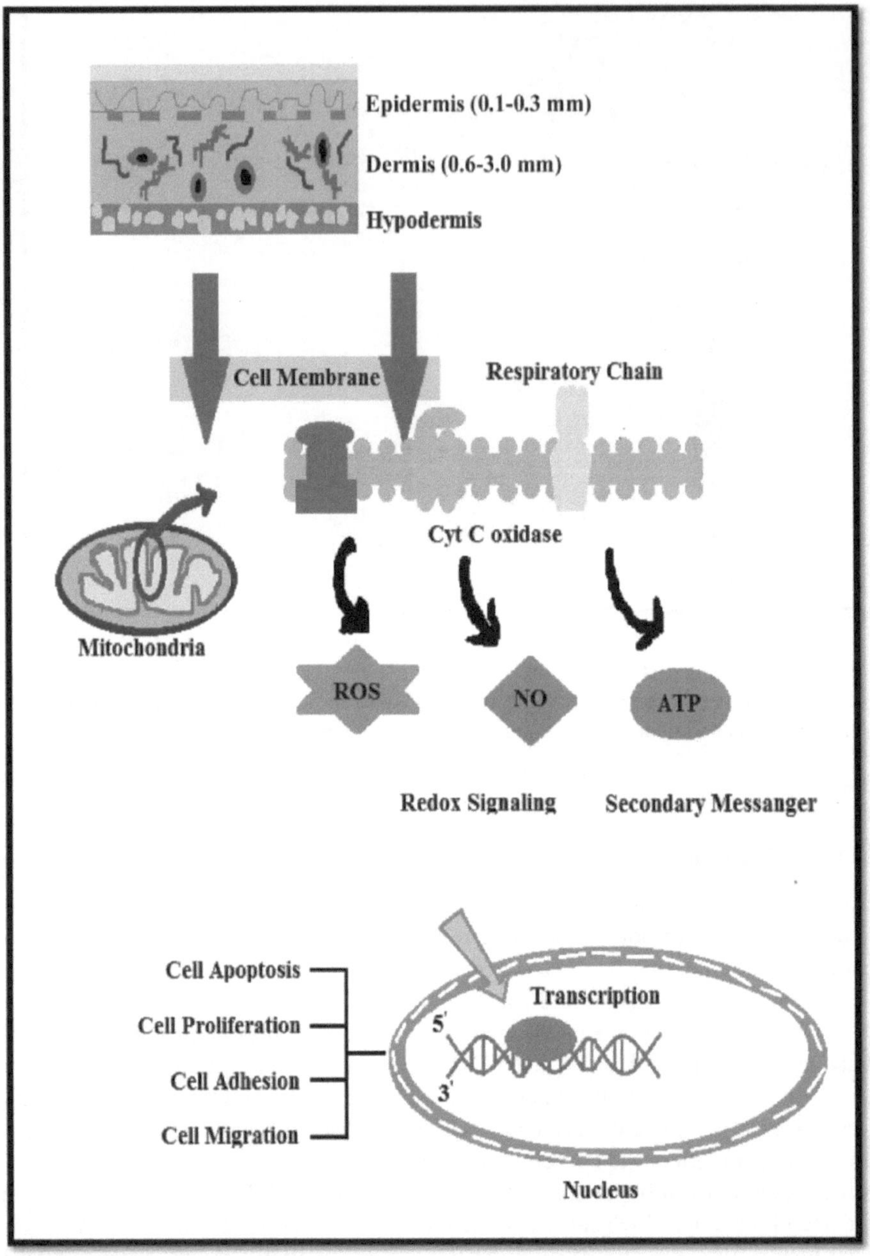

FIGURE 17.2 LLLT's mode of action.

et al., 2012). A literature survey shows that LLLT has few harmful effects, and this may be due to one of the following: wrong selection of dosage and types of the light source, dirt and oily debris on the patient's skin, and an unclean face due to improper make-up removal, causing fewer light penetrations and eventually making the treatment of skin pigmentation less effective (Posten 2005). If LLLT equipment is not being maintained properly, then it also can cause a decline in LLLT's performance and results (Avci et al., 2013).

17.2.3 LLLT System Based on Laser Diodes

This LLLT with laser diodes (LDs) is famous worldwide and manufactured by most of the companies throughout the world, as these are also US Food and Drug Administration (FDA) approved. The LD chips are tiny, and because of their size, they can spot a size range of less than 1 mm^2. Assembly of LDs could be used for treating large areas, as this technique is focused on a point-on-point approach. LDs, as per the name, are known for emitting a defocused elliptical beam. The beam is collimated because of the high angle of divergence. The beam from the handpiece of a system, which is used to emit light for treatment, provides perfect irradiance at the target tissue and is focused within the system, making it possible to use a convergent or a divergent beam. This penetration of the light at the aimed tissue site gives perfect irradiance; emission of the light energy from the handpiece hitting the square meter of the targeted tissue also depends on the space that is present between the lens of the laser system and the aimed target tissue and also makes the divergent beam naturally safer for use. There are many different types of LD-LLLT systems. Some may be battery-operated or have mains-operated control consoles, but both have a similar nature which makes it possible to treat small-size targets, for example, a pen type probe as shown in Figure 17.3 (Calderhead and Tanaka, 2017).

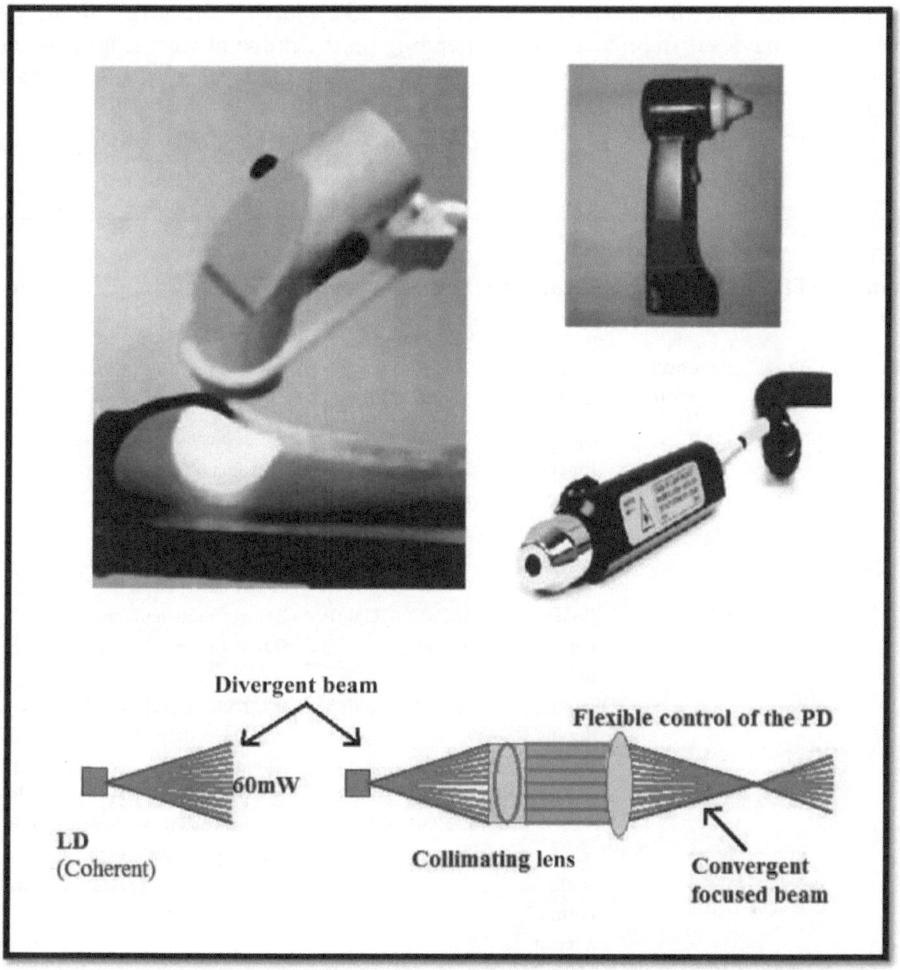

FIGURE 17.3 LLLT system based on laser diode.

17.2.4 LLLT System Based on Laser Emitting Diode

Up to the late 1990s, phototherapy was influenced by light-emitting laser diodes (LEDs) the desired clinical effect at the target tissue could not be proven at that time due to their wide waveband with an unstable and low output power, though they are relatively economical and highly divergent when compared to other sources (Kim and Calderhead, 2011). But in 1998, with the invention of NASA LED at the NASA Space Medicine Laboratory by Prof. Harry Whelan and his teammates, the demand for these LEDs increased, as these were high and stable in terms of output power, quasi-monochromatic, and less divergent than the previous one, making them an important phototherapy source for researchers and clinicians (Whelan et al., 2000).

These newly developed LEDs have higher photon intensities due to the photon interference created by the individual LEDs, leading to a rise in the overall light intensities. For visible red and IR wavelengths and the emissions by LEDs, it was noted that the highest photon intensity inclined towards the target tissue was recorded below the surface because at these wavebands tissue shows the amazing scattering of light and also its phenomenon of interference, where the photons combine (Calderhead, 2007). Then Kendric C. Smith, a US photobiologist, renamed this from 'low-level laser therapy' to 'low-level light therapy' because of it properties, as these LEDs lead towards photoactivation at the cellular level in the absence of heat and without causing any damage due to important photothermal reactions that take place at the target tissue (Whelan et al., 2001). LEDs have a different function depending on the range of wavelength and depth of penetrations related to target tissue, as shown in Table 17.1 (Oh and Jeong, 2019; Avci et al., 2013; Yeh et al., 2010).

TABLE 17.1
Application of LEDs Based on Penetration Depth of Various Wavelengths Relative to the Target Tissue

Wavelength (nm)	Color	Penetration (mm)	Target Tissue	Material and Structure	Applications
200–400	UV	<0.1	Epidermis	AlGaInN	Water purification, biomedical research
400–495	Blue	1–2	Epidermis, Dermis	ZnSe, InGaN, GaN	Seasonal affective disorder, polymerization, acne vulgaris, dental composite, combination therapy,
500–565	Green	0.5–2.0	Epidermis, Papillary dermis	InGaN, AlGaInP, AlGaP	Relaxes body and mind, skin calming
570–590	Yellow	1–2	Papillary dermis	GaAsP, AlGaInP	IPL recovery radiation, photo-aging post–laser dermatitis
630–700	Red	2–3	Adnexa	GaP:Zn—O, GaAl0.35As0.65, GaAs0.6P0.4	Wound healing, sunburn protection mucositis, PDT with the injection of sensitizer, skin rejuvenation, post-surgical recovery, photodynamic therapy
800–1200	Infrared	5–10	Reticular dermis, adnexa	GaAs, AlGaAs	Photorejuvenation, combination therapy, wound healing
Variable	Combine	Variable	Variable	-	Photorejuvenation, acne, decrease in skin melanin, post-procedure recovery

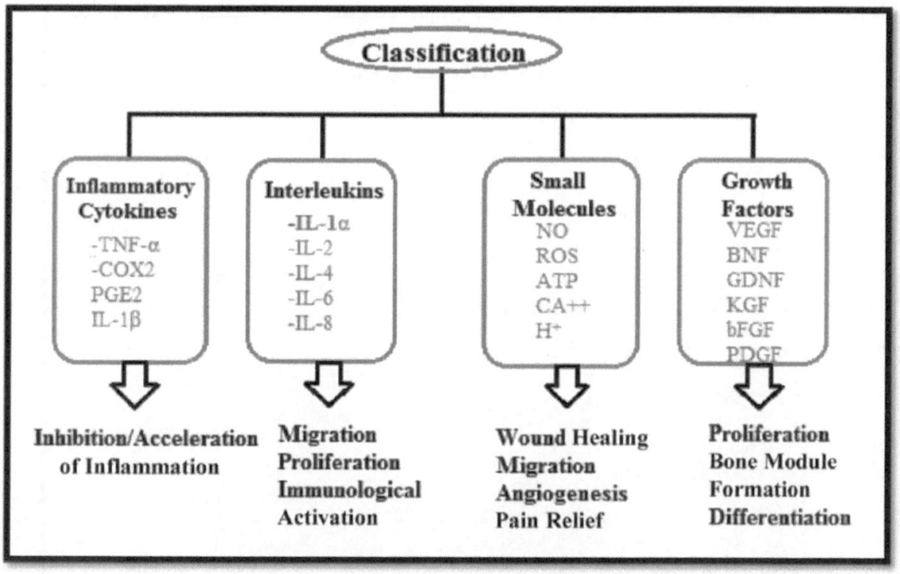

FIGURE 17.4 LLLT systems based on LEDs activate molecular-level mechanisms.

LLLT systems, when linked with LEDs, alter the functions at the molecular level as the light is inclined towards the target tissues. These alterations have solid effects on the working mechanism of various signaling processes, as shown in Figure 17.4 (Kim and Calderhead, 2011).

17.3 LLLT FOR PHOTOREJUVENATION

As one starts to age, fine lines and wrinkles on the skin are the first sign as the collagens present inside the skin break down, forming deeper grooves on the surface, along with other different issues such as hyperpigmentation, sun damage, and enlarged pores. This is where LLLTs come along to treat skin aging through a phototherapy known as photorejuvenation. Low-level light energy is used for irradiation in the epidermal and dermal cells of the skin, as it increases the blood and lymphatic flow, preventing fine lines in that targeted area. The light that is emitted by the LLLT systems happens to form intracellular reactive oxygen species (ROS), and these stimulate collagen formation, as high-level ROS were found to be capable of removing old collagen and triggering the production of new collagen in the targeted tissues. It initiates the formation of fibroblast inside the skin to where the light penetrates and is also responsible for controlling the important signaling molecules like P13K/AKT, MAPK, and TGF-β1/Smads, which leads to the proliferation, extracellular matrix production, and fibroblast migration (Oh and Jeong, 2019). So basically, the function of LLLT is to emit a certain amount of low-level light on the targeted area where the treatment has to be done, enhancing the production of collagen, and this high supply of collagen results in the activation of platelet-derived growth factor (PGDF), which hinders the process of apoptosis, leading to increase in basic fibroblast growth factor (bFGF), transforming growth factor beta (TGF-β), and vascular perfusion. The degradation of collagen reduces is connected with the reduction of matrix metalloproteinases (MMPs), which is the result of the rise in the tissue inhibitors of metalloproteinases (TIMPs) and decrease in the interleukin (IL)-6 within the target cells, as shown in Figure 17.5 (Yeh et al., 2010).

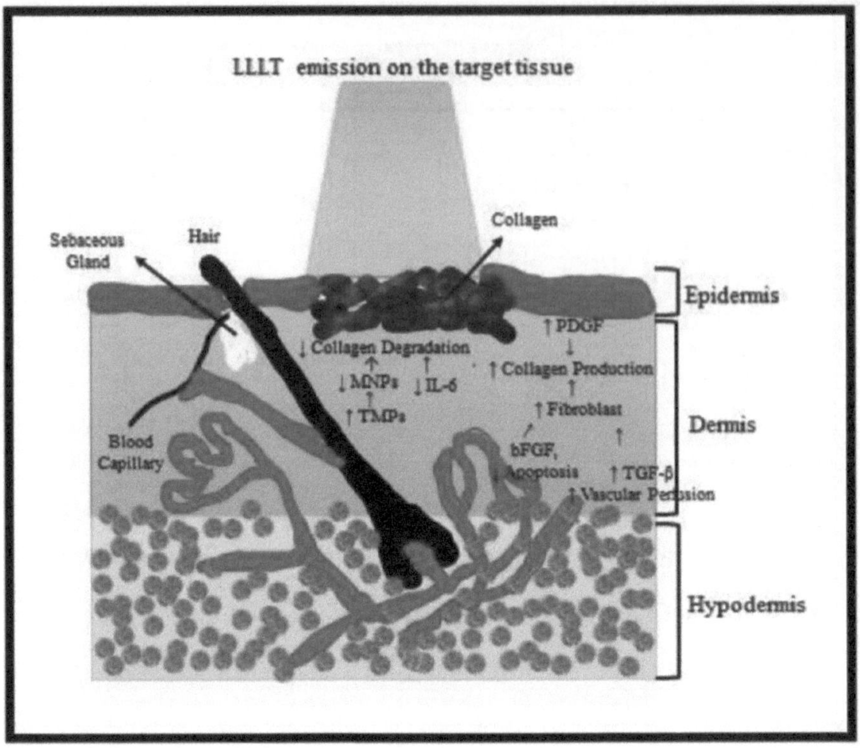

FIGURE 17.5 Photorejuvenation through low-level light therapy.

17.4 CONCLUSION

Phototherapy or photomedicine is one of the most inspiring and interdisciplinary fields in medicine that involves the research and application of photobiology with respect to health and diseases. With aging, the skin of an individual undergoes morphological and physiological changes caused by intrinsic and extrinsic factors that contribute to cellular aging and consequent skin aging. Clinically, it was observed that due to sun radiation, human skin becomes thicker, more flaccid, and more pigmented; there is also an early appearance of skin wrinkles and other skin changes, such as skin cancer. Today LLLT has been used practically worldwide and is reported to have lots of clinical applications. LLLT systems based on laser and LEDs are being used to treat targeted tissues of skin for photorejuvenation or anti-aging. These are used for stimulating repigmentation, decreasing hyperpigmentation, and reducing fine lines and wrinkles by increasing the production of collagen formation in the targeted area where the treatment is being done, and because of their credibility, it is now used by many dermatologists.

BIBLIOGRAPHY

Avci, P., Gupta, A., Sadasivam, M., Vecchio, D., Pam, Z., Pam, N., & Hamblin, M. R. (2013, March). Low-level laser (light) therapy (LLLT) in skin: Stimulating, healing, restoring. *Seminars in Cutaneous Medicine and Surgery*, *32*(1), 41. NIH Public Access.

Barolet, D. (2008, December). Light-emitting diodes (LEDs) in dermatology. *Seminars in Cutaneous Medicine and Surgery*, *27*(4), 227–238. No longer published by Elsevier.

Calderhead, G., & Tanaka, Y. (2017). Photobiological basics and clinical indications of phototherapy for skin rejuvenation. *Photomedicine: Advances in Clinical Practice. Croatia: InTech*, 215–252.

Calderhead, R. G. (2007). The photobiological basics behind light-emitting diode (LED) phototherapy. *Laser Therapy*, *16*(2), 97–108.

Chung, H., Dai, T., Sharma, S. K., Huang, Y. Y., Carroll, J. D., & Hamblin, M. R. (2012). The nuts and bolts of low-level laser (light) therapy. *Annals of Biomedical Engineering*, *40*(2), 516–533.

Grossweiner, L. I., Grossweiner, J. B., & Rogers, B. G. (2005). *The science of phototherapy: An introduction* (pp. 1–328). Dordrecht, The Netherlands: Springer.

Gupta, A., Avci, P., Sadasivam, M., Chandran, R., Parizotto, N., Vecchio, D., . . . Hamblin, M. R. (2013). Shining light on nanotechnology to help repair and regeneration. *Biotechnology Advances*, *31*(5), 607–631.

Harris, D. M. (1991). Editorial comment biomolecular mechanisms of laser biostimulation. *Journal of Clinical Laser Medicine & Surgery*, *9*(4), 277–280.

Karu, T., Pyatibrat, L., & Kalendo, G. (1995). Irradiation with He Ne laser increases ATP level in cells cultivated in vitro. *Journal of Photochemistry and Photobiology B: Biology*, *27*(3), 219–223.

Kim, W. S., & Calderhead, R. G. (2011). Is light-emitting diode phototherapy (LED-LLLT) really effective? *Laser Therapy*, *20*(3), 205–215.

Oh, P. S., & Jeong, H. J. (2019). Therapeutic application of light emitting diode: Photo-oncomic approach. *Journal of Photochemistry and Photobiology B: Biology*, *192*, 1–7.

Ohshiro, T., Calderhead, R. G., & Walker, J. B. (1988). *Low level laser therapy: A practical introduction*. New York: John Wiley & Sons Incorporated.

Oron, U. (2011). Light therapy and stem cells: A therapeutic intervention of the future?. *Interventional Cardiology*, *3*(6), 627.

Posten, W., Wrone, D. A., Dover, J. S., Arndt, K. A., Silapunt, S., & Alam, M. (2005). Low-level laser therapy for wound healing: Mechanism and efficacy. *Dermatologic Surgery*, *31*(3), 334–340.

Trelles, M. A. (2006). Phototherapy in anti-aging and its photobiologic basics: A new approach to skin rejuvenation. *Journal of Cosmetic Dermatology*, *5*(1), 87–91.

Whelan, H. T., Houle, J. M., Whelan, N. T., Donohoe, D. L., Cwiklinski, J., Schmidt, M. H., . . . Stinson, H. (2000, January). The NASA light-emitting diode medical program: Progress in space flight and terrestrial applications. *AIP Conference Proceedings*, *504*(1), 37–43. American Institute of physics.

Whelan, H. T., Smits Jr, R. L., Buchman, E. V., Whelan, N. T., Turner, S. G., Margolis, D. A., . . . Caviness, J. (2001). Effect of NASA light-emitting diode irradiation on wound healing. *Journal of Clinical Laser Medicine & Surgery*, *19*(6), 305–314.

Yeh, N. G., Wu, C. H., & Cheng, T. C. (2010). Light-emitting diodes: Their potential in biomedical applications. *Renewable and Sustainable Energy Reviews*, *14*(8), 2161–2166.

18 Role of Nanotechnology in the Field of Phytopharmaceuticals for the Delivery of Herbal Drugs

Tanvi Jain, Kavita Srivastava, Rajnish Singh, M.J. Ansari, Sachidanand Singh, and P.K. Dutta

CONTENTS

18.1 Introduction and Phytopharmaceuticals: The Need for Today ... 318
18.2 Need for a Novel Drug Delivery System for Herbal Remedies ... 319
18.3 Techniques Used in Nanophytomedicine Formulation ... 319
18.3.1 High-Pressure Homogenization Method ... 319
18.3.2 Complex Coacervation Method ... 320
18.3.3 Solvent-Emulsification-Diffusion Method ... 320
18.3.4 Supercritical Fluid Method ... 320
18.3.5 Self-Assembly Method ... 321
18.3.6 Salting Out Method ... 321
18.3.7 Co-Precipitation Method ... 321
18.3.8 Solvent Displacement Method ... 321
18.4 Classification of Nanophytopharmaceuticals Used in the Medical Field ... 321
18.4.1 Polymeric Nanoparticles ... 321
18.4.2 Solid Lipid Nanoparticles ... 322
18.4.3 Dendrimers ... 322
18.4.4 Colloidal Nanoliposomes ... 322
18.4.5 Microemulsions ... 323
18.5 Limitations in Phytomedicine Product Development ... 323
18.6 Introduction of Nanobiotechnology in Herbal Drug Formulations ... 323
18.7 Future of Nanotechnology in Herbal Drug Delivery ... 324
18.8 Illustration of Herbal Nanocarrier-Based Drug Delivery ... 324
18.8.1 Curcumin ... 324
18.8.2 Paclitaxel ... 325
18.8.3 Resveratrol ... 325
18.8.4 Berberine ... 325
18.8.5 Traditional Chinese Herbs ... 325
18.9 Conclusion ... 327
Bibliography ... 327

DOI: 10.1201/b22917-18

18.1 INTRODUCTION AND PHYTOPHARMACEUTICALS: THE NEED FOR TODAY

Herbal medicines and naturally occurring products were utilized to heal wounds or treat disease in ancient times. Widely renowned allopathic medicines, including herbal products, have thousands of components that work together against various diseases (Yadav et al. 2011).

Both developing and developed countries are more focused on the recognition of phytomedicines primarily due to their natural source and fewer side effects (Sungthongjeen et al. 1999; Sharma et al. 2011). Phytopharmaceuticals need a methodical way to deliver the compounds in a controlled way in order to enhance patient compliance and frequent administration of drugs. This can be attained by coming up with novel drug delivery vehicles for herbal composition. It not only reduces the recurring administration to defeat non-compliance but also helps in increasing the therapeutic rate by decreasing the toxicity and escalating the bioavailability (Currier et al. 2000; Bisht et al. 2007).

Nanotechnology is an area of functional science and technology which is focused on building up the dosage types, providing novel devices, and delivering them in a smaller size, mainly in the range of 1–100 nm. The applications of biological systems used for nanotechnology are shown in Figure 18.1 and are now referred to as nanomedicine (Li et al. 2009; Chen et al. 2009; Jain et al. 2016).

The utilization of herbal plants in the manufacturing of a drug relies on the utility of a strong mechanism, as each component utilizes a coating that enhances the activity of the basic component. Each ingredient plays a vital role in the synthesis of a drug, having inferior bioavailability properties (Li et al. 2006; Jain et al. 2015).

Phytopharmaceuticals were preferred as a feasible drug type for delivery through a nanodelivery system, as they have the following characteristics:

1. Few effective extracts like acetone, petrol, chloroform, and methanolic-based compounds are accessible, which may not be appropriate for drug delivery.
2. These are major parts of drugs, so dosage diminution is required.
3. Drugs in the current markets lack target specificity for many chronic diseases.
4. There are many side effects of the drugs that are currently available in the market.
5. A patient can have non-compliance issues because of less effectiveness and large doses with the currently available formulations.

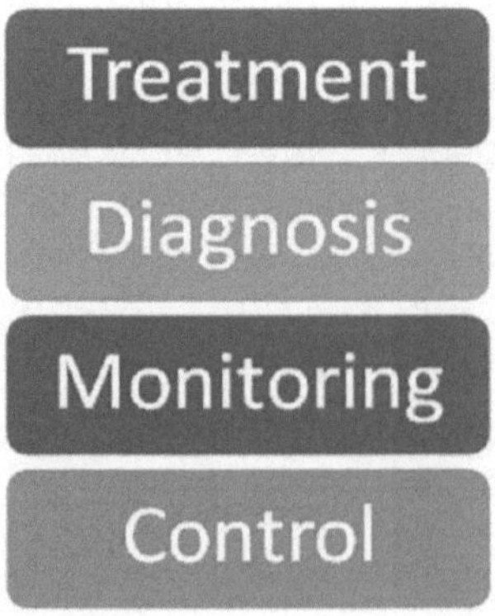

FIGURE 18.1 Applications of biological systems used in nanotechnology.

18.2 NEED FOR A NOVEL DRUG DELIVERY SYSTEM FOR HERBAL REMEDIES

Herbal medications are gaining popularity in the modern world as a means of curing a number of diseases with fewer side effects and greater therapeutic results (Verma et al. 2013). The constitution and biological effects of various medicinal plant products have already been determined by phytochemical and phytopharmacological research. Most biologically active components of extracts, such as terpenoids, flavonoids, and tannins, are highly soluble in water, but they have a low absorption rate, as they are unable to cross lipid membranes because of their larger molecular sizes and have poor absorption, which results in the efficacy and bioavailability being lost. Herbal treatments have been demonstrated to have excellent activity in testing in vitro; however, this has not been replicated with in vivo investigations (Snijesh and Singh 2014). Furthermore, some critical compounds are seldom used, since they are not compatible with the other formulation components or have undesirable effects (Bonifacio et al. 2014). To maximize patient compliance and avoid any repeated administration, phototherapeutics requires a scientific approach to distribute the compounds in a sustained way. Developing a novel drug delivery system (NDDS) for the herbal components can eventually improve the natural mechanism and coping with the challenges that come with using plant medicines (Ansari et al. 2012).

The term "novel drug delivery system" refers to the latest technique in drug delivery that avoids the drawbacks of conventional drug administration methods (Bhardwaj et al. 2021). Treating a disease by modern medicines is done by identifying the affected part inside the body of the patient and delivering the drug directly there. Thus, the drug delivery system can be said to be a method of administering a fixed amount of drug into the body of patient in a way that it reaches the site of action specifically and starts functioning right there only. The aim of a novel medication delivery system is to remove all of the inefficiencies of traditional drug delivery systems. Novel drug delivery can be attained through various methods (Devi et al. 2010).

Nanoparticles are considered to be one of the most essential NDDSs. Herbal drugs have a potential future in phytoformulation research, with the nanosized drug delivery methods improving activity and addressing issues connected with plant medicines. As a result, incorporating nanocarriers as an NDDS into standard therapy is essential to tackling more chronic diseases such as asthma, cancer, diabetes, and many others (Yadav et al. 2011).

Since phytomedicines have less toxicity along with higher therapeutic properties, their demand is increasing day by day in today's world. Applying the techniques of nanotechnology to phytomedicine in order to decrease their size, changing their surface properties, aqueous solubility, and membrane permeability, can lead to more advancement and improvement in the curing properties of phytomedicines (Gunasekaran et al. 2014; Verma et al. 2021).

18.3 TECHNIQUES USED IN NANOPHYTOMEDICINE FORMULATION

The methods for nanophytomedicine formulation include high-pressure homogenization (Sahni et al. 2011; Diab et al. 2012), solvent-emulsification-diffusion (Sahni et al. 2011; Diab et al. 2012), supercritical fluid (Sahni et al. 2011; Kalani and Yunus 2011), complex coacervation (Sahni et al. 2011; Diab et al. 2012), salting out (Sahni et al. 2011), self-assembly (Sahni et al. 2011), and co-precipitation (Sahni et al. 2011). The detailing of all these techniques is provided next, and Figure 18.2 shows the techniques in a summarized form

18.3.1 High-Pressure Homogenization Method

The high-pressure homogenization (HPH) method is often used in pharmaceutical industries for comminution, mixing, and stabilization of solids and nanoparticles. This method is based on the

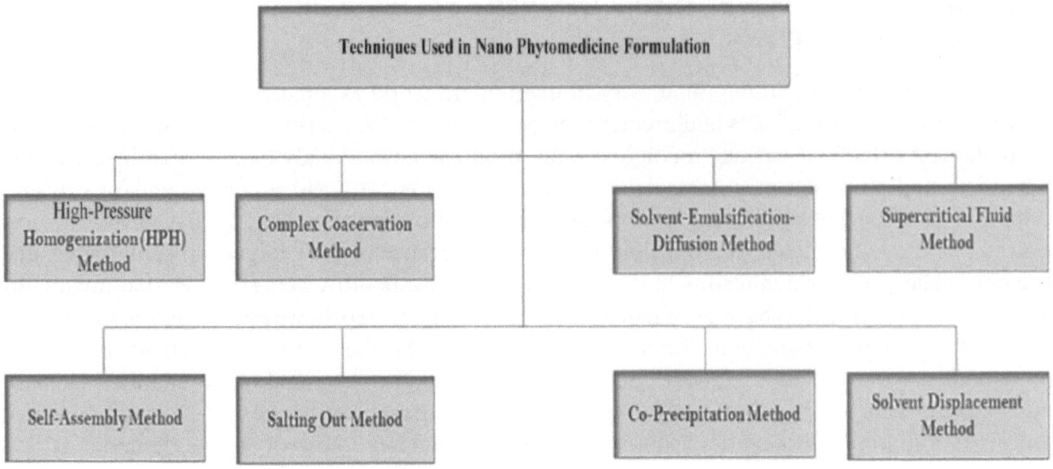

FIGURE 18.2 Techniques used in the preparation of nanophytomedicine formulation.

principle of high-pressure gradient turbulence. In the process of HPH, triglycerides (lipids) are melted, and the drug powder is dissolved homogenously in a molten lipid and then hot aqueous surfactant solution is added to the molten drug-lipid mixture and is continuously mixed by a high-shear mixing device. Afterwards, this hot mixture is compressed at a high pressure (around 100–2000 bar), resulting in the formation of particles of in a size ranging between micrometer and nanometer (Sahni et al. 2011; Mueller et al. 2000).

18.3.2 Complex Coacervation Method

In this method two natural biodegradable polymers (usually alginate and gelatin) with opposite charges are interacted with each other in an aqueous solution, resulting in a polymer-rich dense phase at the bottom and a transparent solution above. This method is efficient in terms of microencapsulation (up to 99%), has a relatively lower cost of processing, and can synthesize products at ambient temperature; hence, it is extensively used in pharmaceuticals, food, textile and agriculture industries (Sahni et al. 2011; Timilsena et al. 2019).

18.3.3 Solvent-Emulsification-Diffusion Method

In this method, oils, polymers, and drugs are emulsified in an aqueous solution in a high-shear mixer containing stabilizers, and water is added at certain intervals of time, causing diffusion of the organic solvent that results in the formation of colloidal nanoparticles. These newly formed nanoparticles have high drug entrapment properties, a certain size, zeta potential, and isopycnic density (Sahni et al. 2011; Gao 2013).

18.3.4 Supercritical Fluid Method

Supercritical fluids may be defined as a solvent with a temperature and pressure higher than their thermodynamic critical point. Supercritical fluid technology is often used in pharmaceutical and food industries. Supercritical fluids do not condense to form liquids; neither do they evaporate, because of their diffusive properties. The most widely used supercritical fluids are supercritical water (H_2O) and supercritical carbon dioxide (CO_2). Supercritical CO2 has properties such as discharge of water, easily recyclable, non-toxic nature, and non-inflammable; hence, it is more widely used for producing nanophytomedicines (Timilsena et al. 2019; Gao 2013; Lopes 2014).

18.3.5 Self-Assembly Method

In this method, pre-existing components such as atoms and molecules organize themselves on their own either by physical or chemical methods in order to form regulated nanoscale structures, without any participation from any external source. This method results in particles of smaller fiber and pore sizes (Sahni et al. 2011; Hegmann et al. 2007).

18.3.6 Salting Out Method

The principle of this method is based on the fact that the solubility of a non-electrolyte substance in water is decreased upon addition of an electrolyte. This method has high efficiency for large-sized molecules, but still cannot be applied to all proteins globally (Sahni et al. 2011; Wu and Lee 1993).

18.3.7 Co-Precipitation Method

The co-precipitation method is a modified version of the complex coacervation method and is used for production of nanophytomedicine by nucleation, growth, coarsening, and agglomeration (at high supersaturating levels). This method enhances the diffusion steadiness and dispersion stability of poor water-soluble drugs (Sahni et al. 2011; Wu and Lee 1993).

18.3.8 Solvent Displacement Method

In this method, a polymer is deposited after the displacement of a semipolar solvent miscible by water in a lipophilic solution, hence resulting in a reduced interfacial aspect between the two phases. This increases the surface area, and small drops of a solvent accumulate without any mechanical stirring (Sahni et al. 2011; Fessi et al. 1989).

18.4 CLASSIFICATION OF NANOPHYTOPHARMACEUTICALS USED IN THE MEDICAL FIELD

This era has become a revolution in the field of nanobiotechnology. Since 1980, there has been a remarkable increase in approved nano-based pharmaceutical products. Various types of nanophytopharmaceuticals are shown in Figure 18.3. These novel nano-based carrier system can either be therapeutic agents or else act as nanovehicles to deliver ample active pharmaceuticals into specific area inside the body.

18.4.1 Polymeric Nanoparticles

Polymeric nanoparticles are made up of biodegradable and biocompatible polymers. Polymeric nanoparticles can be targeted; hence, are often used for controlled drug delivery applications (Khuda-Bukhsh et al. 2010; Mainardes et al. 2006). As compared to conventional formulations, polymeric nanoparticles have a higher solubility for constituents, improved absorption of the constituents, and reduced therapeutic doses. Also, polymeric nanoparticles are stable, non-immunogenic, non-thrombogenic, non-toxic, avoid the reticuloendothelial system, and do not activate neutrophils when used in blood. These nanoparticles are made up of natural (which is preferred) or artificial biodegradable polymers. They can range between 10 and 1,000 nm

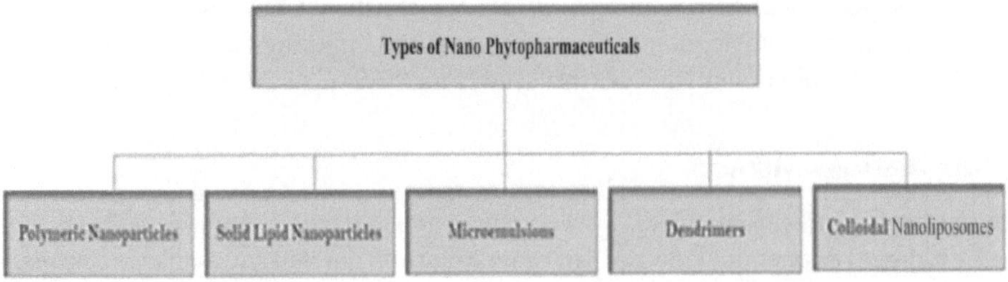

FIGURE 18.3 Types of nanophytopharmaceuticals.

in diameter and can appear as nanocapsules and nanospheres (Ajazuddin 2010; Alexis et al. 2008; Schaffazick et al. 2003; Kumari et al. 2010). Methods used for the production of polymeric nanoparticles include the in situ polymerization method and precipitation of preformed polymers like poly(lactic acid), poly(ε-caprolactone), poly(lactic-co-glycolic acid), and acrylic or methacrylic easters (Schaffazick et al. 2003).

18.4.2 Solid Lipid Nanoparticles

Solid lipid nanoparticles are colloidal particles that contain highly purified lipids or triglycerides, which are mostly solid at room temperature. They are produced by solid lipids and are stabilized by the surfactants. The solid matrix of the lipid particles is used to protect drug molecules from chemical degradation. During this formulation, crystallization can occur, which could result in low encapsulation efficiency and drug release. To overcome this, liquid lipid oil is added to the oil/water emulsion, which contains solid lipids or a mixture of solid lipids. They have higher physiochemical stability and thus provide better protection against degradation of labile drugs. These nanoparticles can be easily produced at large volumes (Pardeike et al. 2009; Souto et al. 2011). Methods used for the production of solid lipid nanoparticles include HPH, solvent emulsification-evaporation, microemulsion, etc. (Naahidi et al. 2013).

18.4.3 Dendrimers

Dendrimers are a class of novel polymers that have a well-defined structure along with high molecular uniformity. Drug delivery by dendrimers can be achieved via two methods. The first method includes imprisoning the hydrophobic drug molecule inside the hydrophobic dendrimer interior in order to make it water soluble. Another method follows the procedure of attaching the drug covalently to the dendrimer's surface (Samad et al. 2009).

Dendritic cells can be used to improve the pharmacokinetic and pharmacodynamic properties of many drugs which have properties like low molecular weight and are protein-based therapeutic agents. One of the most significant properties of dendrimers is that they can act as a particulate system while retaining the properties of polymers (Bai et al. 2006).

18.4.4 Colloidal Nanoliposomes

Liposomes can be described as microscopic vesicles made up of one or more concentric lipid bilayers, which are separated by an aqueous solution. Compounds that are hydrophilic in nature are encapsulated inside the aqueous compartment, and adsorbed lipophiles are inserted in the membrane (Kumari et al. 2010; Akbarzadeh et al. 2013). The composition of these membranes basically includes phospholipids (either synthetic or natural), an antioxidant, or sterols (Chorilli et al. 2013).

Liposomes can be classified on the basis of their size, number of lamellae, and surface charge. On such a basis, liposomes are of various types, namely, unilamellar or multilamellar, oligo, or giant. Unilamellar liposomes are composed of only one bilayer, while multilamellar liposomes consist of more than one concentric lipid bilayer having structures similar to an onion. Unilamellar bilayers are usually found in dilute solutions of surfactants, while multilamellars can be in found highly concentrated systems (Chorilli et al. 2013; Martins et al. 2007).

18.4.5 Microemulsions

The term microemulsion was first coined by Hoar and Schulman in 1943, describing a fluid system obtained as a result of titration, composing one simple emulsion along a medium-chain alcohol, like hexanol or pentanol, which is initially semitransparent and is titrated until it becomes crystal clear (Oliveira and Scarpa 2001).

Hence, microemulsions can be defined as a transparent emulsion, composed of an oil dispersed in an aqueous solution, or vice versa, generating a thermodynamically stable system, resulting in small droplets of internal phase of around the nanometer scale (Oliveira et al. 2001; Oliveira and Scarpa 2001).

18.5 LIMITATIONS IN PHYTOMEDICINE PRODUCT DEVELOPMENT

Due to the instability of herbal extracts in strongly acidic pH, high first-pass metabolism, and other factors (Nikam et al. 2014), drug levels in the blood may fall below the therapeutic concentration, ending in less or no therapeutic effect (Sandhiya and Ubaidulla 2020). There is efficient chloroform, petrol, acetone, and methanolic extracts available for herbal drug formulations, though they may not be ideal for delivery of herbal drugs. Herbal drugs are bulk drugs, and dose minimization is the target. Other concerns are a lack of target specificity of currently marketed herbal formulations for numerous chronic diseases, patient noncompliance owing to high doses, and low efficacy of the presently available formulations (Yadav et al. 2011). Because of these difficulties in phytomedicine product development, herbal remedies were chosen as a viable bioactive compound for delivery via a nanodelivery system.

18.6 INTRODUCTION OF NANOBIOTECHNOLOGY IN HERBAL DRUG FORMULATIONS

Nanoparticles are microscopic particles with sizes ranging from 1 to 100 nm with a variety of shapes, sizes, and structures. They have characteristics including high conductivity, uniformity, and unique optical properties. They are widely used in medication formulation and distribution, therapeutic applications, and imaging because of these characteristics (Salunkhe et al. 2020; Singh et al. 2019). With the possibility of numerous applications, nanotechnologies have evolved fast in recent years. Nanotechnology offers a wide range of applications, including health and medical sciences. Nanosized formulations offer extraordinary efficacy in drug delivery systems than conventional formulations by improving target specificity, biopharmaceutical characteristics, and pharmacokinetic profiles of medicinal agents (Kesharwani et al. 2018). With regard to the stability, effectiveness, biodistribution, bioavailability, and drug release, nanocarrier-assembled drug nanoparticles, such as liposomes, polymeric dendrimers, nanoparticles, nanomicelles, niosomes, metallic nanoparticles, stimuli-responsive nanoparticles, nanostructured lipid carriers, and nanofabricated devices are found to be more effective than the conventional drug delivery systems (Dewanjee et al. 2020). The nanoparticle delivery technique was chosen in herbal drug formulations due to various reasons, like the former can be utilized to target herbal drugs to specific organs, improving solubility, specificity, drug delivery, safety, and efficacy while reducing the number of doses required. Drug delivery by the nanosize particles increases the total surface

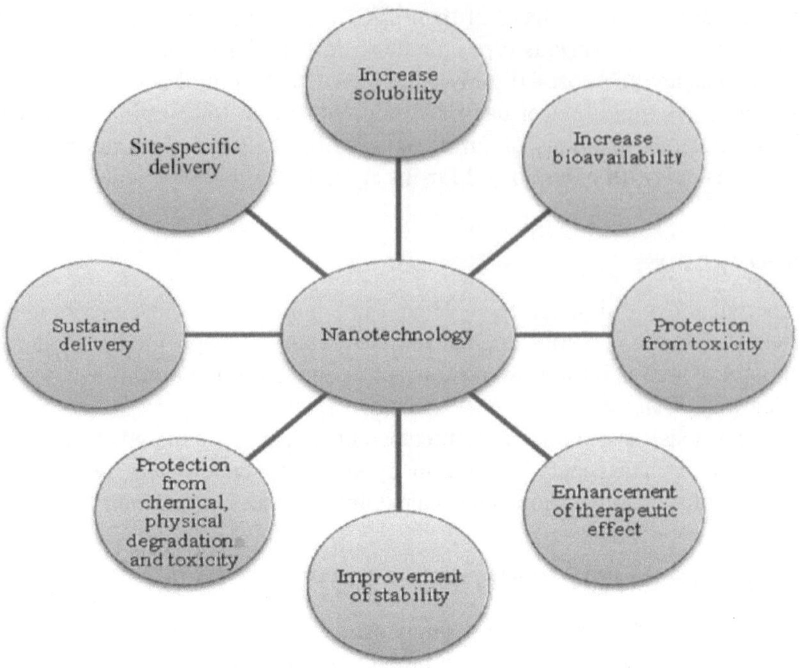

FIGURE 18.4 Role of nanotechnology in herbal drug formulations.

area of the medicines, allowing for faster blood solubility. Toxicity is reduced while therapeutic effects are maintained. Nanoparticles with improved permeability and retention can pass through the blood-brain barrier (BBB) (Chakraborty et al. 2016). An herbal medication containing nanocarriers has received much interest in recent decades due to its future potential and unique qualities that make these materials useful in many spheres of human life. As a result, nanoherbal systems have a potential future for boosting activity and addressing the problems associated with plant medicine (Sandhiya and Ubaidulla 2020). The various formulation used as a phytopharmaceuticals are depicted in Figure 18.4.

18.7 FUTURE OF NANOTECHNOLOGY IN HERBAL DRUG DELIVERY

Nanotechnology applications in herbal drug delivery, including some effective cases with proficient implications, are being researched. Herbal medications have a plethora of useful compounds, including nutrients and antioxidants, that can be employed in targeted nutrition as well. This type of collaborative inquiry between archaic "herbal remedies" and current drug delivery systems has also been highlighted (Yadav et al. 2011; Salunkhe et al. 2020).

Nanocarrier-based drug delivery, on the other hand, has some obstacles, including inadequate drug-loading capacity, toxicity, a lack of cellular uptake ability, and uncertain ligand-tagging ability and biodegradability (Kumar et al. 2017). But formulation scientists have proposed a number of strategies to overcome these drawbacks and bring nanocarrier-based drug delivery into the real world (Dewanjee et al. 2020).

18.8 ILLUSTRATION OF HERBAL NANOCARRIER-BASED DRUG DELIVERY

18.8.1 Curcumin

Curcumin, derived from the *Curcuma longa* L. plant, is one of the most well-known and important natural polyphenols. Antimicrobial (Mahady et al. 2002), antioxidant (Sharma 1976) anticarcinogenic (Kuttan et al. 1985), and anti-inflammatory (Srimal and Dhawan 1973) effects have all been linked to it. Curcumin has been shown to have neuroprotective properties in neurodegenerative disorders such as Parkinson's disease (PD) (Mythri and Bharath 2012) and Alzheimer's disease (Garcia-Alloza et al. 2007). Curcumin has inadequate pharmacokinetics, with low absorption, poor bioavailability, and rapid elimination from the body, despite its outstanding therapeutic qualities. Various nanoparticles like nanoliposomes, nanocapsules, and nanomicelles have been designed to increase the pharmacokinetics and bioavailability of curcumin to overcome these flaws (Bollimpelli et al. 2016).

18.8.2 Paclitaxel

In recent years, nanoparticle delivery technologies have gotten a lot of interest, notably for cancer medicines. Paclitaxel has been produced in several nanodelivery systems as an efficient chemotherapeutic drug with significant advantages over standard-of-care therapy. Paclitaxel was discovered in the bark of a Pacific yew tree (*Taxus brevifolia*). When paclitaxel is encapsulated into lipid-based nanoparticles or conjugated with water-soluble polymers, nanodelivery systems significantly improve the drug's aqueous solubility, permeability, and retention; reduce adverse effects; and improve the drug's pharmacokinetic characteristics (Ma and Mumper 2013).

18.8.3 Resveratrol

Resveratrol is a non-flavonoid polyphenolic molecule found in peanuts, grapes, and other foods. It has many medicinal uses, including cardioprotective, antioxidant, and anti-inflammatory properties (Amri et al. 2012). However, because the resveratrol molecule is unstable and loses its antioxidant effects after a brief exposure to light, a vehicle that can protect the molecule and enhance the efficacy is needed (Müeller et al. 2000). Solid lipid nanoparticles can preserve integrated drugs from degradation and adjust their release rate for enhanced efficiency due to their unique physicochemical and biological features, such as small size and high biocompatibility (Battaglia and Gallarate 2012; Wang et al. 2015; Soldati et al. 2018). The transport of resveratrol to the skin, with solid lipid nanoparticles using *Theobroma grandiflorum* (natural butter) as the lipid core, is possible. Researchers were able to regulate the delivery of the active constituent while increasing the antioxidant activity by using this approach (Soldati et al. 2018).

18.8.4 Berberine

Berberine is an isoquinoline alkaloid present in *Berberis aristata* stem, roots, and bark. Berberine has a variety of pharmacological effects, notably antiarrhythmic properties, antimalarial, anticancer, and antihyperglycemic. Berberine's limited water solubility, on the other hand, affects its oral bioavailability and dissolution rate, limiting its therapeutic application. To boost the drug's bioavailability, berberine nanoparticles were made utilizing the evaporative precipitation of nanosuspension (EPN) and antisolvent precipitation with a syringe pump (APSP) techniques. Berberine nanoparticles created by both the APSP and EPN procedures have showed potential activity against gram-positive and gram-negative bacteria, as well as yeasts. However, nanoparticles produced with the EPN method display better results than the APSP method, as well as the unprocessed drug (Sahibzada et al. 2018).

18.8.5 Traditional Chinese Herbs

Traditional Chinese herbs (TCH) nanoparticles work to promote dispersion and absorption in the body, hence increasing efficacy. TCH such as Szechwan lovage rhizome, Rehmannia root, safflower, peach seed, ground beetle, angelica root, red peony root, gadfly, leech, and earthworm were mixed and prepared using a nanometer collider after drying, grinding, extracting, and crushing into the liquid particles with the help of a ultrasonic wave, filtering, and then nanometerizing into the nanoparticles liquid. TCH nanoparticles exhibited substantial thrombolytic properties, resulting in rapid healing from thrombi diminution and arterial embolism. TCH nanoparticles have substantially stronger thrombolytic effects than their non-nanoparticle types. There are also several studies on integrative analysis, pharmacological efficacy, and pharmacokinetics of traditional Chinese medicine oral prolonged-release formulations (Devi et al. 2010).

Along with the previously mentioned examples, a few more are depicted in Table 18.1 for better understanding about the phytopharmaceuticals.

TABLE 18.1
Nanoformulation Design of Various Herbal Drugs

Active Ingredient	Botanical Name of Sources	Biological Activity	Problem Associated with Herbal Drug Formulation	Types of Nanodelivery	References
Curcumin	*Curcuma longa* L.	Anticancer	Limited absorption, negligible bioavailability, fast elimination	Nanoparticles, nanocapsules, nanomicelles, and nanoliposomes	Bollimpelli et al. (2016)
Berberin	*Berberis aristata*	Antimalarial, antiarrhythmic, antihyperglycemic, anticancer, antimicrobial	Poor water solubility, low bioavailability	Nanosuspension	Sahibzada et al. (2018)
Camptothecin	*Camptotheca acuminata*	Anticancerous	Low bioavailability, poor water solubility	Polymeric nanoparticles	Yang et al. (2016)
Paclitaxel	*Taxus brevifolia*	Anticancer	Low water solubility	Polymeric nanoparticles	Ma and Mumper (2013)
Quercetin	*Santalum album*	Neuroprotective and anti-inflammatory	low bioavailability	Nanocapsules, nanogels, liposomes, nanosuspensions,	Abd El-Rahman and Al-jameel (2014)
Triptolide	*Tripterygium wilfordii* Hook F	Antioxidant, antitumor, anti-inflammatory, antifertility, anticystogenesis, immunosuppressive and anticancer	Poor solubility, toxicity	Solid lipid nanoparticles, liposomes, polymeric micelle, microemulsion	Ren et al. (2021)
Resveratrol	*Vitis vinifera, Arachis hypogaea, Polygonum cuspidatum*	Antioxidant	Low bioavailability	Solid lipid nanoparticles	Soldati et al. (2018)

18.9 CONCLUSION

Recently in the medical field, herbal biomedicines or phytopharmaceuticals have been receiving more attention due to their extraordinary features for targeted drug delivery treatment and created decreased prospects of any infections and side effects. Phytopharmaceuticals possess a noteworthy therapeutic efficiency that must be explored with nanobiotechnology. Although there are various shortcomings such as less bioavailability, limited solubility, slight oral absorption, and possible harmfulness of herbal treatments. Therefore, various nanocarriers can play an important function in order to triumph over these complications. In this chapter, special techniques practiced in the preparation of various nanocarriers and their classifications have been broadly explained along with the use of phytopharmaceuticals and their current trends in the medical field.

BIBLIOGRAPHY

Abd El-Rahman, S., Suhailah, S. and Al-jameel, S. S. 2014. Quercetin nanoparticles: Preparation and characterization. *International Journal of Drugs*, 2(3): 96–103.

Ajazuddin, S. S. 2010. Applications of novel drug delivery system for herbal formulations. *Fitoterapia*, 81(7): 680–689.

Akbarzadeh, A., Rezaei-Sadabady, R., Davaran, S., Joo, S. W., Zarghami, N., Hanifehpour, Y., Samiei, M., Kouhi, M. and Nejati-Koshki, K. 2013. Liposome: Classification, preparation, and applications. *Nanoscale Research Letters*, 8(1): 102.

Alexis, F., Pridgen, E., Molnar, L. K. and Farokhzad, O. C. 2008. Factors affecting the clearance and biodistribution of polymeric nanoparticles. *Molecular Pharmaceutics*, 5(4): 505–515.

Amri, A., Chaumeil, J. C., Sfar, S. and Charrueau, C. 2012. Administration of resveratrol: What formulation solutions to bioavailability limitations. *Journal of Controlled Release*, 158: 182–193.

Ansari, S. H., Islam, F. and Sameem, M. 2012. Influence of nanotechnology on herbal drugs: A review. *Journal of Advanced Pharmaceutical Technology & Research*, 3(3): 142–146.

Bai, S., Thomas, C., Rawat, A. and Ahsan, F. 2006 . Recent progress in dendrimer-based nanocarriers. *Critical Reviews in Therapeutic Drug Carrier Systems*, 23: 437–495.

Battaglia, L. and Gallarate, M. 2012. Lipid nanoparticles: State of the art, new preparation methods and challenges in drug delivery. *Expert Opinion on Drug Delivery*, 9: 497–508.

Bhardwaj, M., Singh, S. and Kumar, A. (2021). Health informatics-A vital strategy to tackle pandemic diseases. *Biomedical Journal of Scientific & Technical Research*, 39(4): 31497–31498.

Bisht, S., Feldmann, G., Soni, S., Ravi, R., Karikar, C. and Maitra, A. 2007. Polymeric nanoparticle-encapsulated curcumin ("nanocurcumin"): A novel strategy for human cancer therapy. *Journal of Nanobiotechnology*, 5: 1–18.

Bollimpelli, V. S., Kumar, P., Kumari, S. and Kondapi, A. K. 2016. Neuroprotective effect of curcumin-loaded lactoferrin nano particles against rotenone induced neurotoxicity. *Neurochemistry International*, 95: 37–45.

Bonifácio, B.V., Silva, P.B., Ramos, M.A., Negri, K.M., Bauab, T.M. and Chorilli, M. 2014. Nanotechnology-based drug delivery systems and herbal medicines: A review. *International Journal of Nanomedicine*, 9: 1–15.

Chakraborty, K., Shivakumar, A. and Ramachandran, S. 2016. Nano-technology in herbal medicines: A review. *International Journal of Herbal Medicine*, 4: 21–27.

Chaudhary, H., Kohli, K. and Kumar, V. 2014. A novel nano-carrier transdermal gel against Inflammation. *International Journal of Pharmaceutics*, 465: 175–186.

Chen, Y., Lin, X., Park, H. and Greever, R. 2009. Study of artemisin in nanocapsules as anticancer drug delivery systems. *Nanomedicine*, 5(3): 316–322.

Chorilli, M., Leonardi, G. R., Oliveira, A. G. and Scarpa, M. V. 2004. Dermocosmetic liposome formulations. *Infarma*, 16(7–8): 75–79.

Chorilli, M., Rimério, T. C., Oliveira, A. G. and Scarpa, M. V. 2007. Study of the stability of small unilamellar liposomes containing caffeine turbidimetric. *Revistabrasileira de fruticulture*, 88(4): 194–199.

Currier, S. J., Johnston, P. D. and Gorelick, K. J. 2000. Complementary and alternative medicine-herbal medicines. *Sci Medicine*, 7: 40–43.

Devi, V. K., Jain, N. and Valli, K. S. 2010. Importance of novel drug delivery systems in herbal medicines. *Pharmacognosy Review*, 4(7): 27–31.

Dewanjee, S., Chakraborty, P., Mukherjee, B. and De Feo, V. 2020. Plant-based antidiabetic nanoformulations: The emerging paradigm for effective therapy. *International Journal of Molecular Science*, 21(6): 2217.

Diab, R., Jaafar-Maalej, C., Fessi, H. and Maincent, P. 2012. Engineered nanoparticulate drug delivery systems: The next frontier for oral administration? *AAPS Journal*, 14: 688–702.

Fessi, H., Puisieux, F., Devissaguet, J. P., Ammoury, N. and Benita, S. 1989. Nano-capsule formation by interfacial polymer deposition following solvent displacement. *International Journal of Pharmaceutics*, 55: 1–4.

Gao, D. 2013. Drug design for cancer: Gold nanoparticleliposome hybrids for drug delivery and monitoring. *Drug Desigining*, 2: 2.

Garcia-Alloza, M., Borrelli, L. A., Rozkalne, A., Hyman, B. T. and Bacskai, B. J. 2007. Curcumin labels amyloid pathology in vivo, disrupts existing plaques, and partially restores distorted neurites in an Alzheimer mouse model. *Journal of Neurochemistry*, 102(4): 1095–1104.

Gunasekaran, T., Haile, T., Nigusse, T., and Dhanaraju, M. D. 2014. Nanotechnology: An effective tool for enhancing bioavailability and bioactivity of phytomedicine. *Asian Pacific Journal of Tropical Biomedicine*, 4(Suppl 1): S1–S7.

Hegmann, T., Qi, H. and Marx, V. M. 2007. Nanoparticles in liquid crystals: Synthesis, self-assembly, defect formation and potential applications. *Journal of Inorganic and Organometallic Polymers and Materials*, 17(3): 483–508.

Jain, T., Kumar, S. and Dutta, P. K. 2015. Chitosan in the light of nanobiotechnology: A mini review. *Journal of Biomedical Technology and Research*, 2(1): 101.

Jain, T., Kumar, S. and Dutta, P. K. 2016. Dibutyrylchitin nanoparticles as novel drug carrier *International Journal of Biological Macromolecules*, 82: 1011–1017.

Kalani, M. and Yunus, R. 2011. Application of supercritical antisolvent method in drug encapsulation: A review. *International Journal of Nanomedicine*, 6: 1429–1442.

Kesharwani, P., Gorain, B., Low, S. Y., Tan, S. A., Ling, E. C. S., Lim, Y. K., Chin, C. M., Lee, P. Y., Lee, C. M., Ooi, C. H., Choudhury, H. and Pandey, M. 2018. Nanotechnology based approaches for anti-diabetic drugs delivery. *Diabetes Research and Clinical Practice*, 136: 52–77.

Khuda-Bukhsh, A. R., Bhattacharyya, S. S., Paul, S. and Boujedaini, N. 2010. Polymeric nanoparticle encapsulation of a naturally occurring plant scopoletin and its effects on human melanoma cell A375. *Zhong Xi Yi Jie He Xue Bao*, 8(9): 853–862.

Kolenyak, F., Oyafuso, M., Kiill, C., Gremiao, M. P. and Chorilli, M. 2013. Nanotechnology-based drug delivery systems for treatment of hyperproliferative skin diseases: A review. *Current Nanoscience*, 9: 159–167.

Kumar, P., Kim, K. H., Bansal, V., Kumar, S., Dilbaghi, N. and Kim, Y. H. 2017. Modern progress and future challenges in nanocarriers for probe applications. *Trends in Analytical Chemistry*, 86: 235–250.

Kumari, A., Yadav, S. K. and Yadav, S. C. 2010. Biodegradable polymeric nanoparticles-based drug delivery system. *Colloids and Surface B: Biointerfaces*, 75(1): 1–18.

Kuttan, R., Bhanumathy, P., Nirmala, K. and George, M. C. 1985. Potential anticancer activity of turmeric (*Curcuma longa). Cancer Letters*, 29(2): 197–202.

Li, D., Zhong, X., Zeng, Z., Jiang, J., Li, L., Zhao, M. M., Yang, X. Q., Chen, J., Zhang, B. S., Zhao, Q. Z., Xie, M. Y., Xiong, H., Deng, Z. Y., Zhang, X. M., Xu, S. Y., Gao, Y. X. 2009. Application of targeted drug delivery system in Chinese medicine. *Journal of Controlled Release*, 138: 103–112.

Li, Y., Dong, L., Jia, A., Chang, X. and Xue, H. 2006. Preparation and characterization of solid lipid nanoparticles loaded traditional Chinese medicine. *International Journal of Biological Macromolecules*, 38(3–5): 296–299.

Lopes, C. M. 2014. Therapeutics delivery: Innovations technology approaches. *Drug Desigining*, 3: 3.

Ma, P. and Mumper, R. J. 2013. Paclitaxel nano-delivery systems: A comprehensive review. *Journal of Nanomedicine & Nanotechnology*, 4(2): 1000164.

Mahady, G. B., Pendland, S. L., Yun, G. and Lu, Z. Z. 2002. Turmeric (*Curcuma longa*) and curcumin inhibit the growth of Helicobacter pylori, a group 1 carcinogen. *Anticancer Research*, 22: 4179–4181.

Mainardes, R. M., Gremião, M. P. D. and Evangelista, R. C. 2006. Thermoanalytical study of praziquatel-loaded PLGA nanoparticles. *Brazilian Journal of Pharmaceutical Sciences*, 42(4): 523–530.

Martins, S., Sarmento, B., Ferreira, D. C. and Souto, E. B. 2007. Lipid-based colloidal carriers for peptide and protein delivery-liposomes versus lipid nanoparticles. *International Journal of Nanomedicine*, 2(4): 595–607.

Mueller, R. H., Maeder, K. and Gohla, S. 2000. Solid lipid nanoparticles (SLN) for controlled drug delivery: A review of the state of the art. *European Journal of Pharmaceuticsand Biopharmaceutics*, 50: 161–177.

Mythri, R. B. and Bharath, M. M., 2012. Curcumin: A potential neuroprotective agent in Parkinson's disease. *Current Pharmaceutical Design*, 18: 91–99.

Naahidi, S., Jafari, M., Edalat, F., Raymond, K., Khademhosseini, A, and Chen, P. 2013. Biocompatibility of engineered nanoparticles for drug delivery. *Journal of Controlled Release*, 166(2): 182–194.

Nikam, A. P., Ratnaparkhiand, M. P. and Chaudhari, S. P. 2014. Nanoparticles: An overview. *International Journal of Research and Development in Pharmacy and Life Sciences*, 3(5): 1121–1127.

Oliveira, A. G. and Scarpa, M. V. 2001. Microemulsion I: Theoretical training of the microemulsion system. *Infarma*, 13(9–10): 73–79.

Oliveira, A. G., Scarpa, M. V. and Cera, L. F. R. 2002. Microemulsions II: Applications for drug delivery. *Infarma*, 14(7–8): 76–80.

Pardeike, J., Hommoss, A., Müller, R. H. 2009. Lipid nanoparticles (SLN, NLC) in cosmetic and pharmaceutical dermal products. *International Journal of Pharmaceutics*, 366 (1–2): 170–184.

Ren, Q., Li, M., Deng, Y., Lu, A. and Lu, J. 2021. Triptolide delivery: Nanotechnology-based carrier systems to enhance efficacy and limit toxicity. *Pharmacological Research*, 165: 105377.

Sahib, S., Sharma, A., Muresanu, D. F., Zhang, Z., Li, C., Tian, Z. R., Buzoianu, A. D., Lafuente, J. V., Castellani, R. J., Nozari, A., Patnaik, R., Menon, P. K., Wiklund, L. and Sharma, H. S. 2021. Nanodelivery of traditional Chinese Gingko Biloba extract EGb-761and bilobalide BN-52021 induces superior neuroprotective effects on pathophysiology of heat stroke. *Progress in Brain Research*, 265: 249–315.

Sahibzada, M., Sadiq, A., Faidah, H. S., Khurram, M., Amin, M. U., Haseeb, A. and Kakar, M. 2018. Berberine nanoparticles with enhanced in vitro bioavailability: Characterization and antimicrobial activity. *Drug Design, Development and Therapy*, 12: 303–312.

Sahni, J. K., Baboota, S. and Ali, J. 2011. Promising role of nanopharmaceuticals in drug delivery. *Pharma Times*, 43: 16–18.

Salunkhe, P., Bhoyar, P., Gode, A. and Shewale, P. S. 2020. Application of nanotechnology to the extraction of herbal components for medicinal uses. *Current Nanomaterials*, 5(1): 4–11.

Samad, A., Alam, Md. I. and Saxena, Kinshuk. 2009. Dendrimers: A class of polymers in the nanotechnology for the delivery of active pharmaceuticals. *Current Pharmaceutical Design*, 15: 2958–2969.

Sandhiya, V. and Ubaidulla, U. 2020. A review on herbal drug loaded into pharmaceutical carrier techniques and its evaluation process. *Future Journal of Pharmaceutical Science*, 6(51).

Schaffazick, S. R., Guterres, S. S., Freitas, L. L. and Pohlmann, A. R. 2003. Physicochemical characterization and physicochemical stability of nanoparticle polymeric systems for drug administration. *Quim Nova*, 26(5): 726–737.

Sharma, A. T., Mitkare, S. S. and Moon, R. S. 2011. Multicomponent herbal therapy: A review. *International Journal of Pharmaceutical Sciences Review and Research*, 6: 185–187.

Sharma, O. P. 1976. Antioxidant activity of curcumin and related compounds. *Biochemical Pharmacology*, 25(15): 1811–1812.

Singh, A., Sengar, R. S., Singh, A. K., Singh, S., Kumar, A. and Ahmad, K. (2019). Impact of genetic engineering on agriculture food crops. *Biotech Today: An International Journal of Biological Sciences*, 9(1): 17–24.

Snijesh, V. P. and Singh, S. (2014). Molecular modeling and network based approach in explaining the medicinal properties of nyctanthes arbortristis, lippia nodiflora for rheumatoid arthritis. *Journal of Bioinformatics and Intelligent Control*, 3(1): 31–38.

Soldati, P. P., Polonini, H. C., Paes, C. Q., Restrepob, J. A. S., Creczynksi-Pasa, T. B., Chaves, M. G. A. M., Brando, A. F., Pittella, F. and Raposo, N. R. B. 2018. Controlled release of resveratrol from lipid nanoparticles improves antioxidant effect. *IFAC-PapersOnLine*, 51(27): 16–21.

Souto, E. B., Severino, P., Santana, M. H. A. and Pinho, S. C. 2011. Solid lipid nanoparticles: Classical methods of laboratory production. *Quim Nova*, 34(10): 1762–1769.

Srimal, R. C. and Dhawan, B. N. 1973. Pharmacology of diferuloyl methane (curcumin), a nonsteroidal anti-inflammatory agent. *Journal of Pharmacy and Pharmacology*, 25: 447–452.

Sungthongjeen, S., Pitaksuteepong, T., Somsiri, A. and Sriamornsak, P. 1999. Studies on pectins as potential hydrogel matrices for controlled-release drug delivery. *Drug Development and Industrial Pharmacy*, 25: 1271–1276.

Timilsena, Y. P., Akanbi, T. O., Khalid, N., Adhikari, B. and Barrow, C. J. 2019. Complex coacervation: Principles, mechanisms and applications in microencapsulation. *International Journal of Biological Macromolecules*, 121: 1276–1286.

Verma, H. and Singh, H. 2013. Herbal drug delivery system: A modern era prospective. *International Journal of Current Pharmaceutical and Research*, 4(3): 88–101.

Verma, N., Sao, P., Srivastava, A. and Singh, S. (2021). Physiological and molecular responses to drought, submergence and excessive watering in plants. In *Harsh Environment and Plant Resilience* (pp. 305–321). Springer, Cham.

Wang, J., Wang, H., Zhu, R., Liu, Q., Fei, J., and Wang, S. 2015. Anti-inflammatory activity of curcumin-loaded solid lipid nanoparticles in IL-1β transgenic mice to the lipopolysaccharide-induced sepsis. *Biomaterials*, 53: 475–483.

Wu, X. Y. and Lee, P. I. 1993. Preparation and characterization of thermal and pH-sensitive nanospheres. *Pharmaceutical Research*, 10(10): 1544–1547.

Yadav, D., Suri, S., Chaudhary, A., Sikander, M., Kardam, H., BEG, M. N., Garg, V., Ahmad, A. and Mohd, A. 2011. A novel approach: Herbal remedies and natural products in pharmaceutical science as nano drug delivery systems. *International Journal of Pharmacy and Technology*, 3(3): 3092–3116.

Yadav, D., Suri, S., Choudhary, A. A., Sikender, M., Hemant, B. and Beg, N. M. 2011. Novel approach: Herbal remedies and natural products in pharmaceutical science as nano drug delivery systems. *International Journal of Pharmaceutical Technology*, 3: 3.

Yang, A., Liu, Z., Yan, B., Zhou, M. and Xiong, X. 2016. Preparation of camptothecin-loaded targeting nanoparticles and their antitumor effects on hepatocellular carcinoma cell line H22. *Drug Delivery*, 23(5): 1699–1706.

19 Phytonanotechnology
Plant-Mediated Green Construction of MNPs and Their Biomedical Implementation against MDR Bacteria

Meenu Gupta, Neeraj Agarwal, and Abhidha Kohli

CONTENTS

19.1 Introduction ... 331
19.2 Pathogenicity and Mechanisms of Multidrug Resistance in Bacteria ... 333
19.3 Plant-Mediated Sustainable Fabrication of MNPs Against MDR Bacteria ... 333
19.4 Antibacterial Activity of Biogenic MNPs ... 333
19.4.1 Mechanism of Antibacterial Activity of MNPs ... 333
19.5 Fabrication of MNPs Using Plant Extracts ... 334
19.5.1 AuNPs ... 334
19.5.2 AgNPs ... 336
19.5.3 PtNPs ... 339
19.5.4 Other MNPs ... 340
19.6 Conclusion, Outlook and Future Aspects ... 342
Ethical Issues ... 342
Bibliography ... 342

19.1 INTRODUCTION

Nanotechnology has resulted in the synthesis of various metal nanoparticles (MNPs) ranging from 0.1 to 100 nm in size. These MNPs have distinctive physiochemical, optical and therapeutic properties. Thus, these MNPs can be suitably manipulated for desired applications.

MNPs can be fabricated through two different approaches: the bottom-up and touch-down method. The bottom-up method brings about the synthesis of MNPs through molecular assembly (chemical and biosynthesis). Chemical synthesis includes many hazardous chemicals like alkyl marcaptan, dimethyl-formamide, tween 80 and polyvinyl-pyrrolidone as stabilizers and formaldehyde and hydrazine as reducing agents

In the top-down approach, making the bulk molecules smaller results in the fabrication of MNPs (mechanical grinding, thermal decomposition, laser ablation) (Kralj and Makovec 2015) See Figure 19.1.

However, many physical and chemical methods for synthesizing MNPs are cost ineffective and have resulted in heavy metal pollution in the environment. Biogenic green synthesis of MNPs

DOI: 10.1201/b22917-19

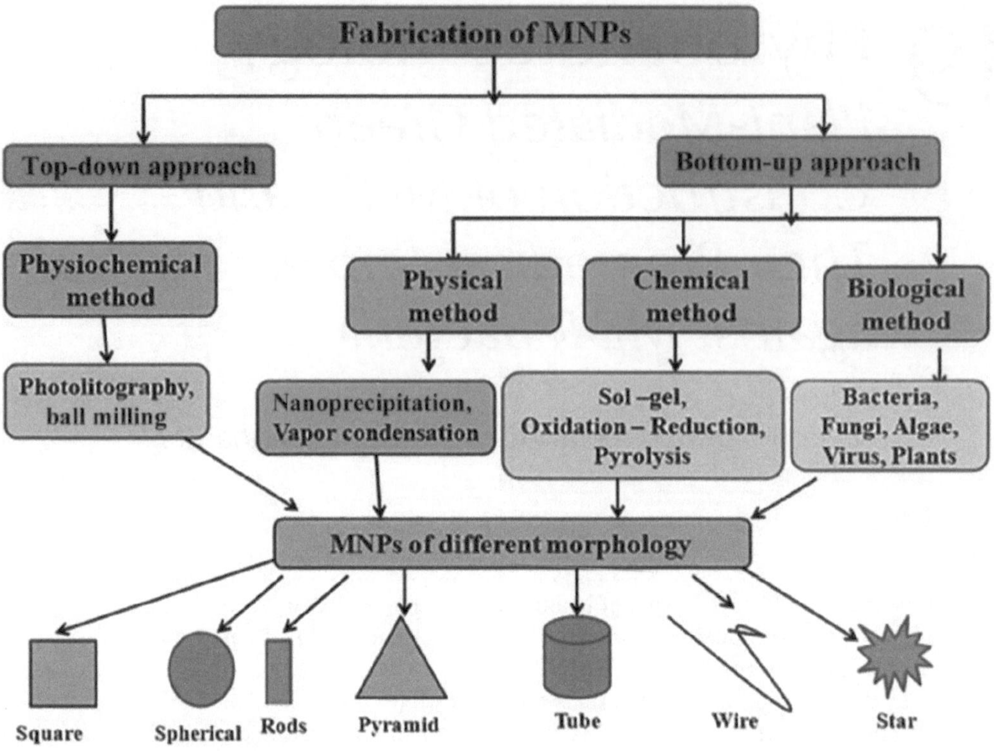

FIGURE 19.1 A summary of different approaches employed for the fabrication of MNPs.

Source: Gupta and Seema, 2021

provides an alternative potential way for convenient, economical, sustainable and eco-friendly synthesis of MNPs (Gupta and Seema, 2021).

The World Health Organization (WHO) released a list of multidrug-resistant MDR and biofilm synthesizing bacteria including *Enterococcus, Staphylococcus aureus, Klebsiella, Acinetobacter baumannii, Pseudomonas aeruginosa* and *Enterobacter* sp. (ESKAPE). All these bacteria can cause chronic disease due to MDR. These opportunistic MDR pathogens have resistance towards the biocidal activity of antibiotics, which has led to an alarming threat in the area of pathogenesis and therapeutics (Pendleton et al., 2013). The WHO has predicted that more people will be killed by bacterial infection than by cancer by 2050 and the cost will be $100 trillion for medical wellness (WHO list 2017). Scientists are trying to find alternative solutions to address this challenge. MNPs (1–100 nm) are known to have potential antibiotic and therapeutic properties (Lashin et al., 2021), and thus can serve as alternative routes. MNPs have a large surface area to volume ratio, which increases their permeability to the plasma membrane of bacteria. Positively charged MNPs also show affinity for negatively charged ligands present on the plasma membrane. MNPs target numerous pathways for bactericidal activities at the same time; thus bacteria cannot easily develop resistance towards MNPs (Singh et al., 2020). Among all the MNPs, silver nanoparticles (AgNPs) have unique chemical and physical properties, low cytotoxicity and a broad surface area, giving them dynamic antimicrobial means to counter gram-negative and gram-positive bacteria (Samuggam et al., 2021). Plant-mediated green construction of MNPs gained a lot of interest recently. MNPs produced using plant extracts are more biocompatible, stable, non-toxic and have diverse biomedical applications (Noruzi, 2015).

19.2 PATHOGENICITY AND MECHANISMS OF MULTIDRUG RESISTANCE IN BACTERIA

Annually around 10,000 tons of antibiotics are manufactured to cure bacterial infections. However, most of them are not effective against MDR bacteria, so there is a continuous requirement of new, novel and potential alternatives as antibacterial agents.

In the process of evolution bacteria have developed several resistance mechanisms against antibacterial agents. These antibiotic-resistance mechanisms include mutation in target proteins, reduction of antibiotic uptake, inactivation of drugs through specific enzymes and efflux pumping of antibacterial agents from bacterial cells—for example, a mutation in one ribosomal protein i.e. RpsL provides resistance against streptomycin in *Escherichia coli* (Alavi and Rai, 2019). The reason for vancomycin resistance is an alteration in the structure of a penta-peptide of the cell wall—an illustration of target bypassing. Activation of a tetracycline efflux pump coded by two genes, TetA and TetA-1, plays a vital role against tetracycline resistance in many gram-negative bacteria (Wang et al., 2014).

19.3 PLANT-MEDIATED SUSTAINABLE FABRICATION OF MNPS AGAINST MDR BACTERIA

From ancient times plants have therapeutic applications in numerous diseases. Thus phytochemical agents and MNPs can be potentially employed against ESKAPE pathogens. Plant-based green synthesis of MNPs has immense possibility for rapid, economic and sustainable mass production of MNPs. Out of all MNPs, gold (AuNPs), silver (AgNPs) and platinum (PtNPs) have gained attention owing to antimicrobial and antagonistic properties for a variety of gram-negative and gram-positive bacteria. AgNPs show biocidal properties against more than 650 microbes, including virus, fungi and bacteria (Ahmed et al., 2016).

The previously mentioned noble metals are known to have unique properties like high reduction potential, high ionization energy, non-reactive in nature, corrosion resistant and a high melting point. Thus, these metals have diverse potential in therapeutics (Pareek et al., 2017; Singh et al., 2020).

Green fabrication of MNPs allowed the synthesized MNPs to interact with biochemical compounds, secondary metabolites and other bioactive compounds required for efficient significant activity. Plant extracts consist of enzymes, vitamins, phenols, polysaccharides, terpenoids, tannins, alkaloids and secondary metabolites, which have remarkable antibiotic properties and play a vital job in the reduction of metal salts and MNPs' stability (Duan et al., 2015). The exact process of phytogenic construction of MNPs is still not understandable. A brief idea of MNP synthesis using plants is shown in Figure 19.2.

19.4 ANTIBACTERIAL ACTIVITY OF BIOGENIC MNPS

Biogenic MNPs are being explored for their efficacy as dynamic antimicrobial agents, especially against MDR bacteria. MNPs with functional bioactive molecules are utilized against MDR bacteria. MNPs have higher permeability and can diffuse inside the cell of pathogenic bacteria, and plant-based bioactive compounds present at the outer surface of MNPs enhance their potential efficacy as antibacterial agents (Khan et al., 2018).

19.4.1 Mechanism of Antibacterial Activity of MNPs

Various explanations have been given to explain the antibacterial activity of MNPs. Possible mechanisms involve alteration in DNA and protein, membrane potential, disruption of cell membrane, nutrient uptake, reactive oxygen species (ROS) production, oxidation of proteins, inhibition of bacterial respiratory chain and by developing asymmetrical pits in the plasma membrane, which leads to the

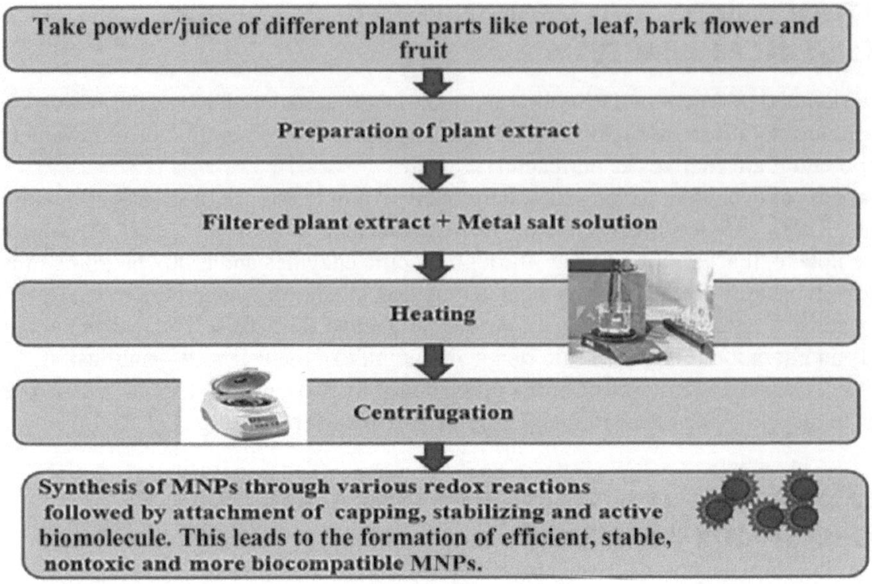

FIGURE 19.2 Systematic representation of the process for plant-mediated MNP synthesis.

Source: Gupta and Seema, 2021

dispersal of lipopolysaccharides and proteins from the bacterial cell, which in turn is responsible for killing the bacteria (Khan et al., 2018). Selenium NPs disrupt the redoxin and glutathione pathway, which induces cytotoxicity monitored in terms of ROS (Menon et al., 2018). ZnNPs interfere with the permeability of the bacterial plasma membrane by inducing oxidative stress and the export of metal ions and hydrogen peroxide also formed from the surface of ZnONPs (Chandra et al., 2020).

19.5 FABRICATION OF MNPS USING PLANT EXTRACTS

19.5.1 AuNPs

AuNPs show surface plasmon resonance (SPR) frequency in the visible region (400–700 nm). Thus these AuNPs can be preferably used in optical applications, as biosensors, bioimaging, in photo thermal treatments and other biomedical applications such as wound healing, tissue engineering, cancer therapy, antimicrobial activity and medical devices (catheters, prostheses, vascular grafts) (Gupta and Seema, 2021). See Table 19.1.

High chemical stability, oxidation resistance, strong antibacterial activity and biocompatibility made them an excellent choice. Synthesis of AuNPs is quick due to the easy reduction of metal salts (Murugan et al., 2015). Table 19.2 summarizes the bactericidal activities of AuNPs fabricated using various plant extracts. AuNPs inhibit bacterial growth by disrupting the cell membrane/cell wall, ribosomes and mitochondria. AuNPs show greater antibacterial activity towards gram-negative bacteria. The thin and negatively charged cell wall of gram-negative bacteria shows stable electrostatic interaction with AuNPs, which enables diffusion of AuNPs, which results into bacterial cell lysis (Sunderam et al., 2019). In another experiment AuNPs show an inhibitory effect against a broad range of bacteria (gram-negative and -positive). The antibacterial activity consists of perforation in the bacterial cell followed by the accumulation of AuNPs. Inside the bacterial cell AuNPs enter the nucleus and inhibit DNA replication and ultimately cell lysis (Chen et al., 2019).

TABLE 19.1
Diverse Applications of AuNPs

MNPs	Application	Reference
AuNPs and its bioconjugates	HIV, TB and cancer treatment; specific delivery of doxorubicin, paclitaxel and methotrexate	Rai et al. (2016)
	Detection of cancer, photoimaging	Singh and Nalwa (2011)
	Antagonistic towards HeLa cells, *Puccinia graminis, Aspergillus flavus, A. niger, Staphylococcus aureus, V. cholera, Staphylococcus epidermidis* and *Staphylococcus typhi*,	Pestovsky and Martínez-Antonio (2017), Ravishankar and Jamuna (2011)
	Radio-treatment	Rai et al. (2016)
	Enhance production of biogenic hydrogen employing artificial waste water and *Clostridium butyricum*	Patrick et al. (2019)
	Increase length and ratio of shoot to root in *Lactucasativa*, enhance seed germination in *Brassica juncea*	Pestovsky and Martínez-Antonio (2017)

TABLE 19.2
Antibacterial Potential of AuNPs Fabricated through Different Plant Extracts

Name of Plant	Plant Organ	Type/Shape/Size of MNPs	Usage	Reference
Convolvulus fruticosus	Extract of flower	AuNPs spherical	Antibacterial	(Gupta and Seema, 2021)
Borago officinalis	Leaf	AuNPs spherical 30–80 nm	Effective against *E. coli*, S. *aeruginosa* and *S. aureus*	(Wahab et al., 2021)
Aloe vera	Leaf	Spherical	Effective against prokaryotes	Das et al. (2017)
Cymbopogon citratus	Leaf	Hexagonal, rod-shaped, spherical and triangular AuNps	Antibacterial	(Singh et al., 2016)
Anacardium occidantale	Leaf	10–60 nm spherical	Antagonistic towards *E. coli* and *Bacilius subtilis*	(Sunderam et al., 2019).
Terrminalia arjuna	Leaf	15–30 nm spherical	Antagonistic towards *Salmonella typhimurium, S. aureus*, and *Pseudomonas aeruginosa*	Dudhane et al. (2019)
Mangifera indica	Seed	46.8 nm spherical	Antagonistic towards *S. aureus* and *E. coli*	Vimalraj et al. (2018)
Mammea	Root	22–50 nm square	Antagonistic towards most common bacteria, *E. coli, B. subtilis, S. aureus* and *P. aeruginosa*	Poojary et al. (2016)
Chenopodium fomosanum	Shell	8 ± 6 nm spherical	Antagonistic towards *S. aureus* and *E. coli*	Chen et al. (2019)
Maytenus royleanus	Stem	30 nm hexagonal	*Leishmania tropica*	Ahmad et al. (2016)
Dimocarpus longan	Juice of fruit	25 nm spherical	Antagonistic towards *S. aureus* and *B. subtilis*	Khan et al. (2016)
Galaxaura elongate		Size range between 3.85 and 77.13 nm. Hexagonal, spherical, few triangular, rod	Antagonistic towards *S. aureus, E. coli, P. aeruginosa*, MRSA and *Klebsiella pneumoniae*	Abdel-Raouf et al. (2017)

(Continued)

TABLE 19.2
(Continued)

Name of Plant	Plant Organ	Type/Shape/Size of MNPs	Usage	Reference
Carica papaya	leaf	2–20 nm, mostly spherical but few triangular, rod and hexagonal	Antagonistic towards *S. aureus, Proteus vulgaricus* and *S. aureus*	Muthukumar et al. (2016)
Plumeria alba	flower	28 ± 5.6 nm and 15.6 ± 3.4 nm spherical	Effective against *E. coli*	Mata et al. (2016)
Acorus calamus	Rhizome	>100 nm spherical	Antagonistic towards *S. aureus* and *E. coli*	Ganesan and Prabu (2019)
Nepenthes khasiana	leaf	50–80 nm spherical	Antagonistic towards *B. subtilis, A. niger, E. coli* and *Candida albicans*	Bhau et al. (2015)
Diospyros	leaf	70–90 nm rectangles	Antagonistic towards *Microsporum gypseum*	Ramesh and Armash (2015)
Gloriosa superba	leaf	10–20 nm spherical and triangular	Antagonistic towards *S. pneumonia, S. aureus, K. pneumonia* and *E. coli*	Gopinath et al. (2016)
Dracocephalum kotschy	leaf	11 nm spherical	Antagonistic towards *P. aeruginosa, Proteus vulgaris* and *E. coli*	Dorosti and Jamshidi (2016)

19.5.2 AgNPs

AgNPs have wide applications in medicine, agriculture, biosensors, wound healing, water purification, nanofabricated devices, electronics, cosmetics, etc. See Table 19.3.

Silver is known to have maximum antibacterial activity with minimum cytotoxicity in humans (Qais et al., 2019). For gram-negative and gram-positive bacteria, the mechanism of antibacterial activity of AgNPs is distinct; gram-positive bacteria have a lower sensitivity to Ag than do gram-negative bacteria (Rajeshkumar and Bharath, 2017). The inhibitory activity of AgNPs towards bacteria includes 1) AgNP attachment to the bacterial cell wall damages the cell membrane, impairing the cell transport system; 2) AgNPs interact with biomolecules and cell organelles inside the bacterial cell, resulting in an antibacterial effect; 3) production of ROS causes cell death; and 4) inside the bacterial cell AgNPs induce mutation, cytotoxicity and genotoxicity (Lee and Jun, 2019). See Figure 19.3. Due to the non-toxicity towards the human cell, AgNPs are an excellent choice as antibacterial agents. The positive charge plays a vital role in the antimicrobial activity of Ag^+, and this positive charge promotes AgNP binding to the bacterial cell membrane, resulting in an increase in local inhibitory agent production and a decrease in pH.

However, the mechanism of AgNP antibacterial activity is yet unknown, but it is thought to be owing to the production of large amounts of singlet oxygen, hydrogen peroxide, hypochlorous acid, superoxide anion and ROS. Inside the bacterial cell AgNPs could produce necrotic agents, apoptotic factors, inhibition of respiratory chain dehydrogenase and production of ROS (Rizzello and Pompa, 2014). Table 19.4 summarizes the antibacterial potential of AgNPs formed using various plant extracts.

TABLE 19.3
Diverse Applications of AgNPs

MNPs	Application	References
AgNPs and its bioconjugates	Peculiar drug transportation in cancer and human oral cancer imaging Inhibitory action towards HIV, HSV-2 virus leukemia, breast cancer (MCF-7), lymphoma tumor L5178Y, B16F10 melanoma Effective against cervical cancer SiHa, malignant melanoma and carcinoma, HSV-1, HSV-2 and HPIV-3 virus, monkey pox virus, influenza virus, *Escherichia coli, Enterobacter aerogenes, Pseudomonas aeruginosa*, MRSA, *Staphylococcus aureus, Streptococcus pyogenes* and *Neisseria gonorrhoeae*	Rai et al. (2016), Ravishankar and Jamuna (2011)
	Condition-dependent inhibition of *Curvularia lunata, Macrophomina phaseolina, Aspergillus terreus, Alternaria sp., Rhizoctonia solani, Aspergillus niger, Penicillium expansum, Rhizopus, Botrytis cinerea, Sclerotinia sclerotiorum, Candida* and *Trichophyton* Enhance production of biohydrogen employing *Clostridium butyricum* and inorganic salts *Triticum aestivum* seed germination is quick. *Borago officinalis* shows an increase in leaf number, plant height and inflorescence, while *Raphanus sativus* and *Arabidopsis* show root elongation.	Pestovsky and Martínez-Antonio (2017) Patrick et al. (2019)

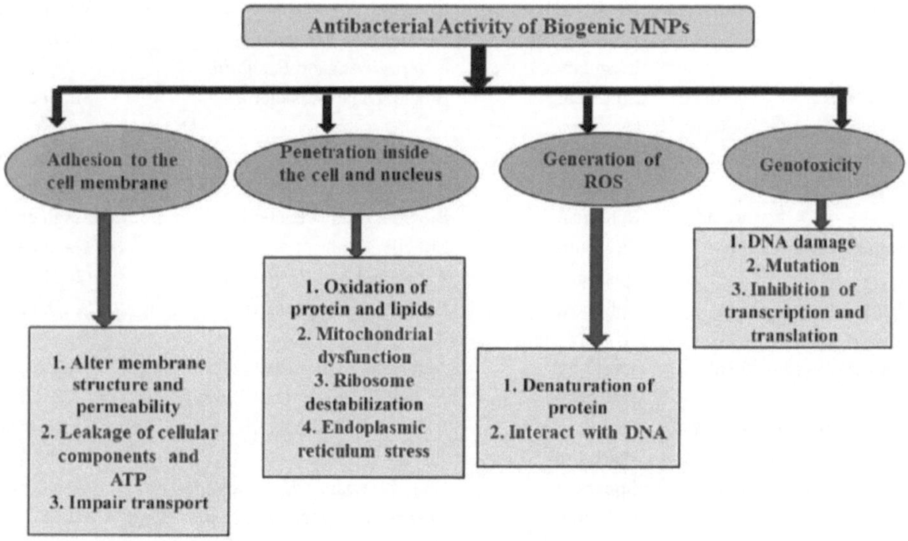

FIGURE 19.3 Mechanism of AgNP-mediated death of bacteria.

TABLE 19.4
Antibacterial Effect of AgNPs Manufactured from Diverse Plant Extracts

S. No	Name of Plant	Plant Organ	Type/Shape/Size of MNPs	Usage	Reference
1	*Nyctanthes arbortristis*	Flower		Biocidal against bacteria	Singh et al. (2016)
2	*Emblica officinalis*	Fruit	15 nm spherical	Effective against *Klebsiella pneumonia, Staphylococcus, Escherichia coli, Bacillus subtilis*	Wahab et al. (2021)
3	*Gardenia jasminoides*	Leaf	Iron rubble like	Biocidal to prokaryotes	Singh et al. (2016)
4	*Curcuma longa*	Leaf	Spherical 15–40 nm	Antagonistic towards *E. coli, Pseudomonas aeruginosa, Streptococcus pyogenes, Candida albicans* and *Staphylococcus aureus*	Maghimaa and Alharbi (2020)
5	*Moringa oleifera*	Flower	Spherical 8 nm	Antagonistic towards *S. aureus* and *K. pneumoniae*	Bindhu et al. (2020)
6	*Artemisia nilagirica*	Leaf	Square/spherical/ hexagonal: 70–90 nm	Effective against *B. subtills, S. aureus, E. coli*	Wahab et al. (2021)
7	*Prosopis farcta*	Leaf	10.8 nm spherical	*S. aureus, P. aeruginosa, E. coli* and *B. subtilis* are all susceptible	Wahab et al. (2021)
8	*Trianthema decandra*	Root	15 nm cubic	Antagonistic toward *P. aeruginosa* and *E. coli*	Wahab et al. (2021)
9	*Panax ginseng*	Root	Spherical	Antibacterial	Singh et al. (2016)
10	*Cleome viscosa*	Fruit	20–50 nm spherical and irregular	Antagonistic towards *S. aureus, K. pneumonia, P. aeruginosa* and *B. subtilis*	Wahab et al. (2021)
11	*Anogeissus latifolia*	Gum powder	Spherical	Biocidal against bacteria	Singh et al. (2016)
12	*Petroselinum crispum*	Leaf	25–30 nm spherical	Effective against *K. pneumonia, S. aureus* and *E. coli*	Wahab et al. (2021)
13	*Red ginseng*	Root	Spherical	Biocidal against bacteria	Singh et al. (2016)
14	*Psidium guajava*	Leaf	25–35 nm spherical	Effective against *E. coli* and *Alcaligenes faecalis*	Wahab et al. (2021)
15	*Abutilon indicum*	Leaf	Spherical	Biocidal against bacteria	Singh et al. (2016)
16	*Curcuma longa* L	Leaf	15–40 nm spherical	Antagonistic towards *E. coli, B. subtilis* and *P. aeruginosa*	Wahab et al. (2021)
17	*Pinus densiflora*	Cones	Oval/triangular shaped	Biocidal against bacteria	Singh et al. (2016)
18	*Moringa olefira*	Leaf	Spherical 9–11 nm	Effective against *K. pneumonia, Enetrococcus faecalis, S. aureus, P. aeruginosa*, and *E. coli.*	Wahab et al. (2021)
19	*Pistacia atlantica*	Seeds	Spherical	Biocidal against bacteria	Singh et al. (2016)
20	*Aegle marmelos*	Fruit	Cubic 159–181 nm	Effective against *Bacillus cereus, Shigella dysenteriae* and *P. aeruginosa*	Wahab et al. (2021)
21	*Ixora coccinea*	Leaf	Spherical 20 nm	Antagonistic towards *S. aureus, Klebsiella, P. aeruginosa* and *E. coli*	Vinay and Chandrasekhar (2019)

S. No	Name of Plant	Plant Organ	Type/Shape/Size of MNPs	Usage	Reference
22	*Terminalia bellerica*	Fruit	Spherical 10 nm	Effective against *K. pneumonia* and *P. aeruginosa*	Wahab et al. (2021)
23	*Geranium graveolens*	Leaf		Antibacterial	Das et al. (2017)
24	*Aloe vera*	Leaf		Antimicrobial	Das et al. (2017)
25	*Ocimum basilicum*	Leaf	13–25 nm spherical	Antagonistic towards to *S. aureus*, *P. aeruginosa* and *E. coli*	Malapermal et al. (2017)
26	*Handelia trichophylla*	Aqueous extract	20–50 nm spherical	Antagonistic towards *B. subtilis, S. aureus, P. aeruginosa* and *E. coli*	Wahab et al. (2021)
27	*Terminalia arjuna*	Bark	130–50 nm Spherical	Biocidal to *E. coli*	Wahab et al. (2021)
28	*Phyllanthus emblica*	Fruit	Ag NPs	Biocidal to prokaryotes	Renuka et al. (2020)
29	*Mentha aquatica*	Leaf	8–14 Spherical	Antagonistic towards *P. aeruginosa, S. aureus, B. cereus* and *E. coli*	Nouri et al. (2020)
30	*Morinda citrifolia*	Leaf	10–60 nm cubic	Antagonistic towards *Enterobacter aerogenes, B. cereus, P. aeruginosa, K. pneumoniae, B. subtilis* and *E. coli*	Wahab et al. (2021)
31	*Rheum ribes*	Leaf	18.2 ± 3.6 nm spherical	Effective against *E. coli,* MRSA and *B. subtilis*	Aygün et al. (2020)
32	*Rheum ribes*		Spherical	Antagonistic towards microbes and cancer	Gupta and Seema (2021)
33	*Elytrariaacaulis*	Leaf	Cuboid	Antibacterial	Gupta and Seema (2021)
34	*Solanum trilobatum*	Fruit	Spherical 50–41.90 nm	Antagonistic towards *E. faecalis, K. pneumoniae, Streptococcus mutans* and *E. coli*	Wahab et al. (2021)
35	*Morinda citrifolia*	Leaves	Cubic 10–60 nm	Antagonistic towards *B. cereus, K. pneumoniae, B. subtilis, P. aeruginosa, E. aerogenes* and *E. coli*	Wahab et al. (2021)
37	*Crocus sativus*	Petal		Antibacterial	Gupta and Seema (2021)
38	*Pimpinell aanisum*	Seed	13–16 nm spherical	Effective against *Salmonella typhi, P. aeruginosa, S. pyogenes, K. pneumoniae* and *Acineotebacter baumannii.*	Wahab et al. (2021)

19.5.3 PtNPs

Besides being antimicrobial PtNPs can be used in various fields like electrocatalysis, catalytic conversions, magnetic nanopowders, biomedicine and agriculture. See Table 19.5.

Song et al. were the first to disclose plant-mediated green synthesis of PtNPs; according to their findings, carboxylic acids, amines and flavonoids are responsible for the reduction of Pt ions to PtNPs. Table 19.6 shows that PtNPs made from plant extracts have good antibacterial potential. The antibacterial activity of PtNPs is attributed to a negative zeta potential, which results in lysis of the bacterial cell (Gopinath et al., 2016).

TABLE 19.5
Diverse Applications of PtNPs

MNPs	Application	Reference
PtNPs and bioconjugates	Reduce oxidative stress, antibacterial, cancer therapy, bone allograph, cancer detection, Parkinson's disease therapeutics, antifungal against *Cladosporium fulvum* and *Colletotrichum acutatum*, prevents bone deprivation	Pestovsky and Martínez-Antonio (2017), s

TABLE 19.6
Antagonistic Effect of PtNPs Constructed Using Different Plant Extracts

Name of Plant	Plant Organ	Type/Shape/Size of MNPs	Usage	Reference
Xanthium strumarium	Leaf	22 nm cubic spherical rectangular	Antagonistic towards *Klebsiella pneumonia, Bacillus subtilis, Staphylococcus aureus* and *Escherichia coli*,	Kumar et al. (2019)
Taraxacum laevigatum	Whole plant	2–7 nm spherical	Antagonistic towards *Pseudomonas aeruginosa* and *B. subtilis*	Tahir et al. (2016)
Jatropha gossypifolia	Leaf	20 nm cubic, hexagonal rectangular	Antagonistic towards *K. pneumonia, Bacillus licheniformis Escherichia coli, Staphylococcus epidermidis, P. aeruginosa* and *S. aureus*	Jeyapaul et al. (2018)
Jatropha glandulifera	Leaf	100 nm spherical, irregular	Antagonistic towards *K. pneumonia, B. licheniformis, E. coli, S. epidermidis, P. aeruginosa* and *S. aureus*	Jeyapaul et al. (2018)
Black cumin	Seed	1–6 nm spherical	Antagonistic towards *Enterobacter aerogenes, Kentucky, E. coli, P. aeruginosa* and *Salmonella*	Aygün et al. (2020)
Dates	Fruit	1.3–2.6 nm spherical	Antagonistic towards *E. coli* and *B. subtilis*	Al-Radadi et al. (2019)

19.5.4 Other MNPs

CuO, FeNPs, ZnO, CdNPs and Cu-doped ZnONPs have been reported from different plants to have biocidal action towards gram-negative and gram-positive bacteria. These plant-based MNPs show higher antibiotic activity and increased biocompatibility in comparison to physiochemically originated MNPs. Plant-derived MNPs were found to have 96.67% activity against bacteria at 30 mM concentrations, whereas chemically manufactured MNPs did not have the same level of activity (Mukherjee et al., 2014). In contrast to chemically generated MNPs, Abeer Mohamad (2015) discovered that biogenic MNPs exhibit outstanding antibacterial activity against *Micrococcus luteus* ATCC 9341, *Salmonella typhimurium* and *Bacillus subtilis* ATCC 6633. Antibacterial activities of different MNPs have been summarized in Table 19.7.

TABLE 19.7
Antibacterial Properties of MNPs Made from Various Plant Extracts

Name of Plant	Plant Organ	Type/Shape/Size of MNPs	Usage	Reference
Allium eriophyllum	Leaves	CuNPs/25–35 nm	Antagonistic towards *Candida albicans, Escherichia coli, Salmonella typhimurium, Pseudomonas aeruginosa, Candida glabrata, Bacillus subtilis, Candida krusei, Candida guilliermondii, Streptococcus pneumonia* and *Staphyloccus. aureus*	Zhao et al. (2020)
Corallocarbus epigaeus	Rhizome	CuNPs/65–80 nm/ spherical	Activity against multidrug resistant *Salmonella*	Selvam et al. (2020)
Mentha pulegium	Leaf/flower mix	ZnONPs/65.02 ± 7.55 nm and CuNPs/26.92 ± 4.7 nm spherical	Activity against resistant *E. coli* and *S. aureus*	Alavi et al. (2021)
Moringa oleifera	Leaf	CuNPs/35.8–49.2 nm	Antagonistic towards *Enterococcus faecalis, Klebsiella pneumoniae, S. aureus* and *E. coli*	Das et al. (2020)
Punica granatum	Fruit extract	CuNPs/56–59 nm	Activity against *S. aureus*	Padma et al. (2018)
Citrus sinensis	Fruit extract	CuNPs/6.93 nm—20.70 nm/spherical	Activity against *E. coli* and *S. aureus*	Jahan et al. (2021)
Artemisia haussknechtii and *Protoparmeliopsis muralis*	Secondary metabolites of usnic acid	CuNPs/21.71 ± 3.37/ spherical	Activity against *S. aureus* (MRSA) *Acinetobacter baumannii* (A52) and *Klebsiella pneumonia* (K38)	Alavi and karimi (2019)
Cinnamomum verum	Bark	ZnONPs	Activity against *E. coli* and *S. aureus*	Hano and Abbasi (2022)
Myristica fragrans	Fruit extract	ZnONPs / 41.23 nm	Antagonistic towards *K. pneumoniae*	Faisal et al. (2021)
Cassia auriculata	Leaf	ZnONPs/ crystalline/1–100 nm	Antagonistic towards *P. aeruginosa, K. pneumonia, B. subtilis* and *Proteus mirabilis*	Ramesh et al. (2021)
Moringa oleifera	Gum powder	AgNPs and ZnONPs/50 and 60 nm	Antagonistic towards *S. aureus*, MRSA, *S. aureus* and *E. coli*	Irfan et al. (2021)
Terminalia belerica	Aqueous extract	Zn, Cu and iron	Antagonistic towards *K. pneumonia, B. subtilis, E. coli, S. aureus* and *Salmonella enterica*	Akhter et al. (2019)
Garcinia pedunculata Roxb	Leaf extract	PdNPs/2–4 nm/ spherical	Antagonistic towards MDR *Cronobacter sakazakii* strain AMD04	Hazarika et al. (2017)
Santalum album	Leaf	PdNPs/10–40 nm/ spherical, crystalline	Antibacterial and catalytic properties	Kalpana et al. (2014)
Filicium decipiens	Leaf	PdNPs/2–22 nm/ spherical, crystalline	Antagonistic to bacteria	Nasrollahzadeh and Sajadi (2016)
Clerodendrum inerme and *Clerodendrum infortuna*	Leaf	Zinc oxide NPs	Antibacterial	Gupta and Seema (2021)

19.6 CONCLUSION, OUTLOOK AND FUTURE ASPECTS

Most of the available tested antibiotics are ineffective against MDR bacteria. So, there is a continuous demand for efficient, affordable, non-toxic antibiotics to neutralize this global medical threat and socioeconomic burden.

Medicinal plants are known to have many bioactive molecules of therapeutic advantage and serve as promising, efficient, eco-friendly routes for the sustainable production of MNPs. These bioactive MNPs are well suited for their potential employment as antibacterial agents. Nanotechnology led to the emergence of a new branch of science called DNA nanotechnology, which will be useful in reducing toxicity of a drug required for efficient and specific drug delivery.

ETHICAL ISSUES

This chapter does not carry any studies with human participants or animals performed by any of the authors. All the articles reviewed for this manuscript have been properly cited.

BIBLIOGRAPHY

Abdel-Raouf, N., Al-Enazi, N. M., & Ibraheem, I. B. (2017). Green biosynthesis of gold nanoparticles using Galaxauraelongata and characterization of their antibacterial activity. *Arabian Journal of Chemistry*, 10, S3029–S3039. doi:10.1016/j.arabjc.2013.11.044

Abeer Mohamad, A. B. (2015). Comparison between biological and chemical synthesis of zinc oxide nanoparticles and its influence over some antibiotics. *International Journal of Pharma and Bio Science*, 6(3), 1357–1364.

Ahmad, A., Syed, F., Imran, M., Khan, A. U., Tahir, K., Khan, Z. U. H., & Yuan, Q. (2016). Phytosynthesis and antileishmanial activity of gold nanoparticles by M aytenusRoyleanus. *Journal of Food Biochemistry*, 40(4), 420–427. doi:10.1111/jfbc.12232

Ahmed, S. Ahmad, M., Swami, B. L., & Ikram, S. (2016). A review on plants extract mediated synthesis of silver nanoparticles for antimicrobial applications: A green expertise. *Journal of Advanced Research*, 7(2016), 17–28. https://doi.org/10.1016/j.jare.2015.02.007

Akhter, S. M. H., Mohammad, F., & Ahmad, S. (2019). Terminaliabelerica mediated green synthesis of nanoparticles of copper, iron and zinc metal oxides as the alternate antibacterial agents against some common pathogens. *BioNanoScience*, 9(2), 365–372.

Alavi, M., Dehestaniathar, S., Mohammadi, S., Maleki, A., & Karimi, N. (2021). Antibacterial activities of phytofabricatedZnO and CuO NPs by Menthapulegium leaf/flower mixture extract against antibiotic resistant bacteria. *Advanced Pharmaceutical Bulletin*, 11(3), 497. https://doi.org/10.34172/apb.2021.057

Alavi, M., & Karimi, N. (2019). Biosynthesis of Ag and Cu NPs by secondary metabolites of usnic acid and thymol with biological macromolecules aggregation and antibacterial activities against multi drug resistant (MDR) bacteria. *International Journal of Biological Macromolecules*, 128,893–901.

Alavi, M., & Rai, M. (2019). Recent advances in antibacterial applications of metal nanoparticles (MNPs) and metal nanocomposites (MNCs) against multidrug-resistant (MDR) bacteria. *Expert Review of Anti-Infective Therapy*. doi:10.1080/14787210.2019.1614914

Al-Radadi, N. S. (2019). Green synthesis of platinum nanoparticles using Saudi's Dates extract and their usage on the cancer cell treatment. *Arabian Journal of Chemistry*, 12(3), 330–349. doi:10.1016/j.arabjc.2018.05.008

Aygün, A., Gülbağça, F., Nas, M. S., Alma, M. H., Çalımlı, M. H., Ustaoglu, B., . . . Şen, F. (2020). Biological synthesis of silver nanoparticles using Rheum ribes and evaluation of their anticarcinogenic and antimicrobial potential: A novel approach in phytonanotechnology. *Journal of Pharmaceutical and Biomedical Analysis*, 179, 113012. doi:10.1016/j.jpba.2019.113012

Bhau, B. S., Ghosh, S., Puri, S., Borah, B., Sarmah, D. K., & Khan, R. (2015). Green synthesis of gold nanoparticles from the leaf extract of Nepenthes khasiana and antimicrobial assay. *Advanced Materials Letters*, 6(1), 55–58. doi:10.5185/amlett.2015.5609

Bindhu, M. R., Umadevi, M., Esmail, G. A., Al-Dhabi, N. A., & Arasu, M. V. (2020). Green synthesis and characterization of silver nanoparticles from Moringaoleifera flower and assessment of antimicrobial and sensing properties. *Journal of Photochemistry and Photobiology B: Biology*, 205, 111836. doi:10.1016/j.jphotobiol.2020.111836

Chandra, H., Kumari, P., Bontempi, E., & Yadav, S. (2020). Medicinal plants: Treasure trove for green synthesis of metallic nanoparticles and their biomedical applications. *Biocatalysis and Agricultural Biotechnology*, 24, 101518. doi:10.1016/j.bcab.2020.101518

Chen, M. N., Chan, C. F., Huang, S. L., & Lin, Y. S. (2019). Green biosynthesis of gold nanoparticles using Chenopodiumformosanum shell extract and analysis of the particles' antibacterial properties. *Journal of the Science of Food and Agriculture*, 99(7), 3693–3702. doi:10.1002/jsfa.9600

Das, P. E., Abu-Yousef, I. A., Majdalawieh, A. F., Narasimhan, S., & Poltronieri, P. (2020). Green synthesis of encapsulated copper nanoparticles using a hydroalcoholic extract of Moringaoleifera leaves and assessment of their antioxidant and antimicrobial activities. *Molecules*, 25(3), 555. https://doi.org/10.3390/molecules25030555

Das, R. K., Pachapur, V. L., Lonappan, L., Naghdi, M., Pulicharla, R., Maiti, S., . . . Brar, S. K. (2017). Biological synthesis of metallic nanoparticles: Plants, animals and microbial aspects. *Nanotechnology for Environmental Engineering*, 2(1), 1–21. doi:10.1007/s41204-017-0029-4

Dorosti, N., & Jamshidi, F. (2016). Plant-mediated gold nanoparticles by Dracocephalumkotschyi as anticholinesterase agent: Synthesis, characterization, and evaluation of anticancer and antibacterial activity. *Journal of Applied Biomedicine*, 14(3), 235–245. doi:10.1016/j.jab.2016.03.001

Duan, H., Wang, D., & Li, Y. (2015). Green chemistry for nanoparticle synthesis. *Chemical Society Reviews*, 44(16), 5778–5792.

Dudhane, A. A., Waghmode, S. R., Dama, L. B., Mhaindarkar, V. P., Sonawane, A., & Katariya, S. (2019). Synthesis and characterization of gold nanoparticles using plant extract of Terminaliaarjuna with antibacterial activity. *International Journal of Nanoscience and Nanotechnology*, 15(2), 75–82.

Faisal, S., Jan, H., Shah, S. A., Shah, S., Khan, A., Akbar, M. T., . . . Syed, S. (2021). Green synthesis of zinc oxide (ZnO) nanoparticles using aqueous fruit extracts of Myristicafragrans: Their characterizations and biological and environmental applications. *ACS Omega*, 6(14), 9709–9722.

Ganesan, R. M., & Prabu, H. G. (2019). Synthesis of gold nanoparticles using herbal Acoruscalamus rhizome extract and coating on cotton fabric for antibacterial and UV blocking applications. *Arabian Journal of Chemistry*, 12(8), 2166–2174. doi:10.1016/j.arabjc.2014.12.017

Gopinath, K., Kumaraguru, S., Bhakyaraj, K., Mohan, S., Venkatesh, K. S., Esakkirajan, M., . . . Arumugam, A. (2016). Green synthesis of silver, gold and silver/gold bimetallic nanoparticles using the Gloriosasuperba leaf extract and their antibacterial and antibiofilm activities. *Microbial Pathogenesis*, 101, 1–11. doi:10.1016/j.micpath.2016.10.011

Gupta, M., & Seema, K. (2021). Living nano-factories: An eco-friendly approach towards medicine and environment. In Bio-manufactured nanomaterials (pp. 95–124). Springer, Cham.

Hano, C., & Abbasi, B. H. (2022). Plant-based green synthesis of nanoparticles: Production, characterization and applications. *Biomolecules*, 12(1), 31.

Hazarika, M., Borah, D., Bora, P., Silva, A. R., & Das, P. (2017). Biogenic synthesis of palladium nanoparticles and their applications as catalyst and antimicrobial agent. *PLoS One*, 12(9), e0184936. doi:10.1371/journal.pone.0184936

Irfan, M., Munir, H., & Ismail, H. (2021). Moringaoleifera gum based silver and zinc oxide nanoparticles: Green synthesis, characterization and their antibacterial potential against MRSA. *Biomaterials Research*, 25(1), 1–8.

Jahan, I., Erci, F., & Isildak, I. (2021). Facile microwave-mediated green synthesis of non-toxic copper nanoparticles using Citrus sinensis aqueous fruit extract and their antibacterial potentials. *Journal of Drug Delivery Science and Technology*, 61, 102172.

Jeyapaul, U., Kala, M. J., Bosco, A. J., Piruthiviraj, P., & Easuraja, M. (2018). An eco-friendly approach for synthesis of platinum nanoparticles using leaf extracts of Jatropagossypifolia and Jatropaglandulifera and its antibacterial activity. *Oriental Journal of Chemistry*, 34(2), 783. doi:10.13005/ojc/340223.

Kalpana, D., Han, J. H., Park, W. S., Lee, S. M., Wahab, R., & Lee, Y. S. (2019). Green biosynthesis of silver nanoparticles using Torreya nucifera and their antibacterial activity. Arabian Journal of Chemistry, 12(7), 1722-1732.

Khan, A. U., Yuan, Q., Wei, Y., Khan, G. M., Khan, Z. U. H., Khan, S., . . . Khan, F. U. (2016). Photocatalytic and antibacterial response of biosynthesized gold nanoparticles. *Journal of Photochemistry and Photobiology B: Biology*, 162, 273–277. doi:10.1016/j.jphotobiol.2016.06.055

Khan, S. A., Noreen, F., Kanwal, S., Iqbal, A., & Hussain, G. (2018). Green synthesis of ZnO and Cu-doped ZnO nanoparticles from leaf extracts of Abutilon indicum, Clerodendruminfortunatum, Clerodendruminerme and investigation of their biological and photocatalytic activities. *Materials Science and Engineering*: C, 82, 46–59. doi:10.1016/j.msec.2017.08.071

Kralj, S., & Makovec, D. (2015). Magnetic assembly of superparamagnetic iron oxide nanoparticle clusters into nanochains and nanobundles. *ACS Nano*, 9(10), 9700–9707. doi:10.1021/acsnano.5b02328

Kumar, P. V., Kala, S. M. J., & Prakash, K. S. (2019). Green synthesis derived Pt-nanoparticles using Xanthium strumarium leaf extract and their biological studies. *Journal of Environmental Chemical Engineering*, 7(3), 103146. doi:10.1016/j.jece.2019.103146

Lashin, I., Fouda, A., Gobouri, A. A., Azab, E., Mohammedsaleh, Z. M., & Makharita, R. R. (2021). Antimicrobial and in vitro cytotoxic efficacy of biogenic silver nanoparticles (Ag-NPs) fabricated by callus extract of Solanumincanum L. *Biomolecules*, 11(3), 341. doi:10.3390/biom11030341

Lee, S. H., & Jun, B. H. (2019). Silver nanoparticles: synthesis and application for nanomedicine. *International Journal of Molecular Sciences*, 20(4), 865. doi:10.3390/ijms20040865

Maghimaa, M., & Alharbi, S. A. (2020). Green synthesis of silver nanoparticles from Curcuma longa L. and coating on the cotton fabrics for antimicrobial applications and wound healing activity. *Journal of Photochemistry and Photobiology B: Biology*, 204, 111806. doi:10.1016/j.jphotobiol.2020.111806.

Malapermal, V., Botha, I., Krishna, S. B. N., & Mbatha, J. N. (2017). Enhancing antidiabetic and antimicrobial performance of Ocimumbasilicum, and Ocimum sanctum (L.) using silver nanoparticles. *Saudi Journal of Biological Sciences*, 24(6), 1294–1305. doi:10.1016/j.sjbs.2015.06.026

Mata, R., Bhaskaran, A., & Sadras, S. R. (2016). Green-synthesized gold nanoparticles from Plumeriaalba flower extract to augment catalytic degradation of organic dyes and inhibit bacterial growth. *Particuology*, 24, 78–86. doi:10.1016/j.partic.2014.12.014

Menon, S., Ks, S. D., Santhiya, R., Rajeshkumar, S., & Kumar, V. (2018). Selenium nanoparticles: A potent chemotherapeutic agent and an elucidation of its mechanism. *Colloids and Surfaces B: Biointerfaces*, 170, 280–292. https://doi.org/10.1016/j.colsurfb.2018.06.006

Mukherjee, S., Chowdhury, D., Kotcherlakota, R., & Patra, S. (2014). Potential theranostics application of biosynthesized silver nanoparticles (4-in-1 system). *Theranostics*, 4(3), 316. doi:10.7150/thno.7819

Murugan, K., Benelli, G., Panneerselvam, C., Subramaniam, J., Jeyalalitha, T., Dinesh, D., . . . & Madhiyazhagan, P. (2015). Cymbopogoncitratus-synthesized gold nanoparticles boost the predation efficiency of copepod Mesocyclopsaspericornis against malaria and dengue mosquitoes. *Experimental Parasitology*, 153, 129–138. doi:10.1016/j.exppara.2015.03.017

Muthukumar, T., Sambandam, B., Aravinthan, A., Sastry, T. P., & Kim, J. H. (2016). Green synthesis of gold nanoparticles and their enhanced synergistic antitumor activity using HepG2 and MCF7 cells and its antibacterial effects. *Process Biochemistry*, 51(3), 384–391. doi:10.1016/j.procbio.2015.12.017

Nasrollahzadeh, M., & Sajadi, S. M. (2016). Pd nanoparticles synthesized in situ with the use of Euphorbia granulate leaf extract: Catalytic properties of the resulting particles. *Journal of Colloid and Interface Science*, 462, 243–251.

Noruzi, M. (2015). Biosynthesis of gold nanoparticles using plant extracts. *Bioprocess and Biosystems Engineering*, 38(1), 1–14. doi.org/10.1007/s00449-014-1251-0

Nouri, A., Yaraki, M. T., Lajevardi, A., Rezaei, Z., Ghorbanpour, M., &Tanzifi, M. (2020). Ultrasonic-assisted green synthesis of silver nanoparticles using Menthaaquatica leaf extract for enhanced antibacterial properties and catalytic activity. *Colloid and Interface Science Communications*, 35, 100252. doi:10.1016/j.colcom.2020.100252

Padma, P. N., Banu, S. T., & Kumari, S. C. (2018). Studies on green synthesis of copper nanoparticles using Punicagranatum. *Annual Research & Review in Biology*, 1–10. doi.org/10.9734/ARRB/2018/38894

Pareek, V., Bhargava, A., Gupta, R., Jain, N., & Panwar, J. (2017). Synthesis and applications of noble metal nanoparticles: A review. *Advanced Science, Engineering and Medicine*, 9, 527–544. doi:10.1166/asem.2017.2027

Patrick, T. S., Naphtaly, C., Moro Ouma Stephanus, M. O., et al. (2019). Application of nanoparticles in biofuels. An overview. *Fuel*, 237, 380–397.

Pendleton, J. N., Gorman, S. P., & Gilmore, B. F. (2013). Clinical relevance of the ESKAPE pathogens. *Expert Rev. Anti-Infective Therapy*, 11, 297–308. https://doi.org/10.1586/eri.13.12

Pestovsky, Y. S., & Martínez-Antonio, A. (2017). The use of nanoparticles and nanoformulations in agriculture. *Journal of Nanoscience and Nanotechnology*, 17(12), 8699–8730. doi:10.1166/jnn.2017.15041

Poojary, M. M., Passamonti, P., & Adhikari, A. V. (2016). Green synthesis of silver and gold nanoparticles using root bark extract of Mammeasuriga: Characterization, process optimization, and their antibacterial activity. *BioNanoScience*, 6(2), 110–120.doi:10.1007/s12668-016-0199-8

Qais, F. A., Shafiq, A., Khan, H. M., Husain, F. M., Khan, R. A., Alenazi, B., . . . Ahmad, I. (2019). Antibacterial effect of silver nanoparticles synthesized using Murrayakoenigii (L.) against multidrug-resistant pathogens. *Bioinorganic Chemistry and Applications*, 2019. doi:10.1155/2019/4649506

Rai, M., Ingle, A.P., Birla, S., Yadav, A., & Santos, C.A.D. (2016). Strategic role of selected noble metal nanoparticles in medicine. *Critical Reviews in Microbiology*, 42 (5), 696–719. doi:10.3109/10408 41x.2015.1018131.

Rajeshkumar, S., & Bharath, L. V. (2017). Mechanism of plant-mediated synthesis of silver nanoparticles: A review on biomolecules involved, characterisation and antibacterial activity. *Chemico-Biological Interactions*, 273, 219–227. doi:10.1016/j.cbi.2017.06.019

Ramesh, P., Saravanan, K., Manogar, P., Johnson, J., Vinoth, E., & Mayakannan, M. (2021). Green synthesis and characterization of biocompatible zinc oxide nanoparticles and evaluation of its antibacterial potential. *Sensing and Bio-Sensing Research*, 31, 100399.

Ramesh, V., & Armash, A. (2015). An in-vitro studies on green synthesis of gold nanoparticles against pathogens and cancer cells. *International Journal of Pharmacological Research*, 5(10). doi:10.7439/ijpr.v5i10.2708

Ravishankar R. V. and Jamuna B. A. (2011). Nanoparticles and their potential application as antimicrobials Science against microbial pathogens, communicating current research and technological advance A, Méndez_Vilas (Ed) 197–209.

Renuka, R., Devi, K. R., Sivakami, M., Thilagavathi, T., Uthrakumar, R., & Kaviyarasu, K. (2020). Biosynthesis of silver nanoparticles using Phyllanthusemblica fruit extract for antimicrobial application. *Biocatalysis and Agricultural Biotechnology*, 24, 101567. doi:10.1016/j.bcab.2020.101567

Rizzello, L., & Pompa, P. P. (2014). Nanosilver-based antibacterial drugs and devices: Mechanisms, methodological drawbacks, and guidelines. *Chemical Society Reviews*, 43(5), 1501–1518.

Samuggam, S., Chinni, S. V., Mutusamy, P., Gopinath, S. C. B., Anbu, P., Venugopal, V., Reddy, L. V., & Enugutti, B. (2021). Green synthesis and characterization of silver nanoparticles using spondiasmombin extract and their antimicrobial activity against biofilm-producing bacteria. *Molecules*, 26, 2681. https://doi.org/10.3390/molecules26092681

Selvam, K., Sudhakar, C., Selvankumar, T., Senthilkumar, B., Selva Kumar, R., & Kannan, N. (2020). Biomimetic synthesis of copper nanoparticles using rhizome extract of corallocarbusepigaeus and their bactericidal with photocatalytic activity. *SN Applied Sciences*, 2(6), 1–7. doi:10.1007/s42452-020-2811-3

Singh, A., Gautam, P. K., Verma, A., Singh, V., Shivapriya, P. M., Shivalkar, S., Sahoo, A. K., & Samanta, S. K. (2020). Green synthesis of metallic nanoparticles as effective alternatives to treat antibiotics resistant bacterial infections: A review. *Biotechnology Reports*, 25, e00427. doi:10.1016/j.btre.2020.e00427

Singh, P., Kim, Y. J., Zhang, D., & Yang, D. C. (2016). Biological synthesis of nanoparticles from plants and microorganisms. *Trends in Biotechnology*, 34(7), 588–599. doi:10.1016/j.tibtech.2016.02.006

Singh, R., & Nalwa, H. S. (2011). Medical applications of nanoparticles in biological imaging, cell labeling, antimicrobial agents, and anticancer nanodrugs. *Journal of Biomedical Nanotechnology*, 7(4), 489–503.

Sunderam, V., Thiyagarajan, D., Lawrence, A. V., Mohammed, S. S. S., & Selvaraj, A. (2019). In-vitro antimicrobial and anticancer properties of green synthesized gold nanoparticles using Anacardiumoccidentale leaves extract. *Saudi Journal of Biological Sciences*, 26(3), 455–459. doi:10.1016/j.sjbs.2018.12.001

Tahir, K., Nazir, S., Ahmad, A., Li, B., Shah, S. A. A., Khan, A. U., . . . Khan, F. U. (2016). Biodirected synthesis of palladium nanoparticles using Phoenix dactylifera leaves extract and their size dependent biomedical and catalytic applications. *RSC Advances*, 6(89), 85903–85916. doi:10.1039/c6ra11409a

Vimalraj, S., Ashokkumar, T., & Saravanan, S. (2018). Biogenic gold nanoparticles synthesis mediated by Mangiferaindica seed aqueous extracts exhibits antibacterial, anticancer and anti-angiogenic properties. *Biomedicine & Pharmacotherapy*, 105, 440–448. doi:10.1016/j.biopha.2018.05.151

Vinay, S. P., & Chandrasekhar, N. (2019). Udayabhanu. Ixoracoccinea extract-mediated green synthesis of silver nanoparticles: Photodegradative and antimicrobial studies. *Int J BiosenBioelectron*, 5(4), 100–105.

Wahab, S., Khan, T., Adil, M., & Khan, A. (2021). Mechanistic aspects of plant-based silver nanoparticles against multi-drug resistant bacteria. *Heliyon*, 7(7), e07448. https://doi.org/10.1016/j.heliyon. e07448

Wang, W., Guo, Q., Xu, X., Sheng, Z.-k., Ye, X., & Wang, M. (2014). High-level tetracycline resistance mediated by efflux pumps Tet(A) and Tet(A)-1 with two start codons. *Journal of Medical Microbiology*, 63(Pt_11), 1454–1459. doi:10.1099/jmm.0.078063-0

WHO Publishes List of Bacteria for Which New Antibiotics Are Urgently Needed. Available online: www.who.int/news/item/27-02-2017-who-publishes-list-of-bacteria-for-which-new-antibiotics-are-urgently-needed (accessed in 2021).

Zhao, H., Su, H., Ahmeda, A., Sun, Y., Li, Z., Zangeneh, M. M., . . . Moradi, R. (2020). Biosynthesis of copper nanoparticles using Allium eriophyllumBoiss leaf aqueous extract; characterization and analysis of their antimicrobial and cutaneous wound-healing potentials. *Applied Organometallic Chemistry*, e5587. doi:10.1002/aoc.5587

20 Past, Present, and Future of Aromatic Plants

Vipin Kumar, Himashu Yadav, and Ramesh Kumar Srivastava

CONTENTS

20.1 Introduction 347
20.2 History 348
20.3 Chemical Compounds of Essential Oils 349
20.4 Pharmaceutical Properties 350
20.4.1 Anticancer 350
20.4.2 Antimicrobial 350
20.5 Uses 351
20.5.1 Food and Flavor 351
20.5.2 Cosmetics 351
20.5.3 Perfumery 352
20.5.4 Aromatherapy 352
20.6 Commercialized Aromatic Plants 353
20.7 Trade Scenario of Aromatic Plants 354
20.8 Government Efforts in the Cultivation of Aromatic Plants 355
20.9 Major Institutes and Industries 355
20.10 Post-Harvest and Distillation Techniques 355
20.10.1 Hydro-Distillation 356
20.10.2 Hydro-Steam Distillation 357
20.10.3 Steam Distillation 357
20.11 Storage 357
20.12 Future Prospects of Aromatic Plants 357
20.13 Conclusion 357
Bibliography 358

20.1 INTRODUCTION

Plants are valuable gifts from nature for human beings, and they have been used since ancient time for food, beverages, cosmetics, traditional medicines, etc. There are different types of plants found in nature, and every plant has its own special importance, including aromatic plants. Due to their special type of aroma, aromatic plants are in great demand in the global market. Recently, the demand for aromatic crops has increased very rapidly. Essential oils are extracted from the aromatic plants by using different techniques. These aromatic crops are widely cultivated in various part of the world, including India. Essential oils are obtained from the various parts of the plants such as seed, fruit, flowers, leaves, stem, wood, bark, root, gum, etc. In rose plants, for example, the oil is extracted from the petals; in cinnamon, from the bark and leaves; in citrus, from the flowers and peel of the fruit; in mint and aromatic grasses, from the leaves; in cumin, fennel and ajwain, oil is extracted from seeds, and in vetiver from the roots.

DOI: 10.1201/b22917-20

The sense of smell plays a significant role in the identification of different types of foods. The term essential oil is similar to fragrance or perfumes because these fragrances are oily in nature, and they represent the essence or the active constituents of the plants. Essential oils of aromatic crops are widely used for manufacturing products in different industries such as pharmaceutical, perfumery, food and flavoring, cosmetics and toiletry-related industries. Many of them have also been used for traditional medicine purposes since ancient times. In aromatherapy, the aromatic compounds of these plants are used for the purpose of mind and body relaxation, headache, refreshing brain and cognitive function and health. India has been considered a treasure house of valuable medicinal and aromatic plant species. The Ministry of Environment and Forests has identified and documented over 9500 plant species, considering their importance in the pharmaceutical industry (Chowti et al. 2018). About 1500 species are known as aromatic plants, in which 500 species are identified for the properties of their essential oil; out of them, only 50 species are used as a commercial source of essential oil (Joy et al. 1998). Global demand for aromatic plants is increasing regularly for food supplements, cosmetics, perfumery, toiletries, plant-based medicines, health products, pharmaceuticals, etc. Hence, the cultivation and value-addition of aromatic plants is to boost the regular supply for business and related industries.

20.2 HISTORY

Aromatic plants have been associated with human life from time immemorial; hence it is difficult to find out when humans first used aromatic plants and their essential oils. According to Sharma (1996) essences obtained from aromatic plants were used in different ways, as mentioned in various ancient literatures of different religious. Perfumery was considered an important trade in ancient times; a person who did this business was known as a perfumer and also called a "gandhikas". Gandhikas created their own blends of perfumes and incense in the form of liquid, sticks, powders, pastilles and pastes (Rajamani et al. 2019). Some references are available in old literature which indicate a relation between humans and aromatic plants. The oldest record dates to 5000–3000 BC in the Alps from Ötzi, the Iceman, who used plant medicines (Inoue and Craker 2014; Christaki et al. 2012). Distillation of rose flowers is mentioned in *Charaka* and *Sushruta samhitas*, believed to have been written in India in 1000 BC. The famous Chinese traveler Fa-Hien described India as the land of aromatic flowers, fruits, woods, roots, resins and grasses. In the Ain-e-Akbari, Abul Fazal mentions Akbar's love of attar and incense, "daily burnt in gold and silver censers". About 5000 years ago Chinese people put their faith in spices for medicinal purpose and preserving flavor and food products. Today about 400 species of aromatic plants are used by Chinese people for their food, flavor and medicinal properties. China is one of the highest essential oil–producing countries and produces about 120 natural essential oils for their own consumption and export in the global market (Xiao 1996).

Later, an apocryphal tale credits the Empress Noor Jahan with discovering the "attar of roses". It is said that she noticed the rose oil floating on the surface of her bathwater and, thus, the legendary attar of rose was born. At Nawabi banquets, guests were welcomed with attar. They were sprinkled with rosewater at the gate and then phayas, small swabs of cotton dipped in attar, would be offered to them on silver trays decorated with flowers. Indian cities like Delhi, Agra, Kannauj, Lucknow, Jaunpur, Ghazipur, Aligarh, Bharatpur, Mysore and Hyderabad emerged as centers of the national and international trade in perfumery and other aromatic compounds, and were known for their quality attars across Asia, Europe and Africa (Rajamani et al. 2019) (Table 20.1).

TABLE 20.1
Major Aromatic Plants in India

S. No.	Common Name	Botanical Name	Major Chemical Compound	Oil Percent on Dry Weight Basis	Plant Part Used	Uses
1.	Mentha	*Mentha arvensis*	Menthol	0.5–1.0	Leaves	Perfumery, flavoring, pharmaceutical
2.	Lemongrasss	*Cymbopogan flexuosus*	Citral	0.8–1.0	Leaves	Perfumery, pharmaceutical
3.	Palmarosa	*Cymbopogan m martini*	Geraniol	0.4–0.5	Flowering shoot	Perfumery, cosmetic
4.	Javagrass/ citronella	*Cymbopogan winterianus*	Citronellal	0.6–0.8	Leaves	Perfumery, pharmaceutical
5.	Vetiver	*Cymbopogan zizanoides*	Khusiminol	0.2–0.4	Root	Perfumery, cosmetic
6.	Basil	*Ocimum basilicum*	Methyl chevicol	0.5–0.8	Flowering shoot	Perfumery, flavoring, pharmaceutical
7.	Geranium	*Pelargonium graveolens*	Geraniol	0.2–0.5	Shoot	Perfumery, aromatherapy
8.	Patchouli	*Pogostemon patcholi*	Patcholal	0.2–0.5	Leaves	Perfumery, pharmaceutical
9.	Chamomile	*Metricaria Chammomila*	CHamazuline	0.8–1.0	Flower	Perfumery, flavoring
10.	Damask rose	*Rosa damascena*	Geraniol	0.02–0.03	Flower	Perfumery, flavoring
11.	Wild marigold	*Tegetes minuta*	Thujene /Muurolol	1.0–1.2	Flower	Perfumery, flavoring
12.	Sandalwood	*Santalum album*	Santalol	0.2–0.5	Heartwood	Perfumery, flavoring, cosmetic
13.	Eucalyptus	*Eucalyptus globulus*	Cineol	8.0–1.0	Leaves	Perfumery, flavoring
14.	Rosemary	*Rosemarinus officinalis*	Ciniol, camphor	0.6–0.8	flowering shoot	Perfumery, cosmetic

20.3 CHEMICAL COMPOUNDS OF ESSENTIAL OILS

Chemical compositions of aromatic plants depend on the variety and climate conditions; however, average percentages of essential oil in some important aromatic plants are as follows: *Mentha arvensis* (70–80% menthol), *Mentha piperita* (30–35% menthol), *Mentha citrate* (40–60% linalool), *Ocimum basilicum* (70–80% linalool, 70–80% methyl chevicol), *Cymbopogon flexuosus* (70–80% citral), *Cymbopogon winterianus* (35–45% citronellal), *Cymbopogon martinii* (80–90% geraniol), *Vetiveria zizanioides* (65–70% veteverol), *Pelargonium graveolens* (20–60% geraniol), *Pogestemon patchouli* (30–40% patchouli), *Rosa damascene* (20–22% citronellol, 25–30% gereniol), *Eucatyptus globules* (70–80% cineol), *Santalum album* (85–90% santalol) and *Lavendula* sp. (25–30% linalyl acetate) (Kumar 2018; Andrade et al. 2014).

20.4 PHARMACEUTICAL PROPERTIES

Aromatic plants' essential oils and secondary metabolites show different pharmaceutical and biological properties, such as antioxidant (Wannes et al. 2010), analgesic and anti-inflammatory (Mendes et al. 2010), antitumor activity (Silva et al. 2008), larvicidal (Rajkumar et al. 2010), anti-cancer, antimicrobial, antiparasitic, insecticidal (Burt 2004; Kaloustian et al. 2008), antifungal (Carmo et al. 2008) and antibacterial (Srivastava et al. 2014).

20.4.1 Anticancer

Aromatic plants' essential oils and their metabolites demonstrate various pharmaceutical activities, and essential oils can be used either in their pure form or mixed with other compounds for the treatment of cancer (Duthie and Brown 1994; Milner 1994). The essential oil cadinol isolated from the leaf of *Neolitsea variabillima* exhibited cytotoxic activity—significantly, α activity against human oral, liver, lung, colon, melanoma and leukemic cancer cells. D the presence of β-caryophyllene and τ-cadinol contributed to the anticancer activities of *N. variabillima* leaf oil. *Zataria multiflora* Bioss is a member of the Lamiaceae family and is one of the most important aromatic plants with immense medicinal properties located in Pakistan, Afghanistan and southern Iran; it is applied as an antiseptic and antitussive in respiratory tract disorders (Azadi et al. 2020).

Furthermore, the anticancer effect of major ingredients in some essential oils of *Z. multiflora* have been repeatedly investigated on different cancer cells. Günes et al. 2020 and Bouhtit et al. 2021 studied anticancer effects of *Z. multiflora* essential oils (ZMEOs). The main constituents of ZMEOs were thymol and carvacrol (monoterpenoids and monoterpenes). Contrary to rosemary extracts, which have been widely investigated for their anticancer effects, the essential oil of rosemary and white poplar have not yet been thoroughly investigated for their biological activities especially antioxidant, antiproliferative and cytotoxic properties. Gezicil et al. 2017 evaluated potential antioxidant properties, lipid peroxidation capacity, in vitro cytotoxic and antiproliferative activities of essential oil extracted from *Populus alba* and *Rosmarinus officinal* against three human cancer cell lines.

20.4.2 Antimicrobial

Essential oils and secondary metabolites are known for their antimicrobial properties. These oils and metabolites are used to control various diseases caused by the fungal and bacterial pathogens (Burt 2004).

Nikolíc et al. 2014 studied the chemical composition of essential oils, as well as antimicrobial and cytotoxic properties of five aromatic crop species belonging to family the Lamiaceae, *viz*., *Mentha piperita, Mentha pulegium, Lavandula angustifolia, Salvia lavandulifolia* and *Satureja montana* over clinical samples of *Pseudomonas aeruginosa, Streptococcus pyogenes, Streptococcus mutans, Streptococcus sanguis, Streptococcus salivarius, Enterecoccus feacalis* and *Lactobacillus acidophilus* along with 58 clinical oral *Candida* spp. In the study significant antimicrobial activity was exhibited by all essential oils against all tested microorganisms; however, *S. montana* essential oil proved to be the most effective in terms of biological activity.

Furthermore, a recent study assessed the antimicrobial activity of six commonly used Brazilian condiments, *viz*., *Ocimum basilicum* L. (basil), *Rosmarinus officinalis* L. (rosemary), *Origanum majorana* L. (marjoram), *Mentha piperita* L. var. *Piperita* (peppermint), *Thymus vulgaris* L. (thyme) and *Pimpinella anisum* L. (anise), against a *Clostridium perfringens* strain (Chouhan et al. 2017). Different components of essential oils obtained from *Alpinia pahangensis*, *Origanum vulgare*, *Origanum dictamnus*, *M. piperita*, *Lavandula hybrida*, *Z. multiflora* and *Hofmeisteria schaffneri* have been tested against *S. aureus*, and all were found to possess potential inhibitory activity (Pérez et al. 2011; Awang et al. 2011). Therefore, essential oils can serve as a powerful tool to check bacterial growth (Stefanakis et al. 2013). According to Skrinjar and Nemet 2009 antibacterial properties of 14 essential oils (clove, oregano, rosemary, pepper, nutmeg, licorice, turmeric, anise seed, cassia bark, fennel, prickly ash, round cardamom, dahurian angelica root and angelica) against four common meat spoilage and pathogenic bacteria (*Listeria monocytogenes, Escherichia coli, Pseudomonas fluorescens* and *Lactobacillus sake*) were studied, and their results showed that individual extracts of clove, rosemary, cassia bark and licorice contained strong antibacterial activity.

20.5 USES

The essential oils of aromatic plants have been commercially used in perfumery, cosmetics and pharmaceuticals because they have different biological properties such as antibacterial, antiviral, antiparasitic, antifungal and antioxidant. Essential oils are used as natural bio products in the agriculture sector in the form of herbicides, insecticide, antifeedants, repellents and growth boosters (Joy et al. 1998). They are also used in cosmetics, food, beverages, soaps, for making scent/perfume, pan masala, toothpaste, floor cleaners, etc.

20.5.1 Food and Flavor

Essential oils can be used as a natural food preservative (Konyak 2018); about 3000 different essential oils are known, yet 300 out of them are commercially used in the flavor and fragrances market. The antimicrobial efficacy of plants depends on several factors, including genetic, culture, post-harvest condition, the method of extracting essential oil, kind of solvent, volume of inoculum, growth phase and intrinsic or extrinsic properties of the food such as pH, fat, protein, water content, antioxidants, preservatives, incubation time/temperature, packaging procedure and physical structure of the food, as well as availability of oxygen concentration (Burt 2004).

20.5.2 Cosmetics

Since ancient time herbs and spices have been used for maintaining and enhancing human health and glory. Sandalwood and turmeric have been used by Indian women since ancient times for skin care; henna for coloring the hair, hand, palms and soles; and natural oils as a fragrance for their bodies. Essential oils are widely used in cosmetics, but their actual mode of action is not fully understood (Shweta et al. 2011). Herbal cosmetics, referred to as products, are formulated using various permissible cosmetic ingredients to form the base in which one or more herbal ingredients are used to provide defined cosmetic benefits only; thus they are called "herbal cosmetics". The demand of herbal medicines is increasing rapidly due to their few side effects The uses of essential oils are determined by their chemical, physical and sensory properties, which differ greatly from oil to oil. Every chemical constituent of oil has its own merit and demerit; on the basis of that property, constituents are used in various herbal cosmetic products. Essential oils can be used in several ways for cosmetic purposes like inhalation, baths, massage, compresses, steam treatments, room fragrance, etc. (Shweta et al. 2011). Cosmetics are commercially available products that are used to protect the appearance of the skin (Mary and Lupo 2001). Numerous herbal plants are available that have different chemical constituents, which are used in cosmetic preparations. Some of those plants commonly called cosmeceuticals are *Aloe vera, Azadirachta indica, Curcuma longa*, coconut oil, sunflower oil, *Rhodiola rosea* (golden root), *Daucus carota, Ginkgo biloba, Lawsonia inermis,*

Camellia sinensis, Acorus calamus, Allium sativum, Alpinia galangal, Avena sativa, Echinacea purpurea, Centella asiatica, Symphytum officinale, Crocus sativus, Vitex negundo, Sesamum indicum, and *Cicer aritinuma* (Joshi and Pawar 2015).

Sandalwood is used in cosmetic products to freshen and revitalize dull skin. Sandalwood paste is known not only for healing skin problems (curative power) but also providing the skin with a glow that is unmatched by any other product. Sandalwood is also used in various beauty products for looking young, such as face packs and scrubs to remove dead cells and regenerate the growth of new cells (Utkarsha et al. 2017). Essential oils are extracted from flowers (e.g. bergamot tree), grass (lemongrass), herbs, fruit (e.g. anise), the peel of citrus fruits, seeds (e.g. nutmeg), leaves (e.g. eucalyptus), bark (e.g. cinnamon), rhizomes (e.g. turmeric) and roots (vetiver)—virtually every part of the plant. They are produced by a process of "expression" (cold-pressure squeezing of the fruit peel) or distillation, which is slow, laborious and expensive (Purohit and Kapsner 1994). Geranium oil is a cleansing, toning and sharpening oil and is so helpful with those problems that come with greasy, over-oily skin, acne, congested skin and eczema. Care should be taken, however, since there is the possibility of contact dermatitis in hypersensitive individuals. It is a very important component of high-grade perfumes due to its strong rose-like odor (Aburjai and Feda 2003).

Essential oils obtained from the genus *Lavandula* have been used both cosmetically and therapeutically for centuries. It is extensively used in all types of soaps, lotions and perfumes, with the most commonly used species being *L. angustifolia*, *L. latifolia*, *L. stoechas* and *L. intermedia*. Among the claims made for lavender oil include properties like antibacterial, antifungal and effective for burns and insect bites (Cavanagh and Wilkinson 2002; Boelens 1995).

20.5.3 Perfumery

Flavors and fragrances have been extracted from plants and animals since ancient times. Several aromatic plants are quite popular for their fragrance. Aromatic plants contain volatile substances, known as essential oils (Parvez et al. 2021). In India, for example, the old literature mentions numerous uses of essences obtained from plants in performing religious rites. India is regarded as the traditional home of oriental perfumes (Sharma 1996). The formulation of the perfume depends upon its composition—if two or more essential oils are being mixed, then the study of suitability of these essential oils for different notes should be studied, which will give the optimum and highest-quality results. Along with essential oils, solvents such as ethanol, methanol, hexane, benzene, etc., are mixed along with other fixatives to increase the longevity of perfume. Also, the proportion of the base note, top note and middle note can be adjusted depending upon the essential oils which are under consideration (Parvez et al. 2021).

20.5.4 Aromatherapy

Aromatherapy is a specialized segment of phytotherapy that uses essential oils extracted from the different parts of aromatic plants, typically administered via inhalation or topical application and massage for several minor, clinical uses (Bagetta et al. 2016). Indeed, during the second half of the past century, since the first article published by Wood HC and Reichut ET in 1880 in the prestigious *Journal of Physiology* entitled "Note on the Action upon the Circulation of Certain Volatile Oils", the interest in aromatherapy for the treatment of several disorders such as anxiety, mood disorders and certain forms of pain experienced a great growth (Bagetta et al. 2016). In particular, the essential oils of two species of the Lamiaceae family, *Melissa officinalis* L. (lemon balm) and *Lavandula officinalis* L. (lavender), are the most used aromatherapeutic treatments for behavioral and psychological symptoms of dementia (Ballard et al. 2002; Burns et al. 2002). A group of 28 demented older patients, 17 of whom were suffering from Alzheimer's disease, were exposed to the aroma of 0.04 mL lemon and 0.08 mL rosemary essential oil in the morning and to the aroma of 0.08 mL lavender and 0.04 mL orange essential oils in the evening (Scuteri et al. 2017; Jimbo et al. 2009). It was demonstrated that the intraplantar administration of bergamot essential oil (BEO), a citrus fruit belonging to the Rutaceae family, significantly attenuates capsaicin-induced nociceptive behavior (Sakurada et al. 2009).

20.6 COMMERCIALIZED AROMATIC PLANTS

Essential oils and aroma chemicals are a major part of industrial products. They are adjuncts of pharmaceuticals, perfumery, cosmetics, soaps, disinfectants, tobacco products, incense sticks, confectionery, ice creams, aerated waters and related products. Some species of aromatic plants are commercially cultivated for industrial use, but most of them are still collected in the wild, e.g. sugandh mantri and wintergreen. The need for renewable sources of industrial products, as well as to protect plant biodiversity, creates an opportunity for farmers to produce such crops (Tables 20.2 and 20.3).

Uttar Pradesh, Rajasthan and Maharashtra state are the major suppliers of aromatic plants and allied products to the national and international market, where the demand for natural drugs/herbal products has increased in recent years. The trade of aromatic plant products is worth hundreds of crores and the scope of employment generation is thousands of persons. The rural persons are engaged in the collection of aromatic plants and allied products from dense forests as well as from

TABLE 20.2
Estimated Cultivation of Different Aromatic Crops in India

S. No.	Crop Name	Total Cultivated Area (Ha)	Total Oil Production (tons)
1.	Menthol mint	300,000	40,000
2.	Peppermint	1200	240
3.	Spearmint	1800	360
4.	Lemongrass	15,000	3000
5.	Palmarosa	10,000	1500
6.	Java citronella	3000	600
7.	Vetiver	2000	50
8.	Tulsi	5000	750
9.	Geranium	1000	30
10.	Lavender	500	100
11.	Rose	5000	3.75

TABLE 20.3
Major Producing States of Different Aromatic Crops in India

S. No.	Crop Name	Main Producing States
1.	Menthol mint	Uttar Pradesh, Bihar, Gujarat, M.P., Punjab
2.	Peppermint	Uttar Pradesh, Madhya Pradesh, Tamil Nadu
3.	Spearmint	Uttar Pradesh, Madhya Pradesh
4.	Lemongrass	Tamil Nadu, Andhra Pradesh, Uttar Pradesh, Kerala, Gujrat
5.	Palmarosa	Andhra Pradesh, Karnataka, Uttar Pradesh, Gujarat
6.	Java citronella	Uttar Pradesh, Assam, Andhra Pradesh, Tamil Nadu
7.	Vetiver	Rajsthan, Uttar Pradesh, Madhya Pradesh, Maharastra
8.	Tulsi	Uttar Pradesh, Bihar, Gujrat, Haryana
9.	Geranium	Uttaranchal, Tamil Nadu, Karnataka, Uttar Pradesh,
10.	Lavender	Jammu and Kashmir, Rajsthan, Uttaranchal,
11.	Rose	Jammu and Kashmir, Andhra Pradesh

domestic cultivation. Traders purchase these products and supply them to wholesalers and industrial units. Industries give employment to the workers, who are primally engaged in processing and packaging. The processed products are transported through the various ways and marketed in the national and international market.

20.7 TRADE SCENARIO OF AROMATIC PLANTS

Aromatic plants and their products are not only a good source of income for farmers and entrepreneurs but also earn valuable foreign exchange by way of export. Sixty-two types of essential oils have a huge and constant demand in the international market. In today's world of a consumer boom, the role of essential oils has increased many-fold. The production and consumption of essential oils are increasing continuously because of their multipurpose application. Essential oils have been used throughout the world in perfumery, the food industry, household industry, condiment industry, beverages and pharmaceuticals. These essential oils are quite expensive due to their global demand in the industrial market (Tables 20.4 and 20.5).

TABLE 20.4
Export of Essential Oils (in US$ million)

S. No.	Crop Name	2010–11	2011–12	2012–13	2013–14	2014–15	2015–16	2016–17	2017–18	2018–19	2019–20
1.	*Mentha piperita*	42.33	74.97	95.08	87.65	77.87	55.68	48.84	70.39	64.24	72.82
2.	*Mentha spicata*	2.89	9.39	12.31	9.54	18.95	20.64	9.33	8.91	13.54	12.82
3.	*Mentha citrate*	0.07	0.06	0.00	0.06	0.07	0.10	0.04	0.06	0.04	0.06
4.	*Cymbopogan flexuosus*	1.72	2.09	2.24	2.70	3.75	5.01	5.97	7.74	10.83	7.06
5.	*Cymbopogan martini*	0.46	0.52	0.20	0.19	0.31	0.41	0.37	0.69	0.85	1.91
6.	*Cymbopogan winterianus*	1.42	2.38	2.00	1.83	1.24	1.41	1.22	1.86	2.99	2.00
7.	*Pogostemon patchouli*	0.18	2.41	6.66	3.62	1.30	0.19	0.21	0.53	0.55	0.48
8.	*Rosa damascene*	0.25	0.47	0.76	2.05	9.81	14.32	7.89	10.41	0.81	0.62
9.	Rose water	0.59	0.71	0.66	0.78	0.80	0.94	1.03	1.22	1.46	1.38
10.	*Eucalyptus globules*	0.65	0.57	0.44	0.46	0.87	1.02	0.99	1.62	1.05	3.81
11	Sandalwood	0.06	0.22	0.06	0.09	0.51	0.11	0.37	0.56	1.61	1.14

TABLE 20.5
Import of Essential Oils (in US$ million)

S. No.	Crop Name	2010–11	2011–12	2012–13	2013–14	2014–15	2015–16	2016–17	2017–18	2018–19	2019–20
1.	*Mentha piperita*	0.78	0.80	0.60	1.02	1.03	1.35	2.30	4.25	520	5061
2.	*Mentha spicata*	0.28	1.16	0.55	0.29	0.50	0.41	0.83	0.66	0.30	1.91
3.	*Mentha citrate*	0.03	0.10	0.13	0.06	0.04	0.05	0.05	0.08	0.10	1.94
4.	*Cymbopogan flexuosus*	0.01	0.12	0.02	0.01	0.02	0.04	0.12	0.15	0.02	0.10
5.	*Cymbopogan martini*	1.19	1.51	0.75	1.24	1.07	1.41	0.90	0.53	0.47	0.42
6.	*Cymbopogan winterianus*	0.02	0.01	0.01	0.00	0.00	0.00	0.00	0.02	0.13	0.02
7.	*Pogostemon patcholi*	11.94	16.39	17.96	14.42	22.04	27.91	24.06	24.43	19.35	24.05
8.	*Rosa damascena*	0.26	0.91	0.39	1.35	7.87	12.32	8.97	7.50	1.21	0.77
9.	Rose water	0.00	0.01	0.01	0.02	0.00	0.09	0.01	0.04	0.00	0.05
10.	*Eucalyptus globulus*	6.28	6.74	6.45	8.44	7.64	7.34	8.02	10.55	11.59	12.95
11.	Sandalwood	8.97	11.11	14.09	16.44	12.42	14.35	19.79	15.33	6.50	10.25

20.8 GOVERNMENT EFFORTS IN THE CULTIVATION OF AROMATIC PLANTS

Realizing the vigorous untapped potential of aromatic plants, the government of India launched a scheme for their development during the successive Five-Year Plans. These programs are being operated in 16 state agricultural universities and three regional research laboratories of the Council of Scientific and Industrial (CSIR). This includes the production of quality planting material, development of herbal gardens, improved agro-techniques for propagation and production of new plant varieties and setup of regional analytical laboratories (Report Planning Commission 2000). The central government and the state government provide financial support in the form of loans and subsidies to the farmers engaged in aromatic crop cultivation through various government bodies/agencies like Ayurveda, Yoga, Unani, Siddha and Homeopathy (AYUSH), Central Institute of Medicinal and Aromatic Plants (CIMAP), National Medicinal Plant Board (NMPB), Agri-Export Zone, National Bank for Agricultural and Rural Development (NABARD), Agricultural and Processed Food Products Export Development Authority (APEDA), National Horticulture Board (NHB), State Aromatic Plant Board, etc.

20.9 MAJOR INSTITUTES AND INDUSTRIES

About 1500 species of aromatic plants are grown in different agro-climatic conditions of the world, and out of them 500 species are a source of raw materials of the perfumery industry, having properties of essential oils. India has a mega-biodiversity and is a hot-spot for many medicinal and aromatic plants, which are now being used for various ailments and pharmaceuticals as well as nutraceuticals. In India, two major scientific councils are working on medicinal and aromatic plants; one is CSIR and the other one is the Indian Council of Agricultural Research (ICAR). CIMAP works under CSIR, which is looking towards an industrial point of view, and the Directorate of Medicinal and Aromatic Plants works under ICAR, which is looking towards farmers' point of view. At the present time demand for aromatic plants is increasing in both developed and developing countries for various reasons. For some people, it is the growing recognition that natural products have fewer or even no side effects; for others, it is their accessibility and affordable costs that tip the scales. After the COVID-19 pandemic, the consumption of medicinal and aromatic plants and their products increased very rapidly (Table 20.6).

There are nearly 1100 aromatic plant product manufacturing industries in Maharashtra, which are totally dependent on the natural forests of Maharashtra and adjoining states for the supply of raw materials of plant parts for manufacturing their medicines/various related aromatic products.

20.10 POST-HARVEST AND DISTILLATION TECHNIQUES

The separation of essential oils from the herbage is done through distillation. There are various methods of distillation in the extraction of essential oils such as water distillation (hydro-distillation), water and steam distillation (hydro-steam), direct steam distillation, distillation with dual technique, etc. Generally, three types of distillation techniques such as steam distillation, hydro-distillation and hydro-steam distillation are used for obtaining essential oils. Most essential oils are extracted from the plants by steam distillation methods. The essential oils of some plants, such as clove and peels of citrus fruits, can be extracted easily. It is quite common to extract the essential oils of citrus fruit, for example, with a special technique in which peels are scraped and the oil collected with a sponge. Tapping is another method used for distillation in some trees, like the commiphora tree (for myrrh), the laurel of Guyana or the camphor tree of Borneo (for borneol).

TABLE 20.6
List of Major Universities/Institutes/Industries Working on Medicinal and Aromatic Plants

S. No.	University/Institute	Address	Working Area
1.	CSIR: Central Institute of Medicinal and Aromatic plants	Picnic Spot Road, Khurram Nagar, Lucknow, 226015	Developed agro-technique, high-yielding variety of MAPs and their value addition
2.	CSIR: Indian Institute of Integrated Medicine	No. 3, Post Bag, Canal Rd, Jammu, Jammu and Kashmir	Developed agro-technique, high-yielding variety and herbal products
3.	Fragrance and Flavour Development Centre, Kannauj	GT Rd, Industrial Estate, Makrand Nagar, Kannauj, Uttar Pradesh	Essential oils
3.	Narendra Deva University of Agriculture and Technology	Kumarganj, Ayodhya	Developed agro-technique and high-yielding variety
4.	Kerala University of Agriculture and Technology	Odakkali, Kerala	Developed agro-technique and high-yielding variety
5.	Govind Pallabh Pant University	Pantnagar	Developed agro-technique and high-yielding variety
6.	CCS Hissar Agriculture University	Hissar, Haryana	Developed agro-technique and high-yielding variety
7.	Jawahar lal Nahru Krishi Vishva Vidhyalaya	Indore, Madhya Pradesh	Developed agro-technique and high-yielding variety
8.	Sri A.K. Jain Aryan International	Nabi Saray, New Delhi	Essential oils
9.	Sri Sai Trading Company	Shahadara, Delhi	Essential oils
10.	Virat Exporters	Patel Nagar, New Delhi	Essential oils
11.	Ajmal Biotech Pvt. Ltd.	Kulaba, Mumbai, Maharastra	Essential oils and products
12.	Sagar Aromatic	East Mumbai, Mharastra	Essential oils
13.	Mentha and Allied Chemical	Barabanki, Uttar Pradesh	Essential oils
14.	Sugandh Vyapar	Lucknow, Uttar Pradesh	Essential oils
15.	Natajel Healthcare Pvt. Ltd.	Haidrabad, Telangana	Essential oils
16.	M.S. Unilink Chemical, Pvt. Ltd.	Barodra, Gujrat	Essential oils
17.	Sai Baba Sugand Bhandar	Bangluru, Karnataka	Essential oils

20.10.1 Hydro-Distillation

Hydro-distillation is the traditional and simplest method of distillation available for the extraction of essential oils from the plants. In this process plant material is placed in a tank filled with water. The hydro-distillation unit is categorized into three main parts, i.e. tank, condenser and separator. The tank is installed on the furnace for heating the water and creating vapor. The tank is connected to the condenser by a vapor pipe. The condenser condenses the vapor through a regular supply of cold water from the overhead tank. The mixture of extracted essential oil and water is collected in the separator. The essential oil leaches up due to high density in comparison to water. This process of distillation takes about 3.5–4 hours to distil of one tank of herbage. In this process, the tank is filled with water at the level of the sieve (Srivastava et al. 2009).

20.10.2 Hydro-Steam Distillation

This is a modified type of hydro-distillation. It is also known as a field distillation unit (FDU). It can be used for the distillation of bulk material from aromatic crops. In this method the distillation unit divided into three main parts, such as a still or tank made of stainless steel, a condenser and receiver-cum-separator. The tank is also divided into two parts with a grid. The perforated grid is salted on 1.6–1.8 feet or more from the actual bottom in the still. The plant material has been filled upon the grid, and the water keeps the grid in place. The water is boiled and the steam passes through the plant material. The vapor and oil mixture is cooled in the condenser and then separated in the separator. The distillation process takes about 4–5 hours.

20.10.3 Steam Distillation

This process is used when a lot of area is under cultivation with hardy and coastal plant material such as root and wood like sandalwood, vetiver, jatamansi, kapoor kachri, nagarmotha, etc. In this method, steam generated in a steam generator or boiler outside of the tank passes through the material, which is in the boiler tank. The pressure of steam is controlled according to the nature of the plant material that is to be distilled. The process of distillation takes about 3–4 hours, and the cost of fuels is very low.

20.11 STORAGE

The essential oil should be store in an aluminum or glass bottle or container for a long time. The container should be filled with the oil without air and stored in cool and dry place.

20.12 FUTURE PROSPECTS OF AROMATIC PLANTS

In perfumery industries, essential oils are in high demand. They are also used to make different products from perfums to mosquito repellent to antifungals. The years 2010–2011 and 2019–2020 have shown increasing exports from India to other countries, as deman both globally and locally continues. Aromatics are in high demand, but some difficulties includee unknown markets and a non-definite marketplace. It is necessary to improve the local market at least at the district level to provide suitable prices for farmers. Due to the high demand for essential oil in the domestic and international market and the availability of high-yielding cultivars, agro-technology, distillation and processing technologies and post-harvest management strategies, the future outlook for aromatics is bright (Rajeswara Rao 1999).

20.13 CONCLUSION

Aromatic plants and their essential oils were only used in a few areas over the last few decades, but at present the uses of essential oil have continued to grow at a global pace. In recent years the consumption of essential oils increased very rapidly because of COVID-19. The international market price of some essential oils increased 1.5–3 times. Due to its uses and benefits, the demand for essential oil in national and international markets has been increasing consistently. To fulfill the requirement and supply a quality product, there is a bright group of aromatic crop growers and related industries. Government should promote the cultivation of aroma crops because these are the highly valuable cash crops for the local area farmers. By cultivating these crops, the socioeconomic condition of the farmers can be improved and we can complete our domestic demands of essential oils.

Table 20.7 shows the comparative economics of some major commercial aromatic crops that provide high returns from cultivation. Due to the low cost of cultivation and high returns, the grower can obtain more income per unit area from aromatic crops compared to other traditional crops.

TABLE 20.7
Economics of Some Commercially Cultivated Aromatic Crops

S. No.	Name of Crop	Cost of Cultivation (Rs/ha)	Gross Income (Rs/ha)	Net Income (Rs/ha/year)
1.	Mentha	30,000	175,000	145,000
2.	Lemongrass	80,000	200,000	120,000
3.	Palmarosa	50,000	150,000	100,000
4.	Java citronella	75,000	200,000	125,000
5.	Vetiver	130,000	230,000	100,000
6.	Tulsi	40,000	10,000	60,000
7.	Geranium	75,000	225,000	150,000
8.	Rose	100,000	700,000	600,000
9.	Rosemary	130,000	350,000	220,000
10.	Marigold	30,000	100,000	70,000

BIBLIOGRAPHY

Aburjai, T., and Feda M.N. 2003. Plants used in cosmetics. *Phytotherapy Research Phytother* 17:987–1000.

Andrade, B.F.M.T., L.N. Barbosa, I.D.S. Probst, and A.F. Junior. 2014. Antimicrobial activity of essential oils. *Journal of Essential Oil Researc* 26(1):34–40.

Awang, K., H. Ibrahim, S.D. Rosmy, M. Mohtar, A.R. Mat, and N.A.M. Ali. 2011. Chemical constituents and antimicrobial activity of the leaf and rhizome oils of Alpinia pahangensis Ridl., an endemic wild ginger from peninsular Malaysia. *Chem. Biodivers* 8:668–673.

Azadi, M., T. Jamali, Z. Kianmehr, G. Kavoosi, and S.K. Ardestani. 2020. In-vitro (2D and 3D cultures) and in-vivo cytotoxic properties of Zataria multiflora essential oil (ZEO) emulsion in breast and cervical cancer cells along with the investigation of immunomodulatory potential. *Journal Ethnopharmacol* 257:112865.

Bagetta, G., M. Cosentino, and T. Sakurada. 2016. Preface. In *Aromatherapy: Basic Mechanisms and Evidence Based Clinical Use*, Oxfordshire, UK: Taylor & Francis.

Ballard, C.G., J.T.O. Brien, K. Reichelt, and E.K. Perry. 2002. Aromatherapy as a safe and effective treatment for the management of agitation in severe dementia: the results of a doubleblind, placebo-controlled trial with Melissa. *Journal of Clinical Psychiatry* 63(7):553–558.

Boelens, M.H. 1995. Chemical and sensory evaluation of lavandula oils. *Perfum Flav* 20:23–51.

Bouhtit, F., M. Najar, D.M. Agha, R. Melki, M. Najimi, and K. Sadki. 2021. New anti-leukemic effect of carvacrol and thymol combination through synergistic induction of different cell death pathways. *Molecules* 26(2):410.

Burns, A., J. Byrne, C. Ballard, and C. Holmes. 2002. Sensory stimulation in dementia. *British Medical Journal* 325(7376):1312–1313.

Burt, S.A. 2004. Essential oils: Their antibacterial properties and potential applications in food: A review. *International Journal of Food Microbiology* 94:223–253.

Carmo, E.S., E.O. Lima, and E.L. Souza. 2008. The potential of Origanum vulgare L. (Lamiaceae) essential oil in inhibiting the growth of some food-related Aspergillus species. *Braz. Journal Microbiol* 39:362–367.

Cavanagh, H.M., and J.M. Wilkinson. 2002. Biological activities of lavender essential oil. *Phytother Res* 16:301–308.

Chouhan, S., K. Sharma, and S. Guleria. 2017. Antimicrobial activity of some essential oils-present status and future perspectives. *Medicines* 4:58.

Chowti, S.P., S. Rudrapur, and B.K. Naik. 2018. Production scenario of medicinal and aromatic crops in India. *Journal of Pharmacognosy and Phytochemistry* 3:274–277.

Christaki, E., E. Bonos, L. Giannenas, and P. Florou-Paneri. 2012. Aromatic plants as a source of bioactive compounds. *Agriculture* 2:228–243.

Duthie, G.G., and K.M. Brown. 1994. Functional foods. In *Reducing the Risk of Cardiovascular Disease*; Goldberg, I., Ed.; London, UK: Chapman & Hall, pp. 19–38.

Gediya, S.K., R.B. Mistry, U.K. Patel, M. Blessy, and H. N. Jain. 2011. Herbal plants: Used as a cosmetics. *J. Nat. Prod. Plant Resour* 1(1):24–32.

Gezicil, S., N. Sekeroglu, and A. Kijjoa. 2017. In vitro anticancer activity and antioxidant properties of essential oils from Populus alba and Rosmarinus officinalis from South Eastern Anatolia of Turkey. *Indian Journal of Pharmaceutical Education and Research* 51(3):498–503.

Günes, B.A., A. Kocyigit, E.M. Guler, and A. Dadak. 2020. In vitro hormetic effect investigation of thymol on human fibroblast and gastric adenocarcinoma cells. *Molecules* 25(14):3270.

Inoue, M., and L. Craker. 2014. Medicinal and aromatic plants-uses and functions. In *Horticulture: Plants for People and Places*; Dixon, G. and D. Aldous, Eds.; Dordrecht: Springer, 2, pp. 645–669.

Jairath, M.S., and N.L. Agarwal. 2005. Marketing of medicinal and aromatic plants in Rajasthan. National Consultative Workshop on Medicinal and Aromatic Plants, held at G.B. Pant University of Agriculture & Technology, Pantnagar, pp. 28–36.

Joy, P.P., J. Thomas, S. Mathew, G. Jose, and J. Joseph. 1998. Aromatic plants.

Jimbo, D., Y. Kimura, M. Taniguchi, M. Inoue, and K. Urakami. 2009. Effect of aromatherapy on patients with Alzheimer's disease. *Psychogeriatrics* 9(4):173–179.

Joshi L.S. and H.A. Pawar. 2015. Herbal Cosmetics and Cosmeceuticals: An Overview, *Natural Products Chemistry & Research* 3:170.

Kaloustian, J., J. Chevalier, C. Mikail, M. Martino, L. Abou, and M.F. Vergnes. 2008. Etude de six huiles essentielles: Composition chimique et activité antibactérienne. *Phytothérapie* 6:160–164.

Konyak, M.B. 2018. Essential oils: Its components and its application in food preservatives. *International Journal of Science and Research* 7(12):207–2011.

Kumar, N. 2018. Introduction to spices, plantation crops, medicinal and aromatic plants. *Medtech*: 265–314.

Mary, P., and M.D. Lupo. 2001. Antioxidants and vitamins in cosmetics. *Clinics in Dermatology* 19:467–473.

Mendes, S.S., R.R. Bomfim, H.C.R. Jesus, P.B. Alves, A.F. Blank, C.S. Estevam, A.R. Antoniolli and S.M. Thomazzi. 2010. Evaluation of the analgesic and anti-inflammatory effects of the essential oil of Lippia gracilis leaves. *J. Ethnopharmacol* 129:391–397.

Milner, J.A. 1994. Functional foods. In *Reducing the Risk of Cancer*; Goldberg, I., Ed.; London, UK: Chapman & Hall, pp. 39–70.

Nikolic, M., K.K. Jovanovic, T. Markovic, D. Markovic, N. Gligorijevic, S. Radulovic, and M. Sokovic. 2014. Chemical composition, antimicrobial and cytotoxic properties of five Lamiaceae essential oils. *Ind. Crops Prod* 61:225–232.

Parameswaran, T.N., A. Kalra, V.K. Mehta, and K. Radhakrishnan. 2000. Chemical control of tip burn and blight of scented geranium (Pelargonium graveolens) caused by Colletotrichum gloeosporioides under South Indian hill conditions. *J Med Arom Plant Sci* 22(1B):666–668.

Parvez, M., T. Nihal, G. Undire, A. Mohammed Raihan, M. Shaikh, and B. Mandake. 2021. Extraction and formulation of perfume from plants: A review. *Journal of Emerging Technologies and Innovative Research (JETIR)* 8(5):663–666.

Perez-Vasquez, A., S. Capella, E. Linares, R. Bye, G. Angeles-López, and R. Mata. 2011. Antimicrobial activity and chemical composition of the essential oil of Hofmeisteria schaffneri. *J. Pharm. Pharmacol* 63:579–586.

Purohit, P., and T.R. Kapsner. 1994. Natural essential oils, functional benefits. *Cosmet Toiletries* 109:51–55.

Purohit, S.S., and S.P. Vyas. 2005. Marketing of medicinal and aromatic plants in Rajasthan. National Consultative Workshop on Medicinal and Aromatic Plants, held at G.B. Pant University of Agriculture Technology, Pantnagar, Uttarakhand, pp. 25–27.

Rajamani, K., L. Nalina, and L. Hegde. 2019. Medicinal and aromatic crops. *AgriMoon*:1–195.

Rajeswara Rao B.R., A.K. Bhattacharya and P.N. Kaul. 1999. Rose-scented geranium (Pelargonium graveolens L.) cultivation and technology for Andhra Pradesh, *Indian Perfum*. 43 (l): 2–8.

Rajkumar, S., and A. Jebanesan. 2010. Chemical composition and larvicidal activity of leaf essential oil from Clausena dentata (Willd) M. Roam. (Rutaceae) against the chikungunya vector, Aedes aegypti Linn. (Diptera: Culicidae). *J. Asia Pac. Entomol* 13:107–109.

Report Planning Commission 2000. niti.gov.in/planningcommission.gov.in/docs/reports/genrep/arep2021/index.php?repts=ar2021

Sakurada, T., H. Kuwahata, and S. Katsuyama. 2009. intraplantar injection of bergamot essential oil into the mouse hindpaw: Effects on capsaicin-induced nociceptive behaviors. *International Review of Neurobiology* 85(18):237–248.

Scuteri, D., L.A. Morrone, L. Rombola, P.R. Avato, A.R. Bilia, M.T. Corasaniti, S. Sakurada, T. Sakurada, and G. Bagetta. 2017. Aromatherapy and aromatic plants for the treatment of behavioural and psychological symptoms of dementia in patients with Alzheimer's disease: Clinical evidence and possible mechanisms. *Hindawi Evidence-Based Complementary and Alternative Medicine* 8.

Sharma, J.R. 1996. Country report: India. Paper presented ASIUMAP, 96:4–9, FAO/RAP, Bangkok.

Shweta, K.G., B.R. Mistry, U.K. Patel, M. Blessy and H.N. Jain. 2011. Herbal Plants: Used as a cosmetics. *J. Nat. Prod. Plant Resour* 1 (1): 24–32.

Silva, S.L., J.S. Chaar, P.M.S. Figueiredo, and T. Yano. 2008. Cytotoxic evaluation of essential oil from Casearia sylvestris Sw on human cancer cells and erythrocytes. *Acta Amazonica* 1:107–112.

Singh, K., B.R. Rajeshwara Rao, S.K. Kothari, C.P. Singh, P.N. Kaul, and S. Kumar. 2000. Cultivation of aromatic crops in south India: Problems and prospects. *Journal of Medicinal and Aromatic Plant Sciences* 22:218–230.

Skrinjar, M.M., and N.T. Nemet. 2009. Antimicrobial effects of spices and herbs essential oils. *Acta Periodica Technologica* 40:195–209.

Srivastava, R.K., S.V. Shukla, and S.S. Dagar. 2009. Cultivation uses of aromatic plants. *IBDC Publication* 1:86–87.

Srivastava, S., D.M. Cahill, X.A. Conlan, and A. Adholeya. 2014. A novel in vitro whole plant system for analysis of polyphenolics and their antioxidant potential in cultivars of Ocimum basilicum, *Journal of Agricultural and Food Chemistry* 62(41):10064–10075.

Stefanakis, M.K., E. Touloupakis, E. Anastasopoulos, D. Ghanotakis, H.E. Katerinopoulos, and P. Makridis. 2013. Antibacterial activity of essential oils from plants of the genus Origanum. *Food Control* 34:539–546.

Utkarsha, C., S.B. Topel, R.B. Gondakar, and Saudager. 2017. Herbal Cosmetics: Review Article. *International Journal of Pharmacy & Technology* 9(2):5908–5919.

Wannes, W.A., B. Mhamdi, J. Sriti, M.B. Jemia, O. Ouchikh, G. Hamdaoui, M.E. Kchouk, and B. Marzou. 2010. Antioxidant activities of the essential oils and methanol extracts from myrtle (Myrtus communis var. italica L.) leaf, stem and flower. *Food Chem. Toxicol* 48:1362–1370.

Xiao, Pei-Gen. 1996. Country report: China. Paper presented ASIUMAP, 4–9 Nov. 96, FAO/RAP, Bangkok.

Yu-Chang, S., H. Kuan-Ping, W.E. I-Chen, and H. Chen-Lung. 2013. Composition and in vitro anticancer activities of the leaf essential oil of Neolitsea variabillima from Taiwan. *Natural Product Communications* 8(4):531–532.

Index

Note: Page numbers in *italics* indicate figures and in **bold** indicate tables on the corresponding pages.

A

acute restraint stress (ARS), cannabinoids in, 231–232
adaptogens
  ashwagandha, 270, **274**
  Bacopa monnieri, 270–271, **274**
  challenges and opportunities in research on, 276
  conclusion on, 276
  Cordyceps sinensis, 272, **275**
  Emblica officinalis, 271, **274**
  ginkgo biloba, 271, **274**
  global commercial value of, 273, 276
  Glycyrrhiza glabra, 272–273, **275**
  health benefits of, *268*
  from Indian traditional medicine, 270–273
  introduction to, 267–269, *268*
  Panax ginseng, 270, **274**
  protective actions of, 269
  Rhodiola rosea, 272, **275**
  Tinospora cordifolia, 273, **275**
AgNPs, 336, **337**, *337*, **338–339**
alkaloids
  as anti-cancer phytopharmaceuticals, **241**, 245
  benzylisoquinoline alkaloids (BIAs), 112–114, *114*
  biosynthetic pathways of, 112–119
  catalyzing the pathway elucidation of, 119–123
  conclusions and future perspectives on, 127
  genomics of, 119–120
  introduction to, 109–110
  metabolomics of, 122–123
  monoterpene indole alkaloids (MIAs), 115, *116*
  multiomics analysis of, 123–127, *124*
  occurrence, diversity, and function of, 110–112, *111*
  proteomics of, 121–122
  purine, 118–119, *119*
  in spices, 21
  transcriptomics of, 120–121
  tropane alkaloids (TAs), 117–118, *118*
Alzheimer's disease, 12
amino acids, 2
amla, 271
angiotensin-converting enzyme (ACE) inhibitors, 143
anthocyandin, 150
anthocyanins, 4
anti-aging, *see* low-level laser (light) therapy
antibody drug conjugates (ADCs), 159
anti-cancer phytopharmaceuticals; *see also* cancer
  from aromatic plants, 350
  camptothecin, 245–246
  cannabinoids, **243**, 249
  carotenoid, 247
  conclusion on, 250
  development, challenges, and opportunities for, 249–250
  diindolylmethane/indole-3-carbinol, **243**, 247
  drug resistance and mechanism of drug resistance and, 240–244, **241–243**
  epigallocatechin, **243**, 248
  epipodophyllotoxin, 245
  flavonoids, 246
  indirubin, **243**, 248–249
  introduction to, 239–240
  isothiocyante, 246–247
  natural herbal compounds, 244–249
  phenolic compounds, 244–245
  resveratrol and stilbenoid, 248
  taxanes, **243**, 248
  triterpenoids, 247
  vinca alkaloid, **241**, 245
  vitamin D and vitamin E, **243**, 248
antimicrobial properties of aromatic plants, 350–351
antioxidants
  action and inhibition of electron transfer by, in breast cancer, 174–175
  from aromatic plants, 188–195, *189*, **190–191**, *194*
  supplementation with, 143
aonla, 83, 91–95, *94*
apigenin, 28, 148
  as aromatic plant, *189*, 192–193
AraCyc, 125
aromatherapy, 352–353
aromatic plants
  apigenin, *189*, 192–193
  carnosic acid, *189*, 194
  carnosol, *189*, 194
  carvacrol, 189, *189*
  chemical compounds of essential oils from, 349
  chrysin, *189*, 193
  classification of, 186–188
  commercialized, **353**, 353–354, **358**
  conclusion and future aspects of, 195, 357
  crop duration of, 187
  curcumin, *189*, 193
  cyanidin, *189*, 193
  diosmetin, *189*, 194
  economic importance of, 188
  eugenol, 188–189, *189*
  future prospects of, 357
  gingerol, *189*, 192
  government efforts in cultivation of, 355
  growth habit of, 186–187
  habitat of, 187
  history of, 348, **349**
  importance of, 186
  introduction to, 185–186, 347–348
  kaempferol, *189*, 192
  luteolin, *189*, 192
  major institutes and industries in, 355, **358**

mechanism of action of natural antioxidant molecules in averting disease risk and, *194*, 194–195
natural antioxidants from, 188–194, ***189***, **190–191**
pharmaceutical properties of, 350–351
post-harvest and distillation techniques for, 355–357
propagation of, 187
quercetin, *189*, 192
rosmanol, *189*, 193
storage of, 357
thymol, 189, *189*
trade scenario of, 354, **354**
artificial selection in breeding, 54–55
asafoetida, 30–31, **41**
ashwagandha, 270, **274**
atherosclerosis, 11
AuNPs, 334, **335–336**
Ayurveda, 20–21
asafoetida in, 30–31
black pepper in, 23
carom seeds in, 30
cinnamon in, 23
garlic in, 25
ginger in, 26
star anise in, 32
turmeric in, 28–29

B

Bacopa monnieri, 270–271, **274**
bael, 84, 95–97, *96*
baicalin, 148
banana, 84, 97–98
belotecan as anti-cancer phytopharmaceutical, **242**
benzylisoquinoline alkaloids (BIAs), 112–114, *114*
berberine, 325, **326**
bilateral common carotid artery occlusion (BCCAO), cannabinoids in, 231
bioactivity molecular networking, 294
Biogenome Project, 120
bioinformatics, 282–283
bioactivity molecular networking, 294
drug discovery tools using, 289, **290–292**
effective target prediction tools for analyzed phytochemicals, 295
evaluation through docking and molecular dynamics simulations, 297–298
ligand-based bioactivity prediction tools, 294–295
network building and pathway advancement, 297
in phytochemical investigations, 283–294
phytomedicinal research tools using, 294–298
plant genomics and EST databases, 295–296
reverse screening, 296–297
biotechnology approach to breeding, 58–60
black pepper, 23–24, **35**
bladder cancer and seaweed metabolites, 258
brahmi, 270–271
brain cancer, cannabinoids in, **230**
breast cancer
antioxidant action and inhibition of electron transfer in, 174–175
binding to proteins/enzymes and blocking substrate/protein partner binding in, 173–174
cannabinoids in, 229, **230**
capsaicin and, 170
causes of, 158–159, *160*
conclusions on, 175
curcumin and, 169–170
gallic acid and, 171
history of medicine and, 163–164
how phytochemicals prevent/treat, 171–175, *172*
incidence of, 158
inhibition of stem cell proliferation in, 173
introduction to, 158–161
molecular mechanisms of, 161–163, *162*
phytomedicine and treatment modalities for, 159, 161
quercetin and, 170
relevance of phytochemicals in therapy for, 164–167, **165–166**, **168**, *169*
resveratrol and, 170–171
seaweeds' cytotoxicity against, 257–258
breeding, medicinal and aromatic plants (MAP)
artificial selection, 54–55
biotechnology approach to, 58–60
conclusions and perspectives on, 60
genetic engineering in, 58–59
genome editing in, 59
hybridization technique for, 55–56
introduction to, 53–54, *54*
marker-assisted selection (MAS), 59–60
mutation, 56, **57**
plant tissue culture, 58
polyploidy, 56–58
traditional approach to, 54–58, **57**

C

caffeine, 118–119, *119*
callus induction of aonla, 94–95
camptothecin, 206, 245–246
in herbal nanocarrier-based drug delivery, **326**
cancer, 205–206; *see also* anti-cancer phytopharmaceuticals
aromatic plants and, 350
challenges for cannabinoids use in, 234
functional role of cannabinoids in, 226–229, *227*, **230**
recent clinical trials of cannabinoids in, 232–233
seaweed metabolites and (*see* seaweed metabolites)
cannabinoids
as anti-cancer phytopharmaceuticals, **243**, 249
in brain cancer, **230**
in breast cancer, 229, **230**
challenges for use in cancer and neurological diseases, 234
conclusion and future prospects for, 234
functional role in cancer, 226–229, *227*, **230**
functional role in neurological diseases, 230–232, *231*, **232**
in glioblastoma, 226–228, *227*
introduction to, 225–226
in lung cancer, 228–229, **230**
recent clinical trials in cancer, 232–233
recent clinical trials in neurological diseases, 233

capsaicin and breast cancer, 170
carbon dioxide (CO_2) concentration in controlled environment agriculture (CEA), 71–72, **72**
cardamom, 31–32, **42**
cardiovascular disease (CVD), 11, 29
carnosic acid, *189*, 194
carnosol, *189*, 194
carom seeds, 30, **40**
carotenoids, 4–5
  as anti-cancer phytopharmaceuticals, 247
chrysin, *189*, 193
cinnamon, 23, **35**
CitrusCyc, 126
clove, 24, **36**
clustered regularly interspaced short palindromic repeats/CRISPR-associated protein 9 (CRISPR/Cas9 system), 59
clustering analysis, 125
cognition and memory, 12
colloidal nanoliposomes, *322*, 322–323
colon cancer and seaweed metabolites, 258
commercialized aromatic plants, **353**, 353–354
common diseases/illnesses
  conclusion and future prospects for phytopharmaceuticals, 209
  current status of phytopharmaceuticals for, 204–205
  defined, 202
  extraction of phytopharmaceuticals for, 202–203, *203–205*
  importance of phytopharmaceuticals for, 202
  introduction to, 201–202
  role of pharmaceuticals in treating, 205–206, **207–209**
  significance of phytopharmaceuticals in treating, 209
complex coacervation method, 320
congestive heart failure, 12
controlled environment agriculture (CEA)
  carbon dioxide (CO_2) concentration in, 71–72, **72**
  conclusion and future perspectives on, 76–77
  hydration, humidity, nutrient concentration factors in, 74–75
  introduction to, 65–66
  lighting in, 72–74, **73**
  limitations of, 76
  market share for, 66
  methods adopted in, 66, 68, **68–69**
  range of facilities for, 66, *67*
  secondary metabolite production in, 69–75, **70**
  synergism among different parameters in, 75–76
conventional drugs, 213–214
  limitations of, 214–215
  natural products as alternative to, 215–217
  pros and cons of, 214
  synergistic effects of seaweed metabolites and anticancer, 259–263, **261**
co-precipitation method, 321
Cordyceps sinensis, 272, **275**
correlation analysis, 125
cosmetics, essential oils for, 351–352
Council of Scientific and Industrial Research (CSIR), 201–202
COVID-19
  aromatic plants and, 186, 355
  spices and, 33–34
C-reactive protein (CRP), 143
crocetin as anti-cancer phytopharmaceutical, **242**
crocins, 28
cumin, 25, **36**
curcumin
  as anti-cancer phytopharmaceutical, 240, **241**
  as aromatic plant, *189*, 193
  breast cancer and, 169–170
  in herbal nanocarrier-based drug delivery, 324–325, **326**
cyanidin, *189*, 193

D

daidzen, *146*, 149–150
Darwin Tree of Life project, 120
delphinidin, 150
dendrimers, 322, *322*
differential analysis, 126
Digitalis purpurea, 12
diindolylmethane (DIM) as anti-cancer phytopharmaceutical, **243**, 247
diosgenin, 27
diosmetin, *189*, 194
DNA Data Bank of Japan (DDBJ), 121
docetaxel, 206

E

E-flux, 126
electrical conductivity (EC) of nutrient solutions, 75
element-based approach in multiomics analysis, 125
Emblica officinalis, 271, **274**
epigallocatechin-3-gallate as anti-cancer phytopharmaceutical, **243**, 248
epipodophyllotoxin, 245
ER signaling, 161, 163
essential oils
  for aromatherapy, 352–353
  cardamom, 32
  carom seed, 30
  chemical compounds of, 349
  clove, 24
  for cosmetics, 351–352
  cumin, 25
  for food and flavor, 351
  future prospects of, 357
  for perfumery, 352
  post-harvest and distillation techniques for, 355–357
  star anise, 32
  storage of, 357
  turmeric, 29
  uses of, 351
EST databases, 295–296
etoposide as anti-cancer phytopharmaceutical, **241**
eugenol, 24, 188–189, *189*
exotic medicinal plants, 11–12
extraction process for phytopharmaceuticals, 202–203, *203–204*

F

fennel, 26–27, **38**
fenugreek, 27–28, **38**
fever, 12
Filipendula ulmaria, 12
5,7-dimethoxyflavone, 147
flavanone, 149
flavones, 147–149
flavonoids, 5, 150–151
  anthocyanidin, 150
  as anti-cancer phytopharmaceuticals, 246
  in clove, 24
  flavanone, 149
  flavones, 147–149
  flavonols, *146*, 146–147
  isoflavone, 149–150
  protective roles in muscle mass preservation and regeneration, 145–150, *146*
  in spices, 21
flavonol, 145–147, *146*
Food and Drug Administration (FDA), 215
food and flavor, essential oils for, 351
Fourier-transform infrared spectroscopy (FT-IR), 284
fruit crops, *see* micropropagation of fruit crops
fucoidan, 262–263
fucosterol, 259
fucoxanthin, 259–260, 261

G

gallic acid and breast cancer, 171
garlic, 25–26, **37**
gas chromatography-mass spectrometry (GC-MS), 283–284
  NIST-2008 dataset, 284
genetic engineering, 58–59, 219
genetic mutation breeding, 56, **57**
genistein, *146*, 149–150
  as anti-cancer phytopharmaceutical, **242**
genome editing, 59
genome-scale analysis, 126–127
genome-wide association studies (GWASs), 120
genomics, 119–120, 295–296
  in multiomics analysis, 123–127, *124*
German Medicines Act (AMG), 3
giloy, 273
ginger, 26, **37**
gingerol, *189*, 192
  as anti-cancer phytopharmaceutical, **241**
ginkgo biloba, 206, 271, **274**
ginseng, 270, **274**
ginsenoside G-Rg3 as anti-cancer phytopharmaceutical, **242**
glioblastoma, cannabinoids in, 226–228, *227*
Global Natural Product Social Molecular Networking (GNPS), 284, 286
glycolytic switch in breast cancer, 163
glycoside-aglycone relationships, 297
Glycyrrhiza glabra, 272–273, **275**
gold nanoparticles, 263
guava, 83, 90–91, *92*
guduchi, 273

H

HER2/ERBB2 signaling, 161
Herbal Ingredients' Targets (HIT) tool, 295
high-performance liquid chromatography (HPLC), 283
high-performance thin layer chromatography (HPTLC), 283
high-pressure homogenization (HPH) method, 319–320
high resolution mass spectrometry (HRMS), 219
high-throughput screening (HTS), 283
homeopathy, 3
hormonal alterations in sarcopenia, 143
hormone replacement therapy (HRT), 10
humidity in controlled environment agriculture (CEA), 74–75
hybridization technique in breeding, 55–56
hydration in controlled environment agriculture (CEA), 74–75
hydro-distillation of essential oils, 356
hydro-steam distillation of essential oils, 357
Hypericum perforatum, 11
hypothalamic-pituitary-adrenal (HPA) axis, 269

I

imaging mass spectrometry (IMS), 123
impaired mitochondrial metabolism in breast cancer, 163
Indian gooseberry, 271
indirubin as anti-cancer phytopharmaceutical, **243**, 248–249
inflammation in sarcopenia, 142–143
International Plant Proteomics Organization (IPPO), 122
investigational new drug applications (INDS), 204
in vitro propagation of mango, 86–90, **89**, *90*
irinotecan as anti-cancer phytopharmaceutical, **242**
isoflavone, 149–150
isoflavonoids, 5, 149–150
isothiocyanate, 246–247
isotope labeling approach, 122–123

J

jamun, 83–84, 97

K

kaempferol, *189*, 192
kapha dosha, 20
kinase inhibition and binding, 173–174
Kyoto Encyclopedia of Genes and Genomes (KEGG), 125–126, 297

L

ligand-based bioactivity prediction tools, 294–295
ligand-based drug design (LBDD), 159
lighting in controlled environment agriculture (CEA), 72–74, **73**
light therapy, *see* low-level laser (light) therapy
liquid chromatography (LC), 219
liquid chromatography-mass spectrometry (LC-MS), 283, 284–286, *285*
low-density lipoprotein (LDL) blood cholesterol, 10

low-level laser (light) therapy
conclusion on, 314
defined, 308
evolution of, 308, *309*
laser diodes-based system of, 311, *311*
laser emitting diode-based system of, **312**, 312–313, *313*
mode of action of, 309–310, *310*
for photorejuvenation, 313, *314*
lung cancer, cannabinoids in, 228–229, **230**
luteolin, 148–149
as aromatic plant, *189*, 192
lycopene, 5
lycopene as anti-cancer phytopharmaceutical, **242**

M

mace, 29–30, **40**
mango, 83, 85–90, **89**, *90*
MapMan, 125–126
marine sources of drugs, 216–217, **217**
marker-assisted selection (MAS), 59–60
mass spectrometry (MS), 284–286, *285*
mathematical model-based approach in multiomics, 126–127
matrix-associated laser desorption ionization (MALDI), 123
medicinal plants, commercial drugs from, 216
Medicinal Plant Transcriptome Project, 121
Melissa officinalis, 12
Metabolic Adjustment by Differential Expression, 126
metabolic genome-wide association studies (mGWASs), 123
metabolomics, 122–123
in multiomics analysis, 123–127, *124*
metal nanoparticles (MNPs)
antibacterial activity of biogenic, 333–334
conclusion, outlook and future aspects for, 342
fabrication using plant extracts, 334–340, **335–341**, *337*
introduction to, 331–332, *332*
other, 340, **341**
pathogenicity and mechanisms of multidrug resistance in bacteria and, 333
plant-mediated sustainable fabrication of, against MDR bacteria, 333, *334*
microbial metabolites, 217, **218**, *218*
microemulsions, *322*, 323
micropropagation of fruit crops
advances in, 85–100
aonla, 91–95, *94*
bael, 95–97, *96*
banana, 97–98
conclusion and future perspective on, 100–101
guava, 90–91, *92*
introduction to, 81–82, **82**
jamun, 97
mango, 85–90, **89**, *90*
papaya, 98–100, **99**, *100*
pharmaceutical importance of fruit crops and, 82–85
mitochondrial dysfunction in sarcopenia, 142
molecular docking, 297–298
molecular dynamics simulations, 297–298
monoterpene indole alkaloids (MIAs), 115, *116*
morin, *146*, 147
motor neuron and muscle fiber deterioration in sarcopenia, 141–142
MS-DIAL, 123
MS-FINDER, 123
multidrug resistant bacteria, 333
plant-mediated sustainable fabrication of MNPs against, 333, *334*
multiomics analysis, 123–127, *124*
multivariate analysis, 125
mutation breeding, 56, **57**
myostatin, 144
MZmine3, 123

N

Nanophytomedicine, *see also* metal nanoparticles (MNPs)
berberine, 325, **326**
camptothecin, **326**
classification of nanophytopharmaceuticals used, 321–323, *322*
complex coacervation method, 320
conclusion on, 327
co-precipitation method, 321
curcumin, 324–325, **326**
future in herbal drug delivery, 324
high-pressure homogenization (HPH) method, 319–320
introduction to, 318, *318*
limitations of, 323
nanobiotechnology in herbal drug formulations and, 323–324, *324*
need for novel drug delivery system for herbal remedies and, 319
paclitaxel, 325, **326**
quercetin, **326**
resveratrol, 325, **326**
salting out method, 321
self-assembly method, 321
solvent displacement method, 321
solvent-emulsification-diffusion method, 320
supercritical fluid method, 320–321
techniques used in, 319–321, *320*
traditional Chinese herbs (TCH), 325–326
triptolide, **326**
National Center for Biotechnology Information (NCBI), 121
natural products, 213–214
as alternative to conventional drugs, 215–217
commercial drugs from medicinal plants, 216
conclusion and future prospects for, 220
disease-specific regulation of molecules and pathways by, 218–220
flow chart of development of, *215*
from marine sources, 216–217, **217**
from microbial metabolites, 217, **218**, *218*
from plants-phytocompounds, 216
significance of, 220
types and targets of, 218, *219*
neurological diseases, cannabinoids in, 230–232, *231*, **232**
challenges for, 234
recent clinical trials of, 233

NIST-2008 dataset, 284
novel drug delivery systems (NDDS), 319
nuclear magnetic resonance (NMR) spectroscopy, 219, 283
nutmeg, 29–30, **40**
nutrient concentration in controlled environment agriculture (CEA), 75

O

1KP project, 121
organogenesis in guava, 90–91
oxygen-glucose deprivation (OGD), cannabinoids in, 231

P

paclitaxel, 206
  in herbal nanocarrier-based drug delivery, 325, **326**
pain, 12
Panax ginseng, 270, **274**
papaya, 84–85, 98–100, **99**, *100*
PathVisio, 125–126
pathway-based approach in multiomics, 125–126
perfumery, essential oils for, 352
phenolic compounds
  in anti-cancer phytopharmaceuticals, 244–245
  in spices, 21
phenylpropanoids, 2
phloroglucinol, 259–260
photorejuvenation, LLLT for, 313, *314*
phytobiotics, 216
phytochemicals; *see also* phytopharmaceuticals
  bioinformatics tools for, 289, **290–292**
  conclusion on, 298
  databases and repositories for, 286–289, **287–288**
  methodological workflow of bioinformatics-based investigations into, 283–294
  usefulness of, 282
phytochemistry
  analysis of MS/MS and LC-MS using GNPS, 284–286, *285*
  identification of components using NIST, 284
  other databases/phytochemical repositories, 286–289, **287–288**
  QSAR in research on, 289, 292–294
  spectroscopic investigations in, 283–284
  workflow in, 283
PhytoClust, 120
phytoestrogens, 10
phytomedicinal research, bioinformatics tools used in, 294–298
PhytoMetaSyn, 121
phytopharmaceuticals
  breast cancer and, 159, 161, 164–167, **165–166**, **168**, *169*
  for common diseases/illnesses, 202–209, *203–205*, **207–209**
  complex mixtures made from secondary plant metabolites, 2–3
  defined, 1–2
  exotic medicinal plants, 11–12
  in fruit crops, 82–85
  high tolerability and strength of, 3
  introduction to, 1–2
  as natural chemicals used to treat illness, 3
  types and medical applications of, 4–11, **6–9**
phytosterols, 10
phytotherapy, 3
  with clinical evidence, 4
picrocrocins, 28
pitta dosha, 20
PlantClusterFinder, 120
plantiSMASH, 120
PlantSEED database, 127
plant tissue culture techniques, 58
polyketides, 2
polymeric nanoparticles, 321, *322*
polyphenols, 10
polyploidy breeding, 56–58
porphyran, 261–262
probiotics, 217, **218**, *218*
protein-protein interactions (PPIs), 297
proteomics, 121–122
  in multiomics analysis, 123–127, *124*
PR signaling, 161, 163
PtNPs, 339, **340**
purine alkaloids, 118–119, *119*

Q

quantitative structure-activity relationships (QSAR), 283, 289, 292–294
quercetin, 145–147, *146*
  as aromatic plant, *189*, 192
  breast cancer and, 170
  in herbal nanocarrier-based drug delivery, **326**

R

reactive oxygen species (ROS), 219, 313
redox enzymes, 171
resveratrol, 10
  as anti-cancer phytopharmaceutical, 248
  breast cancer and, 170–171
  in herbal nanocarrier-based drug delivery, 325, **326**
reverse screening, 296–297
Rhodiola rosea, 272, **275**
rosmanol, *189*, 193
rosmarinic acid, *189*, 193
  as anti-cancer phytopharmaceutical, **241**

S

safety of phytopharmaceuticals, 3
saffron, 28, **39**
salting out method, 321
Salvia officinalis, 12
sarcopenia, 150–151
  cellular derangements and etiology of, 141–143, *142*
  clinical definition of, 141
  current pharmacological and nonpharmacological interventions for, 143–144, *144*
  natural molecules as potential management interventions for, 145
  protective role of flavonoids in muscle mass preservation and regeneration and, 145–150, *146*

satellite cell differentiation and proliferation potential in sarcopenia, 142
seaweed metabolites
  anticancer activity of, 256–258
  anticancer activity of nanoparticles synthesized from, 263
  cytotoxicity against breast cancer, 257–258
  cytotoxicity against colon and bladder cancer, 258
  cytotoxicity against skin cancer, 256
  introduction to, 255–256
  synergistic effects with anticancer drugs, 259–263, **261**
secondary metabolites
  complex mixtures made from, 2–3
  production in controlled environment agriculture (CEA), 68, 69–75, **70**
  in spices, 21–23, **22**, *22*
self-assembly method, 321
Silexin, 206
silver nanoparticles, 263
sinensetin, 149
skin cancer, 205
  seaweeds' cytotoxicity against, 256
Solanum lycopersicum, 11
SolCyc, 125
solid lipid nanoparticles, 322, *322*
solvent displacement method, 321
solvent-emulsification-diffusion method, 320
somatic embryogenesis
  of aonla, 94–95
  of guava, 91
  of papaya, **99**
spices, Indian
  asafoetida, 30–31, **41**
  black, pepper, 23–24, **35**
  cardamom, 31–32, **42**
  carom seeds, 30, **40**
  cinnamon, 23, **35**
  clove, 24, **36**
  COVID-19 and, 33–34
  cumin, 25, **36**
  defined, 19–20
  fennel, 26–27, **38**
  fenugreek, 27–28, **38**
  garlic, 25–26, **37**
  ginger, 26, **37**
  introduction to, 19–20
  nutmeg and mace, 29–30, **40**
  as per Ayurveda, 20–21
  plant secondary metabolites as active pharmaceutical compounds in, 21–23, **22**, *22*
  saffron, 28, **39**
  star anise, 32–33, **43**
  structure of major bioactive compounds in, **35–43**
  summary of, **34**
  turmeric, 28–29, **39**
star anise, 32–33, **43**
steam distillation of essential oils, 357
stem cells, breast cancer, 173
steroids, 2
stilbenoid as anti-cancer phytopharmaceutical, 248
St. John's wort, 11, 74
strength of phytopharmaceuticals, 3
stress, 269
structure-based drug design (SBDD), 159
supercritical fluid method, 320–321
synergism in controlled environment agriculture (CEA), 75–76

T

taxanes as anti-cancer phytopharmaceutical, **243**, 248
teniposide as anti-cancer phytopharmaceutical, **242**
terpenes, 21
terpenoids, 11, 21
thymol, 189, *189*
Tinospora cordifolia, 273, **275**
tolerability of phytopharmaceuticals, 3
topotecan as anti-cancer phytopharmaceutical, **242**
traditional Chinese herbs (TCH), 325–326
transcription activator-like effector nucleases (TALENs), 59
transcriptomics, 120–121
  in multiomics analysis, 123–127, *124*
traumatic brain injury (TBI), cannabinoids in, 230
trigonelline, 27
triptolide in herbal nanocarrier-based drug delivery, **326**
triterpenoids, 247
tropane alkaloids (TAs), 117–118, *118*
turmeric, 28–29, **39**
2-phenethyl isothiocyanate as anti-cancer phytopharmaceutical, **242**

U

ultraviolet (UV)-visible spectroscopy, 283–284

V

vapor pressure deficit (VPD) in controlled environment agriculture (CEA), 74–75
vata dosha, 20
vinblastine, 205
vinca alkaloids, **241**, 245
virtual high-throughput screening (VHTS), 283
vitamin D as anti-cancer phytopharmaceutical, **243**, 248
vitamin E as anti-cancer phytopharmaceutical, **243**, 248

W

Warburg effect in breast cancer, 163
water hyssop, 270–271
Weighted Gene Co-expression Network Analysis (WGCNA), 126
Withania somnifera, 270, **274**

X

XCMS, 123

Z

zinc-finger nucleases, (ZFN), 59
zinc oxide nanoparticles, 263